An Introduction to Fire Dynamics

Second Edition

An Introduction to Fire Dynamics

Second Edition

Dougal Drysdale
University of Edinburgh, UK

JOHN WILEY & SONS

Chichester • Weinheim • New York • Brisbane • Singapore • Toronto

Other Wiley Editorial Offices

John Wiley & Sons, Inc., 605 Third Avenue,
New York, NY 10158-0012, USA

Wiley-VCH Verlag GmbH, Pappelallee 3,
D-69469 Weinheim, Germany

Jacaranda Wiley Ltd, 33 Park Road, Milton,
Queensland 4064, Australia

John Wiley & Sons (Asia) Pte Ltd, 2 Clementi Loop #02-01,
Jin Xing Distripark, Singapore 0512

John Wiley & Sons (Canada) Ltd, 22 Worcester Road,
Rexdale, Ontario M9W 1L1, Canada

Library of Congress Cataloging-in-Publication Data

Drysdale, Dougal.
 An introduction to fire dynamics/Dougal Drysdale. — 2nd ed.
  p. cm.
 Includes bibliographical references and indexes.
 ISBN 0-471-97290-8 (ppc). – ISBN 0-471-97291-6 (pbk)
  1. Fire. 2. Flame. I. Title.
 QD516.D79 1998
 541.3′61 — dc21   98-26200
                CIP

British Library Cataloguing in Publication Data

A catalogue record for this book is available from the British Library

ISBN 0 471 97290 8 (ppc)   0 471 97291 6 (pbk)

Typeset in 10/12pt Times Roman by Laser Words, Madras, India
Printed and bound in Great Britain by Biddles Ltd, Guildford and King's Lynn
This book is printed on acid-free paper responsibly manufactured from sustainable
forestry, in which at least two trees are planted for each one used for paper production.

To Jude

Contents

Preface to the Second Edition

The thirteen years that have elapsed between the appearance of the first and second editions of *Introduction to Fire Dynamics* have seen sweeping changes in the subject and, more significantly, in its application. Fire Engineering – now more commonly referred to as Fire Safety Engineering – was identified in the original preface as 'a relatively new discipline', and of course it still is. However, it is beginning to grow in stature as Fire Safety Engineers around the world begin to apply their skills to complex issues that defy solution by the old 'prescriptive' approach to fire safety. This has been reflected by the concurrent development in many countries of new Codes and Regulations, written in such a way as to permit and promote engineered solutions to fire safety problems. The multi-storey atrium and the modern airport terminal building are but two examples where a modern approach to fire safety has been essential.

Preparing a second edition has been somewhat of a nightmare. I have often said that if the first edition had not been completed in late 1984 it might never have been finished. The increased pace of research in the early 1980s was paralleled by the increasing availability of computers and associated peripherals. The first edition was prepared on a typewriter – a device in which the keyboard is directly connected to the printer. Graphs were plotted by hand. In 1984 I was rapidly being overtaken by the wave of new information, so much so that the first edition was out of date by the time it appeared.

In 1984, the International Association for Fire Safety Science – an organisation which has now held five highly successful international symposia – was still to be launched, and the 'Interflam' series of conferences was just beginning to make an impression on the international scene. The vigour of fire research in the decade after 1985 can be judged by examining the contents of the meetings that took place during this period. The scene has been transformed: the resulting exchange of ideas and information has established fire science as the foundation of the new engineering discipline. This has been largely due to the efforts of the luminaries of the fire research community, including in particular Dr Philip Thomas, the late Prof. Kunio Kawagoe, Prof. T. Akita, Prof. Jim Quintiere and my own mentor, the late Prof. David Rasbash. They perceived the need for organisations such as the IAFSS, and created the circumstances in which they could grow and flourish.

A second edition has been due for over 10 years, but seemed an impossible goal. Fortunately, my friends and colleagues at Worcester Polytechnic Institute came to the rescue. They took the initiative and put me in purdah for four weeks at WPI, with strict instructions to 'get on with it'. Funding for the period was provided by a consortium, consisting of the SFPE Educational Trust, the NFPA, Factory Mutual Research Corporation, Custer Powell Associates, and the Centre for Firesafety Studies at WPI. I am grateful to them all for making it possible, and to Don and Mickey Nelson for making me feel so welcome in their home. Numerous individuals on and off campus helped me to get things together. There was always someone on hand to locate a paper, plot a graph, discuss a problem, or share a coffee. I am grateful to David Lucht, Bob Fitzgerald, Jonathan Barnett, Bob Zalosh and Nick Dembsey for their help. I am indebted to many other individuals who kindly gave their time to respond to questions and comment on sections of the manuscript. In particular, I would like to thank (alphabetically) Paula Beever, Craig Beyler, John Brenton, Geoff Cox, Carlos Fernandez-Pello, George Grant, Bjorn Karlsson, Esko Mikkola, John Rockett and Asif Usmani. Each undertook to review one or more chapters: their feedback was invaluable. Having said this, the responsibility for any errors of fact or omission is mine and mine alone.

It is a sad fact that I managed to carry out over 50% of the revision in four weeks at WPI, but have taken a further two years to complete the task. I would like to thank my colleagues in the Department of Civil and Environmental Engineering for their support and tolerance during this project. This was particularly true of my secretary, Alison Stirling, who displayed amazing *sang-froid* at moments of panic. However, the person to whom I am most indebted is my wife Judy who has displayed boundless patience, tolerance and understanding. Without her support over the years, neither edition would ever have been completed. She finally pulled the pin from the grenade this time, by organising a 'Deadline Party' to which a very large number of friends and colleagues were invited. Missing this deadline was not an option (sorry, John Wiley). It was a great party!

List of Symbols and Abbreviations

a	Absorptivity
A	Arrhenius factor (Chapter 1)
A_f	Fuel bed area (m^2)
A_t	Total internal surface area of a compartment, including ventilation openings (m^2) (Chapter 10)
A_T	Internal surface area of walls and ceiling, excluding the ventilation openings (m^2) (Chapter 10)
A_w	Area of ventilation opening (window or door) (m^2)
b	Plume radius (m) (Section 4.3.1)
b	Conserved variable (Equations (5.18) and (5.19))
B	Spalding's mass transfer number (Equation (5.20))
B	Width of ventilation opening (m) (Chapter 10)
Bi	Biot number $(hL/k)(-)$ (k is the thermal conductivity of the solid).
c_p	Thermal capacity at constant pressure (J/kg.K) or (J/mol.K)
C	Concentration (Chapter 4)
C	Constant (Equations (7.9) and (7.10))
C_d	Discharge coefficient $(-)$ (Chapters 9 and 10)
C_{st}	Stoichiometric concentration (Table 3.1).
d_b	The required depth of clear air above floor level (m) (Chapter 11)
d_q	Quenching distance (mm)
D	Pipe diameter (m) (Sections 2.3 and 3.3)
D	Pool or fire diameter (m) (Chapters 4 and 5)
D	Depth of compartment (Section 10.3.1)
D	Optical density (decibels) (Chapter 11)
D_m	Specific optical density ((b/m).m^3/m^2) (Equation (11.4))
D_0	Smoke potential ((db/m).m^3/g) (Equation 11.5)
D	Diffusion coefficient (m^2/s)
E	Total emissive power of a surface (W/m^2) (Equation (2.4))
E	Constant (Equation (1.14))
E_A	Activation energy (J/mole)

f	Fraction of heat of combustion transferred from flame to surface (Equation (6.16))
f_{ex}	Excess fuel factor (Equation (10.22))
F	Constant (Equation 1.14)
F	Integrated configuration factor ('finite-to-finite area' configuration factor) (Section 2.4.1)
Fo	Fourier number ($\alpha\, t/L^2$)($-$)
g	Gravitational acceleration constant (9.81 m/s^2)
Gr	Grashof number (gl$^3\Delta\rho/\rho\nu^2$) ($-$)
h	Convective heat transfer coefficient (kW/m^2.K)
h_c	Height of roof vent above floor (m) (Chapter 11)
h_0	Height to neutral plane (m) (Figure 10.4)
h_f	Height above neutral plane (m) (Figure 10.4)
h_k	Effective heat transfer coefficient (kW/m^2.K) (Equation (9.6))
H	Height of ventilation opening (m) (Chapter 10)
h	Planck's constant (Equation (2.52))
I	Intensity of radiation (Equations (2.56), (2.79), (10.13))
I	Intensity of light (Equation (11.1))
k	Thermal conductivity (W/m.K)
k'	Rate coefficient (Equation (1.1))
k	Boltzmann's constant (Equations (2.52) and (2.54))
K	'Effective emission coefficient' (m^{-1}) (Equations (2.83), (5.12), (10.27))
l	Flame height or length (m) (Chapter 4)
l	Preheat length (m) (Section 7.2.3)
L	Thickness, or half-thickness (m) as defined locally (Section 2.2)
L	Mean beam length (m) (Section 2.4.2)
L	Lower flammability limit (Chapter 3)
L	Pathlength (m) (Section 11.1)
L_v	Latent heat of evaporation or gasification (J/g)
$\dot{m}$	Rate of mass loss (g/s)
M	Mass of air (kg)
M_f	Mass of fuel (kg)
M_w	Molecular weight
n	An integer
n	Number of moles (Chapter 1)
n_A, n_B	Molar concentration (Equation (1.16))
Nu	Nusselt number (hL/k)($-$) (k is the thermal conductivity of the fluid)
p	Partial pressure (mm Hg, atm)
p^o	Equilibrium vapour pressure (mm Hg)

P	Pressure (atm)
P_f	Perimeter of fire (m) (Section 11.2)
Pr	Prandtl number $(\nu/\alpha)(-)$
$\dot{q}$	Rate of heat transfer (W or kW)
q_f	Fire load (Equation (10.39)) (MJ or kg)
Q	Rate of heat transfer (W or kW)
$\dot{Q}_c$	Rate of heat release (W or kW)
$\dot{Q}_{conv}$	Rate of convective heat release from a flame (Equation (2.76))
$\dot{Q}_c'''$	Rate of heat production per unit volume (W/m^3) (Equation (2.13))
$\dot{Q}^*$	Dimensionless heat release rate $(-)$ (Equation 4.3)
r	Radial distance (m) (Equations (2.58), (4.49))
r	Stoichiometric ratio (fuel/air) (Chapters 1 and 10)
r_c	Height of roof vent above virtual source of the fire (m) (Section 11.2)
r_0	Characteristic dimension (m) (Chapters 6 and 8)
R	Ideal gas constant (Table 1.9) (Chapters 1, 6 and 8)
R	Radius of burner mouth (m) (Section 4.1)
R	Regression rate, or burning rate (Section 5.1.1)
Re	Reynolds number $(ux/\nu)(-)$ (Table 4.4)
S	Surface area (Equation 6.2)
S	Rate of transfer of 'sensible heat', defined in Equation (6.17)
t	Time (seconds)
t_e	Escape time (Chapters 9 and 11)
t_u	Time to achieve untenable conditions (Chapters 9 and 11)
T	Temperature (C or K)
u	Flow velocity (m/s)
U	Upper flammability limit (Chapter 3)
v	Linear velocity or flowrate (m/s)
V	Volume (m^3) (Chapters 1, 2 and 6)
V	Dimensionless windspeed (Section 4.3.4)
V	Flame spread rate (Chapter 7)
W	Width of compartment (m)
W	Mass loss by volatilization (g) (Section 11.1)
x	Distance ($\Delta x =$ thickness) (m)
x_A, x_B	Mole fractions
y	Distance (m)
Y	Mass fraction
z	Distance (m) (e.g. height in fire plume)

Greek symbols

α	Thermal diffusivity $(k/\rho c)$ (m^2/s)
α'	Entrainment constant (Section 4.3.1)
α_A, α_B	Activity (Equation (1.17))
β	Coefficient of expansion (Equations (1.12) and (2.41))
β	Cooling modulus $(-)$ (Equation (6.26))
γ	Energy modulus $(-)$ (Equation (6.30))
γ_A, γ_B	Activity coefficients (Equation (1.17))
γ_i, γ_u	Pettersson's heat transfer coefficient (kW/m^2K) (Equations (10.33) and (10.35))
δ_{cr}	Critical value of Frank-Kamanetskii's δ (Equation (6.13))
δ_h	Thickness of hydrodynamic boundary layer (m) (Section 2.3)
δ_θ	Thickness of thermal boundary layer (m) (Section 2.3)
ΔH	Change in enthalpy (kJ/mol)
ΔH_c	Heat of combustion (kJ/mol or kJ/g)
ΔH_f	Heat of formation (kJ/mol)
ΔU	Change in internal energy (kJ/mol)
ε	Emissivity
θ	Temperature difference (e.g. $T - T_\infty$)
θ	Dimensionless temperature (Chapter 8)
θ	Angle (Equation (2.56), Section 4.3.4 and 7.1)
κ	Absorption coefficient
κ	Constant (Equation (6.9))
λ	Wavelength (μm)
μ	Absolute or dynamic viscosity $(Pa.s, \text{ or } N.s/m^2)$
η_{O_2}	Mole fraction of oxygen (Equations 1.24 and 5.25)
ν	Kinematic viscosity (μ/ρ) (m^2/s)
ρ	Density (kg/m^3)
σ	Stefan-Boltzmann constant $(5.67 \times 10^{-8} \text{ W/m}^2.K^4)$
τ	Slab thickness (m) (Equation (2.21) and Section 6.3.1)
τ	Duration of burning (s) (Equation (10.43) *et seq.*)
τ'	Length of induction period (s)
ϕ	Configuration factor
χ	Factor expressing combustion efficiency (Chapters 1 and 5).
χ	$\dot{m}_{air}/A_w H^{1/2}$

Subscripts

a	Ambient
b	Black body (radiation)
c	Cold (Chapter 2)
c	Convective (with f) (Chapter 6)
C	Combustion
cr	Critical
e	External

f	Fuel
F	Flame
FO	Flashover
g	Gas
h	Hot
i	Ignition or firepoint
l	Liquid (Chapter 5)
L	Loss by gas replacement (Equation 10.23))
m	Mean
max	Maximum
n	Normal to a surface
o	Initial value or ambient value
o	Centreline value (buoyant plume (Chapter 4))
ox	Oxygen
p	Constant pressure
p	Pyrolysis (Chapter 7)
pl	Plate (Equation (8.1.3))
R	Radiative
s	Surface
u	Unburnt gas
W	Wall (Equation (10.23))
x	In the x-direction
∞	Final value

Superscripts

·	Signifies rate of change as in $\dot{m}$
·	Indicates that a chemical species is a free radical (e.g. H, the hydrogen atom) (Chapter 1)
$'$	Single prime (Signifies 'per unit width') (Chapter 4)
$''$	Double prime (Signifies 'per unit area')
$'''$	Per unit volume

List of abbreviations

ASTM	American Society for Testing and Materials
BRE	Building Research Establishment, Garston, Watford, WD2 7JR, England.
BSI	British Standards Institution.
CIB	Counceil Internationale du Bâtiment
DIN	Deutsches Institut für Normung
FMRC	Factory Mutual Research Corporation (Norwood, Ma., USA)
FPA	Fire Protection Association, Loss Prevention Council, Melrose Avenue, Borehamwood, Herts WD6 2BJ, UK
*FRS	Fire Research Station, Building Research Establishment (see above)
ISO	International Organization for Standardisation

NBS   National Bureau of Standards (now NIST)
NFPA   National Fire Protection Association, 1 Batterymarch Park, Quincy, MA
       02269-9101, USA
NIST   National Institute for Standards and Technology (Building and Fire Research
       Laboratory), Gaithersburg, MD 20899, USA

*Frequent references are made in this text to the Fire Research Notes from FRS. These
are obtainable on microfiche on application from the Director at the above address.

1

Fire Science and Combustion

As a process, fire can take many forms, all of which involve chemical reaction between combustible species and oxygen from the air. Properly harnessed, it provides great benefit as a source of power and heat to meet our industrial and domestic needs, but, unchecked, it can cause untold material damage and human suffering. In the United Kingdom alone, direct losses probably exceed £1000 M, while over 800 people die each year in fires (for an analysis of the 1993 figures for England and Wales, see Roy, 1997). In real terms, the direct fire losses may not have increased significantly over the past two decades, but the holding action has been bought by a substantial increase in other associated costs, namely improving the technical capability of the Fire Service and the adoption of more sophisticated fire protection systems.

Further major advances in combating wildfire are unlikely to be achieved simply by continued application of the traditional methods. What is required is a more fundamental approach which can be applied at the design stage rather than tacitly relying on fire incidents to draw attention to inherent fire hazards. Such an approach requires a detailed understanding of fire behaviour from an engineering standpoint. For this reason, it may be said that a study of fire dynamics is as essential to the fire protection engineer as the study of chemistry is to the chemical engineer.

It will be emphasized at various places within this text that although 'fire' is a manifestation of a chemical reaction, the mode of burning may depend more on the physical state and distribution of the fuel, and its environment, than on its chemical nature. Two simple examples may be quoted: a log of wood is difficult to ignite, but thin sticks can be ignited easily and will burn fiercely if piled together; a layer of coal dust will burn relatively slowly, but may cause an explosion if dispersed and ignited as a dust cloud. While these are perhaps extreme examples, they illustrate the complexity of fire behaviour in that their understanding requires knowledge not only of chemistry but also of many subjects normally associated with the engineering disciplines (heat transfer, fluid dynamics, etc.). Indeed, the term 'fire dynamics' has been chosen to describe the subject of fire behaviour as it implies inputs from these disciplines. However, it also incorporates parts of those subjects which are normally associated with the terms 'fire chemistry' and 'fire science'. Some of these are reviewed in the present chapter although detailed coverage is impossible. It is assumed that the reader has a knowledge of elementary chemistry and physics, including thermodynamics: references to relevant texts and papers are given as appropriate.

1.1   Fuels and the Combustion Process

Most fires involve combustible solids, although in many sectors of industry, liquid and gaseous fuels are also to be found. Fires involving gases, liquids and solids will be discussed in order that a comprehensive picture of the phenomenon can be drawn. The term 'fuel' will be used quite freely to describe that which is burning, whatever the state of matter, or whether it is a 'conventional' fuel such as LPG or an item of furniture within a room. With the exception of hydrogen gas, to which reference is made in Chapter 3, all fuels that are mentioned in this text are carbon-based. Unusual fire problems that may be encountered in the chemical and nuclear industries are not discussed, although the fire dynamics will be similar if not identical. General information on problems of this type may be gleaned from the *National Fire Protection Handbook* (NFPA, 1997) and other sources (e.g. Meidl, 1970; Stull, 1977; Lees, 1996).

1.1.1   The nature of fuels

The range of fuels with which we are concerned is very wide, from the simplest gaseous hydrocarbons to solids of high molecular weight and great chemical complexity, some of which occur naturally, such as cellulose, and others which are man-made, e.g. polyethylene and polyurethane (Tables 1.1 and 1.2). All will burn under appropriate conditions, reacting with oxygen from the air, generating combustion products, and releasing heat. Thus, a stream or jet of a gaseous hydrocarbon can be ignited in air to give a flame, which is seen as the visible portion of the volume within which the oxidation process is occurring. Flame is a gas phase phenomenon and, clearly, flaming combustion of liquid and solid fuels must involve their conversion to gaseous form. For burning liquids, this process is normally simple evaporative boiling at the surface,* but for almost all solids, chemical decomposition or *pyrolysis* is necessary to yield products of sufficiently low molecular weight that can volatilize from the surface and enter the flame. As this requires much more energy than simple evaporation, the surface temperature of a burning solid tends to be high (typically 400°C) (Table 1.2). Exceptions to this rule are those solids which sublime on heating, i.e. pass directly from the solid to the vapour phase without chemical decomposition. There is one relevant example, hexamethylenetetramine (or methenamine), which in pill form is used as the ignition source in ASTM D2859-93 (American Society for Testing and Materials, 1993a). It sublimes at about 263°C (Budavari, 1996).

The composition of the volatiles released from the surface of a burning solid tends to be extremely complex. This can be understood when the chemical nature of the solid is considered. All those of significance are polymeric materials of high molecular weight, whose individual molecules consist of long 'chains' of repeated units which in turn are derived from simple molecules known as monomers (Greenwood and Banks, 1968; Billmeyer, 1971; Open University, 1973; Hall, 1981; Friedman 1998). Of the two basic types of polymer (addition and condensation) the addition polymer is the simpler in that it is formed by direct addition of monomer units to the end of a growing

---

* Liquids with very high boiling points (≥250°C) may undergo some chemical decomposition) (e.g. cooking oil).

Table 1.1 Properties of gaseous and liquid fuels[a]

	Formula	Melting point (°C)	Boiling point (°C)	Density (liq) (kg/m³)	Molecular weight
Hydrogen	H_2	−259.3	−252.8	70	2
Carbon monoxide	CO	−199	−191.5	422	28
Methane	CH_4	−182.5	−164	466	16
Ethane	C_2H_6	−183.3	−88.6	572	30
Propane	C_3H_8	−189.7	−42.1	585	44
n-Butane	$n\text{-}C_4H_{10}$	−138.4	−0.5	601	58
n-Pentane	$n\text{-}C_5H_{12}$	−130	36.1	626	72
n-Hexane	$n\text{-}C_6H_{14}$	−95	69.0	660	86
n-Heptane	$n\text{-}C_7H_{16}$	−90.6	98.4	684	100
n-Octane	$n\text{-}C_8H_{18}$	−56.8	125.7	703	114
iso-Octane[b]	$iso\text{-}C_8H_{18}$	−107.4	99.2	692	114
n-Nonane	$n\text{-}C_9H_{20}$	−51	150.8	718	128
n-Decane	$n\text{-}C_{10}H_{22}$	−29.7	174.1	730	142
Ethylene (ethene)	C_2H_4	−169.1	−103.7	(384)	28
Propene	C_3H_6	−185.2	−47.4	519	42
Acetylene (ethyne)	C_2H_2	−80.4	−84	621	26
Methanol	CH_3OH	−93.9	65.0	791	32
Ethanol	C_2H_5OH	−117.3	78.5	789	46
Acetone	$(CH_3)_2CO$	−95.3	56.2	790	58
Benzene	C_6H_6	5.5	80.1	874	78

[a] Data from Lide (1993/94).
[b] 2,2,4-Trimethyl pentane.

Table 1.2 Properties of some solid fuels[a]

	Density (kg/m³)	Heat capacity (kJ/kg.K)	Thermal conductivity (W/m.K)	Heat of combustion (kJ/g)	Melting point (°C)
Natural polymers					
Cellulose	V^b	~1.3	V	16.1	chars
Thermoplastic polymers					
Polyethylene Low density	940	1.9	0.35		
High density	970	2.3	0.44	46.5	130–135
Polypropylene Isotactic	940	1.9	0.24	46.0	186
Syndiotactic					138
Polymethylmethacrylate	1190	1.42	0.19	26.2	~160
Polystyrene	1100	1.2	0.11	41.6	240
Polyoxymethylene	1430	1.4	0.29	15.5	181
Polyvinylchloride	1400	1.05	0.16	19.9	—
Polyacrylonitrile	1160–1180	—	—	—	317
Nylon 66	~1200	1.4	0.4	31.9	250–260
Thermosetting polymers					
Polyurethane foams	V	~1.4	V	24.4	—
Phenolic foams	V	—	V	17.9	chars
Polyisocyanurate foams	V	—	V	24.4	chars

[a] From Brandrup and Immergut (1975); and Hall (1981). Heats of combustion refer to CO_2 and H_2O as products.
[b] V = variable.

polymer chain. This may be illustrated by the sequence of reactions:

$$R^{\cdot} + CH_2{=}CH_2 \rightarrow R.CH_2.CH_2^{\cdot} \tag{1.R1a}$$

$$R.CH_2.CH_2^{\cdot} + CH_2{=}CH_2 \rightarrow R.CH_2.CH_2.CH_2.CH_2^{\cdot} \tag{1.R1b}$$

etc, where $R^{\cdot}$ is a free radical or atom, and $CH_2{=}CH_2$ is the monomer, ethylene. This process is known as polymerization and in this case will give polyethylene which has the idealized structure

$$R\underbrace{(CH_2. \ CH_2)}_{\text{monomer unit}}{}_n R'$$

in which the monomer unit has the same complement and arrangement (although not the same chemical bonding) of atoms as the parent monomer, $CH_2{=}CH_2$: n is the number of repeated units in the chain. In contrast, the process of polymerization which leads to the formation of a condensation polymer involves the loss of a small molecular species (normally H_2O) whenever two monomer units link together. (This is known as a condensation reaction.) Normally, two distinct monomeric species are involved, as in the production of Nylon 66 from hexamethylene diamine and adipic acid: the first stage in the reaction would be:

$$NH_2(CH_2)_6NH_2 + HO.CO.(CH_2)_4.CO.OH$$

$\quad\quad$ hexamethylene $\quad\quad\quad\quad$ adipic acid $\quad\quad\quad\quad\quad\quad$ (1.R2)

$\quad\quad$ diamine

$$\rightarrow NH_2(CH_2)_6NH.CO.(CH_2)_4.CO.OH + H_2O$$

The formula of Nylon 66 may be written in the format used above for polyethylene, namely:

$$H \ (NH.(CH_2)_6.NH.CO.(CH_2)_4.CO)_n OH$$

It should be noted that cellulose, the most widespread of the natural polymers occurring in all higher plants (Section 5.2.2), is a condensation polymer of the monosaccharide D-Glucose. The formulae for both monomer and polymer are shown in Figure 5.11.

An essential feature of any monomer is that it must contain two reactive groups, or centres, to enable it to combine with adjacent units to form a linear chain (Figure 1.1(a)). The length of the chain (i.e. the value of n in the above formulae) will depend on conditions existing during the polymerization process: these will be selected to produce a polymer of the desired properties. Properties may also be modified by introducing branching into the polymer 'backbone'. This may be achieved by modifying the conditions in a way that will induce branching to occur spontaneously (Figure 1.1(b)) or by introducing a small amount of a monomer which has three reactive groups (unit B in Figure 1.1(c)). This can have the effect of producing a cross-linked structure whose physical (and chemical) properties will be very different from an equivalent unbranched, or only slightly branched, structure. As an example, consider the expanded polyurethanes. In most flexible foams the degree of cross-linking is very low, but by increasing it substantially (e.g. by increasing the proportion of trifunctional monomer, B in Figure 1.1(c)), a polyurethane suitable for rigid foams may be produced.

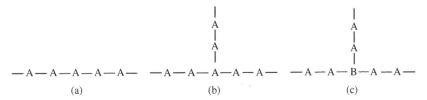

(a)   (b)   (c)

Figure 1.1 Basic structure of polymers: (a) straight chain (e.g. polymethylene, with A = CH$_2$); (b) branched chain, with random branch points (e.g. polyethylene, with A = CH$_2$−CH$_2$, see text); (c) branched chain, involving trifunctional centres (e.g. polyurethane foams in which the straight chains (−A−A− etc.) correspond to a co-polymer of tolylene di-isocyanate and a polymer diol and B is a trihydric alcohol)

With respect to flammability, the yield of volatiles from the thermal decomposition of a polymer is much less for highly cross-linked structures since much of the material forms an involatile carbonaceous char, thus effectively reducing the potential supply of gaseous fuel to a flame. An example of this can be found in the phenolic resins which on heating to a temperature in excess of 500°C, may yield up to 60% char (Madorsky, 1964). The structure of a typical phenolic resin is shown in Figure 1.2. A natural polymer that exhibits a high degree of cross-linking is lignin, the 'cement' that binds the cellulose structures together in higher plants, thus imparting greater strength and rigidity to the cell walls.

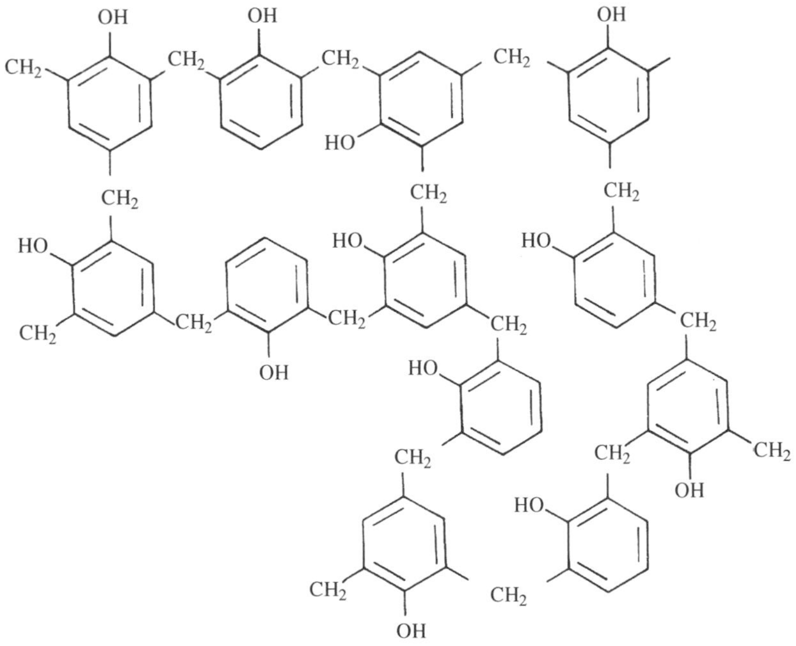

Figure 1.2 Typical cross-linked structure to be found in phenol formaldehyde resins

Synthetic polymers may be classified into two main groups, namely thermoplastics and thermosetting resins (Table 1.2). A third group—the elastomers—may be distinguished on the basis of their rubber-like properties (Billmeyer, 1971; Hall, 1981), but will not be considered further here. From the point of view of fire behaviour, the main difference between thermoplastics and thermosetting polymers is that the latter are cross-linked structures which will not melt when heated. Instead, at a sufficiently high temperature, many decompose to give volatiles directly from the solid, leaving behind a carbonaceous residue (cf. the phenolic resins, Figure 1.2), although with polyurethanes, the initial product of decomposition is a liquid. On the other hand, the thermoplastics will soften and melt when heated, which will modify their behaviour under fire conditions. Fire spread may be enhanced by falling droplets or the spread of a burning pool of molten polymer (Section 9.2.4). This is also observed with flexible polyurethane foams.

1.1.2   Thermal decomposition and stability of polymers

The production of gaseous fuel (volatiles) from combustible solids almost invariably involves thermal decomposition, or pyrolysis, of polymer molecules at the elevated temperatures which exist at the surface (Kashiwagi, 1994; Beyler and Hirschler, 1995). Whether or not this is preceded by melting depends on the nature of the material (Figure 1.3 and Table 1.3). In general, the volatiles comprise a complex mixture of pyrolysis products, ranging from simple molecules such as hydrogen and ethylene, to species of relatively high molecular weight which are volatile only at the temperatures existing at the surface where they are formed, when their thermal energy can overcome the cohesive forces at the surface of the condensed fuel. In flaming combustion most of these will be consumed in the flame, but under other conditions (e.g. pyrolysis without combustion following exposure to an external source of heat or, for some materials, smouldering combustion (Section 8.2)) the high boiling liquid products and tars will condense to form an aerosol smoke as they mix with cool air.

At high temperatures, a few addition polymers (e.g. polymethylmethacrylate) will undergo a reverse of the polymerization process (Equations (1.R1a) and (1.R1b)),

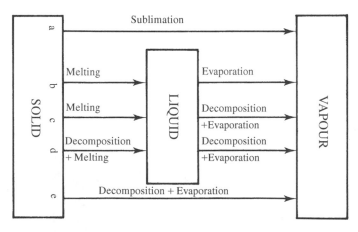

Figure 1.3   Different modes in which fuel vapour is generated from a solid (Table 1.3)

Table 1.3 Formation of volatiles from combustible solids (Figure 1.3)

Designation (Figure 1.3)	Mechanism	Examples
a	Sublimation	Methenamine (see text)
b	Melting and evaporation without chemical change	Low molecular weight paraffin waxes, although the mechanism likely to involve (b) *and* (c)
c	Melting, then decomposition followed by evaporation of low molecular weight products	Thermoplastics; high molecular weight waxes, etc.
d	Decomposition to produce molten products[a] which decompose further to yield volatile species	Polyurethanes
e	Decomposition to give volatile species directly	Cellulose; most thermosetting resins (except polyurethanes)

[a] The initial decomposition may also produce species which can volatilize directly.

Table 1.4 Yield of monomer in the pyrolysis of some organic polymers in a vacuum (% total volatiles) (from Madorsky, 1964)

Polymer	Temperature range (°C)	Monomer (%)
Polymethylene[a]	325–450	0.03
Polyethylene	393–444	0.03
Polypropylene	328–410	0.17
Polymethylacrylate	292–399	0.7
Polyethylene oxide	324–363	3.9
Polystyrene	366–375	40.6
Polymethylmethacrylate	246–354	91.4
Polytetrafluoroethylene	504–517	96.6
Poly α-methyl styrene	259–349	100
Polyoxymethylene	Below 200	100

[a] An unbranched polyethylene.

known as 'unzipping' or 'end-chain scission', to give high yields of monomer in the decomposition products (Table 1.4). This behaviour is a direct result of the chemical structure of the monomer units which favours 'unzipping', in some cases to the exclusion of any other decomposition mechanism (Madorsky, 1964). This may be contrasted with the pyrolysis of polyethylene in which a completely random mechanism appears to operate, again as a consequence of the monomer structure. It may be noted at this point that there is no exact equivalent to the 'unzipping' process in the pyrolysis of condensation polymers (compare Equations (1.R1) and (1.R2)).

In addition to 'end chain scission' and 'random chain scission' which are described above, two other decomposition mechanisms may be identified, namely chain stripping and cross-linking (Wall, 1972; Cullis and Hirschler, 1981; Beyler and Hirschler, 1995). Chain stripping is a process in which the polymer backbone remains intact but molecular species are lost as they break away from the main chain. One relevant example is the thermal decomposition of polyvinyl chloride (PVC) which begins to lose molecular HCl (hydrogen chloride) at about 250°C, leaving behind a char-like residue:

$$R\text{\textparenleft}CH_2-CHCl\text{\textparenright}_n R' \rightarrow R\text{\textparenleft}CH=CH\text{\textparenright}_n R' + nHCl \qquad (1.R3)$$

Polyvinyl chloride   residue

Although the residue will burn at high temperatures (giving much smoke), hydrogen chloride is a very effective combustion inhibitor and its early release will tend to extinguish a developing flame. For this reason it is said that PVC has a very low 'flammability', or potential to burn. This is certainly true for rigid PVC, but the flexible grades commonly used for electrical insulation, for example, contain additives (plasticizers) which make them more flammable. However, even the 'rigid' grades will burn if the ambient conditions are right (Section 5.2.1).

Polymers which undergo cross-linking during pyrolysis tend to char on heating. While this should reduce the amount of fuel available for flaming combustion, the effect on flammability is seldom significant for thermoplastics (cf. polyacrylonitrile, Table 1.2). However, as has already been noted, charring polymers like the phenolic resins have desirable fire properties. These are highly cross-linked in their normal state (Figure 1.2) and it is likely that further cross-linking occurs during pyrolysis.

In subsequent chapters, it will be shown that some of the fire behaviour of combustible materials can be interpreted in terms of the properties of the volatiles, specifically their composition, reactivity and rate of formation. Thermal stability can be quantified by determining how the rate of decomposition varies with temperature. These results may be expressed in a number of ways, the most common arising from the assumption that the pyrolysis proceeds according to a simple kinetic scheme such that

$$\dot{m} = \frac{dm}{dt} = -k'.m \qquad (1.1)$$

where m represents the mass (or more correctly, the concentration) of the polymer. While this is a gross simplification, it does permit k', the rate coefficient, to be determined although this is of little direct value *per se*. However, it allows the temperature dependence of the process to be expressed in a standard form, using the Arrhenius expression for the rate coefficient, i.e.

$$k' = A\exp(-E_A/RT) \qquad (1.2)$$

where E_A is the activation energy (J/mol), R is the universal gas constant (8.314 J/K.mol) and T is the temperature (K). The constant A is known as the pre-exponential factor and in this case will have units of s^{-1}. Much research has been carried out on the thermal decomposition of polymers (Madorsky, 1964; National Bureau of Standards, 1972; Cullis and Hirschler, 1981; Beyler and Hirschler, 1995), but

Table 1.5 Activation energies for thermal decomposition of some organic polymers in vacuum (from Madorsky, 1964)

Polymer	Molecular weight	Temperature range (°C)	Activation energy (kJ/mole)
Phenolic resin	—	332–355	18
Polymethylmethacrylate	15 000	225–256	30
Polymethylacrylate	—	271–286	34
Cellulose triacetate	—	283–306	45
Polyethylene oxide	10 000	320–335	46
Cellulose	—	261–291	50
Polystyrene	230 000	318–348	55
Poly α-methyl styrene	350 000	229–275	55
Polypropylene	—	336–366	58
Polyethylene	20 000	360–392	63
Polymethylene	High	345–396	72

in view of the chemical complexity involved, combined with problems of interpreting data from a variety of sources and experimental techniques, it is not possible to use such information directly in the present context. Some activation energies which were derived from early studies (Madorsky, 1964) are frequently quoted (e.g. Williams, 1974b, 1982) and are included here for completeness (Table 1.5). However, without a knowledge of *A* (the pre-exponential factor) these do not permit relative rates of decomposition to be assessed. Of more immediate value is the summary presented by Madorsky (1964) in which he collates data on relative thermal stabilities of a range of organic polymers, expressed as the temperature at which 50% of a small sample of polymer will decompose in 30 minutes (i.e. the temperature at which the half-life is 1800 s). (This tacitly assumes first-order kinetics, as implied in Equation (1.1).) A selection of these data is presented in Table 1.6. They allowed Madorsky (1964) to make some general comments on polymer stability which are summarized in Table 1.7. It is possible to make limited comparison of the information contained in these tables with data presented in Chapter 5 (Table 5.11) on the heats of polymer gasification (Tewarson and Pion, 1976). However, it must be borne in mind that the data in Table 1.6 refer to 'pure' polymers while those in Table 5.11 were obtained with commercial samples, many of which contain additives which will modify the behaviour.

While at first sight the composition of the volatiles might seem of secondary importance to their ability simply to burn as a gaseous mixture, such a view does not permit detailed understanding of fire behaviour. The reactivity of the constituents will influence how easily flame may be stabilized at the surface of a combustible solid (Section 6.3.2) while their nature will determine how much soot will be produced in the flame. The latter controls the amount of heat radiated from the flame to the surroundings and the burning surface (Sections 2.4.3, 5.1.1 and 5.2.1) and also influences the quantity of smoke that will be released from the fire (Section 11.1.1). Thus, volatiles containing aromatic species such as benzene (e.g. from the carbonaceous residue formed from chain-stripping of PVC, Equation (1.R3)), or styrene (from polystyrene), give sooty

Table 1.6   Relative thermal stability of organic polymers based on the temperature at which the half-life $T_h = 30$ min (from Madorsky, 1964)

Polymer	Polymer unit	T_h (°C)
Polymethylmethacrylate A (MW = 1.5×10^5)	$-CH_2-C(CH_3)-$ $\quad\quad\quad\quad \|$ $\quad\quad\quad CO.OCH_3$	283
Poly α-methyl styrene	$-CH_2-C(CH_3)-$ $\quad\quad\quad\quad \|$ $\quad\quad\quad C_6H_5$	287
Polyisoprene	$-CH_2-C=CH-CH_2-$ $\quad\quad\quad \|$ $\quad\quad\quad CH_3$	323
Polymethylmethacrylate B (MW = 5.1×10^6)	as A	327
Polymethylacrylate	$-CH_2-CH-$ $\quad\quad\quad \|$ $\quad\quad\quad CO.OCH_3$	328
Polyethylene oxide	$-CH_2-CH_2-O-$	345
Polyisobutylene	$-CH_2-C(CH_3)_2-$	348
Polystyrene (polyvinyl benzene)	$-CH_2-CH-$ $\quad\quad\quad\bigcirc$	364
Polypropylene	$-CH_2-CH-$ $\quad\quad\quad \|$ $\quad\quad\quad CH_3$	387
Polydivinyl benzene	$-CH_2-CH-$ $-CH_2-HC$	399
Polyethylene[a]	$-CH_2-CH_2-$	406
Polymethylene[a]	$-CH_2-CH_2-$	415
Polybenzyl	$-CH_2-\langle\bigcirc\rangle-$	430
Polytetrafluoroethylene	$-CF_2-CF_2-$	509

[a] Polyethylene and polymethylene differ only in that polymethylene is a straight chain with no side-chains. It requires very special processing. Polyethylene normally has a small degree of side branching which occurs randomly during the polymerization process.

flames of high emissivity (Section 2.4.3), while in contrast polyoxymethylene burns with a non-luminous flame, simply because the volatiles consist entirely of formaldehyde which does not produce soot. It will be shown later how these factors influence the rates of burning of solids and liquids (Sections 5.1.1 and 5.1.2). In some cases the toxicity of the combustion products is affected by the nature of the volatiles (cf. hydrochloric acid gas from PVC, hydrogen cyanide from wool and polyurethane, etc.) but the principal toxic species (carbon monoxide) is produced in all fires involving carbon-based fuels, and its yield is strongly dependent on the condition of burning and availability of air (see the introduction to Chapter 11).

Table 1.7 Factors affecting the thermal stability of polymers (from Madorsky, 1964)[a]

Factor	Effect on thermal stability	Examples (with values of T_h, °C)	
Chain branching[b]	Weakens	Polymethylene	(415)
		Polyethylene	(406)
		Polypropylene	(387)
		Polyisobutylene	(348)
Double bonds in polymer backbone	Weakens	Polypropylene	(387)
		Polyisoprene	(323)
Aromatic ring in polymer backbone	Strengthens	Polybenzyl	(430)
		Polystyrene	(364)
High molecular weight[c]	Strengthens	PMMA B	(327)
		PMMA A	(283)
Cross-linking	Strengthens	Polyvinyl benzene	(399)
		Polystyrene	(364)
Oxygen in the polymer backbone	Weakens	Polymethylene	(415)
		Polyethylene oxide	(345)
		Polyoxymethylene	(<200)

[a] While these are general observations, there are exceptions. For example, with some polyamides (nylons), stability decreases with increasing molecular weight (Madorsky, 1964).
[b] 'Branching' refers to the replacement of hydrogen atoms linked directly to the polymer backbone by *any* group, e.g. $-CH_3$ (as in polypropylene) or $-C_6H_5$ (as in polystyrene).
[c] Molecular weights of PMMA A and PMMA B are 150 000 and 5 100 000 respectively. Kashiwagi and Omori (1988) found significant differences in the time to ignition of two samples of PMMA of different molecular weights (Chapter 6).

1.2   The Physical Chemistry of Combustion in Fires

There are two distinct regimes in which gaseous fuels may burn, namely: (i) in which the fuel is intimately mixed with oxygen (or air) before burning, and (ii) in which the fuel and oxygen (or air) are initially separate but burn in the region where they mix. These give rise to premixed and diffusion flames, respectively: it is the latter that are encountered in the burning of gas jets and of combustible liquids and solids (Chapter 5). Nevertheless, an understanding of premixed burning is necessary for subsequent discussion of explosions and ignition phenomena (Chapter 6) and for providing a clearer insight into the elementary processes within the flame (Section 3.2).

In a diffusion flame, the rate of burning is equated with the rate of supply of gaseous fuel which, for gas jet flames (Section 4.1), is independent of the combustion processes. A different situation holds for combustible solids and liquids, for which the rate of supply of volatiles from the fuel surface is directly linked to the rate of heat transfer from the flame to the fuel (Figure 1.4). The rate of burning ($\dot{m}''$) can be expressed quite generally as

$$\dot{m}'' = \frac{\dot{Q}_F'' - \dot{Q}_L''}{L_v} \text{ g/m}^2.\text{s} \tag{1.3}$$

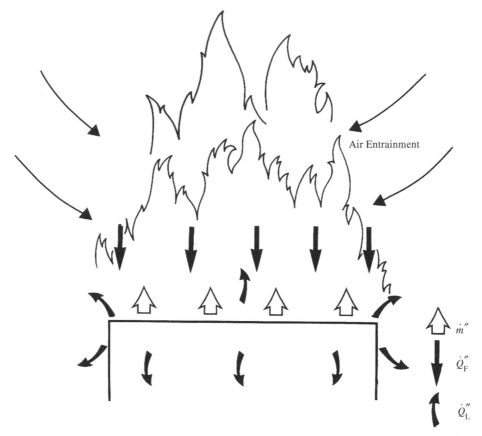

Figure 1.4 Schematic representation of a burning surface, showing the heat and mass transfer processes. $\dot{m}''$, mass flux from the surface: $\dot{Q}_F''$, heat flux from the flame to the surface; $\dot{Q}_L''$, heat losses (expressed as a flux from the surface)

where $\dot{Q}_F''$ is the heat flux supplied by the flame (kW/m^2) and $\dot{Q}_L''$ represents the losses expressed as a heat flux through the fuel surface (kW/m^2). L_v is the heat required to produce the volatiles (kJ/g) which, for a liquid, is simply the latent heat of evaporation (Table 5.8). The heat flux $\dot{Q}_F''$ must in turn be related to the rate of energy release within the flame and the mechanisms of heat transfer involved (see Section 5.1.1 and 5.2.1).

It will be shown later that the rate at which energy is released in a fire ($\dot{Q}_c$) is the most important single factor which characterizes its behaviour (Babrauskas and Peacock, 1992). It is given by an expression of the form:

$$\dot{Q}_c = \chi \cdot \dot{m}'' \cdot A_f \cdot \Delta H_c \text{ kW} \tag{1.4}$$

where A_f is the fuel surface area (m^2), ΔH_c (kJ/g) is the heat of combustion of the volatiles and χ is a factor (<1.0) included to account for incomplete combustion (Tewarson, 1982) (Table 5.12). It is now possible to determine the rate of heat release experimentally using the method of oxygen consumption calorimetry (Section 1.2.3), but Equation (1.4) can still be of value when there is limited information available (see Chapter 5).

Closer examination of Equations (1.3) and (1.4) reveals that there are many contributory factors which together determine $\dot{Q}_c$ including properties relating not only to the material itself (L_v and ΔH_c) but also to the combustion processes within the flame (which in turn determine $\dot{Q}_F''$ and χ). Equation (1.3) emphasizes the importance of the heat transfer terms $\dot{Q}_F''$ and $\dot{Q}_L''$ in determining the rate of supply of fuel vapours to the flame. Indeed, a detailed understanding of heat transfer is a prerequisite to any study of fire phenomena. Consequently, this subject is discussed at some length in Chapter 2, to which frequent reference is made throughout the book. The remainder of this chapter is devoted to a review of those aspects of physical chemistry that are relevant to the understanding of fire behaviour.

1.2.1 The ideal gas law

The release of heat in a fire causes substantial changes in the temperature of the surroundings (Section 10.3) as a result of heat transfer from flames and products of combustion which are formed at high temperatures. Most of the products are gaseous and their behaviour can be interpreted using the ideal gas law:

$$PV = nRT \qquad (1.5)$$

where V is the volume occupied by n moles of gas at a pressure P and temperature T (K). (In the SI system, the mole is the amount of a substance which contains as many elementary particles (i.e. atoms or molecules) as 0.012 kg of carbon-12. In practical terms, the mass of one mole of a substance is the molecular weight expressed in grams. Atomic weights, which may be used to calculate molecular weights, are given in Table 1.8.) R is known as the ideal (or universal) gas constant whose value will depend on the units of P and V (Table 1.9). For simplicity, when the ideal gas law

Table 1.8 Atomic weights of selected elements

	Symbol	Atomic number	Atomic weight
Aluminium	Al	13	27.0
Antimony	Sb	51	121.7
Argon	Ar	18	39.9
Boron	B	5	10.8
Bromine	Br	35	79.9
Carbon	C	6	12.0
Chlorine	Cl	17	35.5
Fluorine	F	9	19.0
Helium	He	2	4.0
Hydrogen	H	1	1.0
Nitrogen	N	7	14.0
Oxygen	O	8	16.0
Phosphorus	P	15	31.0
Sulphur	S	16	32.0

Table 1.9   Values of the ideal gas constant R

Units of pressure	Units of volume	Units of R	Value
N/m^2	m^3	J/K.mol	8.31431^a
atm	cm^3	cm^3.atm/K.mol	82.0575
atm	l	l.atm/K.mol	0.0820575
atm	m^3	m^3.atm/K.mol	8.20575×10^{-5}†

^{a} This is the value applicable to the SI system. However, in view of the variety of ways in which pressure is expressed in the literature. old and new, it is recommended here that the last value (†) is used, with pressure and volume in atmospheres and m^3, respectively.

Table 1.10   Standard atmospheric pressure

Units	Value
Atmospheres	1
Bars	1.01325
Inches of mercury (0°C)	29.9213
Inches of water (4°C)	406.794
kN/m^2 (kPa)	101.325
mm Hg (torr)	760.0

is used, pressure will be expressed in atmospheres as data available in the literature (particularly on the vapour pressures of liquids) are presented in a variety of units, including kN/m^2 (or kPa), mm of mercury (mmHg), and bars. Atmospheric pressure in these and other units is given in Table 1.10.

Equation (1.5) incorporates the laws of Boyle ($PV = $ constant at constant temperature) and Gay-Lussac ($V/T = $ constant at constant pressure), and Avogadro's hypothesis, which states that equal volumes of different gases at the same temperature and pressure contain the same number of molecules (or atoms, in the case of an atomic gas such as helium). Setting $P = 1$ atm, $T = 273.17$ K (0°C) and $n = 1$ mole,

$$V = 0.022\,414 \text{ m}^3 \qquad (1.6)$$

This is the volume that will be occupied by 28 g N$_2$, 32 g O$_2$ or 44 g CO$_2$ at atmospheric pressure and 0°C, assuming that these gases behave ideally. This is not so, but the assumption is reasonable at elevated temperatures. Deviation from ideality increases as the temperature is reduced towards the liquefaction point, but nevertheless, Equation (1.5) can be used in a number of ways to interpret and illustrate features of fire behaviour.

The density, or concentration, of a gas may be calculated: for example, taking the composition of normal air as given in Table 1.11, it can be shown that one mole corresponds to $M_w = 0.028\,95$ kg, so that its density at 0°C (273 K) will be:

$$\rho = \frac{nM_w}{V} = \frac{PM_w}{RT} = 1.292 \text{ kg/m}^3 \qquad (1.7)$$

(see Table 11.6). The composition of a mixture of gases may also be expressed in terms of partial pressures (P_i) of the components, i, so that

Table 1.11 Normal composition of the dry atmosphere[a]

Constituent gas[b]	Mole fraction (%)
Nitrogen (N_2)	78.09
Oxygen (O_2)	20.95
Argon (Ar)	0.93
Carbon dioxide (CO_2)	0.03

[a] From Weast (1974/75). It is convenient for many purposes to assume that air consists only of oxygen (21%) and nitrogen (79%). The molar ratio N_2/O_2 is then $79/21 = 3.76$.
[b] Minor constituents include neon (1.8×10^{-3}%), helium (5.2×10^{-4}%). krypton (1×10^{-4}%) and hydrogen (5×10^{-5}%).

$$P = \sum_i P_i \tag{1.8}$$

where P is the total pressure. As the volume fraction of oxygen in normal air is 0.2095, its partial pressure will be 0.2095 atm. This can be converted into a mass concentration as before: thus at 273 K

$$\frac{PM_w}{RT} = \frac{0.2095 \times 0.032}{273R} \tag{1.9}$$

$$= 0.2993 \text{ kg } O_2/m^3$$

which gives the mass fraction of oxygen in air (Y_{O_2}) as $0.2993/1.2923 = 0.232$, a quantity which is referred to later (e.g. Equation (5.22)).

The effect of increasing the temperature of a volume of gas can be seen by referring to Equation (1.5): if the volume is kept constant then the pressure will rise in direct proportion to the temperature increase (see Section 1.1.5), while if the pressure is held constant, the gas will expand (V increases) and its density fall (Equation (1.7)). In a fire, the high temperatures that are generated in the burning gases give rise to strong buoyant flows, driven by differences in density between the hot gases and the ambient atmosphere (Section 2.3). These create pressure differentials which, although small in absolute terms, are responsible for drawing air into the base of the fire and for the expulsion of flame and hot gases from confined locations (Section 10.2). While these small differences can be taken into account, to a first approximation they may be neglected in the present discussion.

The variation of density (ρ) with temperature (at constant pressure) is given by Equation (1.7), i.e.

$$\rho = \frac{PM_w}{R} \cdot \frac{1}{T} \tag{1.10}$$

where $P \cdot M_w/R$ is constant, thus the product ρT constant. Consequently,

$$\frac{\rho_0 - \rho_\infty}{\rho_0} = \frac{T_\infty - T_0}{T_\infty} \tag{1.11}$$

where the subscripts 0 and ∞ refer to initial (or ambient) and final conditions respectively. As $T_\infty = P \cdot M_w/R \cdot \rho_\infty$, this can be rearranged to give

$$\frac{\Delta\rho}{\rho_\infty} = \beta\Delta T \tag{1.12}$$

where $\beta = R\rho_0/P \cdot M_w = 3.66 \times 10^{-3}$ K^{-1}, at the reference state of 1 atmosphere and 0°C. β is the reciprocal of 273 K and is known as the coefficient of thermal expansion. It was first derived for gases by Gay-Lussac.

If there is any density difference between adjacent masses of air, or indeed any other fluid, relative movement will occur. As the magnitude of this difference determines the buoyant force, the dimensionless group which appears in problems relating to natural convection (the Grashof number, see Section 2.3), can be expressed in terms of either $\Delta\rho/\rho_\infty$ or $\beta\Delta T$ (Table 4.6).

1.2.2   Vapour pressure of liquids

When exposed to the open atmosphere, any liquid which is stable under normal ambient conditions of temperature and pressure (e.g. water, *n*-hexane) will evaporate as molecules escape from the surface to form vapour. (Unstable liquids, such as LPG, will be discussed briefly in Chapter 5.) If the system is closed (cf. Figure 6.8(a)) a state of kinetic equilibrium will be achieved when the partial pressure of the vapour above the surface reaches a level at which there is no further net evaporative loss. For a pure liquid, this is the saturated vapour pressure, a property which varies with temperature according to the Clapeyron-Clausius equation,

$$\frac{\mathrm{d}(\ln p^\circ)}{\mathrm{d}T} = \frac{L_v}{RT^2} \tag{1.13}$$

where p° is the equilibrium vapour pressure and L_v is the latent heat of evaporation (Moore, 1972). An integrated form of this is commonly used, for example

$$\log_{10} p^\circ = (-0.2185E/T) + F \tag{1.14}$$

where E and F are constants, T is in Kelvin and p° is in mm Hg. Values of these for some liquid fuels are given in Table 1.12 (Weast, 1974/5).

The equation may be used to calculate the vapour pressure above the surface of a pure liquid fuel to assess the flammability of the vapour/air mixture (Sections 3.1 and 6.2). The same procedure may be employed for liquid fuel mixtures if the vapour pressures of the components can be calculated. For 'ideal solutions' to which hydrocarbon mixtures approximate, Raoult's law can be used. This states that for a mixture of two liquids, A and B,

$$p_A = x_A \cdot p_A^\circ \quad \text{and} \quad p_B = x_B \cdot p_B^\circ \tag{1.15}$$

where p_A and p_B are the partial vapour pressures of A and B above the liquid mixture, p_A° and p_B° are the equilibrium vapour pressures of pure A and B (given by Equation (1.14)) and x_A and x_B are the respective 'mole fractions', i.e.

$$x_A = \frac{n_A}{n_A + n_B} \quad \text{and} \quad x_B = \frac{n_B}{n_A + n_B} \tag{1.16}$$

Table 1.12   Vapour pressures of organic compounds (Weast, 1974/75)

Compound	Formula	E	F	Temperature range (°C)
n-Pentane	n-C_5H_{12}	6595.1	7.4897	−77 to 191
n-Hexane	n-C_6H_{14}	7627.2	7.7171	−54 to 209
Cyclohexane	c-C_6H_{12}	7830.9	7.6621	−45 to 257
n-Octane	n-C_8H_{18}	9221.0	7.8940	−14 to 281
iso-Octane (2,2,4-Trimethyl pentane)		8548.0	7.9349	−36 to 99
n-Decane	n-$C_{10}H_{22}$	10912.0	8.2481	17 to 173
n-Dodecane	n-$C_{12}H_{26}$	11857.7	8.1510	48 to 346
Methanol	CH_3OH	8978.8	8.6398	−44 to 224
Ethanol	C_2H_5OH	9673.9	8.8274	−31 to 242
n-Propanol	n-C_3H_7OH	10421.1	8.9373	−15 to 250
Acetone	$(CH_3)_2CO$	7641.5	7.9040	−59 to 214
Methyl ethyl ketone	$CH_3CO.CH_2CH_3$	8149.5	7.9593	−48 to 80
Benzene	C_6H_6	8146.5	7.8337	−37 to 290
Toluene	$C_6H_5CH_3$	8580.5	7.7194	−28 to 31
Styrene	$C_6H_5CH = CH_2$	9634.7	7.9220	−7 to 145

Vapour pressures are calculated using the following equation: $\log_{10} p° = (−0.2185E/T) + F$ where $p°$ is the pressure in mm Hg (torr) (Table 1.10), T is the temperature (Kelvin) and E is the molar heat of vaporization. (Note that the temperature range in the table is given in °C.)

where n_A and n_B are the molar concentrations of A and B in the mixture. (These are obtained by dividing the mass concentrations (C_A and C_B) by the molecular weights $M_w(A)$ and $M_w(B)$.) In fact, very few liquid mixtures behave ideally and substantial deviations will be found, particularly if the molecules of A or B are partially associated in the pure state (e.g. water, methanol) or if A and B are of different polarity (Moore, 1972). Partial pressures must then be calculated using the activities of A and B in the solution, thus:

$$p_A = \alpha_A \cdot p_A° \text{ and } p_B = \alpha_B \cdot p_B° \qquad (1.17)$$

where

$$\alpha_A = \gamma_A \cdot x_A \text{ and } \alpha_B = \gamma_B \cdot x_B$$

α and γ being known as the activity and the activity coefficient, respectively. The latter is unity for an ideal solution. Values for specific mixtures are available in the literature (e.g. Perry *et al.*, 1984) and have been used to predict the flashpoints of mixtures of flammable and non-flammable liquids from data on flammability limits (Thorne, 1976). However, this will not be pursued further.

1.2.3   Combustion and energy release

All combustion reactions take place with the release of energy. This may be quantified by defining the heat of combustion (ΔH_c) as the total amount of heat released when unit quantity of a fuel (at 25°C and at atmospheric pressure) is oxidized completely. For a hydrocarbon such as propane (C_3H_8) the products would comprise only carbon dioxide and water, as indicated in the stoichiometric equation:

$$C_3H_8 + 5\,O_2 \rightarrow 3\,CO_2 + 4\,H_2O \qquad (1.R4)$$

in which the fuel and the oxygen are in exactly equivalent — or stoichiometric — proportions. The reaction is exothermic (i.e. heat is produced) and the value
of ΔH_c will depend on whether the water in the products is in the form of liquid or
vapour. The difference will be the latent heat of evaporation of water (44 kJ/mol at
25°C): thus for propane, the two values are:

$$\Delta H_c(C_3H_8) = -2220 \text{ kJ/mol}$$

$$\Delta H_c(C_3H_8) = -2044 \text{ kJ/mol}$$

where the products are liquid water, and water vapour, respectively. In flames and
fires, the water remains as vapour and consequently it is more appropriate to use the
latter value.

The heat of combustion of propane can be expressed as either -2044 kJ/mol or
$-(2044/44) = -46.45$ kJ/g of propane (Table 1.13) where 44 is the gram molecular
weight of C_3H_8. If the reaction is allowed to proceed at constant pressure, the energy
is released as a result of a change in *enthalpy* (ΔH) of the system as defined by
Equation (1.R4). However, heats of combustion are normally determined at constant
volume in a 'bomb' calorimeter, in which a known mass of fuel is burnt completely in
an atmosphere of pure oxygen (Moore, 1972). Assuming that there is no heat loss (the
system is adiabatic), the quantity of heat released is calculated from the temperature rise
of the calorimeter and its contents, whose thermal capacities are accurately known. The
use of pure oxygen ensures complete combustion and the result gives the heat released
at constant volume, i.e. the change in the internal energy (ΔU) of the system defined
by Equation (1.R4). The difference between the enthalpy change (ΔH) and the internal
energy change (ΔU) exists because at constant pressure some of the chemical energy
is effectively lost as work done $(P\Delta V)$ in the expansion process. Thus, ΔH can be
calculated from:

$$\Delta H = \Delta U + P\Delta V \qquad (1.18)$$

remembering that for exothermic reactions both ΔH and ΔU are negative. The work
done may be estimated using the ideal gas law, i.e.

$$PV = nRT \qquad (1.5)$$

where n is the number of moles of gas involved. If there is a change in n, as in
Equation (1.R4), then:

$$P\Delta V = \Delta nRT \qquad (1.19)$$

where $\Delta n = 7 - 6 = +1$ and $T = 298$ K. It can be seen that in this case the correction
is small ($\sim$2.5 kJ/mol) and may be neglected in the present context, although it is
significant given the accuracy with which heats of combustion can now be measured.

Bomb calorimetry provides the means by which heats of formation of many
compounds may be determined. Heat of formation (ΔH_f) is defined as the enthalpy
change when a compound is formed in its standard state (1 atmosphere pressure and

Table 1.13   Heats of combustion^a of selected fuels at 25°C (298 K)

		$-\Delta H_c$ (kJ/mol)	$-\Delta H_c$ (kJ/g)	$-\Delta H_{c,air}$ (kJ/g(air))	$-\Delta H_{c,ox}$ (kJ/g(O$_2$))
Carbon monoxide	CO	283	10.10	4.10	17.69
Methane	CH$_4$	800	50.00	2.91	12.54
Ethane	C$_2$H$_6$	1423	47.45	2.96	11.21
Ethene	C$_2$H$_4$	1411	50.35	3.42	14.74
Ethyne	C$_2$H$_2$	1253	48.20	3.65	15.73
Propane	C$_3$H$_8$	2044	46.45	2.97	12.80
n-Butane	*n*-C$_4$H$_{10}$	2650	45.69	2.97	12.80
n-Pentane	*n*-C$_5$H$_{12}$	3259	45.27	2.97	12.80
n-Octane	*n*-C$_8$H$_{18}$	5104	44.77	2.97	12.80
c-Hexane	*c*-C$_6$H$_{12}$	3680	43.81	2.97	12.80
Benzene	C$_6$H$_6$	3120	40.00	3.03	13.06
Methanol	CH$_3$OH	635	19.83	3.07	13.22
Ethanol	C$_2$H$_5$OH	1232	26.78	2.99	12.88
Acetone	(CH$_3$)$_2$CO	1786	30.79	3.25	14.00
D-Glucose	C$_6$H$_{12}$O$_6$	2772	15.4	3.08	13.27
Cellulose		—	16.09	3.15	13.59
Polyethylene		—	43.28	2.93	12.65
Polypropylene		—	43.31	2.94	12.66
Polystyrene		—	39.85	3.01	12.97
Polyvinylchloride		—	16.43	2.98	12.84
Polymethylmethacrylate		—	24.89	3.01	12.98
Polyacrylonitrile		—	30.80	3.16	13.61
Polyoxymethylene		—	15.46	3.36	14.50
Polyethyleneterephthalate		—	22.00	3.06	13.21
Polycarbonate		—	29.72	3.04	13.12
Nylon 6,6		—	29.58	2.94	12.67

^{a} The initial states of the fuels correspond to their natural states at normal temperature and pressure (298°C and 1 atm pressure). All products are taken to be in their gaseous state—thus these are the net heats of combustion.

298 K) from its constituent elements, also in their standard states. That for carbon dioxide is the heat of the reaction

$$C\ (graphite) + O_2(gas) \rightarrow CO_2(gas) \tag{1.R5}$$

where $\Delta H_f^{298}(CO_2) = -393.5$ kJ/mol. The negative sign indicates that the product (CO$_2$) is a more stable chemical configuration than the reactant elements in their standard states, which are assigned heats of formation of zero.

If the heats of formation of the reactants and products of any chemical reaction are known, the total enthalpy change can be calculated: thus for propane oxidation (Equation (1.R4)):

$$\Delta H_c(C_3H_8) = 3\Delta H_f(CO_2) + 4\Delta H_f(H_2O) - \Delta H_f(C_3H_8) - \Delta H_f(O_2) \tag{1.20}$$

in which $\Delta H_f(O_2) = 0$ (by definition). This incorporates Hess' law of constant heat summation, which states that the change in enthalpy depends only on the initial and final states of the system and is independent of the intermediate steps. In fact, $\Delta H_c(C_3H_8)$ is

Table 1.14   Standard heats of formation of some common gases

Compound	Formula	ΔH_f^{298} (kJ/mol)
Water (vapour)	H_2O	-241.826
Carbon monoxide	CO	-110.523
Carbon dioxide	CO_2	-393.513
Methane	CH_4	-74.75
Propane	C_3H_8	-103.6
Ethene	C_2H_4	$+52.6$
Propene	C_3H_6	$+20.7$
Ethyne	C_2H_2	$+226.9$

easily determined by combustion bomb calorimetry, as are $\Delta H_f(CO_2)$ and $\Delta H_f(H_2O)$, and Equation (1.20) would be used to calculate $\Delta H_f(C_3H_8)$, which is the heat of the reaction

$$3\text{ C (graphite)} + 4\text{ H}_2(\text{gas}) \rightarrow C_3H_8(\text{gas}) \tag{1.R6}$$

Values of the heats of formation of some common gaseous species are given in Table 1.14. Those species for which the values are positive (e.g. ethene and ethyne) are less stable than the parent elements and are known as endothermic compounds. Under appropriate conditions they can be made to decompose with the release of energy. Ethyne, which has a large positive heat of formation, can decompose with explosive violence.

Values of heat of combustion for a range of gases. liquids and solids are given in Table 1.13: these all refer to normal atmospheric pressure (101.3 kPa) and an ambient temperature of 298 K (25°C) and to complete combustion. It should be noted that the values quoted for ΔH_c are the net heats of combustion, i.e. the product water is in the vapour state. They differ from the gross heats of combustion by the amount of energy corresponding to the latent heat of evaporation of the water (2560 kJ/kg at 100°C). Furthermore, it is not uncommon in fires for the combustion process to be incomplete, i.e. χ in Equation (1.4) is less than unity. The actual heat released could be estimated by using Hess' law of constant heat summation if the composition of the combustion products was known. The oxidation of propane could be written as a two-stage process, involving the reactions:

$$C_3H_8 + \tfrac{7}{2}O_2 \rightarrow 3CO + 4\text{ H}_2O \tag{1.R7}$$

and

$$CO + \tfrac{1}{2}O_2 \rightarrow CO_2 \tag{1.R8}$$

Equation (1.R4) can be obtained by adding together (1.R7) and three times (1.R8). Then by Hess' law

$$\Delta H_{R4} = \Delta H_{R7} + 3\Delta H_{R8} \tag{1.21}$$

where ΔH_{R7} is the heat of reaction (1.R7), $\Delta H_{R4} = \Delta H_c(C_3H_8)$ and $\Delta H_{R8} = \Delta H_c(CO)$. As these heats of combustion are both known (Table 1.13), ΔH_{R7} can be shown to be

$$\Delta H_{R7} = 2044 - 3 \times 283 = 1195 \text{ kJ/mol}$$

Consequently, if it was found that the partial combustion of propane gave only H_2O with CO_2 and CO in the ratio 4:1, the actual heat released per mole of propane burnt would be

$$\Delta H = \frac{4\Delta H_c(C_3H_8) + \Delta H_{R7}}{5} \tag{1.22}$$

or

$$\Delta H = \Delta H_c(C_3H_8) - \tfrac{3}{5}\Delta H_c(CO) \tag{1.23}$$

which give the same answer (-1874.2 kJ/mol).

The techniques of thermochemistry provide essential information about the amount of heat liberated during a combustion process that has gone to completion. In principle, a correction can be made if the reaction is incomplete, although the large number of products of incomplete combustion that are formed in fires make this approach cumbersome, and effectively unworkable. Yet, information on the rate of heat release in a fire is often required in engineering calculations (Babrauskas and Peacock, 1992), for example in the estimation of flame height (Section 4.3.2), the flashover potential of a room (Section 9.2.2) or the temperatures achieved in a compartment fire (Section 10.3). Until relatively recently, it was common practice to calculate the rate of energy release using Equation (1.4), taking an appropriate value of $\dot{m}''$ and assuming a value for χ to account for incomplete combustion, but an experimental method is now available by which $\dot{Q}_c$ can be determined. This relies on the fact that the heat of combustion of most common fuels is constant if it is expressed in terms of the oxygen, or air consumed. Taking Equation (1.R4) as an example, it can be said that 2044 kJ are evolved for each mole of propane burnt, or for every five moles of oxygen consumed. The heat of combustion could then be quoted as $\Delta H_{c,ox} = -408.8$ kJ/mol or $(-408.8/32) = -12.77$ kJ/g, where 32 is the molecular weight of oxygen. $\Delta H_{c,ox}$ is given in Table 1.13 for a range of fuels and is seen to lie within fairly narrow limits. Huggett (1980) concluded that typical organic liquids and gases have $\Delta H_{c,ox} = -12.72 \pm 3\%$ kJ/g of oxygen (omitting the reactive gases ethene and ethyne) while polymers have $\Delta H_{c,ox} = -13.02 \pm 4\%$ kJ/g of oxygen (omitting polyoxymethylene). Consequently, if the rate of oxygen consumption can be measured, the rate of heat release can be estimated directly. This method is now widely used both in fire research and in routine testing. It is the basis on which the Cone Calorimeter (Babrauskas, 1992a, 1995b) is founded, and has been adopted in other laboratory-scale apparatuses, such as the Factory Mutual Flammability Apparatus (Tewarson, 1995). The technique is also used in large-scale equipment (Babrauskas, 1992b) such as the Nordtest/ISO Room Fire Test (Sundström, 1984; Nordtest, 1991), the ASTM room (ASTM 1982), and the Furniture Calorimeter (Babrauskas *et al.*, 1982; Nordtest, 1991) (see Figure 9.15(a)). These are designed to study the full-scale fire behaviour of wall lining materials, items of furniture and other commodities. The common feature of all these apparatuses is the system for measuring the rate at which oxygen has been consumed in the fire. This involves a hood and duct assembly, the combustion products flowing through a duct of known cross-sectional area and in which careful measurements are made of temperature and velocity, and of the concentrations of oxygen, carbon monoxide and carbon dioxide. If the combustion process is complete (i.e. the only products are water and carbon dioxide), the rate of heat release may be calculated from the expression

$$\dot{Q}_c = (0.21 - \eta_{O_2}) \cdot V \cdot 10^3 \cdot \rho_{O_2} \cdot \Delta H_{c,ox} \tag{1.24}$$

where V is the volumetric flow of air (m^3/s), ρ_{O_2} is the density of oxygen (kg/m^3) at normal temperature and pressure, and η_{O_2} is the mole fraction of oxygen in the 'scrubbed' gases (i.e. water vapour and acid gases have been removed).

The average value $\Delta H_{c,ox}$ is taken as -13.1 kJ/g(O$_2$) (see Table 1.13), assuming complete combustion to H$_2$O and CO$_2$. Krause and Gann (1980) argue that if combustion is incomplete, i.e. carbon monoxide and soot particles are formed, the effect on the calculated rate of heat release will be small. Their reasoning rests on the fact that if all the carbon was converted to CO, the value used for the heat of combustion ($\Delta H_{c,ox}$) would be no more than 30% too high, while if it all appeared as carbon (smoke particles), it could be no more than 20–25% too low. Given that these factors operate in opposite directions and that in most fires the yield of CO$_2$ is invariably much higher than that of CO, the resulting error is unlikely to be more than 5%. However, this is not sufficiently accurate for current research and testing procedures, and corrections are necessary, based on the amount of carbon monoxide in the products (see Equations (1.R7), (1.R8), *et seq.*) and the yields of carbon dioxide and water vapour. It is not appropriate to examine this any further in this context as the equations are presented in great detail elsewhere (Janssens, 1991b, 1995; Janssens and Parker, 1992).

The heat of combustion may also be expressed in terms of air 'consumed'. Equation (1.R4) may be modified to include the nitrogen complement, thus:

$$C_3H_8 + 5\,O_2 + 5 \times 3.76\,N_2 \rightarrow 3\,CO_2 + 4\,H_2O + 18.8\,N_2 \tag{1.R9}$$

as the ratio of nitrogen to oxygen in air is approximately 3.76 (Table 1.11). Repeating the calculation as before, 2044 kJ are evolved when the oxygen in 23.8 moles of air is consumed. Thus, $\Delta H_{c,air} = 85.88$ kJ/mol or $(85.88/28.95) = 2.97$ kJ/g, where 28.95 is the 'molecular weight' of air (Section 1.2.1). Values quoted in Table 1.13 cover a wide variety of fuels of all types and give an average of 3.03 ($\pm$2%) kJ/g if carbon monoxide and the reactive fuels ethene and ethyne are discounted. A value of 3 kJ/g is a convenient figure to select and is within 12% for the one polymer that appears to behave significantly differently (polyoxymethylene). This may be used to estimate the rate of heat release in a fully developed, ventilation-controlled compartment fire if the rate of air inflow is known or can be calculated, and it is assumed that all the oxygen is consumed within the compartment boundaries (Section 10.3.2).

Stoichiometric equations such as (1.R9) can be used to calculate the air requirements for the complete combustion of any fuel. For example, polymethylmethacrylate has the empirical formula C$_5$H$_8$O$_2$ (Table 1.2), identical to the formula of the monomer. The stoichiometric equation for combustion in air may be written:

$$C_5H_8O_2 + 6\,O_2 + 6 \times 3.76\,N_2 \rightarrow 5\,CO_2 + 4\,H_2O + 22.56\,N_2 \tag{1.R10}$$

which shows that 1 mole of PMMA monomer unit requires 28.56 moles of air. Introducing the molecular weights of C$_5$H$_8$O$_2$ and air (100 and 28.95 respectively), it is seen that 1 g of PMMA requires 8.27 g of air for stoichiometric burning to CO$_2$ and water. (In the same way, it can be shown that 15.7 g of air are required to 'burn' 1 g of propane to completion: see problem 1.11). More generally, we can write

$$1 \text{ kg fuel} + r \text{ kg air} \rightarrow (1 + r) \text{ kg products}$$

where r is the stoichiometric air requirement for the fuel in question. This will be applied in Sections 10.1 and 10.2.

The stoichiometric air requirement can be used to estimate the heat of combustion of any fuel, if this is not known. Taking the example of PMMA, as $\Delta H_{c,air} = 3 \text{ kJ/g}$, then $\Delta H_c(\text{PMMA}) = 3 \times 8.27 = 24.8 \text{ kJ/g}$, in good agreement with the value quoted in Table 1.13.

1.2.4   The mechanism of gas phase combustion

Chemical equations such as (1.R4) and (1.R7) define the stoichiometry of the complete reaction but hide the complexity of the overall process. Thus, while methane will burn in a flame to yield carbon dioxide and water vapour, according to the reaction

$$CH_4 + 2 O_2 \rightarrow CO_2 + 2 H_2O \tag{1.R11}$$

the mechanism by which this takes place involves a series of elementary steps in which highly reactive molecular fragments (atoms and free radicals), such as $H\cdot$, $\cdot OH$ and $\cdot CH_3$ take part (Table 1.15). While these have only transient existences within the flame, they

Table 1.15   Mechanism of the gas-phase oxidation of methane (after Bowman, 1975)

	CH_4	+	M	=	$\cdot CH_3$	+	$H\cdot$	+	M	a
	CH_4	+	$\cdot OH$	=	$\cdot CH_3$	+	H_2O			b
	CH_4	+	$H\cdot$	=	$\cdot CH_3$	+	H_2			c
	CH_4	+	$\cdot O\cdot$	=	$\cdot CH_3$	+	$\cdot OH$			d
	O_2	+	$H\cdot$	=	$\cdot O\cdot$	+	$\cdot OH$			e
	$\cdot CH_3$	+	O_2	=	CH_2O	+	$\cdot OH$			f
	CH_2O	+	$\cdot O\cdot$	=	$\cdot CHO$	+	$\cdot OH$			g
	CH_2O	+	$\cdot OH$	=	$\cdot CHO$	+	H_2O			h
	CH_2O	+	$H\cdot$	=	$\cdot CHO$	+	H_2			i
	H_2	+	$\cdot O\cdot$	=	$H\cdot$	+	$\cdot OH$			j
	H_2	+	$\cdot OH$	=	$H\cdot$	+	H_2O			k
	$\cdot CHO$	+	$\cdot O\cdot$	=	CO	+	$\cdot OH$			l
	$\cdot CHO$	+	$\cdot OH$	=	CO	+	H_2O			m
	$\cdot CHO$	+	$H\cdot$	=	CO	+	H_2			n
	CO	+	$\cdot OH$	=	CO_2	+	$H\cdot$			o
$H\cdot$ +	$\cdot OH$	+	M	=	H_2O	+	M			p
$H\cdot$ +	$H\cdot$	+	M	=	H_2	+	M			q
$H\cdot$ +	O_2	+	M	=	$HO_2\cdot$	+	M			r

This reaction scheme is by no means complete. Many radical-radical reactions, including those of the HO_2 radical, have been omitted.

M is any 'third body' participating in radical recombination reactions (p-r) and dissociation reactions such as a.

are responsible for rapid consumption of the fuel (reactions b-d in Table 1.15). Their concentration is maintained because they are continuously regenerated in a sequence of chain reactions, e.g.

$$CH_4 + {}^{\cdot}OH \rightarrow H_2O + {}^{\cdot}CH_3 \qquad (1.R12)$$

$${}^{\cdot}CH_3 + O_2 \rightarrow CH_2O + {}^{\cdot}OH \qquad (1.R13)$$

(Table 1.15, b and f) although they are also destroyed in chain termination reactions such as p and q (Table 1.15).

The rate of oxidation of methane may be equated to its rate of removal by reactions b-d. This may be written:

$$-\frac{d[CH_4]}{dt} = k_b[CH_4][{}^{\cdot}OH] + k_c[CH_4][H^{\cdot}] + k_d[CH_4][{}^{\cdot}O]$$

$$= (k_b[{}^{\cdot}OH] + k_c[H^{\cdot}] + k_d[{}^{\cdot}O])[CH_4] \qquad (1.25)$$

where the square brackets indicate concentration, and k_b, k_c and k_d are the appropriate rate coefficients (cf. Equation (1.1)). Clearly, the rate of removal of methane depends directly on the concentrations of free atoms and radicals in the reacting system. This is turn will depend on the rates of initiation (reaction a) and termination (m, n, p and q), but will be enhanced greatly if the branching reaction (e) is significant, i.e.

$$O_2 + H^{\cdot} \rightarrow {}^{\cdot}O^{\cdot} + {}^{\cdot}OH \qquad (1.R14)$$

This has the effect of increasing the number of radicals in the system, replacing one hydrogen atom by three free radicals, as can be seen by examining the fate of the oxygen atoms by reactions d, g and j (Table 1.15). In this respect the hydrogen atom is arguably the most important of the reactive species in the system. If other molecules compete with oxygen for H-atoms (e.g. reactions c, i and n), then the branching process, i.e. the multiplication in number of free radicals, is held in check. No such check exists in the H_2/O_2 reaction in which the oxygen molecule is effectively the only gaseous species with which the hydrogen atoms can react (e.g. Dixon-Lewis and Williams. 1977; Simmons, 1995a, 1995b).

$$O_2 + H^{\cdot} \rightarrow {}^{\cdot}O^{\cdot} + {}^{\cdot}OH \qquad (1.R14)$$

$$H_2 + {}^{\cdot}O^{\cdot} \rightarrow H^{\cdot} + {}^{\cdot}OH \qquad (1.R15)$$

$$H_2 + {}^{\cdot}OH \rightarrow H^{\cdot} + H_2O \qquad (1.R16)$$

(Reaction r in Table 1.15 is relatively unimportant in flames.) Consequently, under the appropriate conditions (Section 3.1) the rate of oxidation of hydrogen in air is very high, as indicated by its maximum burning velocity, $S_u = 3.2$ m/s, which is more than eight times greater than that for methane (0.37 m/s) (Table 3.1).

Species that react rapidly with hydrogen atoms, effectively replacing them with atoms or radicals which are considerably less reactive, can inhibit gas-phase oxidation. Chlorine- and bromine-containing compounds can achieve this effect by giving rise to hydrogen halides (HCl or HBr) in the flame. Reactions such as

$$HBr + H^{\cdot} \rightarrow H_2 + Br^{\cdot} \qquad (1.R17)$$

replace hydrogen atoms with relatively inactive halogen atoms and thereby reduce the overall rate of reaction dramatically. For this reason many chlorine- and bromine-containing compounds are found to be valuable for retardants (Lyons, 1970) or chemical extinguishants (see Section 3.5.4).

Returning to the discussion of methane oxidation, it is seen that the molecule CH_2O (formaldehyde) is formed in Reaction (1.R13) (Table 1.15, f) as an intermediate. Under conditions of complete combustion this would be destroyed in Reactions g-i (Table 1.15), but if the reaction sequence was interrupted as a result of chemical or physical quenching, some formaldehyde could survive and appear in the products. Similarly carbon monoxide — the most ubiquitous product of incomplete combustion — will also be released if the concentration of hydroxyl radicals ($\cdot$OH) is not sufficient to allow the reaction

$$CO + \cdot OH \rightarrow CO_2 + H\cdot \tag{1.R18}$$

to proceed to completion. This equation represents the only significant reaction by which carbon monoxide is oxidized and indeed is the principal source of carbon dioxide in any combustion system (e.g. Baulch and Drysdale, 1974).

The complexity of the gas phase oxidation process is shown clearly in Table 1.15 for the simplest of the hydrocarbons, methane, although even this reaction scheme is incomplete (Bowman (1975) lists 30 reactions). The complexity increases with the size and structure of the fuel molecule and consequently the number of partially oxidized species that may be produced becomes very large. Most studies of gas phase combustion have been carried out using well-mixed (*premixed*) fuel/air mixtures (Chapter 3), but in natural fires mixing of fuel vapours and air is an integral part of the burning process (Chapter 4): these flames are known as *diffusion flames*. Consequently, the combustion process is much less efficient, burning occurring only in those regions where there is a sufficient presence of fuel and oxidant. In a free-burning fire in the open, more air is entrained into the flame than is required to burn all the vapours (Steward, 1970; Heskestad, 1986; see Section 4.3.2): despite this, some products of incomplete combustion survive the flame and are released to the atmosphere. Not all the products are gaseous: some are minute carbonaceous particles formed within the flame under conditions of low oxygen and high temperature (e.g. Rasbash and Drysdale, 1982). These make up the 'particulate' component of smoke which reduces visibility remote from the fire (Section 11.1), and can be formed even under 'well-ventilated' conditions, depending on the nature of the fuel. For example, polystyrene produces a great deal of black smoke because of the presence of phenyl groups (C_6H_5) in the polymer molecule (Table 1.6 *et seq.*). This is a very stable structure which is not only resistant to oxidation, but it is ideally suited to act as the building unit for the formation of carbonaceous particles within the flame (Section 11.1.1).

If the supply of air to a fire is restricted in some way — for example if it is burning in an enclosed space, or 'compartment', with restricted ventilation (see Figure 9.3) — then the yield of incompletely burned products will increase (e.g. Rasbash, 1967; Woolley and Fardell, 1982). It is convenient to introduce here the concept of 'equivalence ratio', a term normally associated with premixed fuel/air mixtures. It is given by the expression

$$\phi = \frac{(\text{fuel/air})_{\text{actual}}}{(\text{fuel/air})_{\text{stoich}}} \tag{1.26}$$

where (fuel/air)$_{stoich}$ is the stoichiometric fuel/air ratio. The terms 'lean' and 'rich' refer to the situations where $\phi < 1$ and $\phi > 1$ respectively. In diffusion flames, assuming that the rate of fuel supply is known, a value can only be assigned to ϕ if the rate of air supply into the flame can be deduced or can be measured. This cannot be done meaningfully for free-burning diffusion flames in the open, but the concept has been used to interpret the results of experimental studies of the composition of the smoke layer under a ceiling, at least up to the condition known as 'flashover' (Beyler, 1984b; Gottuk *et al.*, 1992; Pitts, 1994). In these, careful attention was paid to the dependence of the yields of partially burned products (in particular, carbon monoxide) on the 'equivalence ratio' (ϕ) (Section 9.2.1). In general, high yields of carbon monoxide are to be associated with high equivalence ratios: there is competition for the hydroxyl radicals between CO (see (1.R18)) and other partially burned products (e.g. Reaction h in Table 1.15) which causes an effective reduction in the rate of conversion of CO to CO_2 (e.g. Pitts, 1994). This conversion is also suppressed in the presence of soot particles which are now known to react with hydroxyl radicals (Neoh *et al.*, 1984; Puri and Santoro, 1991).

1.2.5   *Temperatures of flames*

In a fire, the total amount of heat that can be released is normally of secondary importance to the rate at which it is released (Section 1.2.3). If the heat of combustion is known, this rate may be calculated from Equation (1.4), provided that $\dot{m}'' \cdot A_f \cdot \chi$ is known. This is seldom the case for fires burning in enclosures but the rate may be estimated if the rate of air inflow ($\dot{m}_{air}$) is known. Then, assuming that all the oxygen is consumed within the enclosure, the rate of heat release is:

$$\dot{Q}_c = \dot{m}_{air} \cdot \Delta H_{c,air} \tag{1.27}$$

The temperatures achieved depend on $\dot{Q}_c$ and the rate of heat loss from the vicinity of the reacting system (see. Section 10.3.2). The only situation where it is reasonable to ignore heat loss (at least to a first approximation) is in premixed burning when the fuel and air are intimately mixed and the reaction rates are high, independent of diffusive or mixing processes. This is the 'adiabatic' model, in which it is assumed that all the heat generated remains within the system, giving rise to an increase in temperature. Taking as an example a flame propagating through a stoichiometric mixture of propane in air (see Figure 3.14), then it is possible to estimate the adiabatic flame temperature, assuming that all the energy released is taken up by the combustion products. From Table 1.13, $\Delta H_c(C_3H_8) = -2044.3$ kJ/mol. The oxidation reaction in air is given by Equation (1.R9): the combustion energy raises the temperature of the products CO_2, H_2O and N_2 whose final temperatures can be calculated if heat capacities of these species are known. These may be obtained from thermochemical tables (e.g. US Department of Commerce, 1971b) or other sources (e.g. Lewis and von Elbe, 1987) (Table 1.16). It is assumed that nitrogen is not involved in the chemical reaction but acts only as 'thermal ballast', absorbing a major share of the combustion energy. The energy released in the combustion of 1 mole of propane is thus taken up by 3 moles of CO_2, 4 moles of H_2O and 18.8 moles of N_2 (Reaction (1.R9)). The total heat capacity of this mixture is 942.5 J/K (per mole of propane burnt) (see Table 1.17), so that the

Table 1.16 Thermal capacities of common gases at 1000 K

	$C_p^{1000\ \mathrm{K}}$ (J/mol.K)
Carbon monoxide (CO)	33.2
Carbon dioxide (CO_2)	54.3
Water (vapour) (H_2O)	41.2
Nitrogen (N_2)	32.7
Oxygen (O_2)	34.9
Helium (He)	20.8

Table 1.17 Heat capacity of combustion products (stoichiometric propane/air mixture)

$$C_3H_8 + 5\ O_2 + 18.8\ N_2 \rightarrow 3\ CO_2 + 4\ H_2O + 18.8\ N_2$$

Species	Number of moles in products[a]	Thermal capacity at 1000 K	
		C_p(J/mol.K)	nC_p(J/K)[a]
CO_2	3	54.3	162.9
H_2O	4	41.2	164.8
N_2	18.8	32.7	614.8
		Total thermal capacity/mole propane $= 942.5$ J/K	

[a]Per mole of propane burnt.

final flame temperature T_f is

$$T_f = 25 + \frac{2044300}{942.5} = 2194°\mathrm{C} \tag{1.28}$$

assuming an initial temperature of 25°C. The result of this calculation is approximate for the following reasons:

(i)   The thermal capacity of each gas is a function of temperature, and for simplicity the values used here refer to an intermediate temperature (1000 K).

(ii) The system is not truly adiabatic as radiation losses from the flame zone and its vicinity will tend to reduce the final temperature and cause the temperature to fall in the post-flame gases (Section 3.3; Figure 3.16).

(iii) At high temperature, the products are partially dissociated into a number of atomic, molecular and free radical species. This can be expressed in terms of the equilibria (Friedman, 1995):

$$H_2O \rightleftharpoons H^{\cdot} + {\cdot}OH \tag{1.R19a}$$

$$H_2O \rightleftharpoons H_2 + \tfrac{1}{2}O_2 \tag{1.R19b}$$

$$CO_2 \rightleftharpoons CO + \tfrac{1}{2}O_2 \tag{1.R19c}$$

As each dissociation is endothermic (absorbs energy rather than releases it), these will tend to depress the final temperature. The effect of dissociation on the calculated temperature is significant above ~1700°C (2000 K). Provided these reactions are sufficient to describe dissociation in the system, Equation (1.R9) can be rewritten:

$$C_3H_8 + 5\ O_2 + 18.8\ N_2 \rightarrow (3 - y)\ CO_2 + y\ CO + (4 - x)\ H_2O + 18.8\ N_2$$

$$+ (x - z)\ H_2 + \left(\frac{x}{2} + \frac{y}{2} - \frac{z}{2}\right) O_2 + z\ H^{\cdot} + z\ ^{\cdot}OH \qquad (1.R20)$$

where the values of x, y and z are unknown but can be determined by calculating the position of these three equilibria using thermodynamic data (e.g. US Department of Commerce, 1971b) provided that the final temperature is known. As this is not the case, a trial value is selected and the corresponding concentrations of $H^{\cdot}$, $^{\cdot}OH$, H_2, CO and O_2 calculated from the appropriate equilibrium constants (Moore, 1974). The heat of Reaction (1.R20) is then calculated from the heats of formation of all species present in the products and the resulting (adiabatic) temperature obtained using the method outlined in Table 1.17. The complete procedure is repeated using the calculated temperature, and then, if necessary, reiterated until two successive iterations give temperatures in satisfactory agreement. The true adiabatic flame temperature will inevitably be less than the value obtained above. The procedure is discussed by Friedman (1995) who shows that the figure for a stoichiometric propane/air mixture is 1995°C.

Despite the approximate nature of the calculation given above, it has value in estimating whether or not particular mixtures are flammable. There is good evidence that there is a lower limiting (adiabatic) temperature below which flame cannot propagate (Section 3.3). The consequence of this is that not all mixtures of flammable gas and air will burn if subjected to an ignition source. The flammable region is well defined and is bounded by the lower and upper flammability limits which can be determined experimentally to within a few tenths of 1% (Sections 3.1.1 and Table 3.1). For propane, the lower limit corresponds to 2.2% propane in air. Assuming that combustion will proceed to completion, the oxidation can be written:

$$0.022\ C_3H_8 + 0.978\ (0.21O_2 + 0.79N_2) \rightarrow \text{products (CO}_2, H_2O, O_2 \text{ and } N_2)$$
$$(1.R21)$$

Dividing through by 0.022 gives:

$$C_3H_8 + 9.335\ O_2 + 35.119\ N_2 \rightarrow 3\ CO_2 + 4\ H_2O + 4.355O_2 + 35.119\ N_2$$
$$(1.R22)$$

As the original mixture was 'fuel lean', the excess oxygen will contribute to the total thermal capacity of the product mixture. Using the method outlined in Table 1.17, the final (adiabatic) flame temperature can be shown to be 1281°C (1554 K), well below that at which the effect of dissociation is significant. If the adiabatic flame temperature is calculated for the limiting mixtures of a number of hydrocarbon gases, they are found to fall within a fairly narrow band (1600 ± 100 K) (see Table 3.1). There is evidence to suggest that the same value also applies to the upper flammability (fuel-rich) limit (Mullins and Penner, 1959; Stull, 1971), but it cannot be derived by the same method as the lower limit because the products will contain a complex mixture of pyrolysis and partially oxidized products from the parent fuel.

It should be noted that the temperature increases reported above will be accompanied by expansion of the system. Using the ideal gas law (Section 1.2.1), it can be seen that a seven-fold increase in temperature (e.g. 300 K to 2100 K) will be accompanied by a seven-fold increase in volume, neglecting any change in the number of moles in the system:

$$\frac{V_2}{V_1} = \frac{n_2}{n_1} \times \frac{T_2}{T_1} \qquad (1.29)$$

where the subscripts 1 and 2 refer to the initial and final states, assuming that $P_1 = P_2$. However, if the volume remains constant there will be a similar, corresponding rise in pressure. Such large increases will be generated very rapidly if a flammable vapour/air mixture is ignited within a confined space. This will almost certainly cause structural damage to a building unless measures have been incorporated to prevent the build-up of pressure. One such technique is the provision of explosion relief in the form of weakened panels in the building envelope, which will fail easily (Drysdale and Kemp, 1982; Bartknecht, 1981; Harris, 1983; Zalosh, 1995).

Problems

1.1 Calculate the vapour densities (kg/m^3) of pure carbon dioxide, propane and butane at 25°C and atmospheric pressure. Assume ideal gas behaviour.

1.2 Assuming ideal gas behaviour, what will the final volume be if 1 m^3 of air is heated from 20°C to 700°C at constant pressure?

1.3 Calculate the vapour pressure of the following pure liquids at 0°C: (a) *n*-octane; (b) methanol; (c) acetone.

1.4 Calculate the vapour pressures of *n*-hexane and *n*-decane above a mixture at 25°C containing 2% *n*-C_6H_{14} + 98% *n*-$C_{10}H_{22}$, by volume. Assume that the densities of pure *n*-hexane and *n*-decane are 660 and 730 kg/m^3 respectively and that the liquids behave ideally.

1.5 Calculate the vapour pressures of iso-octane and *n*-dodecane above a mixture at 20° C containing 5% i-C_8H_{18} and 95% *n*-$C_{12}H_{26}$, by volume. The densities of the pure liquids are 692 and 749 kg/m^3 respectively. Assume ideal behaviour.

1.6 Work out the enthalpy of formation of propane at 25°C (298 K) from Equation (1.20) using data contained in Tables 1.13 and 1.14.

1.7 Given that the stoichiometric reaction for the oxidation of *n*-pentane is:

$$n\text{-}C_5H_{12} + 8\ O_2 \rightarrow 5\ CO_2 + 6\ H_2O$$

calculate ΔH_f^{298} (C_5H_{12}) from data in Tables 1.13 and 1.14.

1.8 Calculate the enthalpy change (20°C) in the oxidation of *n*-pentane to carbon monoxide and water, i.e.

$$n\text{-}C_5H_{12} + 5\tfrac{1}{2}\ O_2 \rightarrow 5\ CO + 6\ H_2O$$

1.9 The products of the partial combustion of *n*-pentane were found to contain CO_2 and CO in the ratio of 4:1. What is the actual heat released per mole of *n*-pentane burnt if the only other product is H_2O?

1.10   Express the result of problem 1.9 in terms of heat released (a) per gram of *n*-pentane burnt; (b) per gram of air consumed.

1.11   Calculate the masses of air required to burn completely 1 g each of propane (C_3H_8), pentane (C_5H_{12}) and decane ($C_{10}H_{22}$).

1.12   Calculate the adiabatic flame temperatures for the following mixtures initially at 25°C assuming that dissociation does not occur:

(a)   stoichiometric *n*-pentane/oxygen mixture;

(b)   stoichiometric *n*-pentane/air mixture;

(c)   1.5% *n*-pentane in air (lower flammability limit, see Chapter 3).

2
Heat Transfer

An understanding of several branches of physics is required in order to be able to interpret fire phenomena (see Di Nenno *et al.*, 1995). These include fluid dynamics, and heat and mass transfer. In view of its importance, the fundamentals of heat transfer will be reviewed in this chapter. Other topics are required in later chapters, but the essentials are introduced in context and key references given (e.g., Landau and Lifshitz, 1987; Tritton, 1988; Kandola, 1995). In this chapter, certain heat transfer formulae are derived for use later in the text, although their relevance may not be immediately obvious. There are many good textbooks which deal with heat transfer in great depth, several of which have been referred to during the preparation of this chapter (e.g. Rohsenow and Choi, 1961; Holman, 1976; Welty *et al.*, 1976; Pitts and Sissom, 1977): it is recommended that such texts are used to provide the detail which cannot be included here.

There are three basic mechanisms of heat transfer, namely conduction, convection and radiation. While it is probable that all three contribute in every fire, it is often found that one predominates at a given stage, or in a given location. Thus, conduction determines the rate of heat flow in and through solids. It is important in problems relating to ignition and spread of flame over combustible solids (Chapters 6 and 7), and to fire resistance, where knowledge of heat transfer through compartment boundaries and into elements of structure is required (Chapter 10). Convective heat transfer is associated with the exchange of heat between a gas or liquid and a solid, and involves movement of the fluid medium (e.g. cooling by directing a flow of cold air over the surface of a hot solid). It occurs at all stages in a fire but is particularly important early on when thermal radiation levels are low. In natural fires, the movement of gases associated with this transfer of heat is determined by buoyancy which also influences the shape and behaviour of diffusion flames (Chapter 4). The buoyant plume will be discussed in Section 4.3.1.

Unlike conduction and convection, radiative heat transfer requires no intervening medium between the heat source and the receiver. It is the transfer of energy by electromagnetic waves, of which visible light is the example with which we are most familiar. Radiation in all parts of the electromagnetic spectrum can be absorbed, transmitted or reflected at a surface, and any opaque object placed in the way will cast a shadow. It becomes the dominant mode of heat transfer in fires as the fuel bed diameter increases beyond about 0.3 m, and determines the growth and spread of fires in compartments. It is the mechanism by which objects at a distance from a fire are heated to the firepoint condition, and is responsible for the spread of fire through open fuel beds (e.g. forests) and between buildings (Law, 1963). A substantial amount of heat released in flames is transmitted by radiation to the surroundings. Most of this radiation is emitted by minute solid particles of soot which are formed in almost all diffusion flames: this is the source of their characteristic yellow luminosity. The effect of thermal radiation from flames, or indeed from any heated object, on nearby surfaces can only be ascertained by carrying out a detailed heat transfer analysis. Such analyses are required to establish how rapidly combustible materials which are exposed to thermal radiation will reach a state in which they can be ignited and will burn.

2.1 Summary of the Heat Transfer Equations

At this point, it is necessary to introduce the basic equations of heat transfer, as it is impossible to examine any one of the mechanisms in depth in isolation from the others. Detailed discussion will follow in subsequent sections, and major review articles may be found in the *SFPE Handbook* (Atreya, 1995; Rockett and Milke, 1995; Tien *et al.*, 1995).

(a) *Conduction (Section 2.2)* Conduction is the mode of heat transfer associated with solids. Although it also occurs in fluids, it is normally masked by convective motion in which heat is dissipated by a mixing process driven by buoyancy. It is common experience that heat will flow from a region of high temperature to one of low temperature; this flow can be expressed as a heat flux, which in one direction is given by:

$$\dot{q}_x'' = -k\frac{\Delta T}{\Delta x} \tag{2.1}$$

where ΔT is the temperature difference over a distance Δx.

In differential form

$$\dot{q}_x'' = -k\frac{\mathrm{d}T}{\mathrm{d}x} \tag{2.2}$$

where $\dot{q}_x'' = (\mathrm{d}q_x/\mathrm{d}t)/A, A$ being the area (perpendicular to the x-direction) through which heat is being transferred. This is known as Fourier's Law of Heat Conduction. The constant k is the thermal conductivity and has units of W/m.K when $\dot{q}''$ is in W/m^2, T is in °C (or K), and x is in m. Typical values are given in Table 2.1. These refer to specific temperatures (0 or 20°C) as thermal conductivity is dependent on temperature. Data on k as a function of T are available for many pure materials (e.g. Kaye and Laby, 1986), but such information for combustible solids and building materials is fragmentary (Abrams, 1979; Harmathy, 1995).

As a general rule, materials which are good thermal conductors are also good electrical conductors. This is because heat transfer can occur as a result of interactions involving free electrons whose movement constitutes an electric current when a voltage is applied. In insulating materials, the absence of free electrons means that heat can only be transferred as mechanical vibrations through the structure of the molecular lattice which is a much less efficient process.

(b) *Convection (Section 2.3)* As indicated above, convection is that mode of heat transfer to or from a solid involving movement of a surrounding fluid. The empirical

Table 2.1   Thermal properties of some common materials[a]

Material	k (W/m.K)	c_p (J/kg.K)	ρ (kg/m^3)	α (m^2/s)	$k\rho c_p$ (W^2.s/m^{4}K^2)
Copper	387	380	8940	1.14×10^{-4}	1.3×10^9
Steel (mild)	45.8	460	7850	1.26×10^{-5}	1.6×10^8
Brick (common)	0.69	840	1600	5.2×10^{-7}	9.3×10^5
Concrete	0.8–1.4	880	1900–2300	5.7×10^{-7}	2×10^6
Glass (plate)	0.76	840	2700	3.3×10^{-7}	1.7×10^6
Gypsum plaster	0.48	840	1440	4.1×10^{-7}	5.8×10^5
PMMA[b]	0.19	1420	1190	1.1×10^{-7}	3.2×10^5
Oak[c]	0.17	2380	800	8.9×10^{-8}	3.2×10^5
Yellow pine[c]	0.14	2850	640	8.3×10^{-8}	2.5×10^5
Asbestos	0.15	1050	577	2.5×10^{-7}	9.1×10^4
Fibre insulating board	0.041	2090	229	8.6×10^{-8}	2.0×10^4
Polyurethane foam[d]	0.034	1400	20	1.2×10^{-6}	9.5×10^2
Air	0.026	1040	1.1	2.2×10^{-5}	—

[a] From Pitts and Sissom (1977) and others. Most values for 0 or 20°C. Figures have been rounded off.
[b] Polymethylmethacrylate. Values of k, c_p and ρ for other plastics are given in Table 1.2.
[c] Properties measured perpendicular to the grain.
[d] Typical values only.

relationship first discussed by Newton is:

$$\dot{q}'' = h\Delta T \ \text{W/m}^2 \tag{2.3}$$

where h is known as the convective heat transfer coefficient. This equation defines h which, unlike thermal conductivity, is not a material constant. It depends on the characteristics of the system, the geometry of the solid and the properties of the fluid including the flow parameters, and is also a function of ΔT. The evaluation of h for different situations has been one of the major problems in heat transfer and fluid dynamics. Typical values lie in the range 5–25 W/m^2.K for free convection and 10–500 W/m^2.K for forced convection in air.

(c) *Radiation (Section 2.4)*   According to the Stefan–Boltzmann equation, the total energy emitted by a body is proportional to T^4, where T is the temperature in Kelvin: the total emissive power is

$$E = \varepsilon \sigma T^4 \ \text{W/m}^2 \tag{2.4}$$

where σ is the Stefan-Boltzmann constant (5.67×10^{-8} W/m^{2}K^4) and ε is a measure of the efficiency of the surface as a radiator, known as the emissivity. The perfect emitter — the 'black body' — has an emissivity of unity. The intensity of radiant energy ($\dot{q}''$) falling on a surface remote from the emitter can be found by using the appropriate 'configuration factor' ϕ which takes into account the geometrical relationship between the emitter and the receiver.

$$\dot{q}'' = \phi \varepsilon \sigma T^4 \tag{2.5}$$

These concepts will be developed in detail in Section 2.4.

2.2   Conduction

While many common problems involving heat conduction are essentially steady state (e.g. thermal insulation of buildings), most of those related to fire are transient and require solutions of time-dependent partial differential equations. Nevertheless, a system of this type will move towards an equilibrium which will be achieved provided that there is no variation in the heat source or in the integrity of the materials involved. Indeed, as the steady state is the limiting condition, it can be used to solve a number of problems, many of which will be discussed in subsequent chapters. Thus, it is worthwhile to consider steady state before examining transient conduction.

2.2.1   *Steady state conduction*

Consider the heat loss through an infinite, plane slab, or wall, of thickness L, whose surfaces are at temperatures T_1 and T_2 ($T_1 > T_2$) (Figure 2.1). In this idealized model the heat flow is unidimensional. Integrating Fourier's equation

$$\dot{q}''_x = -k\frac{\mathrm{d}T}{\mathrm{d}x} \tag{2.2}$$

gives:

$$\dot{q}''_x \int_O^L \mathrm{d}x = -k \int_{T_1}^{T_2} \mathrm{d}T \tag{2.6}$$

provided that k is independent of temperature. This gives:

$$\dot{q}''_x = \frac{k}{L}(T_1 - T_2) \tag{2.7}$$

(If k is a function of temperature, then $k\,\mathrm{d}T$ must be integrated between T_1 and T_2.)

   If the wall is composite consisting of various layers as shown in Figure 2.2. the net heat flux through the wall at the steady state can be calculated by equating the steady state heat fluxes across each layer. Assuming that there is a temperature difference between each surface and the adjacent air, then if h_h and h_c are the convective heat transfer coefficients at the inner and outer (hot and cold) surfaces (Equation (2.3)):

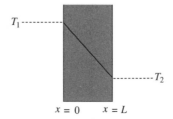

Figure 2.1   The infinite plane slab. In this example, the *surfaces* are at temperatures T_1 and T_2, as indicated

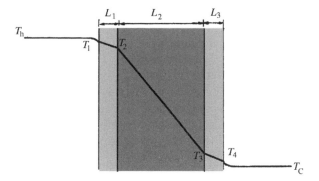

Figure 2.2 The infinite plane composite wall. The temperature of the air in contact with each surface is shown

$$\dot{q}''_x = h_h(T_h - T_1) = \frac{k_1}{L_1}(T_1 - T_2) = \frac{k_2}{L_2}(T_2 - T_3)$$

$$= \frac{k_3}{L_3}(T_3 - T_4) = h_c(T_4 - T_c) \tag{2.8}$$

Although only the air temperatures T_h and T_c are known, the heat flux across each layer gives the following series of relationships:

$$T_h - T_1 = \dot{q}''_x/h_h$$

$$T_1 - T_2 = \dot{q}''_x L_1/k_1$$

$$T_2 - T_3 = \dot{q}''_x L_2/k_2$$

$$T_3 - T_4 = \dot{q}''_x L_3/k_3$$

$$T_4 - T_c = \dot{q}''_x/h_c$$

which, if added together, give, after rearrangement:

$$\dot{q}''_x = \frac{T_h - T_c}{\dfrac{1}{h_h} + \dfrac{L_1}{k_1} + \dfrac{L_2}{k_2} + \dfrac{L_3}{k_3} + \dfrac{1}{h_c}} \tag{2.9}$$

This expression is similar in form to that which describes the relationship between current (I), voltage (V) and resistance (R) in a simple d.c. series circuit:

$$I = \frac{V}{R} \tag{2.10}$$

In Equation (2.9), the flow of heat ($\dot{q}''_x$) is analogous to the current, while potential difference and electrical resistance are replaced by temperature difference and 'thermal resistance' respectively (Figure 2.3). This analogy will be referred to later (e.g. Chapter 10).

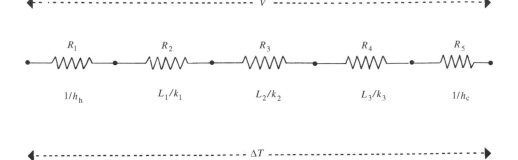

Figure 2.3   The analogy between electrical and thermal resistance

2.2.2   *Non-steady state conduction*

Fires are transient phenomena, and the equation of non-steady state heat transfer must be used to interpret not only the details of fire behaviour, such as ignition and flame spread, but also the gross effects such as the response of buildings to developing and fully developed fires. The basic equations for non-steady state conduction can be derived by considering the flow of heat through a small volume $dx\,dy\,dz$ (Figure 2.4) and the associated heat balance.

Taking flow in the x-direction, the rate of heat transfer through face A is given by:

$$\dot{q}''_x\,dS = -k\frac{\partial T}{\partial x}\,dy\,dz \tag{2.11}$$

where $dS = dy\,dz$, the area of face A. Similarly, the flow of heat *out* through face B is:

$$\dot{q}''_{x+dx}\,dS = -k\left(\frac{\partial T}{\partial x} + \frac{\partial^2 T}{\partial x^2}\,dx\right)dy\,dz \tag{2.12}$$

The difference between Equations (2.11) and (2.12) must be equal to the rate of change in the energy content of the small volume $dx\,dy\,dz$. This is made up of two terms, namely

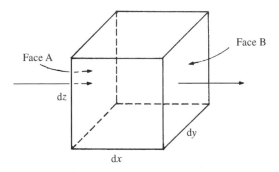

Figure 2.4   Transient heat conduction through an element of volume $dx\,dy\,dz$

heat storage and heat generation, thus:

$$k \left(\frac{\partial^2 T}{\partial x^2} \right) dx\,dy\,dz = \rho c \frac{\partial c}{\partial t} dx\,dy\,dz - \dot{Q}''' dx\,dy\,dz \tag{2.13}$$

where $\dot{Q}'''$ is the rate of heat release per unit volume and ρ and c are the density and heat capacity respectively. This simplifies to:

$$\frac{\partial^2 T}{\partial x^2} = \frac{1}{\alpha} \frac{\partial T}{\partial t} - \frac{\dot{Q}'''}{k} \tag{2.14}$$

where $\alpha = k/\rho c$, the 'thermal diffusivity' of the material (Table 2.1) which is assumed to be constant in the above analysis. In most problems, $\dot{Q}''' = 0$, but Equation (2.14) is used in the development of Frank-Kamenetskii's thermal explosion theory which will be discussed in Section 6.1 in relation to spontaneous ignition. Equation (2.14) would be relevant to any transient heating problem which involves exothermic or endothermic change (e.g. phase changes and chemical decompositon). However, if $\dot{Q}'''$ is zero, Equation (2.14) gives for one dimension:

$$\frac{\partial^2 T}{\partial x^2} = \frac{1}{\alpha} \frac{\partial T}{\partial t} \tag{2.15}$$

The equation for three dimensions is written:

$$\frac{\partial^2 T}{\partial x^2} + \frac{\partial^2 T}{\partial y^2} + \frac{\partial^2 T}{\partial z^2} = \nabla^2 T = \frac{1}{\alpha} \frac{\partial T}{\partial t} \tag{2.16}$$

Fortunately, many problems can be reduced to a single dimension. Thus, Equation (2.15) can be applied directly to conduction through materials which may be treated as 'infinite slabs' or 'semi-infinite solids' (see below). Some problems can be reduced to a single dimension by changing to cylindrical or polar coordinates: such applications will be discussed in the analysis of spontaneous ignition (Sections 6.1 and 8.1).

One of the simplest cases to which Equation (2.15) may be applied is the infinite slab, of thickness $2L$ and temperature $T = T_0$, suddenly exposed on both faces to air at a uniform temperature $T = T_\infty$ (Figure 2.5). Writing $\theta = T - T_\infty$, Equation (2.15) becomes:

$$\frac{\partial^2 \theta}{\partial x^2} = \frac{1}{\alpha} \frac{\partial \theta}{\partial t} \tag{2.17}$$

This must be solved with the following boundary conditions:

$$\frac{\partial \theta}{\partial x} = 0 \text{ at } x = 0 \text{ (i.e. at the mid-plane)}$$

$$\theta = \theta_0 \ (= T_0 - T_\infty) \text{ at } t = 0 \text{ (for all } x)$$

$$\frac{\partial \theta}{\partial x} = -\frac{h}{k} \theta \text{ at } x = \pm L \text{ (at both faces)}$$

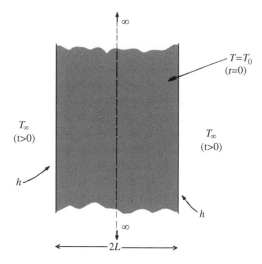

Figure 2.5   Transient heat conduction in a plane infinite slab

where the latter defines the rate of heat transfer through the faces of the slab, h being the convective heat transfer coefficient. The solution is not simple but may be found in any heat transfer text (e.g. Carslaw and Jaeger, 1959):

$$\frac{\theta}{\theta_0} = 2 \sum_{n=1}^{\infty} \frac{\sin \lambda_n L}{\lambda_n L + (\sin \lambda_n L)(\cos \lambda_n L)} \cdot \exp(-\lambda_n^2 \alpha t) \cdot \cos(\lambda_n x) \qquad (2.18)$$

where λ_n are roots of the equation

$$\cot(\lambda_L) = \frac{\lambda_L L}{\text{Bi}} \qquad (2.19)$$

in which Bi is the Biot number (hL/k).

Examination of Equation (2.18) reveals that the ratio θ/θ_0 is a function of three dimensionless groups, the Biot number, the Fourier number (Fo $= \alpha t/L^2$), and x/L, the distance from the centre line expressed as a fraction of the half-thickness. The Biot number compares the efficiencies with which heat is transferred to the surface by convection from the surrounding air, and from the surface by conduction into the body of the solid, while the Fourier number can be regarded as a dimensionless time variable, which takes into account the thermal properties and characteristic thickness of the body. For convenience, the solutions to Equation (2.18) are normally presented in graphical form, displaying in a series of diagrams (each referring to a given value of x/L) the ratio θ/θ_0 as a function of Fo for a range of values Bi. Figure 2.6 shows two such charts, for θ/θ_0 at the surface ($x/L = 1$) and at the centre of the slab ($x/L = 0$).

The form of the temperature profiles within the slab and their variation with time are illustrated schematically in Figure 2.7(a). In a thin slab of a material with high conductivity, the temperature gradients within the slab are much less, and in some circumstances may be neglected (cf. Figure 2.7(b)). This can be seen by comparing the values of θ/θ_0 at the surface and at the mid-plane of the infinite slab for different values of Bi (Figure 2.6). Figure 2.8 shows how the ratio $\theta_{x=0}/\theta_{x=L}$ varies with Bi:

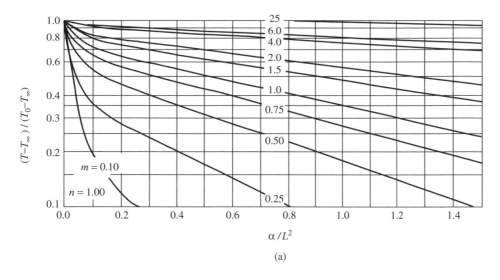

(a)

Flat plate

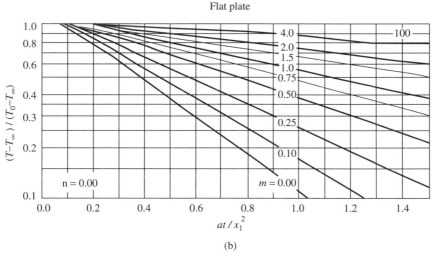

(b)

Figure 2.6 Heisler charts for (a) surface temperature and (b) centre temperature of an infinite slab. $n = x/L$ and $m = 1/Bi$ (Welty *et al.*, 1976). Reproduced by permission of John Wiley & Sons, Inc.

when Bi is vanishingly small, the temperature at the mid-plane of the slab is equal to the surface temperature. If it is less than about 0.1 (i.e. k is large and/or L is small) then temperature gradients within the solid may be ignored (compare Figure 2.7(b)) and the heat transfer problem can be treated by the 'lumped thermal capacity analysis'. Thus, for a 'thin' slab (or for that matter any body which complies with this constraint), the energy balance over the time interval dt may be written:

$$Ah(T_\infty - T)\,dt = V\rho c\,dT \tag{2.20}$$

where A is the area through which heat is being transferred and V is the associated volume. This integrates to give:

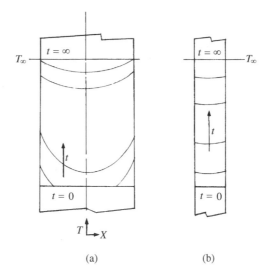

Figure 2.7 (a) Transient temperatures within a thick slab initially at T_0 exposed to an environmental temperature of T_∞: (b) the same for a thin slab. Warren M. Rohsenow, Harry Y. Choi, *Heat, Mass and Momentum Transfer*, © 1961, p. 111. Adapted by permission of Prentice-Hall.Inc., Englewood Cliffs, NJ

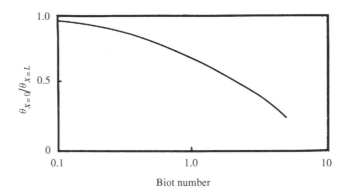

Figure 2.8 Dependence of the ratio $\theta_{x=0}/\theta_{x=L}$ on Bi for Fo $= 1.0$

$$\frac{T_\infty - T}{T_\infty - T_0} = \exp\left(-\frac{2ht}{\tau\rho c}\right) \tag{2.21}$$

where τ, the slab thickness, is equal to $2V/A$ if both faces are heated convectively. A similar model, involving radiant heating on one side and convective cooling at both faces, will be discussed in Section 6.3 in relation to the ignition of 'thin fuels' such as paper and fabrics. This 'lumped thermal capacity' method is a useful approximation, and is relevant to the interpretation of the ignition and flame spread characteristics of thin fuels (Section 7.2). It is also the basis for the response time index (RTI) for sprinklers which will be discussed in Section 4.4.2.

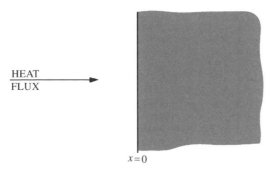

Figure 2.9   Heat transfer to a semi-infinite solid

While the temperature profiles within a 'thick' slab which is being heated symmetrically can be obtained by use of diagrams such as Figures 2.6(a) and 2.6(b), a problem more relevant to ignition and flame spread is that of a slab heated on one side only, with heat losses potentially at both faces. The limiting case is that of the semi-infinite solid subjected to a uniform heat flux (Figure 2.9). 'Thick' slabs will approximate to this model during the early stages of heating, before heat losses from the rear face have become significant. The relationship between heating time and thickness, which defines the limiting thickness to which this model may be applied, can be derived by considering a semi-infinite slab, initially at a temperature T_0, whose surface is suddenly increased to T_∞. Solving Equation (2.17) (where $\theta = T - T_0$) with the boundary conditions:

$$\theta = 0 \text{ at } t = 0 \text{ for all } x$$

$$\theta = \theta_\infty \text{ at } x = 0 \text{ for } t = 0$$

$$\theta = 0 \text{ as } x \to \infty \text{ for all } t$$

gives (Welty *et al.*, 1976)

$$\frac{\theta}{\theta_\infty} = 1 - \text{erf} \frac{x}{2\sqrt{(\alpha t)}} \tag{2.22}$$

where the error function is defined as:

$$\text{erf} \, \xi \equiv \frac{2}{\sqrt{\pi}} \int_0^\xi e^{-\eta^2} \cdot d\eta \tag{2.23}$$

While this cannot be evaluated analytically, it is given numerically in handbooks of mathematical functions, as well as in most heat transfer texts (see Table 2.2).

Equation (2.22) can be used to define the temperature profiles below the surface of a slab of thickness L, heated instantaneously on one face, until the rear face becomes heated to a temperature significantly above ambient (T_0). If this is set arbitrarily as 0.5% of $T_s - T_0$, i.e. $T = T_0 + 5 \times 10^{-3} (T_s - T_0)$ at $x = L$, then substituting in Equation (2.22):

$$1 - \text{erf} \frac{L}{2\sqrt{(\alpha t)}} = 5 \times 10^{-3} \tag{2.24a}$$

Table 2.2 The error function

x	erf x	erfc x
0	0	1.0
0.05	0.056372	0.943628
0.1	0.112463	0.887537
0.15	0.167996	0.832004
0.2	0.222703	0.777297
0.25	0.276326	0.723674
0.3	0.328627	0.671373
0.35	0.379382	0.620618
0.4	0.428392	0.571608
0.45	0.475482	0.524518
0.5	0.520500	0.479500
0.55	0.563323	0.436677
0.6	0.603856	0.396144
0.65	0.642029	0.357971
0.7	0.677801	0.322199
0.75	0.711156	0.288844
0.8	0.742101	0.257899
0.85	0.770668	0.229332
0.9	0.796908	0.203092
0.95	0.820891	0.179109
1.0	0.842701	0.157299
1.1	0.880205	0.119795
1.2	0.910314	0.009686
1.3	0.934008	0.065992
1.4	0.952285	0.047715
1.5	0.966105	0.033895
1.6	0.976348	0.023652
1.7	0.983790	0.016210
1.8	0.989091	0.010909
1.9	0.992790	0.007210
2.0	0.995322	0.004678
2.1	0.997021	0.002979
2.2	0.998137	0.001863
2.3	0.998857	0.001143
2.4	0.999311	0.000689
2.5	0.999593	0.000407

which gives

$$\frac{L}{2\sqrt{(\alpha t)}} \approx 2 \qquad (2.24\text{b})$$

This indicates that a wall, or slab, of thickness L can be treated as a semi-infinite solid with little error, provided that $L > 4\sqrt{(\alpha t)}$. In many fire engineering problems involving transient surface heating, it is adequate to assume 'semi-infinite behaviour' if $L > 2\sqrt{(\alpha t)}$ (e.g. Williams, 1977; McCaffrey *et al.*, 1981). The quantity $\sqrt{(\alpha t)}$ is the characteristic thermal conduction length and may be used to estimate the thickness of the heated layer in some situations (Chapters 5 and 6).

If the above model is modified to include convective heat transfer from a stream of fluid at temperature T_∞ to the surface of the semi-infinite solid (initially at temperature T_0) (Figure 2.9) then Equation (2.17) must be solved with the following boundary conditions:

$$\theta = 0 \text{ at } t = 0 \text{ for all } x$$

$$\frac{\partial \theta}{\partial x} = -\frac{h}{k}(\theta_\infty - \theta_s) \text{ at } x = 0 \text{ for all } t$$

The solution, which is given by Carslaw and Jaeger (1959), is

$$\frac{\theta}{\theta_\infty} = \frac{T - T_0}{T_\infty - T_0} = \text{erfc}\left(\frac{x}{2\sqrt{(\alpha t)}}\right)$$

$$- \exp\left(\frac{xh}{k} + \frac{\alpha t}{(k/h)^2}\right) \cdot \text{erfc}\left(\frac{x}{2\sqrt{(\alpha t)}} + \frac{\sqrt{(\alpha t)}}{k/h}\right) \quad (2.25)$$

(Note that erfc $(\xi) = 1 - \text{erf}(\xi)$ (Table 2.2).)

The variation of surface temperature (T_s) with time under a given imposed heat flux can be illustrated by setting $x = 0$ in Equation (2.25), i.e.

$$\frac{\theta_s}{\theta_\infty} = 1 - \exp\left(\frac{\alpha t}{(k/h)^2}\right) \cdot \text{erfc}\left(\frac{\sqrt{(\alpha t)}}{k/h}\right) \quad (2.26)$$

and plotting θ_s/θ_∞ against time. Figure 2.10 shows that the rate of change of surface temperature depends strongly on the value of the ratio $k^2/\alpha = k\rho c$, a quantity known as the 'thermal inertia'. The surface temperature of materials with low thermal inertia (such as fibre insulating board and polyurethane foam) rises quickly when heated: the relevance of this to the ignition and flame spread characteristics of combustible solids will be discussed in Chapters 6 and 7.

Equation (2.26) is cumbersome to use, but it is possible to derive a simplified expression if the surface is exposed to a constant heat flux, $\dot{Q}_R$. The heat flux (flow) through the semi-infinite solid

$$\dot{q}'' = -k\frac{\partial \theta}{\partial x} \quad (2.27)$$

obeys a differential equation identical in form to Equation (2.15) (Carslaw and Jaeger, (1992)):

$$\frac{\partial^2 \dot{q}''}{\partial x^2} = \frac{1}{\alpha}\frac{\partial \dot{q}''}{\partial t} \quad (2.28)$$

For $x > 0$ and $t > 0$, the boundary condition $\dot{q}'' = \dot{Q}_R''$ at $x = 0$ and $t > 0$ gives the solution:

$$T - T_0 = \theta = \frac{2\dot{Q}_R''}{k}\left\{\left(\frac{\alpha t}{\pi}\right)^{1/2}\exp\left(-\frac{x^2}{4\alpha t}\right) - \frac{x}{2}\text{erfc}\left(\frac{x}{2\sqrt{\alpha t}}\right)\right\} \quad (2.29)$$

The value of θ at the surface ($x = 0$) then becomes:

$$\theta = T - T_0 = \frac{2\dot{Q}_R''}{k}\left(\frac{\alpha t}{\pi}\right)^{1/2} \quad (2.30)$$

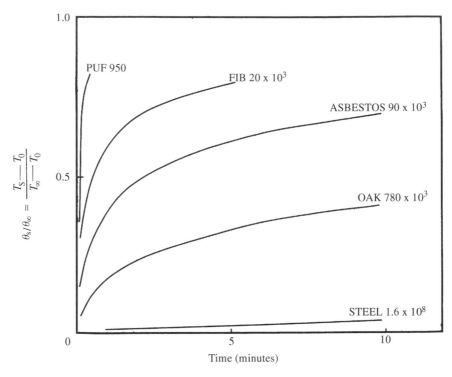

Figure 2.10 Effect of thermal inertia ($k\rho c_p$) on the rate of temperature rise at the surface of a semi-infinite solid. FIB = fibre insulating board: PUF = polyurethane foam. Figures are values of $k\rho c_p$ in $W^2 s/m^4.K^2$. (From Equation (2.26), with $h =$ 20 $W/m^2.K$. T_s = surface temperature)

Although the derivation ignores heat losses from the surface to the surroundings, this equation is nevertheless useful, as will be seen in Chapter 6.

2.2.3   *Numerical methods of solving time-dependent conduction problems*

Where they exist, analytical solutions to transient heat conduction problems are not simple, although they tend to apply only to simple geometries with well-defined boundary conditions (cf. Equations (2.18) and (2.25)). While basic equations can be written for complex geometries and boundary conditions, they may be intractable and require numerical solution. In general, these require lengthy, iterative calculations, but the availability of low-cost computing power now provides the opportunity to undertake complex heat transfer calculations with ease. Simple problems can be solved using spreadsheets, or a few lines of code; more complex problems — particularly if non-uniform temperature distributions are involved — will require either the use of available software packages such as TASEF (Sterner and Wickstrom, 1990) or the preparation of a program specific to the problem.

   As an example of a simple numerical solution, consider a steel plate, or bulkhead, which forms the boundary between two compartments, both initially at a temperature

T_0. Suppose the air in one of the compartments is suddenly increased to a temperature T_h: how rapidly will the temperature of the plate increase, and what will its final temperature be? If it is assumed that the bulkhead behaves as an infinite plate (thus reducing the problem to one dimension) of thickness Δx, and that the temperature of the plate remains uniform at all times (i.e. the Biot number is low), the problem becomes quite simple. Indeed, calculating the *final* temperature is trivial, as it is a steady state problem. Thus, if the final temperature of the steel is T_s, the steady state is expressed as

$$\text{Heat in (kW/m}^2) = \text{Heat out (kW/m}^2)$$

$$h(T_h - T_s) = h(T_s - T_0) \tag{2.31}$$

where h is the convective heat transfer coefficient, assumed to be independent of temperature: radiative heat transfer is neglected. If T_h and T_0 are 100°C and 20°C respectively, the final steel temperature is 60°C, independent of the value of Δx or h.

To calculate how quickly the plate heats up, an analytical solution for the rate of temperature rise may be obtained from the equation:

$$\frac{dT_s}{dt} = \frac{1}{\rho c_p \Delta x} [h(T_h - T_s) - h(T_s - T_0)] \tag{2.32}$$

(which reduces to Equation (2.31) at the steady state ($dT_s/dt = 0$)). This can be integrated to give an analytical solution, as follows:

$$\int_{T_0}^{T_s} \frac{dT_s}{(T_h - 2T_s + T_0)} = \frac{h}{\rho c_p \Delta x} \int_0^t dt \tag{2.33}$$

$$T_s = \frac{1}{2}(T_h + T_0) - \frac{1}{2}(T_h - T_0) \exp\left[-\frac{2ht}{\rho c_p \Delta x}\right] \tag{2.34}$$

Equation (2.34) is shown in Table 2.3 and Figure 2.11(b). (Compare with Equation (2.21).)

For the numerical approach, Equation (2.32) is cast in finite difference form:

$$\Delta T = \frac{1}{\rho c_p \Delta x} [h(T_h - T_s) - h(T_s - T_0)] \Delta t \tag{2.35}$$

This can be rearranged to show that for any time interval Δt:

$$\text{Heat in } - \text{ Heat out} = \text{ Heat stored} \tag{2.36}$$

Writing $\Delta T = T_s(t + \Delta t) - T_s(t)$, Equation (2.35) becomes:

$$h(T_h - T_s(t))\Delta t - h(T_s(t) - T_0)\Delta t = \rho c_p \Delta x(T_s(t + \Delta t) - T_s(t)) \tag{2.37}$$

where $T_s(t)$ and $T_s(t + \Delta t)$ are the temperatures of the steel plate at time t and time $t + \Delta t$, respectively. Equation (2.37) can be rearranged to give $T_s(t + \Delta t)$:

Table 2.3 Comparison of analytical and numerical solutions to the rate of temperature rise of the steel plate shown in Figure 2.11(a) according to Equations (2.34) and (2.38) respectively. $T_h = 100°C$, $T_0 = 20°C$, $h = 0.02$ kW/m²K, $\rho = 7850$ kg/m³, $c_p = 0.46$ kJ/kg K and $\Delta x = 0.005$ m. For the numerical solution, the results for two values of Δt (180 s and 20 s) are shown

Time	Analytical solution (Equation (2.34))	Numerical solution (Equation (2.38))	
t (s)	$T_s(t)$ (°C)	$T_s(t)$ (°C) ($\Delta t = 180$ s)	$T_s(t)$ (°C) ($\Delta t = 20$ s)
0	20.00	20.00	20.00
180	33.15	35.95	33.40
360	41.98	45.54	42.31
540	47.91	51.31	48.23
720	51.88	54.77	52.17
900	54.55	56.86	54.80
1080	56.34	58.11	57.24
1260	57.55	58.86	57.70
1440	58.35	59.32	58.47
1620	58.90	59.59	58.98
1800	59.26	59.75	59.32

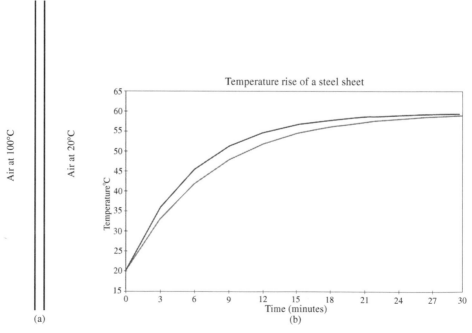

(a)

(b)

Figure 2.11 A comparison between the analytical and numerical solutions for transient, one-dimensional heat transfer through (a) a 5 mm infinite steel sheet with initial temperature 20°C and convective heat transfer coefficient 0.02 kW/m²K (see Table 2.3): (b) lower line: analytical solution (Equation (2.34)); upper line: numerical solution with $\Delta t = 180$ s (Equation (2.38))

$$T_s(t + \Delta t) = T_s(t) + \frac{h\Delta t}{\rho c_p \Delta x}(T_h - 2T_s(t) + T_0) \tag{2.38}$$

expressing the steel temperature at time $t + \Delta t$ in terms of the temperature at the beginning of the timestep, at time t.

This can be solved iteratively, with initial values $t = 0$, and $T_s(0) = T_0$, calculating T_s at times $t = n\Delta t$, where $n = 0, 1, 2, \ldots m$, where m defines the total time of exposure. A spreadsheet or a simple program may be used. The accuracy of the numerical solution can be improved by reducing the timestep, as illustrated in Table 2.3. The numerical and analytical solutions are compared in Figure 2.11(b).

The same technique can be used for more complex problems. A thick plate, or wall, acting as a barrier between the compartments would not be at uniform temperature; there would be a temperature gradient (cf. Figure 2.2 for the steady state). The numerical solution requires that the wall is represented by a series of thin, parallel elements of equal thickness, as shown in Figure 2.12. Transient heat transfer through the plate is then calculated iteratively, by considering adjacent elements and applying the logic of Equation (2.36). Thus the 'internal element' numbered 3 in Figure 2.12, during the heating-up period, gains heat from element 2, and loses heat to element 4. Thus:

$$\frac{k}{\Delta x}(T_2 - T_3) - \frac{k}{\Delta x}(T_3 - T_4) = \rho c_p \Delta x(T_3(t + \Delta t) - T_3(t)) \tag{2.39}$$

which can be arranged to give:

$$T_3(t + \Delta t) = T_3(t) + \frac{k}{\rho c_p (\Delta x)^2}(T_2 - 2T_3 + T_4) \tag{2.40}$$

This equation is the basis for numerical solutions of complex problems such as the heat losses from a fire compartment during the immediate post-flashover stage, when assumptions can be made about the rate of heat release within the compartment (Section 10.3.2) and the temperature variation has to be calculated (Pettersson *et al.*

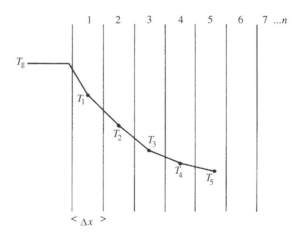

Figure 2.12 Thermal conduction into an infinite, plane slab, divided into equal elements $1, 2, \ldots, n$ for numerical analysis

1976). If necessary, the method can be adapted to two and three dimensions, although for convergence the Fourier number (Fo $= \alpha \Delta t/(\Delta x)^2$) must be less than 0.5 for a one-dimensional problem, 0.25 for two dimensions and 0.16 for three dimensions (e.g. Pitts and Sissom, 1977).

2.3   Convection

'Convection' is associated with the transfer of heat by motion of a fluid. In a free-burning fire, most of the heat released is carried away from the burning surfaces by buoyancy-induced convective flows (see Sections 4.3 *et seq.*), but here we shall consider how it applies to the transfer of heat between a solid surface and the surrounding fluid. If the fluid is flowing as a continuous stream past the surface, the convection is said to be 'forced': an example would be the transfer of heat to the fusible element of a sprinkler head exposed to the flow of hot combustion products from a fire (Section 4.4.2). However, if the movement is created by the buoyant flow of warm fluid next to a hot surface, the convection is then 'natural'. The convective heat transfer coefficient, defined in Equation (2.3) ($h = \dot{q}''/\Delta T$), is known to be a function of the fluid properties (thermal conductivity, density and viscosity), the flow parameters (velocity and nature of the flow) and the geometry of the surface (dimensions and angle to the flow).

The transfer process occurs close to the surface within a region known as the boundary layer whose structure determines the magnitude of h. Consider first an isothermal system in which an incompressible fluid is flowing with a free stream velocity u_∞ across a rigid flat plate, parallel to the flow (Figure 2.13). Given that the layer of fluid next to the plate will be stationary ($u(0) = 0$), there will be a velocity gradient described by an equation $u = u(y)$. At a large distance from the plate, $u = u_\infty = u(\infty)$: by definition, the boundary layer is taken to extend from the surface to the point at which $u(y) = 0.99u_\infty$. For small values of x, i.e. close to the leading edge of the plate (Figure 2.13(a)), the flow within the boundary layer is laminar. This develops into turbulent flow beyond a transition regime, although a laminar sublayer always exists close to the surface (Figure 2.13(b)). As with flow in a pipe, the nature of the flow may be determined by examining the magnitude of the 'local' Reynolds number, $\mathrm{Re}_x = xu_\infty\rho/\mu$, where μ is the absolute viscosity. If $\mathrm{Re}_x < 2 \times 10^5$, then the flow is laminar, while the boundary layer will be turbulent if $\mathrm{Re}_x > 3 \times 10^6$: between these limits, the layer may be laminar or turbulent. This may be compared with pipe flow where laminar behaviour is observed for $\mathrm{Re} < 2300$ ($\mathrm{Re} = Du\rho/\mu$, where D is the pipe diameter).

Figure 2.13 shows the hydrodynamic boundary layer for an isothermal system. Its thickness (δ_h) also depends on the Reynolds number and can be approximated by

$$\delta_\mathrm{h} \approx l \left(\frac{8}{\mathrm{Re}_l}\right)^{1/2} \tag{2.41}$$

for laminar flow, where l is the value of x at which δ_h is measured (Figure 2.13(a)) and Re_l is the local Reynolds number (Kanury, 1975).

If the fluid and the plate are at different temperatures, a thermal boundary layer will exist, as shown in Figure 2.14. The rate at which heat is transferred between the

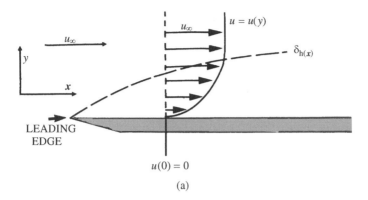

(a)

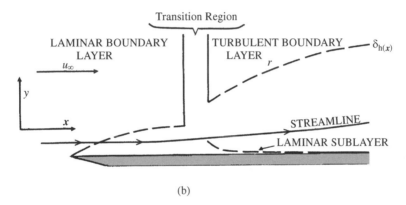

(b)

Figure 2.13 (a) The hydrodynamic boundary layer at the leading edge of a flat plate (isothermal system); (b) as (a) but showing the onset of turbulence and the associated laminar sublayer. Reproduced by permission of Gordon and Breach from Kanury (1975)

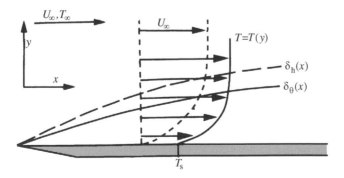

Figure 2.14 The hydrodynamic (– – – –) and thermal (————) boundary layers at the leading edge of a flat plate (non-isothermal system). Reproduced by permission of Gordon and Breach from Kanury (1975)

fluid and the surface will then depend on the temperature gradient within the fluid at $y = 0$, i.e.

$$\dot{q}'' = -k \left(\frac{\partial T}{\partial y} \right)_{y=0} \tag{2.42}$$

where k is the thermal conductivity of the fluid: this is Fourier's law of heat conduction applied to the sublayer of fluid adjacent to the surface. Following Kanury (1975), Equation (2.42) can be approximated by the expression

$$\dot{q}'' \approx \frac{k}{\delta_\theta}(T_\infty - T_s) \tag{2.43}$$

where δ_θ is the thickness of the thermal boundary layer, and T_∞ and T_s are the temperatures of the main stream fluid and the surface of the plate, respectively. The ratio of the thickness of the two boundary layers (δ_θ/δ_h) is dependent on the Prandtl number, $\mathrm{Pr} = \nu/\alpha$, a dimensionless group (see Table 4.6) which relates the 'momentum diffusivity' and 'thermal diffusivity' of the fluid, where $\nu(= \mu/\rho)$ is the kinematic viscosity. These determine the structures of the hydrodynamic and thermal boundary layers respectively. Thus, for laminar flow, Kanury (1975) derives the approximate expression:

$$\frac{\delta_\theta}{\delta_h} \approx (\mathrm{Pr})^{-1/3} \tag{2.44}$$

Combining Equations (2.3), (2.41), (2.43) and (2.44) gives

$$h \approx k/(l(8/\mathrm{Re})^{1/2} \cdot (\mathrm{Pr}^{-1/3})) \tag{2.45}$$

By convention, h is normally expressed as a multiple of k/l, where l is the characteristic dimension of the surface. Thus, Equation (2.45) would be expressed in dimensionless form as

$$\mathrm{Nu} = \frac{hl}{k} \approx 0.35 \, \mathrm{Re}^{1/2} \, \mathrm{Pr}^{1/3} \tag{2.46}$$

where Nu is the local Nusselt number. This has the same form as the Biot number but differs in that k (the thermal conductivity) refers to the fluid rather than the solid. There is considerable advantage to be gained in expressing the convective heat transfer coefficient in this dimensionless form. It allows heat transfer data from geometrically similar situations to be correlated, thus providing the means by which small-scale experimental results can be used to predict large-scale behaviour. This is an example of modelling, a technique to which reference will be made in Section 4.4.5.

Detailed analysis of the boundary layers can be found in most texts on momentum and heat transfer (e.g. Rohsenow and Choi, 1961; Welty *et al.*, 1976). Thus the exact solution to the problem outlined above (laminar flow over a flat plate) is given by

$$\mathrm{Nu} = 0.332 \, \mathrm{Re}^{1/2} \, \mathrm{Pr}^{1/3} \tag{2.47}$$

with which the approximate solution (Equation (2.46)) is in satisfactory agreement. As the Prandtl number does not vary significantly — indeed, it is frequently assumed to be unity in many combustion problems (Kanury, 1975; Williams, 1982) — these equations can be rearranged to show that $h \propto u^{1/2}$ for these conditions. This result is used in Section 4.4.2 in relation to the response of heat detectors to fires.

For turbulent flow, the temperature gradient at $y = 0$ is much steeper than for laminar flow and the Nusselt number is given by

$$\text{Nu} = 0.037 \, \text{Re}^{4/5} \, \text{Pr}^{1/3} \tag{2.48}$$

Expressions for other geometries under both laminar and turbulent forced flow conditions may be found in the literature (e.g. Kanury, 1975; Williams, 1982) (Table 2.4).

In natural or free convection, the hydrodynamic and thermal boundary layers are inseparable as the flow is created by buoyancy induced by the temperature difference between the boundary layer and the ambient fluid. Analysis introduces the Grashof number which is essentially the ratio of the upward buoyant force to the resisting viscous drag:

$$\text{Gr} = \frac{gl^3(\rho_\infty - \rho)}{\rho v^2} = \frac{gl^3 \beta \Delta T}{v^2} \tag{2.49}$$

Table 2.4 Some recommended convective heat transfer correlations[a] (Kanury, 1975; Williams, 1982)

Nature of the flow and configuration of the surface	$\overline{\text{Nu}} = \dfrac{hl}{k}$
Forced convection	
Laminar flow, parallel to a flat plate of length l ($20 < \text{Re} < 3 \times 10^5$)	$0.66 \, \text{Re}^{1/2}\text{Pr}^{1/3}$
Turbulent flow, parallel to a flat plate of length l ($\text{Re} > 3 \times 10^5$)	$0.037 \, \text{Re}^{4/5}\text{Pr}^{1/3}$
Flow round a sphere of diameter l (general equation)	$2 + 0.6 \, \text{Re}^{1/2} \, \text{Pr}^{1/3}$
Natural convection	
Laminar: natural convection at a vertical flat plate of length l (Figure 2.15) ($10^4 < \text{Gr.Pr} < 10^9$)	$0.59 \, (\text{Gr.Pr})^{1/4}$
Turbulent: natural convection at a vertical flat plate of length l ($\text{Gr.Pr} > 10^9$)	$0.13 \, (\text{Gr.Pr})^{1/3}$
Laminar: natural convection at a hot horizontal plate of length l (face up) ($10^5 < \text{Gr.Pr} < 2 \times 10^7$)	$0.54 \, (\text{Gr.Pr})^{1/4}$
Turbulent: natural convection at a hot horizontal plate of length l (face up) ($2 \times 10^7 < \text{Gr.Pr} < 3 \times 10^{10}$)	$0.14 \, (\text{Gr.Pr})^{1/3}$
Vertical parallel plates, separation l: Gr $< 2 \times 10^3$	1
$2 \times 10^3 < \text{Gr} < 2.1 \times 10^5$	$0.2 \, (\text{Gr.Pr})^{1/4}$
$2.1 \times 10^5 < \text{Gr} < 1.1 \times 10^7$	$0.071 \, (\text{Gr.Pr})^{1/3}$

[a] The expressions give average values for the Nusselt number. $\text{Re} = ul/v$, $\text{Pr} = v/\alpha$, $\text{Gr} = gl^3\beta\Delta T/v^2$. Williams (1982) suggests that in most fire problems the Prandtl number can be assumed to be unity ($\text{Pr} = 1$). Note that $\mu (= v\rho)$ and k are temperature-dependent.

[b] Typically, h takes values in the range 5–50 W/m^{2}K and 25–250 W/m^2 K for natural convection and forced convection in air, respectively (Welty *et al.*, 1976).

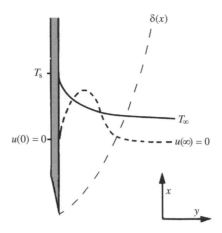

Figure 2.15 Free convective boundary layer at the surface of a vertical flat plate. Reproduced by permission of Gordon and Breach from Kanury (1975)

where g is the gravitational acceleration constant. The convective heat transfer coefficient is found to be a function of the Prandtl and Grashof numbers: thus for a vertical plate (Figure 2.15),

$$\text{Nu} = \frac{hl}{k} = 0.59(\text{Gr} . \text{Pr})^{1/4} \tag{2.50}$$

provided that the flow is laminar ($10^4 < \text{Gr} . \text{Pr} < 10^9$); for turbulent flow ($\text{Gr} . \text{Pr} > 10^9$),

$$\text{Nu} = 0.13(\text{Gr} . \text{Pr})^{1/3} \tag{2.51}$$

Expressions for other configurations are given in Table 2.4.

2.4 Radiation

As indicated earlier, thermal radiation involves transfer of heat by electromagnetic waves confined to a relatively narrow 'window' in the electromagnetic spectrum (Figure 2.16). It incorporates visible light and extends towards the far infra-red, corresponding to wavelengths between $\lambda = 0.4$ and 100 µm. As a body is heated and its temperature rises, it will lose heat partly by convection (if in a fluid such as air) and partly by radiation. Depending on the emissivity and the value of h (the convective heat transfer coefficient), convection predominates at low temperatures ($< c.150$–$200°C$), but above $c.400°C$, radiation becomes increasingly dominant. At a temperature of around $550°C$, the body emits sufficient radiation within the optical region of the spectrum for a dull red glow to be visible. As the temperature is increased further, colour changes are observed which can be used to give a rough guide to the temperature (Table 2.5). These changes are due to the changing spectral distribution with temperature, illustrated in Figure 2.17(a) for an ideal emitter, i.e. a black body. These curves are described by Planck's distribution law which embodies the fundamental concept of the quantum theory, i.e. that electromagnetic radiation is discontinuous, being emitted

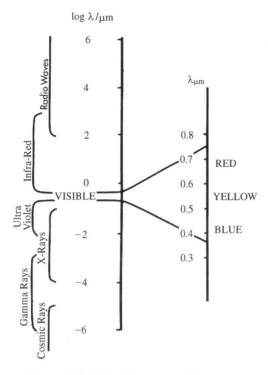

Figure 2.16   The electromagnetic spectrum

Table 2.5   Visual colour of hot objects

Temperature (°C)	Appearance
550	First visible red glow
700	Dull red
900	Cherry red
1100	Orange
1400	White

in discrete amounts known as 'quanta':

$$E_{b,\lambda} = \frac{2\pi c^2 h \lambda^{-5}}{\exp(ch/\lambda kT) - 1} \tag{2.52}$$

where $E_{b,\lambda}$ is the total amount of energy emitted per unit area by a black body within a narrow band of wavelengths (between λ and $\lambda + d\lambda$), c is the velocity of light, h is Planck's constant, k is Boltzmann's constant, and T is the absolute temperature. The maximum moves to shorter wavelengths as the temperature increases according to Wien's Law:

$$\lambda_{\max} T = 2.9 \times 10^3 \ \mu\text{m.K} \tag{2.53}$$

Thus, at 1000 K, the maximum is at 2.9 μm, as shown in Figure 2.17(a).

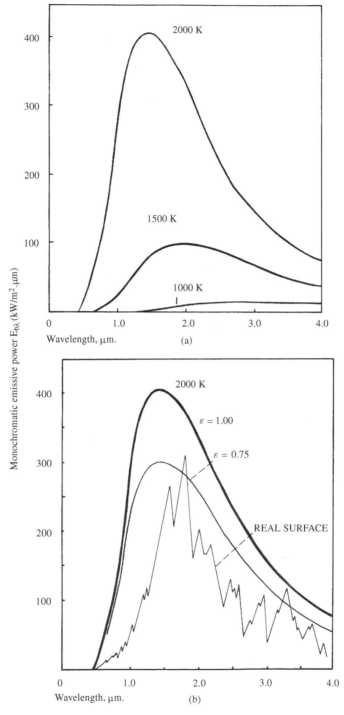

Figure 2.17   (a) Black body emissive power as a function of wavelength and temperature: (b) comparison of emissive power of ideal black bodies and grey bodies with that of a 'real' surface. (Adapted from Gray and Muller (1974) by permission)

Integrating Equation (2.52) between $\lambda = 0$ and $\lambda = \infty$ gives the total emissive power of a black body as:

$$E_b = \int_0^\infty E_{b,\lambda} . \mathrm{d}\lambda = \frac{2\pi^5 k^4 T^4}{15 c^2 h^3} \tag{2.54}$$

Comparing this with the relationship derived semi-empirically by Stefan and Boltzmann (Equation (2.4)) (with $\varepsilon = 1$)) shows that the constant σ is a function of three fundamental physical constants, c, h and k. Consequently its value is known very accurately ($\sigma = 5.67 \times 10^{-8}$ W/m^2.K^4).

The emissivity of a real surface is less than unity ($\varepsilon < 1$) and can depend on wavelength. Thus it should be defined as:

$$\varepsilon_\lambda = \frac{E_\lambda}{E_{b,\lambda}} \tag{2.55}$$

where E_λ is the emissive power of the real surface between λ and $\lambda + \mathrm{d}\lambda$. The variation of monochromatic emissive power with λ for a fictitious 'real body' is shown in Figure 2.17(b). However, it is found convenient to introduce the concept of a 'grey body' (or an 'ideal, non-black' body) for which ε is independent of wavelength: while this is an approximation, it permits simple use of the Stefan-Boltzmann equation (Equation (2.4)). Typical values of ε for solids are given in Table 2.6. Kirchhoff's law states that these are equal to their absorptivities, as dictated by the first law of thermodynamics: thus a black body is a perfect absorber, with $a = 1$.

$$E = \varepsilon \sigma T^4 \tag{2.4}$$

Equation (2.4) gives the total radiation emitted by unit area of a grey surface into the hemisphere above it. It can be used without modification to calculate radiative heat loss from a surface but as the radiation is diffuse, calculation of the rate of heat transfer to nearby objects requires a method of calculating the amount of energy being radiated in any direction. To enable this to be done, the intensity of normal radiation (I_n) is defined as 'the energy radiated per second per unit surface area per unit solid angle from an element of surface within a small cone of solid angle with its axis normal to the surface'. Lambert's cosine law can then be used to calculate the emission intensity

Table 2.6   Emissivities $(\varepsilon)^a$

Surface	Temperature (°C)	Emissivity
Steel, polished	100	0.066
Mild steel		0.2–0.3
Sheet steel, with rough oxide layer	24	0.8–0.9
Asbestos board	24	0.96
Fire brick	1000	0.75
Concrete tiles	1000	0.63

^{a} From Pitts and Sissom (1977) and Weast (1974/75).

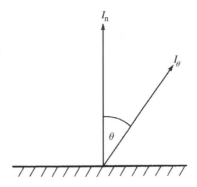

Figure 2.18   The intensity of normal radiation (I_n)

in a direction θ to the normal (Figure 2.18), i.e.

$$I = I_n \cos \theta \tag{2.56}$$

which applies only to diffuse emitters. The relationship between I_n and E may be found by considering the thermal radiation emitted from a small element of surface area dA_1 through the solid angle $d\omega$ obtained by rotating the vectors defined by the angles θ and $\theta + d\theta$ through an angle of $360°$ with the normal to the surface as the axis (Figure 2.19). From the definition of I_n and Lambert's cosine law:

$$dE = I_n \cos \theta \, dA_1 \cdot d\omega \tag{2.57}$$

where the differential solid angle $d\omega$ is, by definition,

$$d\omega = dA_2/r^2 \tag{2.58}$$

and

$$dA_2 = 2\pi r \sin \theta \cdot r \, d\theta \tag{2.59}$$

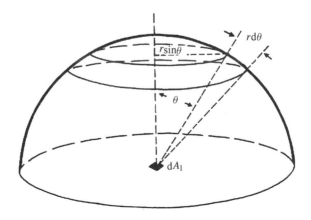

Figure 2.19   Derivation of the relationship between I_n and E (Equations (2.56)–2.62))

Substituting Equations (2.58) and (2.59) into Equation (2.57) gives

$$dE = 2\pi I_n \sin\theta \cos\theta \, d\theta \, dA_1 \tag{2.60}$$

Expressing this as a heat flux from dA_1 and integrating from $\theta = 0$ to $\theta = \pi/2$, gives

$$E = 2\pi I_n \int_0^{\pi/2} \sin\theta \cos\theta \, d\theta \tag{2.61}$$

$$= \pi I_n \tag{2.62}$$

This equation relates the emissive power to the intensity of normal radiation (see also Tien *et al.*, 1995).

2.4.1 Configuration factors

Equation (2.4) gives the total heat flux emitted by a surface. In order to calculate the radiant intensity at a point distant from the radiator, a geometrical — or 'configuration' — factor must be used. Consider two surfaces, 1 and 2, of which the first is radiating with an emissive power E_1 (Figure 2.20). The radiant intensity falling on a small element of surface dA_2 on surface 2 is obtained by calculating the amount of energy from a small element of surface dA_1 that is transmitted through the solid angle subtended by dA_2 at dA_1:

$$d\dot{q} = I_n \, dA_1 \cos\theta_1 \cdot \frac{dA_2 \cos\theta_2}{r^2} \tag{2.63}$$

The incident radiant flux at dA_2 is then

$$d\dot{q}'' = \frac{d\dot{q}}{dA_2} = I_n \, dA_1 \cos\theta_1 \frac{\cos\theta_2}{r^2} \tag{2.64}$$

But $(dA_1 \cos\theta_1)/r^2$ is the solid angle subtended by dA_1 at dA_2. Integrating over A_1, and setting $I_n = E/\pi$,

$$\dot{q}'' = E \cdot \int_0^{A_1} \frac{\cos\theta_1 \cos\theta_2}{\pi r^2} \cdot dA_1 \tag{2.65}$$

$$= \phi E \tag{2.66}$$

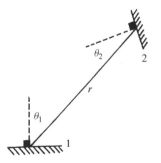

Figure 2.20   Derivation of the configuration factor ϕ (Equations (2.63)–(2.66))

where ϕ is known as the configuration factor. Values may be derived for various shapes and geometries from tables and charts in the literature (McGuire, 1953; Rohsenow and Choi, 1961; Hottel and Sarofim, 1967; Tien *et al.*, 1995).

Figure 2.21 is a diagram from which the configuration factor ϕ may be derived for the geometry shown in Figure 2.22(a), i.e. a receiving element dA lying on the perpendicular to one corner of a parallel rectangle. This method of presentation allows advantage to be taken of the fact that configuration factors are additive. Thus the element dA in Figure 2.22(b) views four rectangles, A, B, C and D. The configuration

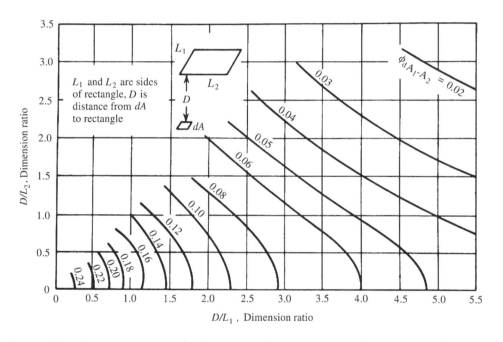

Figure 2.21 Configuration factor ϕ for direct radiation from a rectangle to a parallel small element of surface dA on a perpendicular to one corner (Figure 2.22(a)) (Hottel, 1930). Reproduced by permission of John Wiley & Sons, Inc.

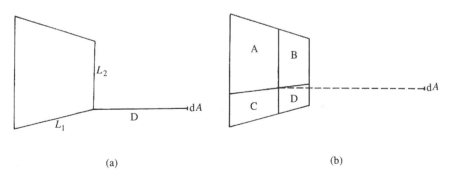

(a)

(b)

Figure 2.22 (a) Receiver element dA on perpendicular from corner of parallel rectangle (see Figure 2.21);
(b) illustrating that configuration factors are additive (Equation (2.67))

Table 2.7   Values of $\phi(\alpha, S)$ for various values of α and S^*

α	$S=1$	$S=0.9$	$S=0.8$	$S=0.7$	$S=0.6$	$S=0.5$	$S=0.4$	$S=0.3$	$S=0.2$	$S=0.1$
2.0	0.178	0.178	0.177	0.175	0.172	0.167	0.161	0.149	0.132	0.102
1.0	0.139	0.138	0.137	0.136	0.133	0.129	0.123	0.113	0.099	0.075
0.9	0.132	0.132	0.131	0.130	0.127	0.123	0.117	0.108	0.094	0.071
0.8	0.125	0.125	0.124	0.122	0.120	0.116	0.111	0.102	0.089	0.067
0.7	0.117	0.116	0.116	0.115	0.112	0.109	0.104	0.096	0.083	0.063
0.6	0.107	0.107	0.106	0.105	0.103	0.100	0.096	0.088	0.077	0.058
0.5	0.097	0.096	0.096	0.095	0.093	0.090	0.086	0.080	0.070	0.053
0.4	0.084	0.083	0.083	0.082	0.081	0.079	0.075	0.070	0.062	0.048
0.3	0.069	0.068	0.068	0.068	0.067	0.065	0.063	0.059	0.052	0.040
0.2	0.051	0.051	0.050	0.050	0.049	0.048	0.047	0.045	0.040	0.032
0.1	0.028	0.028	0.028	0.028	0.028	0.028	0.027	0.026	0.024	0.021
0.09	0.026	0.026	0.026	0.026	0.025	0.025	0.025	0.024	0.022	0.019
0.08	0.023	0.023	0.023	0.023	0.023	0.023	0.022	0.022	0.020	0.017
0.07	0.021	0.021	0.021	0.021	0.020	0.020	0.020	0.019	0.018	0.016
0.06	0.018	0.018	0.018	0.018	0.018	0.017	0.017	0.017	0.016	0.014
0.05	0.015	0.015	0.015	0.015	0.015	0.015	0.015	0.014	0.014	0.013
0.04	0.012	0.012	0.012	0.012	0.012	0.012	0.012	0.012	0.011	0.010
0.03	0.009	0.009	0.009	0.009	0.009	0.009	0.009	0.009	0.009	0.008
0.02	0.006	0.006	0.006	0.006	0.006	0.006	0.006	0.006	0.006	0.006
0.01	0.003	0.003	0.003	0.003	0.003	0.003	0.003	0.003	0.003	0.003

* $S = L_1/L_2$ and $\alpha = (L_1 \times L_2)/D^2$ (see Figure 2.21). From McGuire (1953). Reproduced by permission of The Controller, HMSO. © Crown copyright.

factors for each can be read from Figure 2.21 (or from Table 2.7 (McGuire, 1953)) and the total configuration factor obtained as their sum:

$$\phi_{total} = \phi_A + \phi_B + \phi_C + \phi_D \tag{2.67}$$

This may be used to estimate heat fluxes on surfaces exposed to radiation from a fire. In the UK, permissible building separation distances are calculated on the basis that the exterior of one building must not be exposed to a heat flux of more than 1.2 W/cm^2 (1.2×10^4 W/m^2) if an adjacent building is involved in fire (Law, 1963). This level of radiant flux is commonly assumed to be the minimum necessary for the pilot ignition of wood (Section 6.3). The radiating surfaces of a building are taken to be windows and exterior woodwork, and the separation distance is worked out on the basis of the *maximum* heat flux to which an adjacent building may be exposed. Figure 2.23 illustrates the locus of a given value of ϕ for a building on fire. The individual radiators (windows, etc.) may achieve temperatures of up to 1100°C (1373 K), corresponding to a maximum emissive power of 20 W/cm^2 if $\varepsilon = 1$, unless the fire load is small (in which case it does not burn for a sufficient length of time for these temperatures to be achieved) or the fire is fuel-controlled (Chapter 10). Law (1963) assumes that the radiating areas will have an emissive power of 17 W/cm^2 unless the latter conditions hold: then 8.5 W/cm^2 is used. It should be noted that in the absence of external combustible cladding, only openings are considered as 'radiators'. If large flames are projecting from the openings, as will occur particularly with underventilated compartment fires (Section 10.6), much higher heat fluxes may be expected on target surfaces (Lougheed and Yung, 1993).

To illustrate how configuration factors may be used, consider the side of a building, 5.0 m long by 3.0 m high, with two windows, each 1 m by 1 m, located symmetrically,

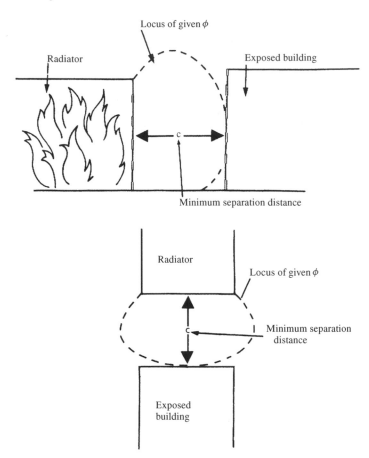

Figure 2.23 Locus of a given configuration factor for a particular radiator (Law, 1963). (Reproduced by permission of The Controller, HMSO © Crown copyright)

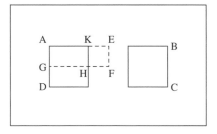

Figure 2.24 Calculation of the configuration factor for the face of a building with two windows, symmetrically located (see Equation (2.68))

as shown in Figure 2.24. To calculate the maximum heat flux at a distance of 5 m from the wall if the building is on fire, only the rectangle ABCD enclosing the windows need be considered. At 5 m distance on the axis of symmetry where the heat flux will be greatest (see Figure 2.23)

$$\phi_{ABCD} = 4\phi_{AKHG} = 4(\phi_{AEFG} - \phi_{KEFH}) \tag{2.68}$$

Using Table 2.7, it is found that:

$$\phi_{AEFG} = 0.009 \text{ and } \phi_{KEFH} = 0.003$$

Therefore

$$\phi_{ABCD} = 4 \times 0.006 = 0.024$$

Assuming that the emissive power of each window is 17 W/cm^2 (after Law, 1963)

$$\dot{q}''_{max} (5 \text{ m}) = 0.024 \times 17 \text{ W/cm}^2$$

$$= 0.41 \text{ W/cm}^2$$

A simpler solution is possible for symmetrical geometries similar to Figure 2.24. Since only 67% of the area ABCD is radiating, the 'average emissive power' for area ABCD is 0.67×17 W/cm^2 = 11.4 W/cm^2. Then, as the configuration factor for AEFG is 0.009,

$$\dot{q}''_{max} (5 \text{ m}) = 4 \times 0.009 \times 11.4 = 0.41 \text{ W/cm}^2$$

It must be emphasized that the configuration factor defined by Equations (2.65) and (2.66), i.e.

$$\phi = \int_0^{A_1} \frac{\cos\theta_1 \cos\theta_2}{\pi r^2} \cdot dA_1 \tag{2.69}$$

allows the radiant heat *flux* at a point to be calculated at a distance r from a radiator. In Schaum's terminology (Pitts and Sissom, 1977) this would be a 'finite-to-infinitesimal area' configuration factor, and is useful in certain problems relating to ignition (Chapter 6) and for evaluating situations in which people might be exposed to levels of radiant heat (Table 2.8). However, if it is required to calculate the energy exchange between two surfaces, then Equation (2.63) must be integrated twice, over

Table 2.8 Effects of thermal radiation

Radiant heat flux (kW/m^2)	Observed effect
0.67	Summer sunshine in UK[a]
1	Maximum for indefinite skin exposure
6.4	Pain after 8 s skin exposure[b]
10.4	Pain after 3 s exposure[a]
12.5	Volatiles from wood may be ignited by pilot after prolonged exposure (see Section 6.3)
16	Blistering of skin after 5 s[b]
29	Wood ignites spontaneously after prolonged exposure[a] (see Section 6.4)
52	Fibreboard ignites spontaneously in 5 s[a]

[a] D.I. Lawson (1954)
[b] S.H. Tan (1967).
The data quoted for human exposure are essentially in agreement with information given by Purser (1995) and Mudan and Croce (1995).

the areas of both surfaces, A_1 and A_2 (see Figure 2.20). The rate of radiant heat transfer to surface 2 from surface 1 is then given by:

$$\dot{Q}_{1,2} = F_{1,2} A_1 \varepsilon_1 \sigma T_1^4 \tag{2.70}$$

where

$$F_{1,2} = \frac{1}{A_1} \int_{A_1} \int_{A_2} \frac{\cos \theta_1 \cos \theta_2}{\pi r^2} dA_1 \, dA_2 \tag{2.71}$$

and is called the 'integrated configuration factor', or 'finite-to-finite area configuration factor'. It is commonly given the symbol F and the product $F_{1,2}A_1$ is known as the 'exchange area'. By symmetry

$$F_{1,2} A_1 = F_{2,1} A_2 \tag{2.72}$$

Values of the integrated configuration factor are available in the literature in the form of charts and tables. From Figure 2.25 values of $F_{1,2}$ can be deduced for radiation exchange between two parallel rectangular plates. Figure 2.26 can be applied to plates at right angles to each other. As with the configuration factor ϕ, F's are additive and can be manipulated to obtain an integrated configuration factor for more complex situations such as that shown in Figure 2.27. Their application to fire problems is discussed in detail by Steward (1974a) and Tien *et al.* (1995).

It is important to remember that radiative heat transfer is a two-way process. Not only will the receiver radiate but also the emitting surface will receive radiation from its

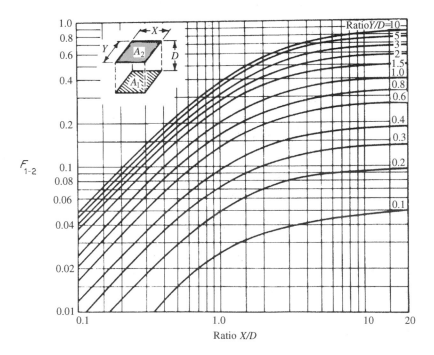

Figure 2.25   View factor for total radiation exchange between two identical, parallel, directly opposed flat plates (Hamilton and Morgan, 1952)

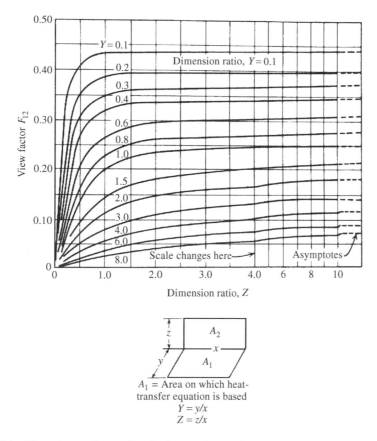

Figure 2.26 View factor for total radiation exchange between two perpendicular flat plates with a common edge (Hamilton and Morgan, 1952)

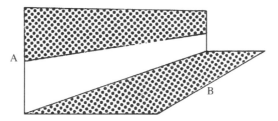

Figure 2.27 View factors for surfaces A and B can be calculated from Figure 2.26 (see text)

surroundings, including an increasing contribution from the receiver as its temperature rises. This can best be illustrated by an example. Consider a vertical steel plate, 1 m square, which is heated internally by means of electrical heating elements at a rate corresponding to 50 kW (Figure 2.28(a)). The final temperature of the plate (T_p) can be calculated from the steady-state heat balance, Equation (2.73).

$$50\,000 = 2\varepsilon\sigma(T_p^4 - T_0^4) + 2h(T_p - T_0) \tag{2.73}$$

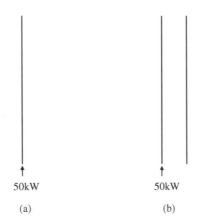

(a)   (b)

Figure 2.28   (a) Heat losses from a vertical, internally heated flat plate (Equations (2.73) and (2.74)); (b) heat losses and radiation exchange between two vertical, flat plates, one of which is internally heated (Equations (2.75)–(2.78))

where T_0 is the ambient temperature, 25°C. The factor of 2 appears because the plate is losing heat from both surfaces: it is assumed that the plate is sufficiently thin for heat losses from the edges to be ignored. Equation (2.73) can be reduced to:

$$2\varepsilon\sigma T_{\rm p}^4 + 2hT_{\rm p} - (50\,000 + 596h) = 0 \tag{2.74}$$

as $2\varepsilon\sigma T_0^4 \ll 50\,000$. Equation (2.74) may be solved for $T_{\rm p}$ with $\varepsilon = 0.85$ and $h = 12$ W/m².K, using the Newton-Raphson method (Margenau and Murphy, 1956) to give $T_{\rm p} = 793$ K (520°C). If a second steel plate, 1 m square but with no internal heater, is suspended vertically 0.15 m from the first (Figure 2.28(b)) then, ignoring reflected radiation, the following two steady state equations can be written:

For plate 1:

$$50\,000 + A_2 F_{2,1}\varepsilon^2\sigma T_2^4 + (1 - A_2 F_{2,1})\varepsilon\sigma T_0^4 = 2A_1 h(T_1 - T_0) + 2A_1\varepsilon\sigma T_1^4 \tag{2.75}$$

and for plate 2:

$$A_1 F_{1,2}\varepsilon^2\sigma T_1^4 + (1 - A_1 F_{1,2})\varepsilon\sigma T_0^4 = 2A_2 h(T_2 - T_0) + 2A_2\varepsilon\sigma T_2^4 \tag{2.76}$$

Of the two terms expressing radiative heat gain on the left-hand sides of Equations (2.75) and (2.76), the first contains ε^2, which is equivalent to the product: (emissivity of emitter) × (absorptivity of receiver). The second refers to radiation from the surroundings at ambient temperature and can be ignored. From Figure 2.25, $A_1 F_{1,2} = A_2 F_{2,1} \approx 0.75$. The above equations then become:

$$9.639T_1^4 + 2.4 \times 10^9 T_1 - 3.072T_2^4 - 5.715 \times 10^{12} = 0 \tag{2.77}$$

and

$$3.072T_1^4 - 9.639T_2^4 - 2.4 \times 10^9 T_2 + 7.152 \times 10^{11} = 0 \tag{2.78}$$

These give $T_1 = 804$ K (531°C) and $T_2 = 526$ K (253°C), thus illustrating the results of cross-radiation in confined situations. This general effect is even more significant at temperatures associated with burning, and is extremely important in fire growth and spread, particularly in spaces such as ducts, ceiling voids and even gaps between items of furniture (Section 9.2.4).

2.4.2   Radiation from hot gases and non-luminous flames

Only gases whose molecules have a dipole moment can interact with electromagnetic radiation in the 'thermal' region of the spectrum (0.4–100 μm). Thus, homonuclear diatomic molecules such as N_2, O_2 and H_2 are completely transparent in this range, while heteronuclear molecules such as HCl, CO, H_2O and CO_2 absorb (and emit) in certain discrete wavelength bands (Figure 2.29). Such species do not exhibit the continuous absorption which is characteristic of 'black' and 'grey' bodies (Figure 2.17) and absorption (and emission) occur throughout the volume of the gas and consequently the radiative properties depend on its depth or 'path length'.

Consider a monochromatic beam of radiation of wavelength λ passing through a layer of gas (Figure 2.30). The reduction in intensity as the beam passes through a thin layer dx is proportional to the intensity $I_{\lambda x}$, the thickness of the layer (dx) and the

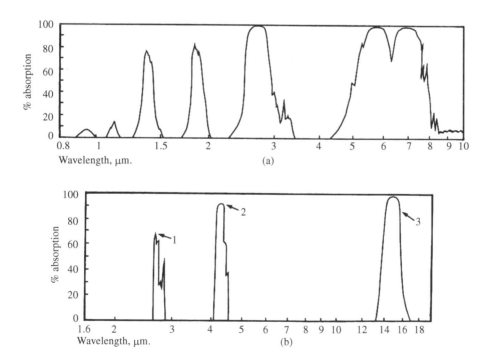

Figure 2.29   Absorption spectra of (a) water vapour, 0.8–10 μm, at atmospheric pressure and 127°C: thickness of layer 104 cm; (b) carbon dioxide, 1.6–20 μm, at atmospheric pressure: curve 1, thickness 5 cm; curves 2 and 3, thickness 6.3 cm. Adapted from Kreith (1976)

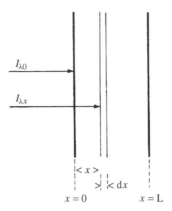

Figure 2.30   Absorption of monochromatic radiation in a layer of absorbing medium

concentration of absorbing species within that layer (C), i.e.

$$dI_\lambda = \kappa_\lambda C I_{\lambda x}\, dx \tag{2.79}$$

where κ_λ, the constant of proportionality, is known as the monochromatic absorption coefficient. Integrating from $x = 0$ to $x = L$ gives:

$$I_{\lambda L} = I_{\lambda 0} \exp(-\kappa_\lambda C L) \tag{2.80}$$

where $I_{\lambda 0}$ is the incident intensity at $x = 0$. This is known as the Lambert-Beer law. The monochromatic absorptivity is then:

$$a_\lambda = \frac{I_{\lambda 0} - I_{\lambda L}}{I_{\lambda 0}} = 1 - \exp(-\kappa_\lambda C L) \tag{2.81}$$

which, by Kirchhoff's law, is equal to the monochromatic emissivity, ε_λ, at the same wavelength λ. Equation (2.81) shows that as $L \to \infty$, a_λ and ε_λ approach a value of unity.

A volume of gas containing carbon dioxide and water vapour does not behave as a 'grey' body as the emissivity is strongly dependent on wavelength (Figure 2.29). Hottel and Egbert (1942) developed an empirical method by which an 'equivalent grey body' emissivity of a volume of hot gas containing these species could be worked out. (Other gases were included in the original work, but only CO_2 and H_2O are relevant here.) The procedure is based on a series of careful measurements of the radiant heat output from hot carbon dioxide and water vapour (separately and together) at various uniform temperatures and partial pressures with different geometries of radiating gas. As emissivity at a single wavelength is known to depend on both the concentration of the emitting species and the 'path length' through the radiating gas as viewed by the receiver (see Figure 2.30), Hottel's first step was to determine the effective total emissivity of CO_2 and water vapour as a function of temperature for a range of values of the product pL where p is the (partial) pressure of the emitter and L is the mean equivalent beam length, which depends on the geometry of the volume of gas (see

Table 2.9 Mean equivalent beam length (L) for a gaseous medium emitting to a surface (Gray and Muller, 1974)[a]

Shape	L
Right circular cylinders	
1. Height = diameter (D), radiating to:	
(a) centre of base	$0.7D$
(b) whole surface	$0.6D$
2. Height = $0.5D$, radiating to:	
(a) end	$0.43D$
(b) side	$0.46D$
(c) whole surface	$0.45D$
3. Height = $2D$, radiating to:	
(a) end	$0.60D$
(b) side	$0.76D$
(c) whole surface	$0.73D$
Sphere, diameter D, radiating to entire surface	$0.64D$

[a] These values correspond to the optically thick limit (see Tien *et al.*, 1995). They will be approximately 10% higher for an optically thin gas.

Table 2.9). The results are shown in Figure 2.31, in which the values of the product pL have been converted from Hottel's original units of atmosphere-feet to atmosphere-metres (Edwards, 1985; Tien *et al.*, 1995). If the partial pressure of the emitting species and the mean beam length are known, then the effective 'grey body' emissivity can be obtained at any temperature up to $c.3000$ K. While these diagrams apply to gas mixtures at a total pressure of 1 atmosphere, an effect known as pressure broadening, which depends on the partial pressures of emitting species, influences the emission and must be taken into account in any accurate analyses. In addition, a correction for the overlap of the 4.4 µm band of CO_2 and the 4.8 µm band of H_2O is necessary. These modifications are described in detail in most heat transfer texts (e.g. Gray and Muller, 1974; Welty *et al.*, 1976; Edwards, 1985), but will not be considered here; the example that follows is included in order to introduce the concepts and illustrate the fact that non-luminous flames have very low emissivities. Errors arising from neglect of the corrections are not significant in this context.

Consider a small fire involving a 0.3 m diameter pool of methanol. The flame, which will be non-luminous, can be approximated by a cylinder of height 0.3 m. The mean beam length for radiation falling at the centre of the surface of the pool then becomes 0.7×0.3 m $= 0.21$ m (Table 2.9). Furthermore, if the composition of the flame is taken to be that produced by burning a limiting methanol/air mixture (6.7% methanol in air) then the partial pressures of carbon dioxide and water vapour will be $p_c = 0.065$ atm and $p_w = 0.13$ atm respectively: these are likely to be overestimates. Multiplying these by the mean beam length (giving 0.014 and 0.027 atm.m respectively) and assuming a uniform temperature of 1200°C (Rasbash *et al.*, 1956) ε_c and ε_w can be evaluated from Figures 2.31(a) and (b) respectively: thus, approximately

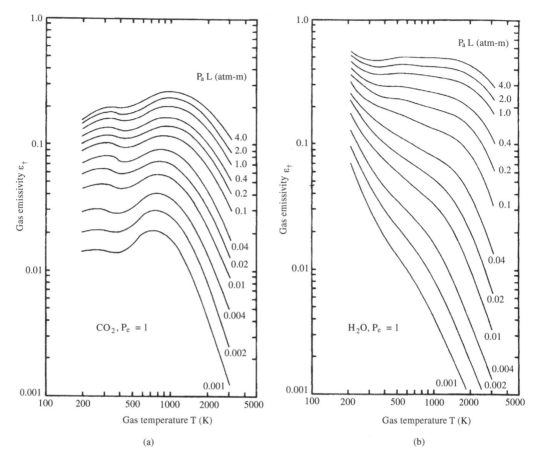

Figure 2.31 (a) Emissivity of carbon dioxide at 1 atmosphere total pressure and near zero partial pressure. (b) Emissivity of water vapour at 1 atmosphere total pressure and near zero partial pressure. From Edwards, 1985. Reproduced by permission of the Society of Fire Protection Engineers

$\varepsilon_c = 0.04$ and $\varepsilon_w = 0.04$. The resultant emissivity will be:

$$\varepsilon_g = 0.04 + 0.04 = 0.08 \qquad (2.82)$$

although this must be regarded as highly approximate as the assumptions were somewhat arbitrary. However, it is not too dissimilar to the emissivities observed by Rasbash *et al.* (1956). The low emissivity of the flame has a significant effect on the burning behaviour of this particular fuel. This will be discussed in Section 5.1.1.

Substantial progress has been made in developing methods of calculating the emissivity of non-luminous combustion gases (de Ris, 1979). While a discussion of these techniques is beyond the scope of the present text, it is worth mentioning that they reveal that the empirical method outlined above gives acceptable values of emissivity up to about 1000 K. Above this temperature and particularly at long path lengths, Hottel's method apparently underestimates the emissivity, probably as a result of overestimating the overlap correction. However, these methods apply to gases of uniform temperature and composition. If, as in a flame, the outer regions are cooler than the inner, then there will be some re-absorption of radiation and a consequent reduction of the emissive power. The effect is much more pronounced with luminous flames and layers of hot smoke (Grosshandler and Modak, 1981; Orloff *et al.*, 1979).

2.4.3 Radiation from luminous flames and hot smoky gases

With few exceptions (e.g. methanol and paraformaldehyde) liquids and solids burn with luminous diffusion flames. The characteristic yellow luminosity is the net effect of emission from minute carbonaceous particles, known as "soot," with diameters of the order of 10–100 nm which are formed within the flame, mainly on the fuel side of the reaction zone (Section 4.1). These may be consumed as they pass through the oxidative region of the flame, but otherwise they will escape from the flame tip to form smoke (Section 11.1). The propensity of different fuels — gases, liquids and solids — to produce soot can be assessed by measuring the laminar flame 'smoke point', i.e. the minimum height of a laminar diffusion flame (Section 4.1) at which smoke is released (Section 11.1 and Table 11.1). The smaller the value of the smoke point, the greater the tendency for soot to form in the flame. While within the flame, individual soot particles attain high temperatures and each will act as a minute black (or 'grey') body. The resulting emission spectrum from the flame will be continuous, and the net emissive power will be a function of the particle concentration and the flame thickness (or mean beam length, L). The smoke point has been found to relate inversely to the proportion of the heat of combustion that is lost by radiation from the flame. This holds for a range of fuels (gases, liquids and solids) (Markstein, 1984; de Ris and Cheng, 1994) and for both laminar and turbulent flames (Markstein, 1984). Generally speaking, as the presence of soot particles in the flame provides the mechanism for radiative heat loss, the 'sootier' the flame, the lower its average temperature — see Table 2.10 (de Ris, 1979) and Table 5.4 (Rasbash *et al.*, 1956).

The emissivity can be expressed in terms of a relationship identical in form to Kirchhoff's law for monochromatic absorptivity, Equation (2.81), thus:

$$\varepsilon = 1 - \exp(-KL) \tag{2.83}$$

where K is an effective emission coefficient, proportional *inter alia* to the concentration, which can be determined as the 'particulate volume fraction', f_v (Section 11.1.1). A few empirical values of K are available in the literature (Table 2.10) and permit approximate values of emissive power to be calculated, provided the flame temperature

Table 2.10 Emissivities (ε) and emission coefficients (K) for four thermoplastics (de Ris, 1979)

Fuel[a]	Flame temperature (K)	Emissivity (ε)	Emission coefficient K (m^{-1})	Carbon appearing as soot (%)
Polyoxymethylene	1400	0.05	—	0
PMMA	1400	0.26	1.3	0.30
Polypropylene	1350	0.59	1.8	5.5
Polystyrene	1190	0.81	5.3	18

[a] The fuel beds were 0.305 m square, except for the PMMA experiments (0.305 m × 0.311 m).

is known or can be measured. However this simple method for calculating emissivity cannot be used for large fires as it contains an implicit assumption that the flame is uniform with respect to both temperature and soot concentrations. There is now clear evidence that this is not so. This has come from more detailed studies of the structure of the flames, mapping not only the temperature fields but also the local concentrations of the soot particles.

Theory indicates that provided the soot particle diameter is less than the radiation wavelength (mostly $\lambda > 1$ μm (10^3 nm)), the emission coefficient will be proportional to the 'soot volume fraction' (f_v), which is the proportion of the flame volume occupied by particulate matter. This can be determined using sophisticated optical techniques. Markstein (1979) and Pagni and Bard (1979) obtained data on f_v as a function of height above horizontal slabs of burning PMMA which showed clearly that f_v decreases with height from a maximum close to the fuel surface (Brosmer and Tien, 1987). Furthermore, because of non-uniformity of temperature there is some attenuation of radiation from large flames because the outside perimeter is at a lower temperature, and the cooler soot, while still radiating, will absorb radiation from the hotter regions within. This makes accurate modelling of thermal radiation from large pool fires very difficult, particularly as the 'mean beam length' cannot be assumed equal to the pool — or tank — diameter. Simple examples of calculation of heat flux at a distance are given in Chapter 4, but these will inevitably overestimate the heat flux as they do not take into account the non-uniformity of temperature.

For smaller fires ($D < 1$ m, perhaps), emission from the soot particles is superimposed on emission from the molecular species H_2O and CO_2. This is shown clearly for the wood crib flames of Figure 2.32. Hydrocarbons (gases, liquids and solids) are much sootier, and the black-body background will tend to dominate the radiation, particularly

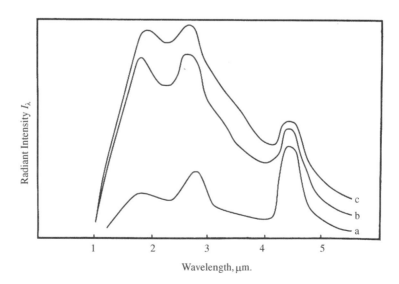

Figure 2.32 Spectra of flames at different thicknesses above wooden cribs (Hägglund and Persson, 1976a). The emission 'peaks' are due to H_2O, H_2O and CO_2 respectively, reading from left to right. Compare with Figure 2.29. Reproduced by permission of the Swedish Fire Protection Ass./FoU-Brand

if the hydrocarbons have aromatic character (e.g. polystyrene). Good progress has been made towards modelling the radiant output from flames of this nature in which both the soot emission and the molecular emissions are taken into account. However, further discussion of this is beyond the scope of this text, and the reader is referred elsewhere (de Ris, 1979; Mudan, 1984; Tien *et al.*, 1995).

For convenience and simplicity, it is sometimes assumed that a luminous flame behaves as a 'grey body', i.e. the emissivity is independent of the wavelength. However, the dominance of the CO_2 and H_2O emissions in the early stages of a fire provides an opportunity to design infra-red detectors which can distinguish a flame unambiguously from a hot surface. This can be done by using sensors which can compare the intensity of emission at 4.4 μm with that at *c*.3.8 μm, outside the CO_2 band: a significantly stronger signal at 4.4 μm will be recognized as the presence of flame, at least during the early stages when the flame is still relatively 'thin'.

If the flame is thick ($L > 1$ m) and luminous (e.g. hydrocarbon flames), it is common to assume black-body behaviour, i.e. $\varepsilon = 1$. This was not found to be the case for 2 m thick flames from free-burning wood crib fires (Figure 2.32).

Whether the emissivity of a flame is assumed or calculated on the basis of an empirical or theoretical equation, Equation (2.4) is still applicable. If the flame temperature is known, then the emissive power can be calculated using the mean equivalent beam length (Table 2.9), but before the radiant heat flux at a distance can be estimated, a configuration factor must be derived. This is generally obtained by assuming that the flame can be approximated by a simple geometrical shape, such as a rectangle of height between 1.5 and 2 times the fuel bed diameter (Section 4.3.2) and working out the appropriate configuration factor using Figure 2.21 or Table 2.7. This type of model has been used to predict levels of radiant heat at various locations in a petrochemical plant during an emergency, such as a tank fire or emergency flaring (Robertson, 1976; Lees, 1996).

Radiation from hot smoke is now known to be an important contributory factor to the development of fire within enclosed spaces. During the growth period of a compartment fire, hot smoky gases accumulate under the ceiling, radiating to the lower levels and thereby enhancing the onset of fully developed burning (Chapter 9). The smoke layer is non-homogeneous and re-absorption of radiation in the lower layers is significant. This has been modelled successfully by Orloff *et al.* (1979) and others, and will be discussed further in Chapter 9.

Problems

2.1 Consider a steel barrier, 5 mm thick, separating two compartments which are at temperatures of 100°C and 20°C respectively. Calculate the rate of heat transfer through the barrier under steady state conditions if the thermal conductivity of the steel is 46 W/m.K and the convective heat transfer coefficient is 8 W/m^2.K.

2.2 Using the example given in Problem 2.1, calculate the temperatures of the two exposed surfaces (the 'hot' and 'cold' sides of the barrier). Check the Biot number to demonstrate that this is an example of a 'thermally thin' material. What would the surface temperatures be if the barrier was 50 mm thick? Calculate the Biot number for this situation — what are your conclusions?

2.3 Calculate the total steady-state heat loss through a 200 mm external brick wall which measures 8 m × 4 m high and contains one single-glazed window, 3 m by 1.5 m, located centrally. The glass is 3 mm thick. Ignore the effects of the window frame. The inside and outside temperatures are 25°C and 0°C respectively. Assume a convective heat transfer coefficient of 8 W/m² K on both sides of the wall.

2.4 What would the steady-state heat loss through the wall described in Problem 2.3 be if the following modifications are made:

   (a)   The window is replaced by a double-glazed unit comprising two 2 mm thick sheets of glass separated by a 2 mm air gap. (Assume that heat transfer across the gap is by conduction through the air.)

   (b)   The brick is lined on the inside by 10 mm fibre insulating board and on the outside by 12 mm pine boards.

   (c)   The wall was constructed as a cavity wall, filled with polyurethane foam. Assume that the cavity and both courses of brick are 100 mm across.

2.5 A fire in a room rapidly raises the temperature of the surface of the walls and maintains them at 1000°C for a prolonged period. On the other side of one wall is a large warehouse, whose ambient temperature is normally 20°C. If the wall is solid brick, 200 mm thick, and retains its integrity, what would be the steady-state temperature of the surface of the wall on the warehouse side? Assume that the thermal conductivity of the brick is independent of temperature and that the convective heat transfer coefficient at the warehouse wall is 12 W/m².K.

2.6 What would the steady-state temperature at the inner surface of the warehouse wall (described in the previous problem) be if several sheets of fibre insulation board (total thickness 0.3 m) were stacked vertically against the wall? (Assume perfect contact between the boards and the wall.)

2.7 Take the model described in Problem 2.1, with both sides of the barrier at 20°C. Assume that the temperature of the air on one side is suddenly increased to 150°C. Calculate numerically the temperature of the steel after 10 minutes using a timestep of 60 s. Compare your answer with the value of T_s taken from the analytical solution derived in Section 2.2.3 (Equation (2.34)).

2.8 A vertical sheet of cotton fabric, 0.6 mm thick, is immersed in a stream of hot air at 150°C. Calculate how long it will take to reach 100°C if the heat transfer coefficient is $h = 20$ W/m² K and ρ and c are 300 kg/m³ and 1400 J/kg respectively. Assume the initial temperature to be 20°C.

2.9 What temperature will a heavy cotton fabric, 1.0 mm thick, reach in 5 seconds if it is exposed to a hot air stream at 500°C? Use the data given in Problem 2.8.

2.10 Vertical slabs of Perspex (PMMA) are exposed on both sides to a stream of hot air at 200°C. Using the Heisler charts (Figure 2.6), derive the surface temperature after 30 seconds for thicknesses of: (a) 3 mm, (b) 5 mm, (c) 10 mm. Assume that the initial temperature is 20°C.

2.11 Estimate the mid-plane temperatures for the slabs of PMMA described in Problem 2.10, for the same conditions.

2.12 Which of the following could be treated as semi-infinite solids if exposed to a convective heat flux for 30 seconds: (a) 6 mm PMMA, (b) 40 mm PUF, (c) 10 mm fibre insulating board?

2.13 Using Equation (2.26), calculate the surface temperature of a thick slab of (a) yellow pine, (b) fibre insulating board, after 15 s exposure to a steady stream of air at 300°C. Take the heat transfer coefficient to be 15 W/m²K. Assume the slabs to behave as semi-infinite solids, initially at a temperature of 20°C.

2.14 A horizontal steel plate measuring 25 × 25 cm lies horizontally on an insulating pad and is maintained at a temperature of 150°C. What would the rate of heat loss be if the surface was cooled by:

    (a)   natural convection (Nu = 0.54 (GrPr)$^{1/4}$)?

    (b)   a laminar flow of air ($u = 5$ m/s) (Nu = 0.66 Re$^{1/2}$Pr$^{1/3}$)?

    (c)   by a turbulent flow of air ($u = 40$ m/s) (Nu = 0.037 Re$^{4/5}$Pr$^{1/4}$)?

(Ignore heat losses from the edge of the plate and to the insulating pad.) The air temperature is 20°C and the kinematic viscosity of air is $v = 18 \times 10^{-6}$m²/s. Would radiative heat losses be significant?

2.15 Calculate the configuration factor for an element of surface at S, parallel to the radiator ABCD, as shown in Figure 2.33. Dimensions are as follows: $a = 2$ m, $b = 1$ m, $c = 2.5$ m, $d = 0.5$ m and $e = 6$ m.

2.16 A building is totally involved in fire. One wall, measuring 20 m long by 10 m high, has four symmetrically placed windows, each 6 m long by 2 m high (see Figure 2.34). Assuming that the windows act as black body radiators at 1000°C, calculate the radiant heat flux 10 m from the building, where the radiant heat flux is a maximum. Assume that the receiver is parallel to the face of the building.

2.17 Taking the building fire described in Problem 2.16, what will the radiant heat flux be at a point 10 m from the building at ground level on a line perpendicular to the mid-point of the base of the wall? Assume that the receiver is parallel to the face of the building.

2.18 A steel plate, measuring 1 m × 1 m, is held vertically. It contains electrical heating elements which are capable of raising its temperature as high as 800°C. Given that

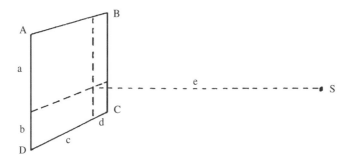

Figure 2.33

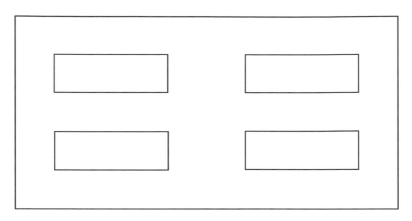

Figure 2.34

the convective heat transfer coefficient can be calculated from:

Nu $= 0.59$ (GrPr)$^{1/4}$ ($10^4 <$ GrPr $< 10^9$)

Nu $= 0.13$ (GrPr)$^{1/3}$ (GrPr $> 10^9$),

calculate the total rate of heat loss from the plate when its temperature is maintained at 200°C, 400°C, 600°C and 800°C, in each case comparing the radiative and convective components. Assume that the emissivity of the steel is 0.85 and that the atmospheric temperature is 20°C. The kinematic viscosity of air is approximately 18×10^{-6} m^2/s at 20°C: use this value for the question, although it is known to increase with temperature (e.g. see Welty *et al.*, 1976).

2.19 The plate described in Problem 2.18 is heated electrically at a rate corresponding to 25 kW. Using the data provided, calculate the steady-state temperature of the plate. (Hint: use Newton's method to solve the equation.)

3
Limits of Flammability and Premixed Flames

In premixed burning, gaseous fuel and oxidizer are intimately mixed prior to ignition. Ignition requires that sufficient energy is supplied in a suitable form, such as an electric spark, to initiate the combustion process which will then propagate through the mixture as a flame (or 'deflagration') (Chapter 1). The rate of combustion is typically high, determined by the chemical kinetics of oxidation rather than by the relatively slow mixing of fuel and oxidizer which determines the structure and behaviour of diffusion flames (Chapter 4). However, before premixed flames are discussed further, it is appropriate to examine flammability limits in some detail and identify the conditions under which mixtures of gaseous fuel and air, or any other oxidizing atmosphere, will burn.

3.1 Limits of Flammability

3.1.1 Measurement of flammability limits

Although it is common practice to refer to gases and vapours such as methane, propane and acetone as 'flammable', their mixtures with air will only burn if the fuel concentration lies within well-defined limits, known as the lower and upper flammability (or 'explosive') limits. For methane, these are 5% and 15% by volume (i.e. molar proportions) respectively. The most extensive review of the flammability of gases and vapours is that of Zabetakis (1965) which, despite its age, remains the standard reference. It is based largely on a collection of data obtained with an apparatus developed at the US Bureau of Mines (Coward and Jones, 1952) (Figure 3.1). Although there are certain disadvantages in this method, these data are considered to be the most reliable that are available. Alternative methods do exist (e.g. Sorenson *et al.*, 1975; Hirst *et al.*, 1981/82), but none has been used extensively enough to provide a challenge to the Bureau of Mines apparatus.

In this method the experimental criterion used to determine whether or not a given mixture is flammable, is its ability to propagate flame. The apparatus, which is shown schematically in Figure 3.1, consists of a vertical tube 1.5 m long and 0.05 m internal

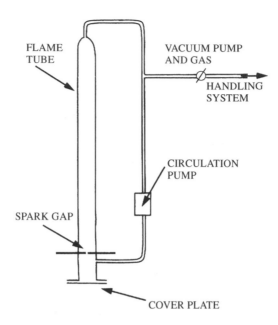

Figure 3.1   The essential features of the US Bureau of Mines apparatus for determining limits
of flammability of gases and vapours (Coward and Jones, 1952). (Not to scale.)
The circulation pump is necessary to ensure rapid and complete mixing of the
gases within the flame tube

diameter, into which premixed gas/air mixtures of known compositions can be introduced. An ignition source, which may be a spark or a small flame, is introduced to
the lower end of the tube which is first opened by the removal of a cover plate. The
mixture is deemed flammable if flame propagates upward by at least 75 cm. The limits
are established experimentally by a process of 'bracketing' and defined as:

$$L = \tfrac{1}{2}(L_w + L_r) \tag{3.1a}$$

$$U = \tfrac{1}{2}(U_w + U_r) \tag{3.1b}$$

where L_w and U_r are the greatest and least concentrations of fuel *in air* that are
non-flammable, and L_r and U_w are the least and greatest concentrations of fuel in air
that are flammable (Zabetakis, 1965). The limits are normally expressed in terms of
volume percentage at 25°C although they are functions of temperature and pressure
(Section 3.1.3). Flammability limit data are given in a number of publications (Fire
Protection Association, 1972; Lewis and von Elbe, 1987; NFPA, 1997) but most can
be found in Zabetakis' review (1965) (Table 3.1).

Earlier studies showed that the tube diameter has an effect on the result, although it
is small for the lower limit if the diameter is 5 cm or more (Figure 3.2). The closing
of the limits in narrower tubes can be explained in terms of heat loss to the wall:
indeed, if the diameter is reduced to the quenching diameter, flame will be unable to
propagate even through the most reactive mixture (Section 3.3). The limits quoted in
the literature (Zabetakis, 1965) refer to upward propagation of flame. These are slightly
wider than the limits for downward propagation (Figure 3.2) which may be determined

Table 3.1 Flammability data for gases and vapours

	Lower flammability limit $(L)^a$			$\dfrac{L}{C_{st}}$	Upper flammability limits $(U)^a$		$\dfrac{U}{C_{st}}$	S_u^b	Minimum ignition energy[b]	Minimum quenching distance[b]
	% Vol	g/m³	kJ/m³		% Vol	g/m³		(m/s)	(mJ)	(mm)
Hydrogen	4.0[c]	3.6	435	0.13	75	67	2.5	3.2	0.01	0.5
Carbon monoxide	12.5	157	1591	0.42	74	932	2.5	0.43	—	—
Methane	5.0	36	1906	0.53	15	126	1.6	0.37	0.26	2.0
Ethane	3.0	41	1952	0.53	12.4	190	2.2	0.44	0.24	1.8
Propane	2.1	42	1951	0.52	9.5	210	2.4	0.42	0.25	1.8
n-Butane	1.8	48	2200	0.58	8.4	240	2.7	0.42	0.26	1.8
n-Pentane	1.4	46	2090	0.55	7.8	270	3.1	0.42	0.22	1.8
n-Hexane	1.2	47	2124	0.56	7.4	310	3.4	0.42	0.23	1.8
n-Heptane	1.05	47	2116	0.56	6.7	320	3.6	0.42	0.24	1.8
n-Octane	0.95	49	2199	0.58	—	—	—	—	—	—
n-Nonane	0.85	49	2194	0.58	—	—	—	—	—	—
n-Decane	0.75	48	2145	0.56	5.6	380	4.2	0.40	—	—
Ethene	2.7	35	1654	0.41	36	700	5.5	>0.69	0.12	1.2
Propene	2.4	46	2110	0.54	11	210	2.5	0.48	0.28	—
Butene-1	1.7	44	1998	0.50	9.7	270	2.9	0.48	—	—
Acetylene	2.5	29	1410		(100)	—	—	1.7	0.02	
Methanol	6.7	103	2141	0.55	36	810	2.9	0.52	0.14	1.5
Ethanol	3.3	70	1948	0.50	19	480	2.9	—	—	—
n-Propanol	2.2	60	1874	0.49	14	420	3.2	0.38	—	—
Acetone	2.6	70	2035	0.52	13	390	2.6	0.50	1.1	—
Methyl ethyl ketone	1.9	62	1974	0.52	10	350	2.7	—	—	—
Diethyl ketone	1.6	63	2121	0.55	—	—	—	—	—	—
Benzene	1.3	47	1910	0.48	7.9	300	2.9	0.45	0.22	1.8

[a] Data from Zabetakis (1965). Mass concentration values are approximate and refer to 0°C ($L(g/m^3) \approx 0.45\,M_w L$ (vol %)).

[b] Data from various sources including Kanury (1975) and Lees (1996). There is uncertainty with some of these data (Harris, 1983; Lees, 1996).

[c] See Section 3.5.4.

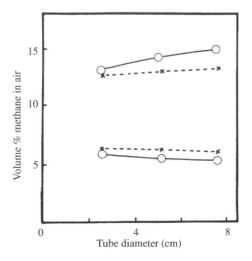

Figure 3.2 Variation of observed flammability limits for methane/air mixtures. ○, upward propagation; ×, downward propagation (Linnett and Simpson, 1957)

using a modified apparatus. The difference arises because the buoyant movement of the burnt gases acts in opposition to the downward propagating flame, creating instability. However, the same behaviour can be observed in unconfined mixtures. Following central ignition, flame will propagate spherically while at the same time the growing volume of burnt gas will rise, causing distortion. Even in this situation it is possible to observe upward (and horizontal) propagation without flame travelling downwards through a near-limit mixture (e.g. Sapko *et al.*, 1976; Roberts *et al.*, 1980; Hertzberg, 1982). Indeed, the existence of the limits can be explained in terms of buoyancy-induced convective flows (Section 3.3).

Very small amounts of energy are sufficient to ignite flammable vapour/air mixtures. Figure 3.3 shows how the minimum spark energy capable of igniting a mixture varies with composition. The minimum on this curve—known as the minimum ignition energy (MIE)—corresponds to the most reactive mixture, which normally lies just on the fuel-rich side of stoichiometry. Typical values of MIE are quoted in Table 3.1. As no mixture can be ignited by a spark of energy less than the minimum ignition energy, it is possible to design certain items of low power electrical equipment which are intrinsically safe and may be used in locations where there is a risk of a flammable atmosphere being formed (British Standards Institution, 1977; National Fire Protection Association, 1993b). This can be achieved by designing the equipment and circuits in such a way that even the worst fault condition cannot ignite a stoichiometric mixture of the specified gas in air.

Limits of ignitability which vary with the strength of the ignition source can be distinguished from limits of flammability (Figure 3.3). The latter must be determined using an ignition source which is sufficiently large to ignite near-limit mixtures. However, as the limits vary significantly with temperature (Figure 3.4), flame may propagate in a mixture which is technically 'non-flammable' under ambient conditions if the ignition source is large enough to cause a local rise in temperature. Thus the criterion for flammability in the US Bureau of Mines apparatus is propagation of flame at least half-way up the flame tube. At this point it is assumed that the flame will be propagating into a mixture which has not been affected by the ignition source.

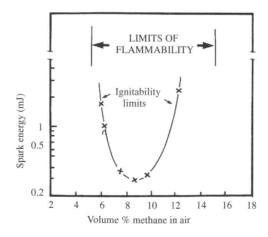

Figure 3.3 Ignitability curve and limits of flammability for methane/air mixtures at atmospheric pressure and 26°C (Zabetakis, 1965)

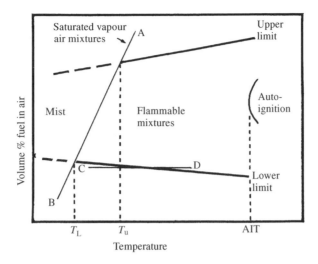

Figure 3.4 Effect of temperature on the limits of flammability of a flammable vapour/air mixture at a constant initial pressure (Zabetakis, 1965)

Various criticisms can be made of the US Bureau of Mines apparatus, relating mainly to procedure. In its original form, it is unsatisfactory for examining the effects that small quantities of chemical extinguishants (e.g. the halons) have on the limits, as the method used to prepare the mixture is cumbersome. Furthermore, heavier-than-air mixtures tend to 'slump' from the tube when the cover plate is removed. This will affect the local concentration of vapour at the lower end of the tube where the ignition source is located.

The need to make accurate measurements of the flammability limits of mixtures containing chemical extinguishants has led to the examination of a new method which relies not on flame propagation as the criterion of flammability, but on pressure rise inside a spherical steel vessel, 6 litres in volume (Figure 3.5). This is a very sensitive indicator as outside the limit the pressure rise is effectively zero, provided that the energy dissipated by the ignition source is not excessive. In general, data obtained with this apparatus agree quite well with flame tube results for the lower limit as shown in Table 3.1, although the upper limit values are less satisfactory. However, the lower flammability limit obtained for hydrogen in air using the pressure criterion ($\sim$8%) is much higher than that based on observations of flame propagation ($\sim$4%). This behaviour is likely to be unique, arising from the high molecular diffusivity of hydrogen combined with buoyancy effects to produce finger-like flamelets capable of propagating vertically, but consuming little fuel (Hertzberg, 1982; Lewis and von Elbe, 1987). Further study is required to establish whether or not the new method is a more relevant and reliable means of measuring the limits.

3.1.2   Characterization of the lower flammability limit

When expressed as volume percentage, the flammability limits of the members of a homologous series decrease monotonically with increasing molecular weight, or carbon

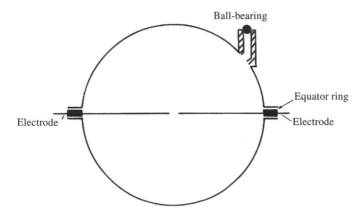

Figure 3.5   The essential features of the spherical pressure vessel used to determine limits of flammability of gases and vapours (Hirst *et al.*, 1981/82). The equator ring also carries the gas inlet port and the outlet port to a vacuum line. Pressure rise can be detected by means of a pressure transducer or a low pressure relief valve (as shown)

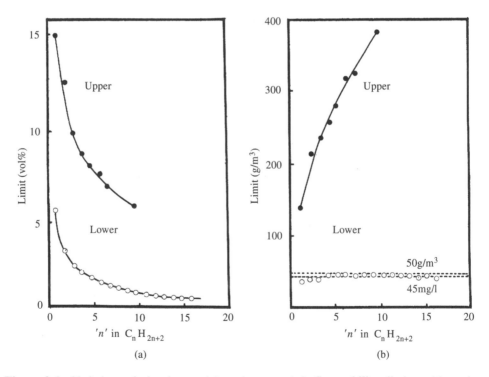

Figure 3.6   Variation of the lower (○) and upper (●) flammability limits with carbon number for the *n*-alkanes. Limits expressed as (a) percentage by volume; (b) mass concentration (g/m^3). (Data from Zabetakis, 1965)

number (Figure 3.6(a)). However, a different picture is found if the limits are converted to mass concentration. Figure 3.6(b) shows that the lower flammability limits of the C_4–C_{10} straight chain alkanes correspond to $\sim$48 g/m^3 (see also Table 3.1). As the heat of combustion of the alkanes per unit mass is approximately constant ($\sim 45 \pm 1$ kJ/g, see Table 1.13), the lower limit is defined as an 'energy density' of $\sim$2160 kJ/m^3, or 48.4 kJ per mole of the lower limit mixture. Examination of the lower limits of a range of hydrocarbons and their oxygenated derivatives suggests that the concept of a 'critical energy density' may be more widely applicable. Most lie in the range 2050 ± 150 kJ/m^3, the exceptions being the alkenes and alkynes, although the higher homologues in the alkene series (e.g. butene-1) tend to fall within the same spread. It should be borne in mind that the uncertainty in the measured values of the lower limits from which these figures are derived is likely to be at least $\pm 5\%$.

These various criteria offer methods of assessing whether or not mixtures of gases in air will be flammable. For a mixture of alkanes in air at ambient temperature, the lower flammability limit corresponds to:

$$\sum_{i}^{n} C_i = 48 \text{ g/m}^3 \tag{3.2a}$$

where C_i is the mass concentration of component i at 25°C. This can also be expressed as:

$$\sum_{i} \frac{C_i}{48} = \sum_{i} \frac{l_i}{L_i} = 1 \tag{3.2b}$$

where l_i is the percentage composition (molar proportion) of component i in the vapour/air mixture, and L_i is the corresponding value for its lower flammability limit. If the lower flammability limit of the mixture of hydrocarbons in air is L_m, then writing $l_i = L_m.f_i$, where f_i is the proportion of hydrocarbon i in the original hydrocarbon mixture (by volume),

$$L_m \sum_{i} \frac{f_i}{L_i} = 1 \tag{3.2c}$$

or in the form normally quoted

$$L_m = \frac{100}{\sum_{i} \dfrac{P_i}{L_i}} \tag{3.2d}$$

in which P_i is the percentage composition of component i in the original mixture such that $\Sigma_i P_i = 100$. This is Le Chatelier's law (Le Chatelier and Boudouard, 1898; Coward and Jones, 1952). As an example, it can be used to calculate the lower flammability limit of a mixture of hydrocarbon gases containing 50% propane, 40% n-butane and 10% ethane: thus

$$L_m = \frac{100}{\dfrac{50}{2.1} + \dfrac{40}{1.8} + \dfrac{10}{3.0}} = 2.0\%$$

If the adiabatic flame temperatures for the limiting mixtures are calculated using the method outlined in Section 1.2.5, the results suggest that — at least for the alkanes — there is a limiting flame temperature of 1500–1600 K, below which flame cannot exist (Table 3.1). Although the unsaturated hydrocarbons give lower values, the concept of a limiting temperature may be used to check whether or not lean gas mixtures are flammable. For example, consider the mixture comprising 2.5% butane, 20% carbon dioxide and 77.5% air. The components are in the ratio 1:8:31; thus the overall reaction may be expressed as:

$$C_4H_{10} + 8\ CO_2 + 31(0.21\ O_2 + 0.79\ N_2) = 12\ CO_2 + 5\ H_2O + 0.02\ O_2 + 24.49\ N_2$$

The thermal capacity of the final products is 1659 J/K per mole of butane. Combining this with the heat of combustion of butane (2650 kJ/mol) gives a temperature rise of 1597 degrees, i.e. an adiabatic flame temperature of 1890 K, well above the limiting value. While little reliance can be placed on the absolute value of this figure, it is sufficiently large to indicate that the mixture should be considered flammable. Of course, this type of calculation can be turned round to estimate how much carbon dioxide would be required to 'inert' a stoichiometric butane/air mixture (i.e. render it non-flammable under ambient conditions) (Section 1.2.5).

It is interesting to note that for many of the gases and vapours listed in Table 3.1, the ratio of L to the percentage concentration in the stoichiometric mixture is approximately constant, $L \approx 0.55\ C_{st}$ at 25°C. This would suggest that about 45% of the heat of combustion released during stoichiometric burning would have to be removed to quench (extinguish) the flame.

3.1.3 *Dependence of flammability limits on temperature and pressure*

As the temperature is increased, the limits widen, as illustrated schematically in Figure 3.4. The line AB in this diagram identifies the saturated vapour pressure so that the region to its left corresponds to an aerosol mist or droplet suspension. The limits are continuous across this boundary: thus the lower limit for tetralin (1,2,3,4-tetrahydronaphthalene, $C_{10}H_{12}$) mist at 20°C corresponds to a concentration of 45–50 g/m³, in close agreement with the lower limit concentrations observed for hydrocarbon gases and vapours (Table 3.1). However, if the droplet diameter is increased above ~10 μ, the lower limit appears to decrease: coarse droplets will tend to fall into an upward-propagating flame, effectively increasing the local concentration (Figure 3.7) (Burgoyne and Cohen, 1954).

A vapour/air mixture which is non-flammable under ambient conditions may become flammable if its temperature is increased: compare points C and D in Figure 3.4, which refer to the same mixture at different temperatures. The lower limit decreases with rising temperature simply because less combustion energy need be released to achieve the limiting flame temperature (T_{lim}): consequently, a lower concentration of fuel in air will be sufficient for flame to propagate through the mixture. In terms of the changes in enthalpy,

$$\frac{L_{25}}{100} \cdot \Delta H_c = c_p(T_{lim} - 25) \tag{3.3a}$$

$$\frac{L_T}{100} \cdot \Delta H_c = c_p(T_{lim} - T) \tag{3.3b}$$

where L_{25} and L_T are the lower limits (% by volume) at 25°C and T°C respectively.

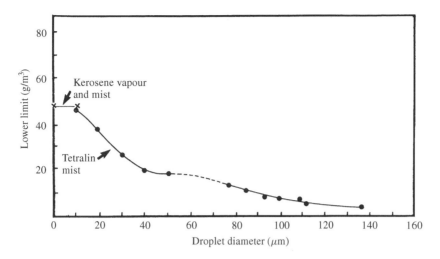

Figure 3.7   Variation of the lower flammability limit of tetralin mist as a function of droplet diameter (Burgoyne and Cohen, 1954, by permission)

ΔH_c is the heat of combustion (J/mol) and c_p is the thermal capacity of the products (J/K). Dividing Equation (3.3b) by (3.3a) gives

$$\frac{L_T}{L_{25}} = \frac{T_{\lim} - T}{T_{\lim} - 25}$$

$$= 1 - \frac{T - 25}{T_{\lim} - 25} \tag{3.3c}$$

Taking $T_{\lim} = 1300°C$ (Zabetakis, 1965) Equation (3.3c) can be rearranged to give the lower limit at any temperature T:

$$L_T = L_{25}(1 - 7.8 \times 10^{-4}(T - 25)) \tag{3.3d}$$

While this is only an approximation as the temperature dependence of ΔH_c and c_p are neglected, it agrees satisfactorily with the empirical relationship quoted by Zabetakis *et al.* (1959),

$$L_T = L_{25}(1 - 7.21 \times 10^{-4}(T - 25)) \tag{3.3e}$$

which is based on work by Burgess and Wheeler. The upper limit is found to obey a similar relationship, namely:

$$U_T = U_{25}(1 + 7.21 \times 10^{-4}(T - 25)) \tag{3.4}$$

provided that the mixture does not exhibit cool flame formation at temperature T. Equation (3.4) suggests that the upper limit is also determined by a limiting temperature criterion, a view expressed by Mullins and Penner (1959). It is not possible to calculate the adiabatic flame temperature at the upper limit by the simple method outlined above as the products will contain a complex mixture of pyrolysed and partially oxidized species. However, Stull (1971) has shown theoretically that the flame temperature at the upper limit is approximately the same as the lower limit (Section 1.2.3).

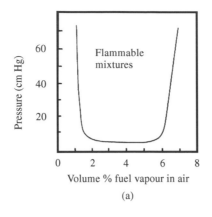

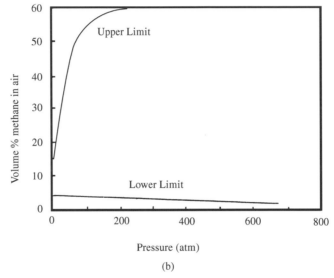

Figure 3.8 Variation of flammability limits with pressure: (a) gasoline vapour in air at reduced pressures (reprinted with permission from Mullins and Penner, copyright 1959 Pergamon Press); (b) methane in air at super-atmospheric pressures (Zabetakis, 1965)

Pressure has little effect on the limits in the sub-atmospheric range, provided that it is not less than 75–100 mm Hg (approximately 0.1 atm) (Figure 3.8(a)). Thus it is possible to determine flammability limits at reduced pressures and apply the results to ambient conditions. In this way the flammability limits of vapours which would be supersaturated at atmospheric pressure can be derived. This is entirely consistent with the thermal nature of the limit. Thus, *n*-decane is quoted as having $L_{25} = 0.75\%$, corresponding to a vapour pressure of 5.7 mm Hg under a standard atmosphere (760 mm Hg). At 25°C the saturated vapour pressure of *n*-decane is 1.77 mm Hg (Table 1.10) or 0.24% by volume at a total pressure of 760 mm Hg. Thus, under normal conditions the vapour/air mixture at the surface of a pool of *n*-decane at 25°C will be non-flammable (Section 6.2.2). However, if the atmospheric pressure is reduced

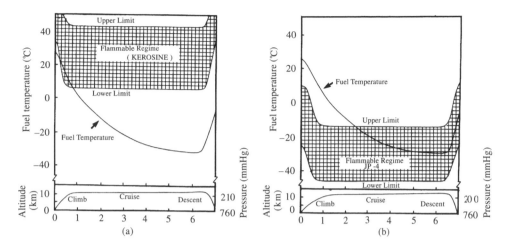

Figure 3.9 Changing regimes of flammability during aircraft flight (a) kerosene. (b) JP-4 fuel. (Reproduced with permission from Ministry of Aviation, 1962)

to 236 mm Hg and the liquid temperature remains constant, the mixture becomes flammable as the decane vapour now constitutes 0.75% of the total pressure. Such dramatic changes in pressure occur in fuel tanks of aircraft following take-off. With kerosene as the fuel, the free space above the liquid surface will contain a vapour/air mixture which is non-flammable at sea level but will become flammable when the aircraft climbs above a certain height. Of course, at high altitude the temperature of the fuel will gradually fall as the ambient temperature will be low: eventually the vapour pressure of the liquid will decrease until the mixture is no longer flammable (Figure 3.9(a)) (see Equation (1.14)). However, while the mixture lies within the limits of flammability, there is a potential explosion hazard. Highly flammable liquids such as JP-4, whose vapour pressure lies above the upper flammability limit under ambient conditions, will not present a hazard during the early stages of flight. However, during a flight at high altitude, its temperature will decrease and the hazard may develop as the plane descends and ambient pressure increases (Figure 3.9(b)).

Substantial increases of pressure above atmospheric produce significant changes in the upper limit (Figure 3.8(b)) while the lower limit is scarcely affected. Thus, at an initial pressure of 200 atm, the upper and lower limits for methane/air mixtures are approximately 60% and 4% respectively (cf. 15% and 5% at 1 atm).

3.1.4 Flammability diagrams

Information is readily available in the literature on flammability limits of vapour/air mixtures (e.g. Table 3.1) but in some circumstances it is necessary to know the regimes of flammability associated with a more complex combination of gases, such as hydro-carbon, oxygen and nitrogen. Similarly, it may be necessary to record the effects of adding flame suppressants to flammable vapour/air mixtures, presenting the results in a way that would be useful to a fire protection engineer or a plant operator. As an example, consider the three-component mixture of methane, oxygen and nitrogen.

The flammable regime which must be established experimentally can be presented on a triangular diagram as shown in Figure 3.10, or, as the third component (e.g. oxygen) is a dependent variable, displayed using rectangular coordinates as shown in Figure 3.11.

Thus in Figure 3.10, each axis represents one of the three component gases and the region of flammability is defined by the locus of points which correspond to the limits. Thus, the mixture marked M1 is non-flammable. 'Air' corresponds to the line running from the top apex, C (fuel = 100%), to the lower axis at the point A where the fuel concentration is zero and O_2 and N_2 are in the ratio 21:79. (By drawing a line from 21% on the oxygen axis, parallel to the methane axis (OC), the concentration of nitrogen can be read directly). The 'air line' CA intersects the envelope of the flammability region at two points corresponding to 5% and 15% methane, i.e., the lower and upper limits of methane in air. The flammability limits of methane in pure oxygen can be obtained from the diagram by examining the intersections of the flammability envelope with the axis OC on which $N_2 = 0\%$. These are 5% and 60% respectively. It is significant that the lower flammability limit of methane in oxygen is the same as that in air. This is because the heat capacities of nitrogen and oxygen are very similar (Table 1.16) and at the fuel-lean limit the excess oxygen acts only as thermal 'ballast' (Section 1.2.3).

Another important observation that can be made from these diagrams is that there is a minimum oxygen concentration below which methane will not burn. The corresponding O_2/N_2 mixture is given by the line CL which forms a tangent with the tip of the flammability region. Oxygen concentrations falling to the right of this line (less than

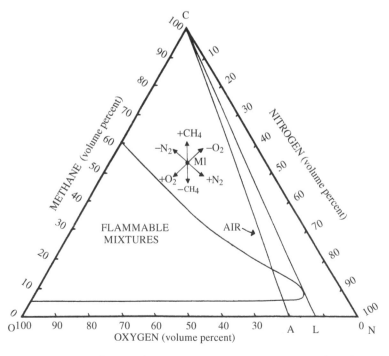

Figure 3.10   Flammability diagram for the three-component system methane/oxygen/nitrogen at atmospheric pressure and 26°C (after Zabetakis, 1965). Points on the line CA correspond to methane/air mixtures

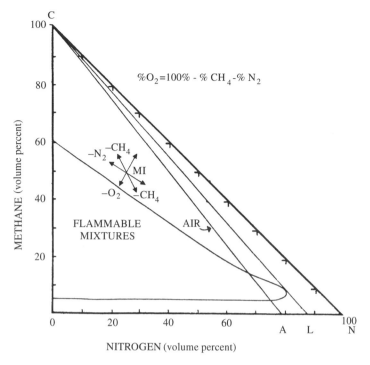

Figure 3.11 Flammability diagram for $CH_4/O_2/N_2$ at atmospheric pressure and 26°C. The information contained in this diagram is identical to that contained in Figure 3.10 (after Zabetakis, 1965)

about 13%) will not support methane combustion at ambient temperatures. Figure 3.11 displays exactly the same information as shown in Figure 3.10. This method of presentation has the advantage that it requires conventional graph paper.

Such diagrams may be used to decide how spaces containing mixtures of flammable gas, oxygen and nitrogen may be inerted safely. The point corresponding to M1 in Figures 3.10 and 3.11 corresponds to a non-flammable mixture consisting of 50% CH_4, 25% O_2 and 25% N_2. If this represents a mixture which is normally flowing through an item of chemical plant, then part of the shutdown procedure would be to replace the mixture by air. However, as a general principle, one should avoid flammable mixtures inside the equipment. If the flowing mixture was gradually replaced by air, then there would be a period of time when a flammable mixture would exist in the system until CH_4 fell below 5% (straight line joining M1 and A). A similar effect would occur if the flow of fuel was simply turned off. The optimum method would be to reduce the oxygen flow and/or increase the nitrogen flow until the O_2/N_2 ratio corresponds to (or lies to the right of) the line CL in Figure 3.10. Then the methane flow can be stopped safely, and the item purged with air as soon as the concentration of CH_4 falls below 5%.

This type of plot may also be used to compare the effect of adding different gases to vapour/air mixtures. Figure 3.12 shows the reduction in the limits of methane in air caused by the addition of a number of gases, including helium, nitrogen, water vapour and carbon dioxide. Of these, carbon dioxide is clearly the most effective in rendering

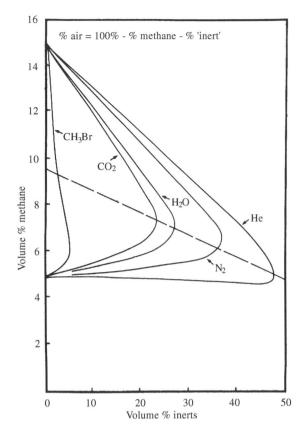

Figure 3.12 Flammability limits of various methane/air/inert gas mixtures at atmospheric pressure and 26°C (after Zabetakis, 1965)

the mixture non-flammable, which can be understood in terms of the different heat capacities of the four gases (Table 1.16). On the other hand, the very high efficiency of methyl bromide is due to chemical inhibition involving bromine: the unusual shape of this particular curve indicates that methyl bromide is itself combustible (Wolfhard and Simmons, 1955).

3.2   The Structure of a Premixed Flame

A premixed flame can be studied experimentally by stabilizing it on a gas burner. The simplest is the Bunsen burner operating with full aeration (Figure 3.13(a)): the characteristic blue cone is a premixed flame which, although fixed in space, is propagating against the gas flow. However, the porous disc burner developed by Botha and Spalding (1954) is more suitable for experimental work as it produces a stationary flat flame, ideal for measurement (Figure 3.13(b)). By use of suitable probes, temperature and concentration profiles through the flame can be obtained (e.g. Fristrom and Westenberg, 1965). These are similar to those illustrated in Figure 3.14, which shows principally the variation of temperature through the flame. The leading edge is located

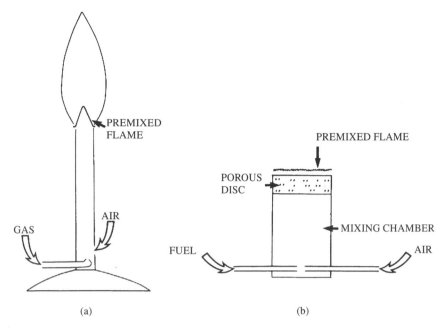

Figure 3.13 (a) Premixed flame on a Bunsen burner with full aeration. (b) Flat premixed flame stabilized on a porous disc (e.g. Botha and Spalding, 1954)

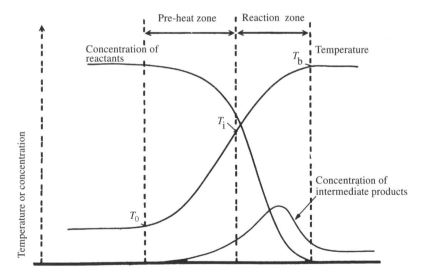

Figure 3.14 Temperature and concentration profiles through a plane combustion wave (reproduced by permission of Academic Press from Lewis and von Elbe, 1987)

at $x = 0$. Three distinct zones may be identified, as follows:

(i) a pre-heat zone in which the temperature of the unburnt gases rises to some arbitrary value, T_i (see below);

(ii) the reaction zone in which most of the combustion takes place; and

(iii)   the post-flame region, characterized by high temperature and radical recombina-
        tion, leading to local equilibrium. Cooling will subsequently occur.

Of these, zone (ii) is the visible part of the 'flame' and is about 1 mm thick for
common hydrocarbon fuels at ambient pressure, but less for highly reactive species
such as hydrogen and ethylene.

The thickness of the pre-heat zone ((i) above) can be estimated from an analysis
of the temperature profile (Figure 3.15). If it is assumed that no oxidation occurs at
temperatures below T_i (a convenient but fictitious 'ignition temperature'), the following
quasi-steady state equation may be written to describe conduction of heat ahead of the
leading edge of the flame (which lies at $x = 0$):

$$k \left(\frac{d^2 T}{dx^2} \right) - \rho_u S_u c_p \left(\frac{dT}{dx} \right) = 0 \tag{3.5}$$

where ρ_u is the density of the unburnt gas at the initial temperature T_0 and S_u is the
rate at which unburnt gas flows into the flame (the burning velocity, see Section 3.4).
This equation should be compared with Equation (2.15), with $dt = dx/S_u$. Integration
between the limits $x = -\infty$ to $x = 0$ (i.e. from T_0 to T_i) gives:

$$k \left(\frac{dT}{dx} \right) = \rho_u S_u c_p (T_i - T_0) \tag{3.6}$$

Setting dT/dx equal to $(T_i - T_0)/\eta_0$ as shown in Figure 3.15, where η_0 is taken to be
a first-order approximation to the thickness of the pre-heat zone, gives:

$$\eta_0 = \frac{k}{\rho_u S_u c_p} \tag{3.7a}$$

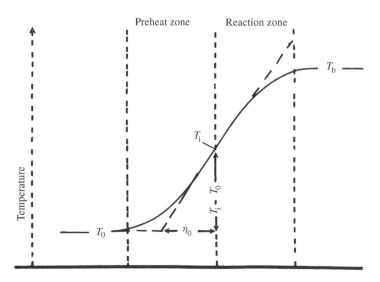

Figure 3.15   Analysis of the temperature profile in the pre-heat zone of the combustion wave
(reproduced by permission of Academic Press from Lewis and von Elbe, 1987)

(Lewis and von Elbe, 1987). The derived value of the zone thickness depends on how the leading edge of the pre-heat zone is defined. Gaydon and Wolfhard (1979) identify it as the point at which $(T - T_0) = 0.01 \times (T_i - T_0)$; Equation (3.5) must be integrated twice to give:

$$\eta_0' = \frac{4.6k}{\rho_u S_u c_p} \qquad (3.7b)$$

While this is very approximate—ambient values of k, ρ_u, S_u and c_p give $\eta_0' \approx 0.3$ mm—the derivation emphasizes the existence and importance of the pre-heat zone. If a premixed flame comes close enough to a solid surface to disturb the pre-heat zone, then heat will be transferred to the surface, and the flame will be cooled (Section 3.3).

However, in free propagation through a quiescent mixture, the velocity with which a flame will travel will depend on the efficiency with which heat is transferred ahead of zone (ii). (This general concept can be applied to almost all types of flame or fire spread, and will be discussed at length in Chapter 7.) While a comprehensive analysis of premixed flame propagation must incorporate the four conservation equations of mass, momentum, energy and chemical species (e.g. Williams, 1965), an approximate solution may be gained for an infinite, plane adiabatic wave from the energy equation on its own (cf. Equation (2.14)), i.e.

$$k\left(\frac{d^2T}{dx^2}\right) - \rho u c_p \left(\frac{dT}{dx}\right) - \dot{Q}''' = 0 \qquad (3.8)$$

where $\dot{Q}'''$ is the rate of heat release per unit volume and u is the linear flowrate of gas in the combustion zone. If the reaction rate is strongly temperature dependent (i.e. $\dot{Q}''' \propto \exp(E_A/RT)$ where the activation energy, E_A, is large) it is possible to neglect combustion in the 'pre-heat' zone (zone (i)), where the temperature is less than the fictitious ignition temperature (T_i in Figure 3.15). This done, zones (i) and (ii) can be treated separately to derive expressions for the temperature gradient at $x = 0$ (where $T = T_i$) which are then equated to each other. Using this procedure, which is attributed to Zeldovich and Frank-Kamenetskii (e.g., see Kanury (1975)), an expression for S_u, the 'fundamental burning velocity', can be derived:

$$S_u = \left(\frac{2k}{\rho_0^2 c_p^2 (T_F - T_0)} \cdot \dot{Q}'''_{ave}\right)^{0.5} \qquad (3.9)$$

In this equation the zero subscript refers to initial conditions, T_F is the flame temperature and $\dot{Q}'''_{ave}$ is the average rate of heat release in the reaction zone (ii). The quantity S_u is the velocity with which the flame propagates into the unburnt mixture, provided that there is no turbulence in the system (Section 3.5.5).

A more fundamental analysis of flame propagation confirms that for any given mixture there is one and only one burning velocity—an eigenvalue—and that mass diffusion must also be taken into account (e.g. Dixon-Lewis, 1967). Thus, it is found that burning velocity is a maximum for a slightly fuel-rich mixture, compatible with experimental observation. This occurs because highly mobile hydrogen atoms diffuse ahead of the reaction zone and thus contribute significantly to the mechanism of propagation. Their concentration is higher in the fuel-rich flame.

The relationship between the fundamental burning velocity S_u and the parameters k, c_p, T_F and T_0, indicated in Equation (3.9), is compatible with the observations which will be discussed in Section 3.4. It also predicts that the burning velocity is directly proportional to the square root of the reaction rate, which is incorporated in $\dot{Q}'''$: consequently, $S_u \propto (\exp(-E_A/RT))^{0.5}$, or $S_u \propto \exp(-E_A/2RT)$. This result will be used in Section 3.3.

3.3   Heat Losses from Premixed Flames

The existence of flammability limits is not predicted by the theories outlined in the previous section. Spalding (1957) and Mayer (1957) pointed out that this discrepancy could be resolved by incorporating heat losses into the model. To examine Mayer's argument, consider the temperature profile through the adiabatic flame front shown in Figure 3.16: if there is any heat loss, there will be a decrease in temperature and a consequent fall in the rate of heat release.

Mayer begins his analysis by comparing the temperature profiles normal to the flame front of adiabatic and non-adiabatic flames (Figure 3.16). Without heat losses, the flame temperature reaches the adiabatic value, T_F^a but in reality it is likely to deviate from the adiabatic curve in zone (ii), thereafter passing through a maximum and decaying slowly in the 'post-flame gases' (Figure 3.16). Because the reaction rate is strongly dependent on temperature (proportional to $\exp(-E_A/RT)$) any reduction in T_F will be accompanied by a substantial decrease in the rate of heat generation. A flame will be unable to propagate if the rate of heat loss exceeds the rate of heat production.

Applying a crude 'lumped thermal capacity' model to the propagating flame, the adiabatic case can be described by:

$$\rho_0 S_u^a c_p (T_F^a - T_0) = \rho_0 S_u^a \Delta H_c \tag{3.10}$$

which rearranges to give:

$$\rho_0 S_u^a (c_p T_0 + \Delta H_c) - \rho_0 S_u^a c_p T_F^a = 0 \tag{3.11}$$

where T_F^a is the adiabatic flame temperature and ΔH_c is the heat of combustion.

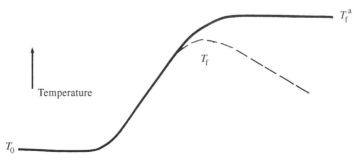

Figure 3.16   Temperature profiles through adiabatic (——) and non-adiabatic (– – – –) premixed flames (after Mayer, 1957)

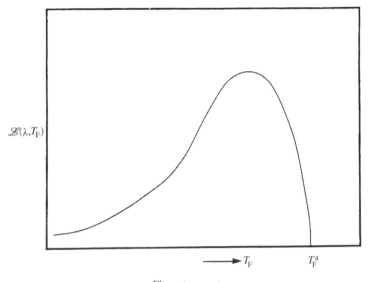

Flame temperature

Figure 3.17   Relationship between heat loss and premixed flame temperature according to
Equation (3.15) (after Mayer, 1957)

For the non-adiabatic flame a heat loss term $\mathscr{L}(\lambda, T_F)$ must be included, thus:

$$\rho_0 S_u(c_p T_0 + \Delta H_c) - \rho_0 S_u c_p T_F - \mathscr{L}(\lambda, T_F) = 0 \qquad (3.12)$$

where λ is a 'heat loss parameter'. The following relationships are substituted into
Equation (3.12):

$$S_u = B \exp(-E/2RT) \qquad (3.13)$$

where B is assumed to be a constant (see Section 3.2.1), and

$$\Delta H_c = c_p(T_F^a - T_0) \qquad (3.14)$$

(from Equation (3.10)), to give:

$$\mathscr{L}(\lambda, T_F) = \rho_0 c_p B(T_F^a - T_F) \exp(-E/2RT_F) \qquad (3.15)$$

The form of this relationship is illustrated in Figure 3.17. It gives the rate of heat loss
which is being experienced by a flame at a given temperature. Thus, when $\mathscr{L} = 0$,
the temperature is equal to the adiabatic flame temperature, T_F^a. The relationship also
indicates that there is a maximum heat loss which the flame is able to sustain: this can
be used to interpret flame quenching and limits of flammability.

(a) *Convective heat losses and quenching diameters*   The ability of a flame to propagate along a narrow duct or tube will depend on the extent of heat losses to the walls.
Consider a flame propagating through a flammable mixture contained within a narrow,

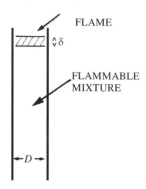

Figure 3.18   Propagation of a premixed flame in a tube or duct (Mayer, 1957)

circular pipe of internal diameter D (Figure 3.18). If the flame thickness is approximated by $\delta \approx k/\rho_0 S_u c_p$ (see Equation (3.7a)) the heat transferred by convection from the flame to the walls per unit surface area of flame ($\pi D^2/4$) is given by:

$$\dot{q}_{conv} = h(T_F - T_0) \cdot \frac{\pi D \delta}{\pi D^2/4} \tag{3.16}$$

$\pi D \delta$ is the area of contact between the flame and the pipe, and the heat transfer coefficient is

$$h = \text{Nu} \cdot k/D \tag{3.17}$$

(Section 2.3). For this configuration Nu = 3.65 (Mayer, 1957). Writing $\alpha = k/\rho_0 c_p$, as before (Section 2.2.2)

$$\dot{q}_{conv} = 14.6 \frac{k\alpha}{D^2 S_u}(T_F - T_0) \tag{3.18}$$

This indicates that $\dot{q}_{conv}$ will increase with decreasing pipe diameter, but decrease with increasing T_F, as S_u is strongly temperature dependent (Equation (3.13)).

If $\dot{q}_{conv}$ for a pipe, diameter D, is plotted on the same diagram as $\mathscr{L}$ (Figure 3.19), any intersection of the curves represents a quasi-equilibrium situation, which defines the temperature of the flame that will propagate through the mixture confined by the pipe. Although there are normally two intersections, only the right-hand one corresponds to stable propagation. Consider the intersection at b_1. A slight decrease in temperature gives $\dot{q}_{conv} > \mathscr{L}$ which results in cooling of the system, while a slight increase causes a continuing rise in temperature as $\mathscr{L} > \dot{q}_{conv}$. If the same arguments are applied to the intersection a_1, it can be seen that any perturbation is self-correcting and the system will always return to the starting point (a_1).

The three convective heat loss curves shown in Figure 3.19 correspond to three pipe diameters which decrease in the sequence D_1, D_2 and D_Q, and which identify the limiting pipe diameter (D_Q) below which $\dot{q}_{conv} > \mathscr{L}$ for all values of T_F. If $D < D_Q$, flame cannot propagate as heat losses to the walls of the tube are too great. However, while the heat loss mechanism can account qualitatively for this phenomenon, it is likely that the surface will also be responsible for the loss of free radicals from the reaction zone by destroying those which migrate to the surface. At present it is

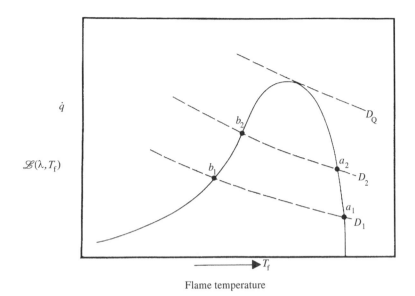

Figure 3.19   Quenching of a flame in a narrow pipe (Mayer, 1957). $D_1 > D_2 > D_Q$ (schematic)

not possible to quantify the relative importance of these two mechanisms in physical quenching. Conceptually, D_Q is related to the quenching distance which is of relevance to the design of flame arresters and explosion proof equipment.

(b) *Radiative heat losses*   In the previous paragraphs, it was tacitly assumed that radiative heat losses were negligible compared with convective loss to the tube walls. If the flammable vapour/air mixture is unconfined so that a propagating flame will not come into contact with any surfaces, then radiation will be the *only* mechanism by which heat can be lost. Radiative loss from the reaction zone occurs in two ways, namely:

(i)   indirectly, by conduction into the post-flame gases which cool by radiation ($\dot{q}_{\mathrm{rad(i)}}$); and
(ii)   directly, by radiation from the reaction zone ($\dot{q}_{\mathrm{rad(ii)}}$).

Of these, the latter is relatively unimportant as the reaction zone is very thin and the average concentrations of the emitting species (principally CO_2 and H_2O, see Section 2.4.2) are low. The following expression, derived from Equation (3.8) with $\dot{Q}''' = 0$, may be written for the post-flame gases:

$$\rho_0 S_u c_p \left(\frac{\mathrm{d}T}{\mathrm{d}x} \right) - k \left(\frac{\mathrm{d}^2 T}{\mathrm{d}x^2} \right) = -R_l(T, C_n) \qquad (3.19)$$

where R_l, the volumetric heat loss, is a function of temperature and concentration ($n = CO_2$ and H_2O). Close to the downstream edge of the flame, where the temperature is a maximum, the flow or convective term on the left-hand side of Equation (3.19) dominates the conduction term, which is ignored, to allow the temperature gradient at this 'hot boundary' to be approximated by

$$\frac{dT}{dx} = -\frac{R_l(T, C_n)}{\rho_0 S_u c_p} \tag{3.20}$$

The heat loss from the reaction zone to the post-flame gases then becomes

$$\dot{q}_{rad(i)} = -k_f\left(\frac{dT}{dx}\right) = k_f\frac{R_l(T, C_n)}{\rho_0 S_u c_p} \tag{3.21}$$

Using Hottel's charts for the emissivity of water vapour and carbon dioxide (Section 2.4.2), Mayer showed that R_l could be expressed as:

$$R_l = 1.7 \times 10^{-6}(p_{CO_2} + 0.18 p_{H_2O})T^2 \tag{3.22}$$

where the partial pressures are in atmospheres.

Following the argument developed above for convective heat losses, steady propagation of flame through an unconfined flammable mixture can be represented by the stable intersection of the curves $\dot{q}_{rad}$ and $\mathscr{L}$ (Figure 3.20), where $\dot{q}_{rad} = \dot{q}_{rad(i)}$. As in this case it is composition that is changed, changes in both the $\mathscr{L}$ versus T_F and $\dot{q}_{rad}$ versus T_F curves must be considered, as illustrated schematically in Figure 3.20. The intersection between $\mathscr{L}$ and $\dot{q}_{rad}$ defines the temperature of the steady-state flame propagating through a particular flammable vapour/air mixture. If the concentration of the vapour is reduced then clearly $\mathscr{L}$ will change but so will $\dot{q}_{rad}$, as the partial pressures of the products CO_2 and H_2O will also be reduced. The limit mixture corresponds to

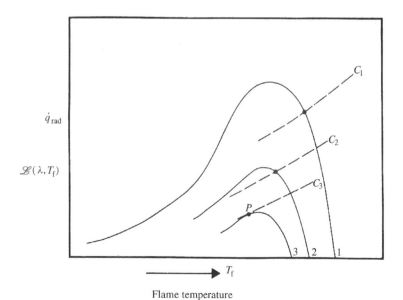

Figure 3.20   Flammability limit of an unconfined premixed flame with radiation losses (Mayer, 1957). $C_1 > C_2 > C_3$ (schematic)

that in which there is a tangency condition between the corresponding curves — the point marked '*P*' in Figure 3.20.

While this model shows how the existence of the limits may be explained in terms of the rates of heat production and loss, other factors are likely to be significant. Thus, it has been suggested that buoyancy is capable of creating sufficient instability at the leading edge of an upward-propagating flame to cause extinction in a limiting mixture (Lovachev *et al.*, 1973; Hertzberg, 1982). Further refinement will be necessary before there is a detailed understanding of the importance of these contributory mechanisms.

3.4 Measurement of Burning Velocities

The fundamental burning velocity (S_u) is defined as the rate at which a plane (i.e., flat) combustion wave will propagate into a stationary, quiescent flammable mixture of infinite extent. Normally the maximum value is quoted for a given flammable gas as this is a measure of its reactivity and to some extent determines the violence of any confined explosion in which it might be involved (Bartknecht, 1981; Harris, 1983; Lewis and von Elbe, 1987).

Burning velocity must be distinguished from 'flame speed' which is a measure of the rate of movement of flame with respect to a fixed observer. To illustrate the difference, consider a flammable mixture confined to a tube or duct of length *l*, one end of which is closed (Figure 3.21). Following ignition at the closed end, flame will propagate along the duct, reaching the open end in time *t*. The average flame speed is then *l*/*t*. However, this is substantially greater than the burning velocity as the flammable mixture ahead of the flame front is set into motion by the expansion of the burnt gas. Flame speed cannot be converted into a burning velocity simply by taking the rate of movement of the unburnt mixture into account; the flame front is not planar, there will be heat losses to the walls and (most significantly) the unburnt gas is likely to be in a turbulent state as it flows along the duct (see Section 3.5.5).

It is difficult to devise an experiment for measuring S_u in which interaction between the flame and the apparatus does not influence the result. Several methods are available (e.g., Gaydon and Wolfhard, 1979) although in most, it is necessary to correct the final result to take such interactions into account. Perhaps the simplest method of estimating burning velocity is that using a device similar to the Bunsen burner but which has the capability of producing laminar flows (Re < 2300) of a gas/air mixture whose composition can be varied. Within certain limits of mixture composition and flowrate, a flame can be established which takes the form of a cone sitting at the

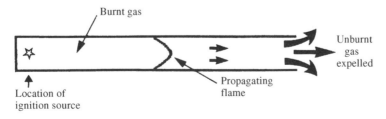

Figure 3.21 Propagation of premixed flame through a flammable mixture in a duct following ignition (☆) at the closed end

open end of the vertical burner tube (Figure 3.13(a)). The flame front is in a state of quasi-equilibrium with the flowing mixture, adopting the configuration in which the local burning velocity is equal to the local flowrate vector perpendicular to the flame front. If S_u is the burning velocity, U is the average (laminar) flowrate parallel to the tube axis, and θ is the half-angle of the 'cone' (Figure 3.22), then

$$S_u = U \sin \theta \qquad (3.23)$$

However, this method underestimates S_u by at least 25% for the following reasons:

(a)   no account is taken of the velocity distribution across the diameter of the tube;

(b)   heat transfer from the flame zone to the unburnt gas will cause the flowlines to diverge from being parallel to the tube axis close to the flame front (i.e., the measured value of θ is too small);

(c)   heat losses from the edge of the cone to the burner rim, while stabilizing the flame, also affect the flame shape; and

(d)   the flame front is not planar.

Corrections can be made for (a) and (b) (Andrews and Bradley, 1972; Gaydon and Wolfhard, 1979): it is more difficult to compensate for the effects of (c) and (d), although their effects may be reduced by using a wider tube.

   Botha and Spalding (1954) developed a burner that was capable of producing flat, laminar flames and that allowed the amount of heat transferred from the flame to the burner to be measured. The burner comprised a water-cooled, sintered metal disc through which a fuel/air mixture of known composition could be made to flow at selected linear flowrates (Figure 3.13(b)). Flat flames could be stabilized, the heat transferred to the burner decreasing as the flowrate was increased and the flame establishing itself at greater distances from the burner surface. This is shown schematically in Figure 3.23.

   The rate of heat transfer to the burner was obtained by measuring the increase in temperature of the flow of cooling water. This was determined as a function of the flowrate for each premixture studied and a value for the burning velocity obtained by extrapolating the heat loss to zero (Figure 3.24). Botha and Spalding obtained 0.42 m/s as the maximum burning velocity for propane/air mixtures, which may be compared with $S_u = 0.3$ m/s obtained using the 'cone angle method'.

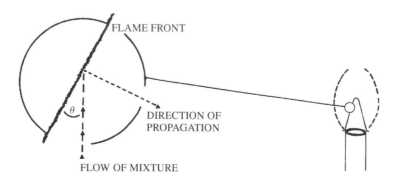

Figure 3.22   An approximate determination of burning velocity: the cone angle method

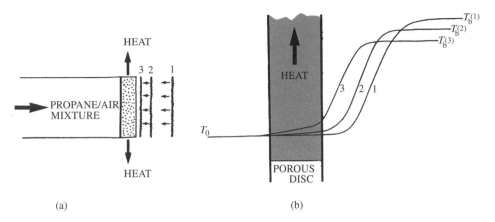

Figure 3.23 Determination of burning velocity: Botha and Spalding's porous burner: (a) showing different positions of the flame front corresponding to different flowrates of fuel/air mixture through the porous disc; (b) showing schematically the temperature distributions for different positions of the flame front in (a) (after Botha and Spalding, 1954, by permission)

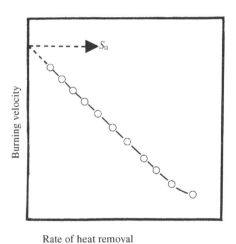

Figure 3.24 Determination of burning velocity (Botha and Spalding, 1954). Extrapolation of burning rate to zero heat loss (reproduced by permission)

Several other techniques are available for measuring burning velocity (Gaydon and Wolfhard, 1979). The most widely accepted is the 'spherical bomb' method of Lewis and co-workers (Lewis and von Elbe, 1987). The mixture is contained within a sphere capable of withstanding explosion pressures and ignited centrally by means of a spark. Analysis of the rate of pressure rise and the rate of propagation of the spherical flame front allows S_u to be calculated. The results obtained by this method are in good agreement with those of Botha and Spalding (1954) for propane ($S_u = 0.42$ m/s).

Values of fundamental burning velocities of a limited number of flammable gases and vapours are quoted in Table 3.1. Andrews and Bradley (1972) have demonstrated the need for careful evaluation of existing data. They examined the available data on the burning velocity of methane in air and concluded that $S_u(CH_4)$ was somewhat higher than the normally accepted value of 0.36 m/s, and much closer to the figure for the higher alkanes (0.42 m/s). However, this seems to go against common experience of the behaviour of methane: not only is it less reactive than other hydrocarbons, but gas explosions involving methane *tend* to be less severe than those involving higher alkanes such as propane (Bartknecht, 1981).

It has been suggested that all values of S_u need to be revised upwards in the light of Andrews and Bradley's (1972) re-evaluation of the data on methane (Harris, 1983), but more research is needed to resolve this matter.

3.5   Variation of Burning Velocity with Experimental Parameters

Many studies have been made of the variation of burning velocity with experimental parameters such as flammable gas concentration and temperature (Kanury, 1975; Gaydon and Wolfhard, 1979; Lewis and von Elbe, 1987). The main conclusions are outlined in the following sections.

3.5.1   Variation of mixture composition

The burning velocity of fuel/air mixtures is a maximum for mixtures slightly on the fuel-rich side of stoichiometric (Section 3.2). Mixtures close to the flammability limits have finite burning velocities and there is no evidence that S_u tends to zero at the limit. This is consistent with the concept that the limit represents criticality at which the rate of heat generation within the flame cannot sustain the heat losses (Mayer, 1957; Spalding, 1957) (Section 3.3) and with observations that a limiting flame temperature exists (White, 1925). However, it is also consistent with the concept of aerodynamic instability in which buoyancy-induced flows are sufficient to cause flame extinction at the limit (Hertzberg, 1982).

Figure 3.25 shows typical results for the variation of burning velocity with mixture composition for methane/air and propane/air mixtures. Values quoted in the literature (see Table 3.1) refer to the maximum values measured, i.e. to slightly fuel-rich mixtures. It might be anticipated that the maximum would be observed for the stoichiometric mixture for which the flame temperature is a maximum, but the fact that it does not suggests that flame propagation cannot be explained entirely in terms of heat transfer (Section 3.2).

Burning velocity is increased if the proportion of oxygen in the atmosphere is increased. Thus the maximum burning rate for methane/air mixtures increases from about 0.37 m/s to over 3.25 m/s as nitrogen in the air is replaced by oxygen (Figure 3.26). This is consistent with our observations of the temperatures achieved in the combustion zone (Chapter 1), if Equation (3.13) holds (Sections 3.2 and 3.3).

The mechanism by which a flammable vapour/air mixture may be rendered non-flammable by adding a gas such as N_2 or CO_2 which is 'inert' from the combustion standpoint can be interpreted in terms of a limiting flame temperature. However, if the nitrogen in the air is replaced completely by another gas (e.g. CO_2 or He) then

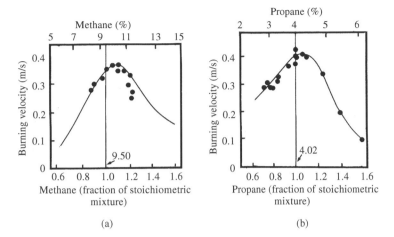

Figure 3.25   Variation of burning velocity with composition: (a) methane/air mixtures; (b) propane/air mixtures (reproduced by permission of Academic Press from Lewis and von Elbe, 1987)

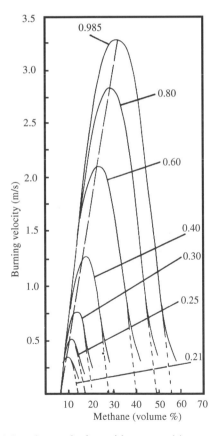

Figure 3.26   Variation of burning velocity with composition: methane in oxygen/nitrogen mixtures (Lewis and von Elbe, 1987). Numbers refer to the ratio $O_2/(O_2 + N_2)$

the burning velocity will be changed. For example, as the heat capacity of CO_2 is over 60% higher than that of N_2 at 1000 K, replacing N_2 by CO_2 will result in a reduction in the burning velocity because the flame temperature will be significantly less. On the other hand, the burning velocity of a stoichiometric mixture of a flammable gas in a 21/79 mixture of O_2/He is much higher than in air because helium has a low thermal capacity and a much higher thermal conductivity (see Equation (3.9), Section 3.2). Oxygen/helium mixtures are used in certain diving applications: in such an environment an explosion resulting from a leak of flammable gas or vapour would be more violent than if the atmosphere were normal air.

3.5.2   Variation of temperature

The burning velocities quoted in Table 3.1 refer to unburnt mixtures at ambient temperature (20–25°C). However, S_u is increased at higher temperatures, as illustrated for methane, propane and ethylene in Figure 3.27. Zabetakis (1965) quotes the following expression for methane, propane, n-heptane and iso-octane in the range 200–600 K:

$$S_u = 0.1 + 3 \times 10^{-6} T^2 \qquad \text{m/s} \tag{3.24}$$

At 300 K, this gives 0.37 m/s, close to the currently accepted value of S_u for methane, but somewhat less than that for propane (0.42 m/s). At temperatures above 800 K, mixtures of gaseous fuel and air will undergo slow oxidation and pyrolysis ('preflame reactions'), thus changing the chemical composition of the mixture. Under these conditions the measured burning velocity will be less than that based on an extrapolation from lower temperatures (e.g. from Equation (3.24) or Figure 3.27). If the temperature is sufficiently high, the flammable mixture may ignite spontaneously (see Figure 3.4). Some auto-ignition temperatures — which refer to near-stoichiometric mixtures for which the AIT is a minimum — are shown in Table 6.3; the phenomenon is discussed in Section 6.1.

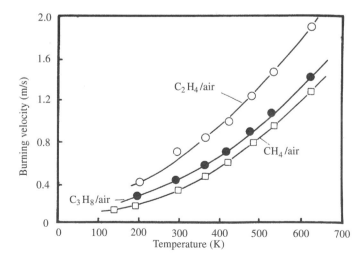

Figure 3.27   Variation of burning velocity with temperature according to Dugger *et al.* (1955). Reproduced by permission of Gordon and Breach from Kanury (1975)

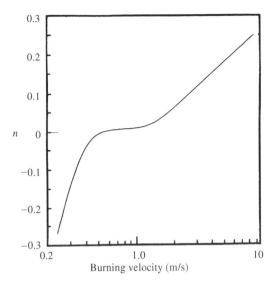

Figure 3.28 Influence of pressure on flame speed (Lewis, 1954; Kanury, 1975). n refers to the exponent in the proportionality $S_u \propto p^n$, where p is the initial pressure

3.5.3 Variation of pressure

There is no simple relationship between burning velocity and pressure. Lewis (1954) assumed that a proportionality of the form $S_u \propto p^n$ would hold, where p is the pressure, and determined the value for n for a range of gases and oxygen concentrations using the spherical bomb method described above. He found that n depended strongly on the value of S_u, being zero for burning velocities in the range 0.45–1.0 m/s (Figure 3.28). For $S_u < 0.45$ m/s, the dependence (i.e. n) is negative, while for $S_u > 1.0$ m/s the dependence is positive. Thus, S_u for methane/air mixtures will decrease with pressure while that for methane/O_2 will increase. Note, however, that the effect is small. Doubling the pressure of a stoichiometric methane/oxygen increases the burning velocity by a factor of only 1.07.

3.5.4 Addition of suppressants

A flammable mixture may be rendered non-flammable by the addition of a suitable suppressant. Additives such as nitrogen and carbon dioxide act as inert diluents, increasing the thermal capacity of the mixture (per unit mass of fuel) and thereby reducing the flame temperature, ultimately to below the limiting value when flame propagation will not be possible (Section 3.1.4). This is illustrated in Figure 3.29(a) which shows the variation of flame temperature, determined by infra-red radiance measurements, as nitrogen is added to stoichiometric methane/air mixtures (Hertzberg *et al.*, 1981). The limiting flame temperature (1500–1600 K) corresponds with 35–38% N_2, in agreement with values calculated on the basis that the lower flammability limit is determined by a critical limiting temperature of this magnitude. Consequently, it is anticipated that the burning velocity at the limit will be similar to that of a limiting methane/air mixture.

However, if chemical inhibitors are present in the unburnt vapour/air mixture, there will be significant reduction in burning velocity without a corresponding reduction in flame temperature. Halogen-containing species are particularly effective in this respect. For example, Simmons and Wolfhard (1955) found that the addition of 2% methyl bromide to a stoichiometric mixture of ethylene and air reduced the burning velocity from 0.66 m/s to 0.25 m/s. These species exert their influence by inhibiting the oxidation chain reactions, reacting with the chain carriers (in particular hydrogen atoms) and replacing them by relatively inert atoms or radicals. As the branching reaction:

$$O_2 + H^{\cdot} = {}^{\cdot}O^{\cdot} + OH^{\cdot}$$

is largely responsible for maintaining the high reaction rate (Section 1.2.2), any reduction in hydrogen atom concentration will have a very significant effect on the overall reaction rate. The accompanying heat capacity changes are relatively small so that there is a relatively minor reduction in the flame temperature. This is shown in Figure 3.29(b) in which flame temperature is plotted against concentration of Halon 1301 (bromotrifluoromethane); 4% 1301 is sufficient to render a stoichiometric CH_4/air mixture non-flammable, although the flame temperature at the limit is still high (>1800 K). However, there is some dispute over the interpretation of these data. Hertzberg's experiments involved spark ignition of the gaseous mixtures inside a 3.66 m diameter sphere.

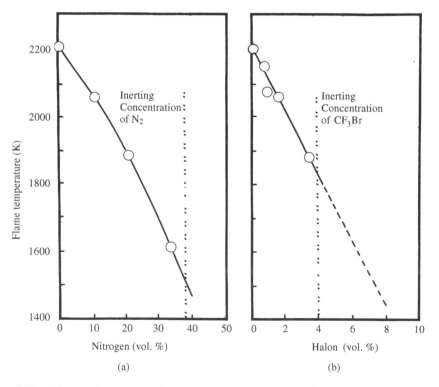

Figure 3.29 Measured premixed flame temperature during explosions in a 3.7 m diameter sphere, stoichiometric methane/air mixtures with addition of (a) nitrogen and (b) CF_3Br (Hertzberg, 1982). Reproduced by permission of University of Waterloo Press

Table 3.2 Peak concentrations of various halons in *n*-hexane/air mixtures (Hirst *et al.*, 1981/82)

Halon number	Common name	Formula	Boiling point (°C)	Peak concentration (%)
1211	BCF	CF_2ClBr	−4.0	8.1
1301	BTM	CF_3Br	−57.6	8.0
1202	DDM	CF_2Br_2	24.4	5.4
2402	DTE	$C_2F_4Br_2$	47.5	5.2

There is evidence to suggest that the halon is very efficient at suppressing ignition by a small spark: somewhat higher concentrations are required to suppress flame propagation when a larger source of ignition — such as a flame — is used. Hertzberg (1982) suggests that 8% of Halon 1301 is necessary to inert a stoichiometric methane/air mixture under these conditions. If this is correct, it would suggest that chemical inhibition may be of less importance than is currently assumed (at least for CF_3Br).

Sawyer and Fristrom (1971) have used the effect on S_u as a means of assessing the relative efficiencies of a range of inhibitors. However, normal practice is to determine the effect of an inhibitor on the flammability limits of suitable gases or vapours. The 'peak concentration' is determined from a flammability diagram such as those shown in Figures 3.12 and 3.30, and refers to the minimum concentration of the agent which is capable of rendering the most reactive vapour/air mixture non-flammable. Some typical values are shown in Table 3.2. Unfortunately, many of these inhibitors (particularly BCF and BTM) have now been banned as they have been shown to survive for long enough in the atmosphere to be harmful to the ozone layer (Montreal Protocol, 1987): they are now used only where there is an unacceptably high risk, and there is no alternative. In view of this problem, halon replacements have been sought:

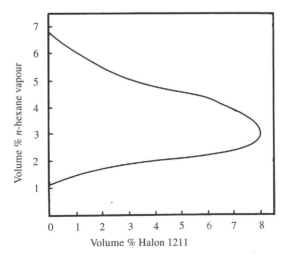

Figure 3.30 Flammability envelope for the addition of CF_2Br Cl to a stoichiometric *n*-hexane/air mixture (Hirst *et al.*, 1981/82)

a review of the current situation appears in the latest edition of the *NFPA Handbook* (Di Nenno, 1997).

3.5.5   The effect of turbulence

On the basis of the original definition (Section 3.4), the term 'fundamental burning velocity' cannot be applied to a turbulent mixture. Turbulence in the unburnt gas increases the rate at which flame will propagate through a mixture, but while this is of considerable importance regarding the behaviour of gas explosions, the effect is difficult to quantify. Damkohler and others have shown a relationship between turbulent burning velocity and Reynolds number of the unburnt mixture for particular pipe and duct flow situations (Damkohler, 1940; Rasbash and Rogowski, 1960). The mechanism is understood to involve an increase in the efficiency of the transport processes (transfer of heat *and* reactive species) as a result of eddy mixing at the flame front. As these control the rate of propagation (Section 3.2), the rate of burning in turbulent mixtures is high.

In pipes or ducts of a sufficient length (see Figure 3.21), the generation of pipe flow turbulence following ignition at a closed end may cause sufficient flame acceleration to produce a detonation, i.e. a shock wave merged with the combustion wave, propagating at supersonic velocities. This can only occur for mixtures which lie within the limits of 'detonability' which are analogous to and lie within the limits of flammability for flammable gases and vapours (Lewis and von Elbe, 1987). For detonation to develop in pipes or ducts, a minimum 'run-up' length can be identified (Health and Safety Executive, 1980). This may be as much as 60 pipe diameters for alkanes, but is substantially less for more reactive gases such as ethylene and hydrogen. Bends or obstacles will reduce the run-up length, regardless of the nature of the gas involved as they create flame acceleration. This has been studied in some detail (Rasbash and Rogowski, 1962; Rasbash, 1986).

Turbulence also increases the violence of gas explosions in large spaces in which there are obstacles ahead of the flame front (e.g. Rasbash, 1986). Such a situation exists in chemical plant, and in typical modules found on offshore oil production platforms. The initial explosion which led to the loss of the Piper Alpha platform in 1988 (Cullen, 1989) produced high overpressures on account of flame acceleration due to the turbulence induced in the unburnt gas ahead of the flame front. The effect has been demonstrated experimentally (Harrison and Eyre, 1987) and modelled with a considerable degree of success using computational fluid dynamics (Hjertager, 1993). A similar effect occurs if an enclosure is subdivided into compartments linked through open doors. A gas/air explosion will create turbulence in the adjacent space as the unburnt mixture is pushed through the openings (Figure 3.31). This is sometimes referred to as 'pressure piling': very rapid and unpredictable rates of pressure rise can be generated in this way. The high pressure achieved in the Ronan Point gas explosion (Rasbash, 1969) was attributed to this mechanism (Rasbash *et al.*, 1970). It is very difficult to design an adequate explosion prevention system for these circumstances (e.g. Harris, 1983).

The mechanism by which the ignition of a large cloud of gas or vapour in the open can give rise to flame acceleration and pressure effects has not yet been established. These events are commonly referred to as 'unconfined vapour cloud explosions'

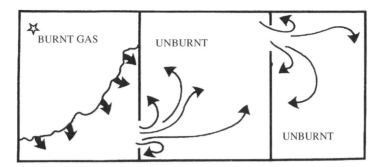

Figure 3.31 Development of an explosion in a multi-chambered compartment, showing development of turbulence ahead of the flame front. Ignition at ☆

(Strehlow, 1973; Gugan, 1979; Zalosh, 1995) although some degree of obstruction seems to be necessary to cause acceleration of the flame. It has been suggested that such 'clouds' can detonate under certain conditions (e.g. Burgess and Zabetakis, 1973) but this is by no means proven. These events should be distinguished from BLEVES (boiling liquid, expanding vapour explosions), in which the flammable material is released suddenly following the violent rupture of a pressurized storage vessel which has been exposed to fire for a prolonged period (e.g. Feyzin in 1966, Crescent City, Illinois in 1970 (Strehlow, 1973)). BLEVES are discussed briefly in Section 5.1.

Problems

3.1 Calculate the lower flammability limit of a mixture containing 84% methane, 10% ethane and 6% propane.

3.2 Given that the lower flammability limit of n-butane (n-C_4H_{10}) in air is 1.8% by volume, calculate the adiabatic flame temperature at the limit. (Assume initial temperature to be 20°C.)

3.3 Calculate the lower flammability limit of propane in a mixture of (a) 21% oxygen + 79% helium, and (b) 21% oxygen + 79% carbon dioxide, assuming a limiting adiabatic flame temperature of 1600 K. (Initial temperature 20°C).

3.4 Using the result of Problem 3.2, calculate by how much a stoichiometric n-butane/air mixture would have to be diluted by (a) nitrogen (N_2), (b) carbon dioxide (CO_2), to render the mixture non-flammable.

3.5 Calculate by how much a stoichiometric propane/air mixture would have to be diluted by (a) carbon dioxide (CO_2), (b) bromotrifluoromethane (CF_3Br, Halon 1301), to render the mixture non-flammable. Assume that the limiting adiabatic flame temperature is 1600 K and that the heat capacity of CF_3 Br is 101 J/mol K at 1000 K. It is found experimentally that only 5% CF_3Br is required to inert a stoichiometric propane/air mixture. Explain why this differs from your answer.

3.6 Calculate the lower and upper flammability limits of propane at 200°C and 400°C.

3.7 Calculate the range of temperatures within which the vapour/air mixture above the liquid surface in a can of *n*-hexane at atmospheric pressure will be flammable.

3.8 Calculate the range of ambient pressures within which the vapour/air mixture above the liquid surface in a can of *n*-decane (n-$C_{10}H_{22}$) will be flammable at 25°C.

3.9 Given that the lower and upper flammability limits of butane in air and in oxygen are 1.8% and 8.4%, and 1.8% and 49%, respectively, and that the 'limiting oxygen index' for butane is 13%, sketch the flammability limits for the $C_4H_{10}/O_2/N_2$ system, using rectangular coordinates.

3.10 By inspection of Equation (3.9) *et seq.* how will S_u vary with thermal conductivity (k), thermal capacity (c) and temperature? Compare your conclusions with the empirical results discussed in Section 3.5.

4
Diffusion Flames and Fire Plumes

The principal characteristic of the diffusion flame is that the fuel and oxidizer (air) are initially separate and combustion occurs in the zone where the gases mix. The classical diffusion flame can be demonstrated using a simple Bunsen burner (Figure 3.13(a)) with the air inlet port closed. The stream of fuel issuing from the burner chimney mixes with air by entrainment and diffusion and, if ignited, will burn wherever the concentrations of fuel and oxygen are within the appropriate (high temperature) flammability limits (Section 3.1.3). The appearance of the flame will depend on the nature of the fuel and the velocity of the fuel jet with respect to the surrounding air. Thus, hydrogen burns with a flame which is almost invisible, while all hydrocarbon gases yield flames which have the characteristic yellow luminosity arising from incandescent carbonaceous particles formed within the flame (Section 2.4.3). Laminar flames are obtained at low flowrates. Careful inspection reveals that just above the burner rim, the flame is blue, similar in appearance to a premixed flame. This zone exists because some premixing can occur close to the rim where flame is quenched (Section 3.3a). At high flowrates, the flame will become turbulent (Section 4.2), eventually 'lifting off' when flame stability near the burner rim is lost due to excess air entrainment at the base of the flame. The momentum of the fuel vapour largely determines the behaviour of these types of flame which are often referred to as 'momentum jet flames'. A typical example is an emergency flare which acts to relieve the pressure in an item of chemical plant.

In contrast, flames associated with the burning of condensed fuels (i.e. solids and liquids) are dominated by buoyancy, the momentum of the volatiles rising from the surface being relatively unimportant. If the fuel bed is less than 0.05 m in diameter, the flame will be laminar, the degree of turbulence increasing as the diameter of the fuel bed is increased, until for diameters greater than 0.3 m buoyant diffusion flames with fully developed turbulence are observed (Section 5.1.1).

This chapter deals principally with flames from burning liquids and solids, although much of our knowledge comes from studies of flames produced on flat, porous bed gas burners such as those used by McCaffrey (1979), Cox and Chitty (1980) and Zukoski (1981a): these are designed to give a low momentum source of fuel vapour. The relative importance of momentum (or inertia) and buoyancy in the flame will determine the type of fire, and the Froude number (Fr) may be used as a means of classification. It is a measure of the relative importance of inertia and buoyancy in the system, and is conveniently expressed as:

$$\mathrm{Fr} = U^2/gD \qquad (4.1)$$

where U is the velocity of the gases, D is a characteristic dimension (e.g. the diameter of the burner) and g is the acceleration due to gravity. Turbulent jet flames have high Froude numbers, based on the exit velocity of the fuel from a pipe or orifice. With natural fires, the initial velocity of the vapours in general cannot be measured, but can be derived from the rate of heat release* ($\dot{Q}_c$). Assuming a circular fuel bed of diameter D (area $\pi D^2/4$), fuel density ρ, and heat of combustion of the fuel vapour of ΔH_c, the initial velocity of the fuel vapours can be expressed as:

$$U = \frac{\dot{Q}_c}{\Delta H_c \rho (\pi D^2/4)} \qquad (4.2)$$

Comparing the above two expressions, it can be seen that the Froude number is proportional to $\dot{Q}_c^2/D^5$, a scaling criterion that will be encountered below. Indeed, a 'dimensionless heat release rate' ($\dot{Q}_c^*$), introduced in the 1970s by Zukoski (1975) and others, is the square root of a Froude number expressed in terms of the heat release rate of a fire. It is used to classify fire types and correlate aspects of fire behaviour

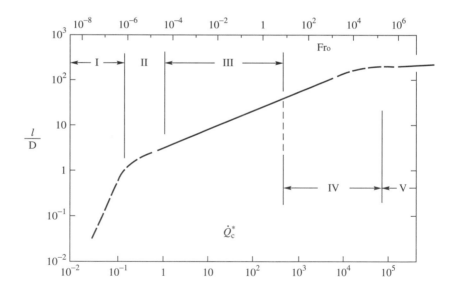

Figure 4.1   Schematic diagram showing flame length (l) as a function of the fuel flow rate parameters, expressed as $\dot{Q}_c^*$. The extreme right hand-region (V) corresponds to the fully turbulent jet fire (cf. Figure 4.7), dominated by the momentum of the fuel. Regions I and II correspond to buoyancy-driven turbulent diffusion flames (cf. Figure 4.8). (Adapted from Zukoski, 1986). Reprinted by permission

---

* The term 'rate of energy release' is more satisfactory in this context, but 'rate of heat release' has come to be the accepted terminology.

(McCaffrey, 1995), such as flame height (Figures 4.1 and 4.17). It is given by:

$$\dot{Q}_c^* = \frac{\dot{Q}_c}{\rho_\infty c_p T_\infty \sqrt{gD} \cdot D^2} \qquad (4.3)$$

Heskestad (1981) recommends the use of an alternative form of modified Froude number which takes into account the stoichiometry of the reaction (see Equation (4.39)).

4.1 Laminar Jet Flames

When a jet of gas issues into a still atmosphere, air is entrained as a result of shear forces between the jet and the surrounding air (cf. Section 4.3.1). The resulting flame will be laminar provided that the Reynolds number at the origin is less than ~2000. However, the shear forces cause instability in the gas flow which gives rise to flame flicker (Gaydon and Wolfhard, 1979). For hydrocarbon diffusion flames on a Bunsen burner, the flickering has a frequency of 10–15 Hz. This can be virtually eliminated if the surrounding air is made to move concurrently with and at the same linear velocity as the gas jet. Burke and Schumann (1928) chose to work with this arrangement in their classic study of laminar diffusion flames. They enclosed the burner tube inside a concentric cylinder carrying the flow of air: by varying the relative diameters of the tubes they were able to establish 'over-ventilated' and 'under-ventilated' flames as shown in Figure 4.2. These studies established that combustion occurred in the

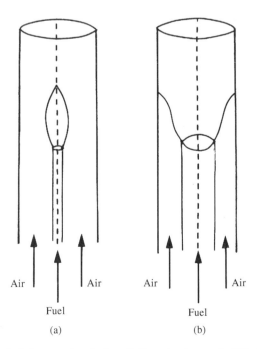

Figure 4.2 Burke and Schumann's study of the structure of diffusion flames: (a) over-ventilated and (b) under-ventilated flames

fuel/air mixing zone and suggested that the flame structure could be analysed on the assumption that the burning rate was controlled by the rate of mixing rather than by the chemical kinetics.

The rate of diffusion of one gas into another can be described by Fick's law which for one dimension is:

$$\dot{m}_i'' = -D_i \frac{dC_i}{dx} \tag{4.4}$$

where $\dot{m}_i''$ and C_i are the mass flux and concentration of species i respectively and D_i is the diffusion coefficient for species i in the particular gas mixture. It is analogous to Fourier's law of conductive heat transfer in which the heat flux is proportional to the temperature gradient, $\dot{q}_c'' = -k(dT/dx)$; here, mass flux is proportional to the concentration gradient. Transient mass transfer in three dimensions requires solution of the equation

$$\nabla^2 C_i = \frac{1}{D_i} \left(\frac{\partial C_i}{\partial t} \right) \tag{4.5}$$

which can be compared with Equation (2.16). As with heat transfer problems, solution of this basic partial differential equation is made easier if the problem is reduced to a single space dimension. This is possible for the diffusion flames illustrated in Figure 4.2 since the model can be described in cylindrical coordinates, i.e.

$$\frac{\partial C(r, y)}{\partial t} = D \left[\frac{\partial^2 C(r, y)}{\partial r^2} + \frac{1}{r} \cdot \frac{\partial C(r, y)}{\partial r} \right] \tag{4.6}$$

where the concentration $C(r, y)$ is a function of radial distance from the axis of symmetry (r) and height above the burner rim (y) (the subscript 'i' has been dropped for convenience). Normally this would be applied to the 'infinite cylinder' in which there would be no diffusion parallel to the axis, but here we have a flowing system in which time can be expressed as a distance travelled vertically (y) at a known velocity (u). Thus, as $t = y/u$, Equation (4.6) can be rewritten:

$$u\frac{\partial C(r, y)}{\partial y} = D \left[\frac{\partial^2 C(r, y)}{\partial r^2} + \frac{1}{r} \cdot \frac{\partial C(r, y)}{\partial r} \right] \tag{4.7}$$

The solution to this equation will give concentration (e.g. of fuel in air) as a function of height and radial distance from the burner axis (see Figure 4.3). (Axial diffusion will occur but is neglected in this approximate model.) Burke and Schumann (1928) suggested that the flame shape would be defined by the envelope corresponding to $C(r, y) = C_{\text{stoich}}$, where C_{stoich} is the stoichiometric concentration of fuel in air, but to obtain a solution to the above equation the following additional assumptions were necessary:

(i)   the reaction zone (i.e. where $C(r, y) = C_{\text{stoich}}$) is infinitesimally thin;
(ii) rate of diffusion determines the rate of burning; and
(iii) the diffusion coefficient is constant.

Assumptions (i) and (ii) are effectively equivalent and, combined with the basic assumption regarding flame shape, imply that reaction is virtually instantaneous wherever the concentration is stoichiometric. This is a gross oversimplification as reaction

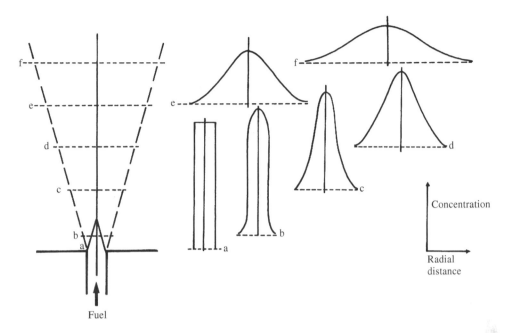

Figure 4.3   Concentration profiles in a jet emerging into an infinite quiescent atmosphere (after Kanury, 1975)

will occur wherever the mixture is within the limits, which will be wide at the high temperatures encountered in flames (see Figure 3.4). Moreover, diffusion coefficients vary considerably with both temperature and composition of the gas mixture. Nevertheless, the resulting analytical solution to Equation (4.7) accounts very satisfactorily for the shapes of both over-ventilated and under-ventilated flames, as illustrated in Figure 4.4, thus establishing the validity of the proposed basic structure.

A much simpler model was developed by Jost (1939) in which the tip of a diffusion flame was defined as the point on the flame axis ($r = 0$) at which air is first found ($y = l$ in Figure 4.5). He used Einstein's diffusion equation, $x^2 = 2Dt$ (where x is the average distance travelled by a molecule in time t) to establish the (average) time it would take a molecule from the air to diffuse from the rim of the burner to its axis, i.e. $t = R^2/2D$, where R is the radius of the burner mouth. Considering the concentric burner system of Burke and Schumann in which air and fuel are moving concurrently with a velocity u (and air is in excess, Figure 4.2(a)), in time t the gases will flow through a distance ut. Thus the height of the flame (l), according to the above definition, will be

$$l = \frac{uR^2}{2D} \tag{4.8}$$

which, if expressed in terms of volumetric flowrate, $\dot{V} = \pi R^2 u$, gives:

$$l = \frac{\dot{V}}{2\pi D} \tag{4.9}$$

This equation predicts that the flame height will be proportional to the volumetric flowrate and independent of the burner radius, of which the latter is essentially correct.

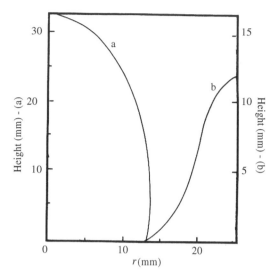

Figure 4.4   Shapes of (a) over-ventilated and (b) under-ventilated diffusion flames according to Equation (4.7), for $C(r, y) = C_{stoich}$. (Reprinted from Burke and Schumann, *Ind. Eng. Chem.*, **20**, 998. Published 1928 American Chemical Society)

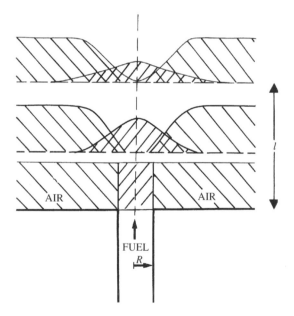

Figure 4.5   Jost's model of the diffusion flame (Jost, 1939)

However, buoyancy influences the height of the laminar flame and the dependence is closer to $\dot{V}^{0.5}$. The predicted inverse dependence on the diffusion coefficient is not observed strictly, but this is not unexpected as D varies considerably with temperature and with mixture composition. Furthermore, a change in the stoichiometry would be expected to alter the flame height, but this is not incorporated into Jost's model.

The limited success of these simple diffusion models indicates that the underlying assumptions are essentially correct. They can be expressed in a different format by identifying the tip of the flame with the height at which combustion is complete, implying that sufficient air is entrained through the jet boundary in the time interval $t = l/u$ to burn all the fuel issuing from the mouth of the burner during the same period. While this is a useful concept, it is an oversimplification, as will be seen below (Section 4.3.2).

Laminar jet flames are unsuitable for studying the detailed structure of the diffusion flame. Wolfhard and Parker developed a method of producing flat diffusion flames using a burner consisting of two contiguous slots, one carrying the fuel gas and the other carrying the oxidant (Figure 4.6(a)) (Gaydon and Wolfhard, 1979). Provided that there is a concurrent flow of nitrogen surrounding the burner, this arrangement yields a stable, vertical flame sheet on which various types of measurement can be made on both sides of the combustion zone. Using this device, valuable information has been obtained on the spatial concentrations of combustion intermediates (including free radicals) which has led to a better understanding of the chemical processes within the flame. This type of work has given an insight into the mechanism of smoke (or 'soot') formation in diffusion flames (Kent *et al.*, 1981). However, like the jet flames, this type

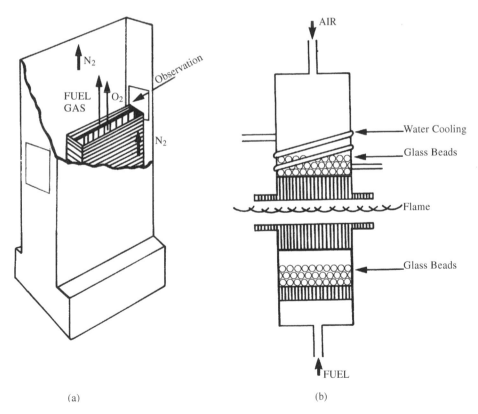

(a)   (b)

Figure 4.6   (a) The Wolfhard–Parker burner for producing flat diffusion flames. (b) The counterflow diffusion flame apparatus. (Reproduced with permission from Gaydon and Wolfhard. 1979)

of flat flame is affected by the presence of the burner rim. This can be avoided by using a counter-flow diffusion flame burner in which a flat flame is stabilized in the stagnant layer where diametrically opposed flows of fuel and oxidant meet (Figure 4.6(b)) (see Gaydon and Wolfhard, 1979). This system has been widely used to examine the stability and extinction of diffusion flames of gaseous and solid fuels (e.g. Williams, 1981).

4.2   Turbulent Jet Flames

In the previous section it was pointed out that the height of a jet flame will increase approximately as the square root of the volumetric flowrate of the fuel, but this is true only in the laminar regime. Above a certain jet velocity, turbulence begins, initially at the flame tip, and the flame height decreases with flowrate to a roughly constant value for the fully turbulent flame (Figure 4.7). This corresponds to high values of $\dot{Q}_c^*$

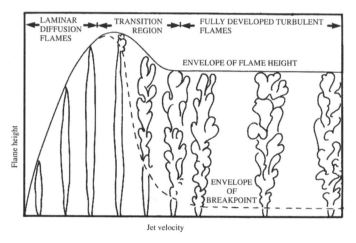

Figure 4.7   Height of momentum jet flames as a function of nozzle velocity, showing transition to turbulence (Hottel and Hawthorne, 1949). © 1949 Williams and Wilkins Co., Baltimore

Table 4.1   Flame temperatures (Lewis and von Elbe, 1987)

Fuel	Fuel concentration in air (%)	Flame temperature[a] (°C)
Hydrogen	31.6	2045
Methane	10.0	1875
Ethane	5.8	1895
Propane	4.2	1925
Butane	3.2	1895
Ethylene	7.0	1975
Propylene	4.5	1935
Acetylene	9.0	2325

[a] Determined by the sodium D-line reversal method (see Gaydon and Wolfhard, 1979). Valid for $T_0 = 20°C$.

(Equation (4.3)). The transition from a laminar to a turbulent flame is observed to occur at a nozzle Reynolds number significantly greater than 2000 (Hottel and Hawthorne, 1949) as it is the local Reynolds number ($Re = ux/v$) within the flame which determines the onset of turbulence. Re decreases significantly with rise in temperature as a result of the variation in kinematic viscosity (v). Turbulence first appears at the tip of the flame, extending further down towards the burner nozzle as the jet velocity is increased, although never reaching it (Figure 4.7). The decrease in flame height from the maximum inside the laminar region to a constant value in the fully-turbulent regime can be understood qualitatively in terms of increased entrainment of air by eddy mixing which results in more efficient combustion.

Hawthorne, Weddel and Hottel (1949) derived the following expression theoretically, relating the turbulent flame height l_T to the diameter of the burner jet, d_i, the flame temperature T_F (K) (Table 4.1), the initial temperature T_i (K), and the average molecular weights of air (M_{air}) and the fuel issuing from the jet (M_f):

$$\frac{l_T}{d_i} = \frac{5.3}{C_f} \left[\frac{T_F}{mT_i} \left(C_f + (1 - C_f)\frac{M_{air}}{M_f} \right) \right]^{1/2} \tag{4.10}$$

where m is the molar ratio of reactants to products (both inclusive of nitrogen) for the stoichiometric mixture, and $C_f = (1 + r_i)/(1 + r)$ in which r is the stoichiometric molar air/fuel ratio, and r_i is the initial air/fuel ratio, taking into account situations in which there is air in the initial fuel mixture.

This refers to the fully turbulent momentum jet flame in which buoyancy effects are neglected (high Froude number). It is in good agreement with measurements made on the turbulent flames for a range of gases (Lewis and von Elbe, 1987; Kanury, 1975) and shows that the flame height is linearly dependent on nozzle diameter, but independent of the volumetric flowrate. Because combustion is more efficient in these than in laminar diffusion flames, their emissivity tends to be less as a result of the lower yield of carbonaceous particles. The magnitude of the effect depends on the nature of the fuel: for methane and propane, c.30% of the heat of combustion may be lost by radiation from a laminar diffusion flame, while this may be reduced to only 20% for a turbulent flame (Markstein, 1975, 1976; Delichatsios and Orloff, 1988). The effect is even greater for flames from fuels which have a greater tendency to produce soot, such as ethylene (ethene) and acetylene (ethyne) (Delichatsios and Orloff, 1988).

4.3   Flames from Natural Fires

Turbulent jet flames are associated with high Froude numbers which correspond to values of $\dot{Q}_c^*$ of the order of 10^6 (McCaffrey, 1995), indicating that the momentum of the fuel stream is dominating the behaviour. In natural fires, buoyancy is the predominant driving force, consistent with values of $\dot{Q}_c^*$ which are around six orders of magnitude lower (Figure 4.1). These flames have a much less ordered structure and are more susceptible to external influences (such as air movement) than jet flames. Corlett (1974) drew attention to the existence of a layer of pure fuel vapour above the centre of the surface of a burning liquid when the pool diameter was between 0.03 and 0.3 m (see also Bouhafid *et al.*, 1988). The flames from this size of fire are essentially laminar, but become increasingly turbulent as the diameter is increased. The turbulence aids mixing at low level, but the layer close to the surface will still be fuel-rich. Flame shapes are illustrated in Figure 4.8 which shows a progression of increasing fire sizes, corresponding to *decreasing* values of $\dot{Q}_c^*$.

Porous bed gas burners have been used by several authors to study this type of flame (Corlett, 1968, 1970; Chitty and Cox, 1979; McCaffrey, 1979; Zukoski *et al.*, 1981a, b; Cetegen *et al.*, 1984; Hasemi and Tokunaga, 1984; Cox and Chitty, 1985; and others). The system has the advantage over fires involving combustible solids and liquids in that the fuel flowrate is an independent variable and the flame can be maintained indefinitely for experimental purposes (e.g. Smith and Cox, 1992). McCaffrey (1979) showed that the 'fire plume' above a 30 cm square burner consisted of three distinct regimes (see Figure 4.9), namely:

(i)   the near field, above the burner surface, where there is persistent flame and an accelerating flow of burning gases (the flame zone);

(ii)   a region in which there is intermittent flaming and a near-constant flow velocity (the intermittent zone); and

(iii)   the buoyant plume which is characterized by decreasing velocity and temperature with height.

While these are inseparable in the fire plume, it is appropriate to consider the buoyant plume on its own since its properties are relevant to other aspects of fire engineering, including fire detection (Section 4.4.2) and smoke movement and control (Sections 11.2 and 11.3). In the next two sections, we shall be discussing the unbounded plume; interactions with ceilings and walls will be considered in Section 4.3.4.

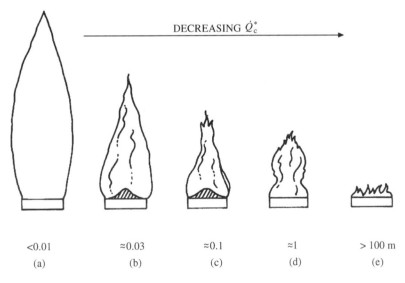

Figure 4.8 Classification of natural diffusion flames as 'structured' (b and c) and 'unstructured' (a, d and e) according to Corlett (1974). (e) flame would be classified as a 'mass fire'. The shaded areas indicate fuel-rich cores

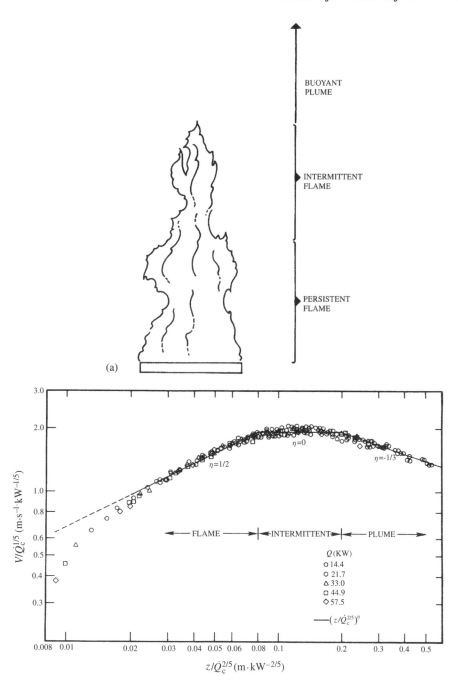

(a)

Figure 4.9   (a) Schematic diagram of the fire plume showing McCaffrey's three regimes. (b) Variation of upward velocity (V) with height (z) above the burner surface, plotted as $V/\dot{Q}_c^{1/5}$ versus $z/\dot{Q}_c^{2/5}$ (Table 4.2), where $\dot{Q}_c$ is the nominal rate of heat release (kW) (McCaffrey, 1979)

4.3.1 The buoyant plume

The concept of buoyancy was introduced in Section 2.3 in relation to natural convection. If a density difference exists between adjacent masses of fluid as a result of a temperature gradient, then the force of buoyancy will cause the less dense fluid to rise with respect to its surroundings. The buoyancy force (per unit volume), which is given by $g(\rho_\infty - \rho)$, is resisted by viscous drag within the fluid, the relative magnitude of these opposing forces being expressed as the Grashof number (Equation (2.49)). The term 'buoyant plume' is used to describe the convective column rising above a source of heat. Its structure is determined by its interaction with the surrounding fluid. Intuitively, one would expect the temperature within the plume to depend on the source strength (i.e. the rate of heat release) and the height above the source: this may be confirmed by theoretical analysis.

   The mathematical model of the simple buoyant plume is based on a point source as shown in Figure 4.10(a) (Yih, 1952; Morton *et al.*, 1956; Thomas *et al.*, 1963; Heskestad, 1972; Williams, 1982; Heskestad, 1995). The ideal plume in an infinite, quiescent atmosphere would be axisymmetric and extend vertically to a height where the buoyancy force has become too weak to overcome the viscous drag. Under certain atmospheric conditions, a temperature inversion can form which will effectively trap a rising smoke plume, halting its vertical movement and causing it to spread laterally at that level. The same effect can be observed in confined spaces: the commonest example is the stratification of cigarette smoke at relatively low levels in a warm room, under quiescent conditions. In high spaces, such as atria, the temperature at roof level may be sufficient to prevent smoke from a fire at ground level reaching smoke detectors at the ceiling. Heskestad (1989) has reviewed the work of Morton *et al.* (1956) and shown how their original expressions may be adapted to give the minimum rate of heat

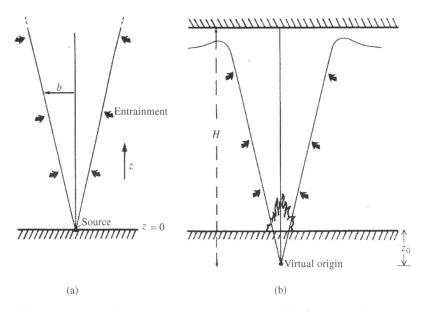

(a)   (b)

Figure 4.10   The buoyant plume (a) from a point source, and (b) from a 'real source', showing interaction with a ceiling

release $\dot{Q}_{conv,min}$ (kW) necessary to ensure that the plume reaches the ceiling (height H (m)):

$$\dot{Q}_{conv,min} = 1.06 \times 10^{-3} H^{5/2} \Delta T_a^{3/2} \qquad (4.11)$$

where T_a (the ambient temperature in K) is assumed to increase linearly with height, ΔT_a being the increase in ambient temperature between the level of the fire source and the ceiling. As an example, Heskestad uses this equation to illustrate that in a 50 m high atrium with $\Delta T_a = 5$ K, a minimum convective heat output of 210 kW will be required before smoke will reach a detector mounted on the ceiling.

Cooling of the plume occurs as a result of dilution with ambient air which is entrained through the plume boundary. The decrease in temperature with height is accompanied by broadening of the plume and a reduction in the upward flow velocity. The structure of the plume may be derived theoretically through the conservation equations for mass, momentum and energy, but a complete analytical solution is not possible: simplifying assumptions have to be made. It is likely that detailed solutions would develop Gaussian-like radial distributions of excess temperature (ΔT), density deficit ($\Delta \rho$) and upward velocity (u) through horizontal sections of the plume as a function of height. Morton *et al.* (1956) and others (see Zukoski, 1995) assumed Gaussian distributions and self-similarity of the radial profiles, but $\Delta \rho$ and ΔT cannot be self-similar unless the plume is 'weak', i.e. $T_0/T \simeq 1$. This is clear from the following relationship derived from the ideal gas law:

$$\frac{\Delta \rho}{\Delta \rho_0} = \left(\frac{\Delta T}{\Delta T_0} \right) \left(\frac{T_0}{T} \right) \qquad (4.12)$$

Self-similarity cannot hold for 'strong' plumes for these variables. Accordingly, it is normal to assume self-similarity either between u and $\Delta \rho$ (Yih, 1952; Morton *et al.*, 1956; Thomas *et al.*, 1963) or between u and ΔT (Zukoski *et al.*, 1981a; Cetegen *et al.*, 1984).

For the present argument, a more simple approach is appropriate in which 'top hat profiles' are assumed, i.e. T, ρ and u are assumed constant across the plume (radius b) at any specified height (Morton *et al.*, 1956; Heskestad, 1972). Starting with relationships derived from the conservation equations, a simple dimensional analysis may be applied to obtain the functional relationships between temperature and upward flow velocity on the one hand and source strength and height on the other. For conservation of momentum, the following proportionality may be written for an axisymmetric plume (of radius b at height z above a point source) in an infinite atmosphere (density ρ_∞) if viscous forces are neglected and temperature differences are small:

$$\frac{d}{dz}(\rho_0 u_0^2 b^2) \propto g(\rho_0 - \rho_\infty)b^2 \qquad (4.13)$$

where u_0 and ρ_0 are the vertical flow velocity and density on the plume axis at height z above the point source (Figure 4.10(a)). Similarly, for the conservation of mass:

$$\frac{d}{dz}(\rho_0 u_0 b^2) \propto \rho v b \propto \rho u_0 b \qquad (4.14)$$

in which the increase in mass flow with height is due to entrainment of air through the plume boundary. The entrainment velocity (v) is assumed to be directly proportional to u_0, i.e. $v = \alpha' u_0$, where α' is the entrainment constant which Morton *et al.* (1956) estimated to be about 0.09 for still air conditions. Any wind, or other air movement, will deflect the plume and effectively increase the entrainment constant (see also Section 4.3.5).

Finally, the conservation of energy may be represented by the following:

$$c_p \rho_0 u_0 b^2 \Delta T_0 \propto \dot{Q}_{conv} \tag{4.15}$$

where ΔT_0 is the temperature excess over ambient on the axis at height z and $\dot{Q}_{conv}$ is the convective heat output from the source, i.e. the source strength. Heat losses from the plume rising from a pure heat source (e.g. by radiation) are assumed to be negligible.

Heskestad (1972, 1975) assumed that the variables b, u_0 and ΔT_0 are directly proportional to simple powers of z, the height, i.e.

$$b \propto z^s; \quad u_0 \propto z^m \text{ and } \Delta T_0 \propto z^n \tag{4.16}$$

By substituting these three relationships into Equations (4.13), (4.14) and (4.15) and solving for s, m and n, assuming consistency of units, it can be shown that:

$$b \propto z \tag{4.17}$$

$$u_0 \propto A^{1/3} \dot{Q}_{conv}^{1/3} z^{-1/3} \tag{4.18}$$

$$\Delta T_0 \propto (A^{2/3} T_\infty / g) \dot{Q}_{conv}^{2/3} z^{-5/3} \tag{4.19}$$

where $A = g/c_p T_\infty \rho_\infty$ and T_∞ is the ambient air temperature (see also Heskestad, 1995). Applying the buoyant plume model of Morton *et al.* (1956), Zukoski *et al.* (1981a) and Cetegen *et al.* (1984) developed relationships which are in close agreement with Equations (4.17)–(4.19), using self-similarity between velocity (u) and temperature excess (ΔT) profiles. Using a dimensionless heat release rate Q_Z^* of the form

$$\dot{Q}_Z^* = \frac{\dot{Q}_c}{\rho_\infty C_p T_\infty Z^2 \sqrt{gZ}} \tag{4.20}$$

they derived the following expressions:

$$b = C_l z \tag{4.21}$$

$$u_0 = C_v (gz)^{1/2} \dot{Q}_z^{*1/3} \tag{4.22}$$

$$\Delta T_0 = C_T \dot{Q}_z^{*2/3} T_\infty \tag{4.23}$$

The constants C_l, C_v and C_T were derived from data of Yokoi (1960). These relationships have been reviewed (Beyler, 1986b) and compared with experimental data which have become available on the rate of entrainment into fire plumes (e.g. Cetegen *et al.* (1984)). Beyler recommends the following expression for the centreline temperature rise at height z, assuming that $T_\infty = 293$ K:

$$\Delta T_0 = 26 \frac{\dot{Q}_{conv}^{2/3}}{z^{5/3}} \tag{4.24}$$

or, if it is not possible to correct for radiative losses:

$$\Delta T_0 = 22 \frac{\dot{Q}_c^{2/3}}{z^{5/3}} \tag{4.25}$$

The above correlations hold remarkably well in the 'far field' of a fire plume (i.e. in the buoyant plume above the flames), particularly where the plume is 'weak' in the sense that the ratio $\Delta T / T_\infty$ is small. However, it is necessary to introduce a correction for the finite area of the source by identifying the location of a 'virtual origin' (Figure 4.10(b)), defined as the point source which produces a (buoyant) plume of identical entrainment characteristics to the real plume.

In early work, it was assumed that for a real fire, the virtual origin would lie approximately $z_0 = 1.5 A_f^{1/2}$ m below a fuel bed of area A_f. This was based on the assumption that the plume spreads with an angle of $c.15°$ to the vertical (Figure 4.10(b)) (Morton *et al.*, 1956; Thomas *et al.*, 1963). However, studies by Heskestad (1983), Cetegen *et al.*, (1984), Cox and Chitty (1985) and others (Zukoski, 1995) have shown that the location of the virtual source is dependent on the rate of heat release, as well as the diameter (or area) of the fire. Heskestad (1983, 1995) recommends the use of the formula

$$\frac{z_0}{D} = -1.02 + 0.083 \frac{\dot{Q}_c^{2/5}}{D} \tag{4.26}$$

as giving a good mean of the other correlations (Figure 4.11), although others have been proposed (e.g. Zukoski (1995) presents a correlation based on flame height rather than heat release rate). In the above equations, 'the height above the point source' should be replaced by $(z - z_0)$ where z remains the height above the fuel surface.

By implication, the above discussion refers to flat fuel beds (e.g. pool fires, gas burners, etc., on which the correlations are based), and Equation (4.26) will not apply if the 'fire' is three-dimensional, in that the fuel bed has a significant vertical extension (e.g. with wood cribs and fires in multi-tier storage arrays). This is discussed by You and Kung (1985) who quote correlations for two-, three- and four-tier storage (see also Heskestad, 1995).

The proportionalities indicated in Equations (4.17)–(4.19) provide the basis for the scaling laws which can be used to correlate data and compare behaviour in situations according to the principles of similarity (see Section 4.4.4). As a simple example, consider the temperature at a height H_1 directly above a source of convective heat output $\dot{Q}_{c1}$. The same temperature will exist at a height H_2 on the centreline of the buoyant plume from a similar source of heat output $\dot{Q}_{c2}$, provided that

$$\dot{Q}_{c2} = \dot{Q}_{c1} \left(\frac{H_2'}{H_1'} \right)^{5/2} \tag{4.27}$$

where H_1' and H_2' refer to the heights above the respective virtual origins (Figure 4.10(b)). In this way, the product $\dot{Q}_{conv}^{2/3} z^{-5/3}$ in Equation (4.19) ($z = H'$) is constant. Specific examples of the application of this type of analysis are given

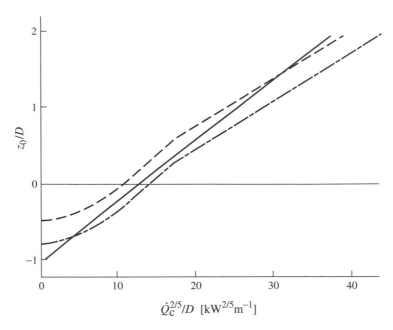

Figure 4.11   Correlations for the virtual origin. Solid line: Heskestad (1986) (Equation (4.26)); dashed line and dotted line: Cetegen *et al.* (1984) with and without a flush floor, respectively. (After Heskestad, 1995 by permission of the Society of Fire Protection Engineers.)

in Sections 4.3.4 and 4.4.3. In principle, a similar relationship should hold for the concentration of smoke particles. Heskestad (1972) quotes

$$C_0 \propto A^{-1/3} \dot{m} \dot{Q}_{conv}^{-1/3} z^{-5/3} \tag{4.28}$$

where C_0 is the centreline concentration of combustion products and $\dot{m}$ is the rate of burning, expressed as a mass flow. However, as $\dot{Q}_{conv} \propto \dot{m}$ for a given fuel, Equation (4.28) can be rewritten

$$C_0 \propto A^{-1/3} \dot{Q}_{conv}^{2/3} z^{-5/3} \tag{4.29}$$

showing that the concentration of smoke follows ΔT (compare Equations (4.19) and (4.29)), i.e. if the term $\dot{Q}_{conv}^{2/3} H^{-5/3}$ is maintained constant, the concentration of smoke particles will be the same (for a given fuel bed). This is relevant to the operation of smoke detectors in geometrically similar locations of different heights.

This section has concentrated on axisymmetric plumes from square or circular sources, but other geometries are encountered. Yokoi (1960) and Hasemi (1988) have considered the plumes arising from sources which have length/breadth ratios significantly greater than one (see Section 4.3.2). In general, in the far field, rectangular sources can be approximated by a virtual point source, but the extreme situation — the 'line source' — has relevance to certain problems. For these, Zukoski (1995) has shown how the heat release rate can be specified per unit length and has derived a set of equations similar to (4.21)–(4.23) using a modified version of the heat release group $\dot{Q}_l^*$:

$$\dot{Q}_l^* = \frac{\dot{Q}_c/L}{\rho_\infty c_p T_\infty (gZ)^{1/2} Z} \qquad (4.30)$$

where $\dot{Q}_c/L$ is the rate of heat release per unit length of source. There are few data available to enable correlations to be tested. An example of where the line plume is important is in the spill plume which emerges from a shop at ground level in an atrium or multi-storey shopping mall. The volume of smoke that has to be managed by the smoke control system is determined by the rate of entrainment into the plume: it is important to be able to calculate this in the design of a smoke control system.

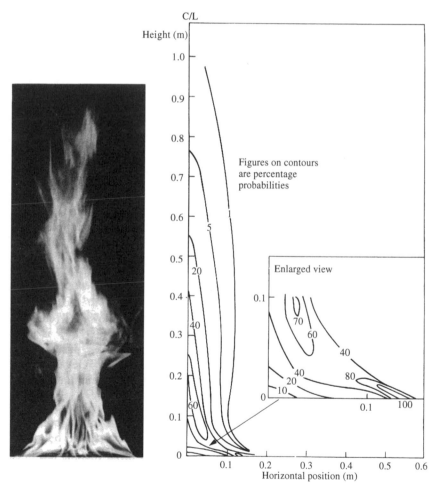

Figure 4.12 Intensity of combustion within a buoyant diffusion flame, shown as probability contours and compared with a typical instantaneous photograph of the flame. 0.3 m square porous burner, $\dot{Q}_c = 47$ kW. Visual flame height. 1.0–1.2 m (Chitty and Cox, 1979). (Reproduced by permission of The Controller, HMSO. © Crown copyright)

4.3.2   The fire plume

The subdivision of the fire plume into three regions was discussed briefly in the introduction to Section 4.3. Flame exists in the near field and the intermittent zone, although it is persistent only in the former. This is illustrated by results of Chitty and Cox (1979) who mapped out regimes of 'equal combustion intensity' throughout a methane diffusion flame above a 0.3 m square porous burner. Using an electrostatic probe, they determined the fraction of time that flame was present at different locations within the fire plume and found that the most intense combustion (defined as flame being present for more than 50% of the time) occurs in the lower region, particularly near the edge of the burner (Figure 4.12). Bouhafid *et al.* (1988) report contours of temperature and concentrations of CO, CO_2 and O_2 near the base of the flames above a kerosene pool fire, 0.15 m in diameter, which show similarities to Chitty and Cox's map of combustion intensity (Figure 4.12). The low probability recorded immediately above the central area of the burner, or pool, is consistent with the presence of a cool fuel-rich zone above the fuel surface: see Figure 4.8 (Corlett, 1974).

Visual estimates of average flame height are 10–15% greater than the vertical distance on the flame axis to the point where flame intermittency is 50%, as determined photographically (Zukoski *et al.*, 1981a, b). The motion of the intermittent (oscillating) flames occupies a considerable proportion of the fire plume (Figures 4.13 and 4.14) and is quite regular, exhibiting a frequency (f) which is a function of $D^{-1/2}$, where

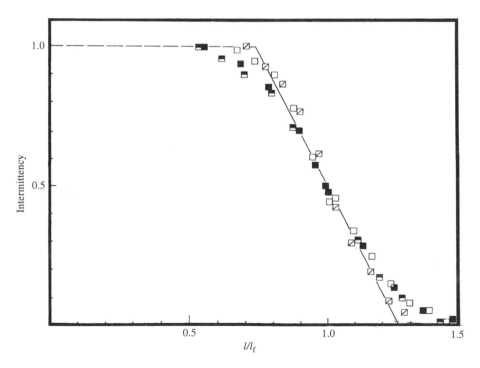

Figure 4.13   Intermittency of a buoyant diffusion flame on the axis of a 0.19 m porous burner. ■, $\dot{Q}_c = 21.1$ kW, $l_f = 0.65$ m; ▰, $\dot{Q}_c = 42.2$ kW; $l_f = 0.90$ m; □, $\dot{Q}_c = 63.3$ kW, $l_f = 1.05$ m; ▨, $Q_c = 84.4$ kW, $l_f = 1.16$ m (Zukoski *et al.*, 1981b)

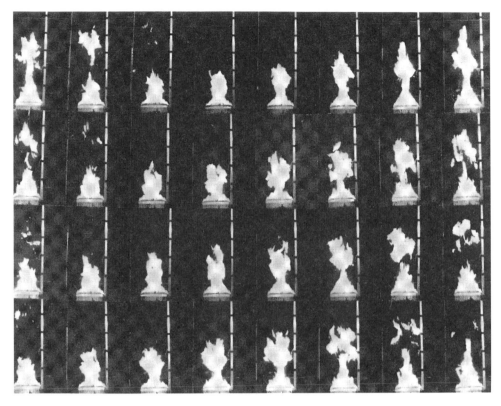

Figure 4.14 Intermittency of a buoyant diffusion flame burning on a 0.3 m porous burner. The sequence represents 1.3 s of cine film, showing 3 Hz oscillation (McCaffrey, 1979)

D is the fire diameter (Figure 4.15). Zukoski (1995) suggests

$$f = (0.50 \pm 0.04)(g/D)^{1/2} \text{ Hz} \qquad (4.31)$$

which is essentially in agreement with observations made by Pagni (1990) and Hamins *et al.* (1992). The phenomenon is illustrated in Figure 4.14, which shows 1.3 s of a cine film of the flame on the 0.3 m square gas burner used by McCaffrey (1979) and Chitty and Cox (1979). The oscillation frequency is 3 Hz, similar to that observed by Rasbash *et al.* (1956) for a 0.3 m diameter petrol fire. The oscillations are generated by instabilities at the boundary layer between the fire plume and the surrounding air although they have their origins low in the flame, close to the surface of the fuel (Weckman and Sobiesiak, 1988). These give rise to disturbances, the largest taking the form of axisymmetric vortex-like structures (Figure 4.16). Zukoski *et al.* (1981a, b) have suggested that these play a significant part in determining the rate of air entrainment into the flame. The observed oscillations are a result of these structures rising upwards through the fire plume and burning out, thus exposing the upper boundary of the next vortex structure which becomes the new flame tip. This behaviour produces the 'flicker' which is characteristic of the diffusion flames of small fires and which

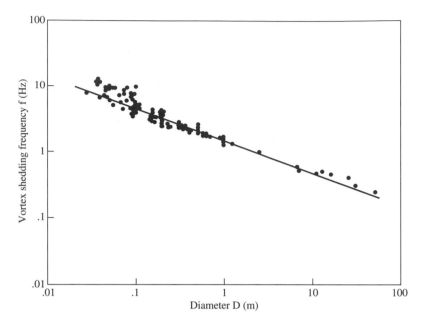

Figure 4.15 Variation of flame oscillation frequency with fire diameter for pool and gas burner fires (Pagni, 1990). (From Cox, 1995, with permission)

Figure 4.16 Schematic diagram of the axisymmetric vortex-like structures in the buoyant diffusion flame (after Zukoski *et al.*, 1981a, by permission)

may be used to distinguish infra-red emission from a flame and that from a steady background source (Bryan, 1974; Middleton, 1983).

It is sometimes necessary to know the size of a flame above a burning fuel bed, as this will determine how the flame will interact with its surroundings, in particular whether it will reach the ceiling of a compartment or provide sufficient radiant heat to ignite nearby combustible items. The basic parameters which determine height were first derived by Thomas *et al.* (1961) who applied dimensional analysis to the problem of the free-burning fire, i.e. one in which 'the pyrolysis rate and energy release rate are affected only by the burning of the fuel itself and not by the room environment' (Walton and Thomas, 1995). They assumed that buoyancy was the driving force and that air for combustion of the fuel volatiles was entrained through the flame envelope. The tip of the flame was defined as the height at which sufficient air had entered the flame to burn the volatiles, and the following functional relationship was derived:

$$\frac{l}{D} = f\left(\frac{\dot{m}^2}{\rho^2 g D^5 \beta \Delta T}\right) \tag{4.32}$$

in which l is the flame height above the fuel surface, D is the diameter of the fuel bed, $\dot{m}$ and ρ are the mass flowrate and density of the fuel vapour, ΔT is the average excess temperature of the flame and g and β are the acceleration due to gravity and the expansion coefficient of air, respectively. The group $g\rho\Delta T$ is indicative of the importance of buoyancy, which is introduced into the analysis in terms of the Grashof number (Equation (2.49)). The dimensionless group in Equation (4.32) is a Froude number and may be compared with $\dot{Q}^*$ (Equation (4.3)). It contains the elements of Froude modelling (Section 4.4.4), in which the rate of heat release must scale with $D^{5/2}$.

Data on flame heights can be correlated by using either $(\dot{Q}_c/D^{5/2})$, which has dimensions kW/m$^{5/2}$, or the dimensionless group $\dot{Q}^*$, defined above (Equation (4.3)). An example is shown in Figure 4.17 in which the data obtained by Zukoski *et al.* (1981a) with three gas burners of different diameters are compared with data (Terai and Nitta, 1975; McCaffrey, 1979) and correlations of others (Steward, 1970; Thomas *et al.*, 1961; You and Faeth, 1979). The flame heights (l) are normalized against the diameter of the fuel bed, or burner, and $\log(l/D)$ plotted against $\log(\dot{Q}_c/D^{5/2})^*$. Figure 4.17 also shows $\log(l/D)$ as a function of $\log(\dot{Q}^*)$. For large values of l/D (>6), the slope of the line is 2/5, indicating that the flame height is virtually independent of the diameter of the burner, or fuel bed, i.e.

$$\frac{l}{D} \propto \left(\frac{\dot{Q}_c}{D^{5/2}}\right)^{2/5} \propto \frac{\dot{Q}_c^{2/5}}{D} \tag{4.33}$$

giving (from the data on visible flame heights of Zukoski *et al.* (1981a)):

$$l = 0.23\dot{Q}_c^{2/5} \text{ m} \tag{4.34}$$

if $\dot{Q}_c$ is in kW. This corresponds to values of $\dot{Q}^*$ greater than c.5.

However, Thomas *et al.* (1961) found their data for wood crib fires to give values of l/D between 3 and 10 which correlated as follows:

$$\frac{l}{D} \propto \left(\frac{\dot{Q}_c}{D^{5/2}}\right)^{0.61} \tag{4.35}$$

or

$$l \propto \frac{\dot{Q}_c^{0.61}}{D^{0.5}} \tag{4.36}$$

---

* Logically, $\dot{Q}_{conv}$ should be used in these correlations, but it is impossible to make reliable corrections for radiative losses.

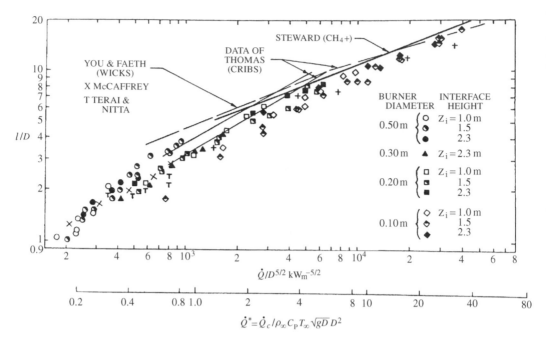

Figure 4.17 Dependence of flame height on heat release parameters (Zukoski, *et al.* 1981a).
'Interface height' refers to the vertical distance from the fire source to the lower
boundary of the ceiling layer. Additional data of Thomas *et al.*, (1961), Steward
(1970), You and Faeth (1979), McCaffrey (1979) and Terai and Nitta (1975). By
permission

corresponding approximately to a two-thirds power law in the range $0.5 < \dot{Q}^* < 7$.
However, for $l/D < 2$, the relationship between l/D and $(Q_c/D^{5/2})$ appeared to be
almost linear ($\dot{Q}^* < 0.5$). There is now strong evidence that the slope is changing
rapidly in this range of values of $\dot{Q}^*$, and a square law is more appropriate when $\dot{Q}^*$
falls below 0.2 (Figure 4.18) (Zukoski, 1985):

$$\frac{l}{D} \propto \left(\frac{\dot{Q}_c}{D^{5/2}} \right)^2 \propto \dot{Q}^{*2} \tag{4.37}$$

Zukoski (1985) draws attention to this fact, and identifies several regimes in the relationship of l/D to $\dot{Q}^*$ in which different power laws apply. These are summarized in
Figure 4.1. Most of the fires of interest in the context of buildings are identified with
$\dot{Q}^* < 5$, but in the chemical and process industries, a much wider range is encountered,
from large-scale pool fires ($\dot{Q}^* < 1$) to fully turbulent jet fires with $\dot{Q}^* > 10^3$.

In general, the above is in essential agreement with results of McCaffrey (1979)
and Thomas *et al.* (1961). Steward (1970) obtained a substantial amount of data on
turbulent diffusion flames and carried out a fundamental analysis of the flame structure
based on the conservation equations. One interesting conclusion that he derived from
the study was that within its height the momentum jet diffusion flame entrains a much
greater quantity of air (400% excess) than is required to burn the fuel gases. For the
buoyant diffusion flame, Heskestad (1983) has correlated data from a wide variety of

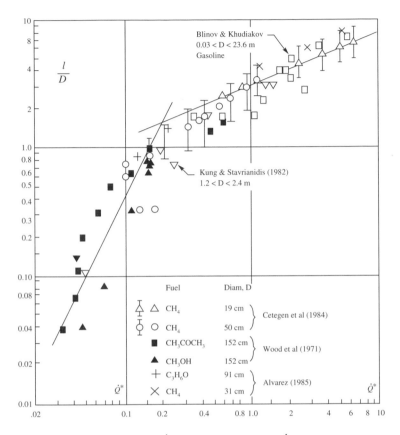

Figure 4.18   Correlation of l/D with $\dot{Q}^*$ for small values of $\dot{Q}^*$. A range of fuels are represented here, and include methane (Cetegen *et al.*, 1984), methanol and acetone
(Wood *et al.*, (1971)) and gasoline (Blinov and Khudiakov, (1957)). (Adapted
from Zukoski, (1986) by permission)

sources, including pool fires (Section 5.1), using the equation

$$\frac{l}{D} = 15.6N^{1/5} - 1.02 \tag{4.38}$$

in which the non-dimensional number N is derived from a modified Froude number
(Heskestad, 1981) and is given by:

$$N = \left(\frac{c_p T_\infty}{g\rho_\infty^2 (\Delta H_c/r)^3} \right) \frac{\dot{Q}_c^2}{D^5} \tag{4.39}$$

where c_p is the specific heat of air, ρ_∞ and T_∞ are the ambient air density and
temperature, respectively, ΔH_c is the heat of combustion, and r is the stoichiometric
ratio of air to volatiles. Given that most of the terms in Equation (4.39) are known

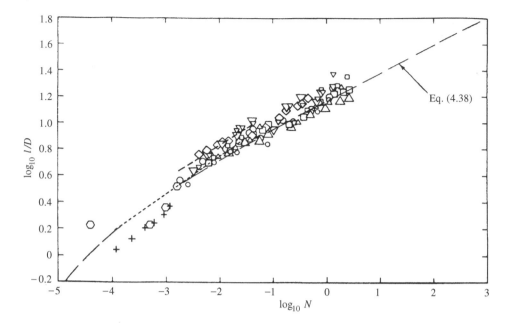

Figure 4.19   Correlation of flame height data from measurements by Vienneau (1964). ($\circ$, methane; $\circ$, methane + nitrogen; $\triangledown$, ethylene; $\triangledown$, ethylene + nitrogen; $\square$, propane; $\square$, propane + nitrogen; $\triangle$, butane; $\triangle$, butane + nitrogen; $\diamond$ hydrogen): D'Sousa and McGuire (1977) (- - - - -, natural gas); Blinov and Khudiakov (1957) ($\bigcirc$, gasoline); Hägglund and Persson (1976b) (+, JP-4 fuel); and Block (1970) ($- - - -$, Equation (4.38)). (From Heskestad (1983) by permission)

($\Delta H_c/r \approx 3000$ kJ/kg, see Section 1.2.3), Equation (4.38) can be rewritten:

$$l = 0.23\dot{Q}_c^{2/5} - 1.02\ D \tag{4.40}$$

$\dot{Q}_c$ in kW, and l and D in m. The correlation is very satisfactory (Figure 4.19), although it has not been tested outside the range $7 < \dot{Q}_c^{2/5}/D < 700$ kW$^{2/5}$/m. It can be expressed in terms of $\dot{Q}^*$ as follows ($0.12 < \dot{Q}^* < 1.2 \times 10^4$) (McCaffrey, 1995):

$$l/D = 3.7\dot{Q}^{*2/5} - 1.02 \tag{4.41}$$

It captures the change of slope which occurs around $\dot{Q}^* = 1$.

This correlation breaks down for flames with $l/D < 1$, corresponding to values of $\dot{Q}^* < 0.2$. In very low Froude number fires, such as large mass fires (e.g. Figure 4.8, $D > 100$ m (Corlett, 1974)), the flame envelope breaks up and a number of separate, distinct 'flamelets' are formed. The heights of these flames are much less than the fuel bed diameter (Zukoski, 1995).

Orloff and de Ris (1982) carried out a very detailed study of the radiation characteristics of flames above porous burners 0.1–0.7 m in diameter. They used the radiation measurements to define the outline of the flame which allowed them to calculate the flame volume (V_f). This proved to be directly proportional to the rate of heat release in the range studied (25–250 kW), for two gaseous fuels (methane and propene) and polymethylmethacrylate, yielding the relationship:

$$\dot{Q}''' = \frac{\dot{Q}_c}{V_f} = 1200 \text{ kW/m}^3 \text{ or } 1.2 \text{ MW/m}^3$$

where $\dot{Q}'''$ is the 'power density' of the flame. These authors draw attention to the fact that the principles of Froude modelling predict that $\dot{Q}'''$ should be proportional to $\dot{Q}_c^{-1/5}$, but this weak dependence does not reveal itself over this range of data. However, Cox (1995) quotes $\dot{Q}''' = 0.5$ MW/m^3, which may be more consistent with fires two orders of magnitude greater than those studied by Orloff and de Ris (1982).

Average temperatures and gas velocities on the centreline of axisymmetric buoyant diffusion flames have been measured by McCaffrey (1979) for methane burning on a 0.3 m square porous burner. The results clearly delineated the three regions of the fire plume, for each of which there were identifiable correlations between temperature (expressed as $2g\Delta T/T_\infty$), gas velocity (normalized as $u_0/\dot{Q}_c^{1/5}$) and $(z/\dot{Q}_c^{2/5})$, where z is the height above the burner surface. These are summarized in Table 4.2 and Figures 4.9 and 4.20. It can be seen that the average temperature is approximately constant in the upper part of the near field (persistent flaming) ($\Delta T = 800°C$ in these flames), but falls in the region of intermittent flaming to ~320°C at the boundary of the buoyant plume. Thus, one would expect the temperature at the average flame height

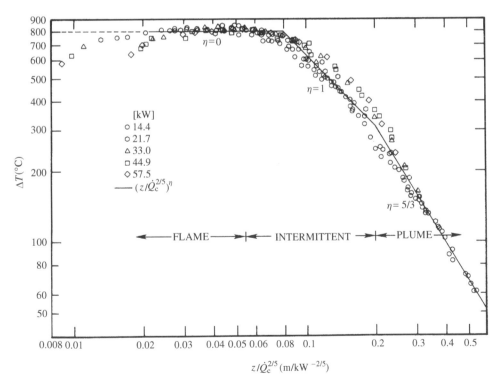

Figure 4.20 Variation of centreline temperature rise with height in a buoyant methane diffusion flame. Scales as $z/\dot{Q}_c^{2/5}$ (Table 4.2) (McCaffrey, 1979, by permission)

as defined by Zukoski *et al.*, (1981a,b) to lie in the region of 500–600°C. In fact, a temperature of 550°C is sometimes used to define maximum vertical reach, e.g. of flames emerging from the window of a flashed-over compartment (Bullen and Thomas, 1979) (see Section 10.2).

The average centreline velocity within the near field is independent of fire size ($\dot{Q}_c$) but increases as $z^{1/2}$ to a maximum velocity which is independent of z in the intermittent region (Table 4.2). McCaffrey (1979) found that this maximum was directly proportional to $\dot{Q}_c^{1/5}$, an observation which is significant in understanding the interaction between sprinklers and fire plumes. If the fire is too large ('strong source'), the downward momentum of the spray or the terminal velocity of the droplets may be insufficient to overcome the updraft and water will not penetrate to the fuel bed. This is discussed further in Section 4.4.3.

Table 4.2 Summary of centreline data for a buoyant methane diffusion flame on a 0.3 m square porous burner (McCaffrey, 1979) (Figure 4.9) These refer to values of $\dot{Q}^*$ in the range 0.25–1.0

Centreline velocity:
$$\frac{u_0}{\dot{Q}^{1/5}} = k\left(\frac{z}{\dot{Q}^{2/5}}\right)^{\eta}$$

Centreline temperature:
$$\frac{2g\Delta T_0}{T_0} = \left(\frac{k}{C}\right)^2 \left(\frac{z}{\dot{Q}^{2/5}}\right)^{2\eta-1}$$

Region[a]	k	η	$z/\dot{Q}^{2/5}$ $(m/kW^{2/5})$	C
Flame	6.8 m$^{1/2}$/s	1/2	<0.08	0.9
Intermittent	1.9 m/kW$^{1/5}$.s	0	0.08–0.2	0.9
Plume	1.1 m$^{4/3}$/kW$^{1/3}$.s	−1/3	>0.2	0.9

[a] See Figure 4.9(a).

The above discussion refers to axisymmetric fire plumes, where the burner/burning surface is square, or circular. There are very limited data on the behaviour of flames from surfaces of other shapes, for example rectangular sources with one side significantly longer than the other. This has been studied by Hasemi and Nishihata (1989) who found that the flame height data could be correlated with a modified $\dot{Q}^*$, given by:

$$\dot{Q}^*_{\text{mod}} = \frac{\dot{Q}_c}{\rho_0 c_p T_0 g^{1/2} A^{3/2} B} \tag{4.42}$$

where A and B are the lengths of the shorter and the longer sides of the rectangular fuel bed, respectively. When $A = B$, this becomes identical to the original definition of $\dot{Q}^*$ (Equation (4.3)), while for the line fire ($B \rightarrow \infty$), expressing the rate of heat release in terms of unit length of burner (or fire) gives:

$$\dot{Q}^*_l = \frac{\dot{Q}_l}{\rho_0 c_p T_0 g^{1/2} A^{3/2}} \tag{4.43}$$

where $\dot{Q}_l$ is the rate of heat per unit length (kW/m). Hasemi and Nishihata's flame height correlation is shown in Figure 4.21.

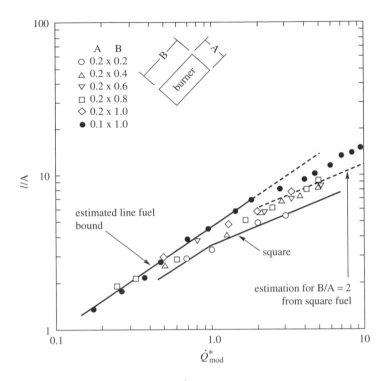

Figure 4.21 Relationship between l/A and $\dot{Q}^*_{\text{mod}}$ (Equation (4.42)), from Hasemi and Nishihata (1989). Reprinted by permission

4.3.3 Upward flows

The vertical movement of the buoyant gases in the fire plume causes air to be entrained from the surrounding atmosphere (see Equation (4.13) *et seq.*). Not only does this provide air for combustion of the fuel vapours, but it dilutes and cools the fire products as they rise above the flame into the far field, causing a progressive increase in the volume of 'smoke' generated by the fire. In the open (and in the early stages of a fire in a compartment), this will be clear air at normal temperatures: the amount entrained will quickly dominate the upward flow, even below the maximum height of the flame (Heskestad, 1986). It is necessary to estimate this flow to be able to calculate the rate of accumulation of smoke under a ceiling, or the extraction rate that will be required to maintain the smoke layer at or above a certain critical level (see Chapter 11).

Following Heskestad (1995), the upward mass flow at any level in a weak plume (i.e. $\Delta T_0/T_\infty \ll 1$) may be written:

$$\dot{m}_{\text{ent}} = E'\rho_\infty u_0 b^2 \qquad (4.44)$$

where u_0 is the centreline velocity, b is the radius of the plume where $u = 0.5u_0$, and E' is a proportionality constant. Using Equations (4.21) and (4.22) for b_u and u_0, this becomes for height z:

$$\dot{m}_{\text{ent}} = E\rho_\infty z^2 \sqrt{gz}\dot{Q}^{*1/3}_z \qquad (4.45)$$

or

$$\dot{m}_{ent} = E \left(\frac{g\rho_\infty^2}{c_p T_\infty} \right)^{1/3} \dot{Q}_c^{1/3} z^{5/3} \tag{4.46}$$

where $E = E'C_vC_l^2$, and z is the height above the (virtual) source. Yih (1952) deduced a value of $E = 0.153$ from measurements of the flow above a point source, although subsequently, Cetegen *et al.* (1984) found $E = 0.21$ to give good agreement with a range of experimental data on the weak plume. They also concluded that it gave a reasonable approximation to the flow in the strongly buoyant region above the flame tip, provided that z was the height above the virtual origin. This was based on results of experiments in which natural gas was burned under a hood from which the fire products were extracted at a rate sufficient to maintain the smoke layer at a constant level. The mass flow into the layer could then be equated to the extract rate. By varying the distance between the burner and the hood, and the rate of burning of fuel, data were gathered on the mass flow as a function of $\dot{Q}^*$ and z. Heskestad (1986) converted Equation (4.46) (with $E = 0.21$) to

$$\dot{m} = 0.076\dot{Q}_c^{1/3} z^{5/3} \text{ kg/s} \tag{4.47}$$

for standard conditions (293 K, 101.3 kPa)*, but noted that their analysis was based on similarity between excess temperature and upward velocity. He showed that improved agreement was obtained if the analysis was based on similarity between upward velocity and density deficit, which gave:

$$\dot{m} = 0.071\dot{Q}_c^{1/3} z^{5/3}[1 + 0.026\dot{Q}_c^{2/3} z^{-5/3}] \text{ kg/s} \tag{4.48}$$

Figure 4.22 compares these two interpretations of the plume mass flow results of Cetegen *et al.* (1984). Although there is a significant scatter, Equation (4.48) appears to give better agreement with the experimental data.

The results against which the above equations have been tested were obtained in a carefully controlled, draught-free environment. The amount entrained is influenced significantly by any air movement, created artificially by, for example, an air conditioning system (Zukoski *et al.*, 1981a), or naturally, if the fire is burning in a confined space and induces a directional flow from an open door, etc. (Quintiere *et al.*, 1981). If a fire is burning against a wall, or in a corner, the entrainment is also affected (see below).

4.3.4   Interaction of the fire plume with compartment boundaries

With the unconfined axisymmetric plume, there are no physical barriers to limit vertical movement or restrict air entrainment across the plume boundary, but in a confined space the fire plume can be influenced by surrounding surfaces. Thus, if an item is burning against a wall, the area through which air may be entrained is reduced (Figure 4.23); similarly, if the fire plume impinges on a ceiling, it will be deflected

---

*In a comprehensive review, Beyler (1986b) noted that the constant derived from Zukoski's work (0.076) was not found by all investigators. Values range from 0.066 (Ricou and Spalding, 1961) to 0.138 (Hasemi and Tokunaga, 1984).

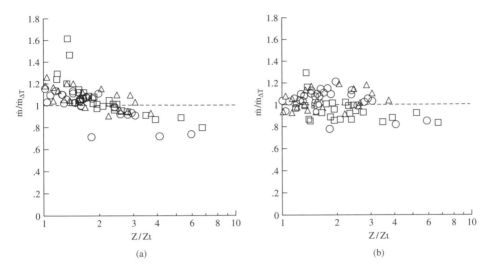

Figure 4.22 Plume mass flows above flames measured by Cetegen *et al.* (1984): (a) according to Equation (4.47), assuming similarity between ΔT and upward velocity; (b) according to Equation (4.48), assuming similarity between $\Delta \rho$ and upward velocity. In (a), Cetegen's formula for the virtual origin is used; in (b), Heskestad's (Equation (4.26). (Heskestad (1986), by permission of the Combustion Institute)

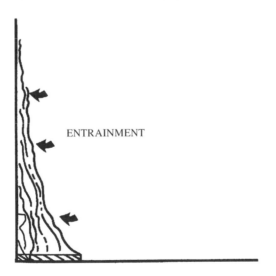

Figure 4.23 Flame deflection near a vertical surface

horizontally to form a ceiling jet, again with restricted entrainment (Figure 4.24). The consequences regarding flame height (or length) and plume temperatures need to be examined. However, additional effects must be considered, the most important relating to heat transfer to the surfaces involved and how quickly these surfaces (if combustible) will ignite and contribute to the fire growth process if given such exposure to flame. This is a topic which is directly relevant to our understanding of fire development in a room (Chapter 9).

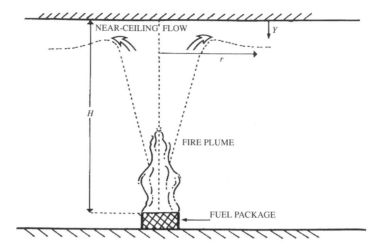

Figure 4.24   The fire plume and its interaction with a ceiling (after Alpert, 1972)

(a) *Walls*

If the fire is close to a wall, or in a corner formed by the intersection of two walls, the resulting restriction on free air entrainment will have a significant effect (Figure 4.25). In a buoyant plume, temperature decreases less rapidly with height as the rate of mixing with ambient air will be less than for the unbounded case (Hasemi and Tokunaga, 1984). Surprisingly, relatively few measurements have been made on the effect on flame height. The convention used in the past was to assume that a fire burning against a wall could be modelled as if there were an identical 'imaginary' fire source mirrored on the other side of the wall (Figure 4.26). The flame height was then calculated as if the real and imaginary fires were burning unhindered in the open, producing a combined flame of height significantly greater than would be predicted if the 'real' fire were burning in the open. However, Hasemi and Tokunaga (1984) found evidence that this was not the case for experimental gas fires ($0.4 < Q^* < 2.0$), and that the flame height was similar to that predicted as if the 'real' fire were burning in the open, despite a significant reduction (c.40%) in the amount of air entrained (Zukoski *et al.*, 1981a). This has been confirmed by recent results of Back *et al.* (1994). On the other hand, if the fire is in a corner, the flame height is increased by about 20% ($0.6 < \dot{Q}^* < 4.0$).

It is interesting to note that for the fire both beside a wall and in a corner, the maximum temperatures as a function of height were estimated more closely by the 'imaginary fire source model' than by assuming free burning of the 'real fire' in the open. This suggests that the turbulent structure of the fire plume is altered, enhancing burnout of the fuel vapours. When entrainment is not axisymmetric, the fire plume will be deflected towards the restricting wall(s) as a result of the directional momentum of the inflowing air (Figure 4.23). This effect will enhance upward flame spread on sloping and vertical combustible surfaces (Section 7.2.1) as well as encouraging fire spread to vertical surfaces from adjacent burning items (Section 9.2.4). This is a consequence of high rates of heat transfer which, in some cases, may exceed 100 kW/m² , depending on the exact circumstances (Back *et al.*, 1994; Foley and Drysdale, 1995).

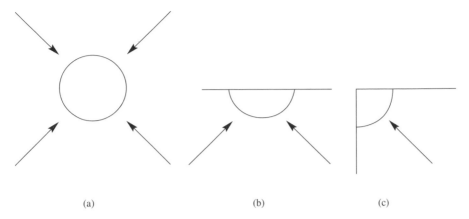

(a)   (b)   (c)

Figure 4.25   Plan view of a fire (a) free burning; (b) burning against a wall; and (c) burning in a corner. The arrows signify the direction of air entrainment into the flame

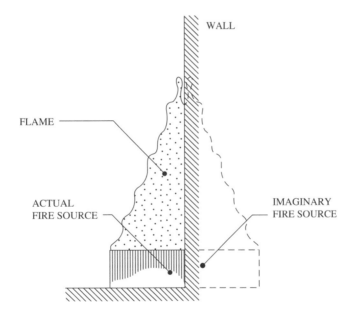

Figure 4.26   Concept of the 'imaginary fire source' (after Hasemi and Tokunaga, 1984)

(b) *Ceilings*

If the vertical extent of a fire plume is confined by a ceiling, the hot gases will be deflected as a horizontal ceiling jet, defined by Evans (1995) as 'the relatively rapid gas flow in a shallow layer beneath the ceiling surface which is driven by the buoyancy of the hot combustion products'. It spreads radially from the point of impingement and provides the mechanism by which combustion products are carried to ceiling-mounted fire detectors. To enable the response of heat and smoke detectors to be analysed, the rate of development and the properties of the ceiling jet must be known. Although the time lag which is associated with the response of any detector can be considered in terms of a transport time lag and a delay to detector operation (Newman, 1988; Mowrer, 1990), it is convenient to consider first the steady-state fire and the resulting temperature distribution under the ceiling. Alpert (1972) has provided correlations based on a series of large-scale tests, carried out at the Factory Mutual Test Centre, in which a number of substantial fires were burned below flat ceilings of various heights, H (Table 4.3). Temperatures were measured at different locations under the ceiling (Figure 4.24): it was found that at any radial distance (r) from the plume axis, the vertical temperature distribution exhibited a maximum (T_{max}) close to the ceiling, at $Y \not> 0.01H$ (see Figure 4.24). Below this, the temperature fell rapidly to ambient (T_∞) for $Y \not> 0.125H$. These figures are valid only if horizontal travel is unconfined and a static layer of hot gases does not accumulate beneath the ceiling. This will be achieved to a first approximation if the fire is at least $3H$ distant from the nearest vertical obstruction; however, if confinement occurs by virtue of the fire being close to a wall, or in a corner, the horizontal extent of the free ceiling from the point of impingement would presumably have to be much greater for this condition to hold.

Alpert (1972) showed that the maximum gas temperature (T_{max}) near the ceiling at a given radial position r could be related to the rate of heat release ($\dot{Q}_c$ kW) by the steady-state equations:

$$T_{max} - T_\infty = \frac{5.38(\dot{Q}_c/r)^{2/3}}{H} \tag{4.49}$$

if $r > 0.18\,H$, while for $r \leq 0.18\,H$ (i.e. within the area where the plume impinges on the ceiling),

$$T_{max} - T_\infty = \frac{16.9\dot{Q}_c^{2/3}}{H^{5/3}} \tag{4.50}$$

where $\dot{Q}_c$ (kW) is the rate of heat release. Inspection of Equations (4.49) and (4.50) reveals a lower rate of entrainment into the ceiling jet than into the vertical fire plume: in the latter, the temperature varies as $H^{-5/3}$, while in the horizontal ceiling jet it varies as $r^{-2/3}$. This is consistent with the knowledge that mixing between a hot layer

Table 4.3   Summary of the fire tests from which Equations (4.49) and (4.50) were derived (Alpert, 1972)

Fuel	Fuel array size (m)	Fire intensity (MW)	Ceiling height (m)
Heptane spray	3.7 m diameter	7.0–22.8	4.6–7.9
Heptane pan[a]	0.6 × 0.6	1.0	7.6
Ethanol pan	1.0 × 1.0	0.67	8.5
Wood pallets[a]	1.2 × 1.2 × 1.5	4.9	6.1
Cardboard boxes	2.4 × 2.4 × 4.6	3.9	13.7
Polystyrene in cardboard boxes	2.4 × 2.4 × 4.6	98	13.7
PVC in cardboard boxes	2.4 × 2.4 × 4.6	35	13.4
Polyethylene pallets	1.2 × 1.2 × 2.7	4.2–11.4	15.5

[a] Located in a 90° corner.

moving on top of a cooler fluid is relatively inefficient. The process is controlled by the Richardson number (Zukoski, 1995), which is the ratio of the buoyancy force acting on the layer to the dynamic pressure of the flow. i.e.

$$\mathrm{Ri} = \frac{g(\rho_0 - \rho_{\text{layer}})h}{\rho_{\text{layer}}V^2}$$

where ρ_{layer} and h are the density and the depth of the layer, respectively, and V is its velocity with respect to the ambient layer below. Mixing is suppressed at high values of Ri. (A detailed analysis of the turbulent ceiling jet has been developed by Alpert (1975).)

If the fire is by a wall, or in a corner, the temperature will be greater, due not only to the lower rate of entrainment into the vertical plume, but also to the restriction under the ceiling where the flow is no longer radial. This can be accounted for in Equations (4.49) and (4.50) by multiplying $\dot{Q}_c$ by a factor of two or four, respectively. The dependence of T_{max} on r and H for a 20 MW fire according to these equations is shown in Figure 4.27. Such information may be used to assess the response of heat detectors to steady burning or slowly developing fires (Section 4.4.2).

If the ceiling is sufficiently low (or the fire sufficiently large) for direct flame impingement, that part of the flame that is deflected horizontally will become part

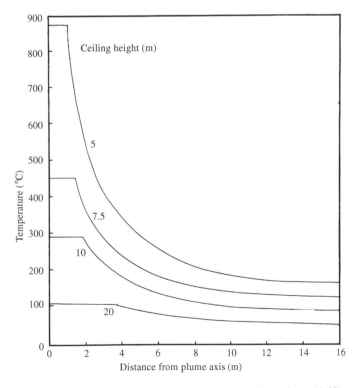

Figure 4.27 Gas temperatures near the ceiling according to Equations (4.49) and (4.50), for a large-scale fire ($\dot{Q}_c = 20$ MW) for different ceiling heights (see Figure 4.24). Note that the formulae are unlikely to apply for the 5 m ceiling because of flame impingement (after Alpert, 1972)

of the ceiling jet. When only the upper portion of a flame is deflected, it is 'fuel lean', in that the flow of burning gas contains excess air (Section 4.3.2), and the fact that the rate of air entrainment is now reduced is of little consequence. The total flame length (vertical distance plus horizontal deflection) for a flame impinging on an uncon-fined ceiling is reported by Gross (1989) to be less than the vertical flame height in the absence of the ceiling. This is consistent with some observations of Kokkala and Rinkenen (1987), but is relevant *only* to the fuel-lean situation. If the burning gases are fuel-rich, as will occur if the fire is large in relation to the height of the ceiling (e.g. see Figure 4.28), considerable flame extension can occur, depending on the configuration. Babrauskas (1980a) drew attention to this after reviewing the existing information on flames under non-combustible ceilings, suggesting that the length of the horizontal part of the flame (h_r) could be related to the 'cut-off height', h_c in Figure 4.29. On the basis of the amount of air that had to be entrained into the horizontal flame to complete the burning of the fuel gases, he argued that the ratio h_r/h_c depended strongly on the

Table 4.4 Flame extension under ceilings: $\dot{Q}_c = 0.5$ MW and $H = 2$ m (Babrauskas, 1980a,b) (Figure 4.29)

Configuration	h_r/h_c
Unrestricted plume, unbounded ceiling	1.5
Full plume, quarter ceiling[a]	3.0
Quarter plume, quarter ceiling	12
Corridor	Dependent on width

[a] Fire located in a corner but not close enough for flame attachment as in Figure 4.26.

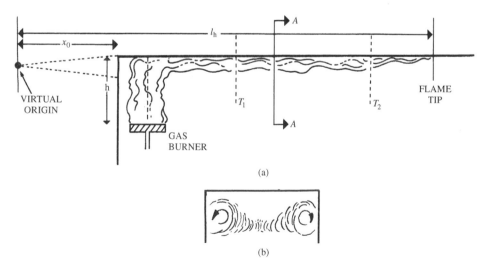

(a)

(b)

Figure 4.28 (a) Deflection of a flame beneath a model of a corridor ceiling (longitudinal section) showing the location of the 'virtual origin'. T_1 and T_2 identify the locations of the vertical temperature distributions shown in Figure 4.30. (b) Transverse section A-A. Not to scale (after Hinkley *et al.*, 1968). (Reproduced by permission of The Controller, HMSO. © Crown copyright)

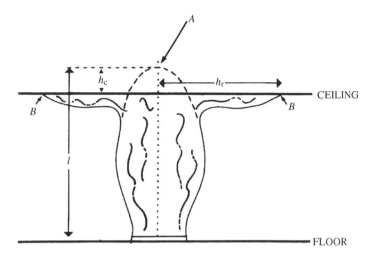

Figure 4.29 Deflection of a flame beneath a ceiling, illustrating Babrauskas' 'cut-off height' $h_c \cdot A$ = location of the flame tip in the absence of a ceiling; B = limit of the flame deflected under the ceiling (after Babrauskas, 1980a, by permission)

configuration (Table 4.4): the greatest extension will occur when a fire is confined in a corridor-like configuration (cf. Figure 4.28).

The phenomenon of flame extension was first investigated by Hinkley *et al.* (1968) who studied the deflection of diffusion flames, produced on a porous bed gas burner, by an inverted channel with its closed end located above the burner (Figure 4.28). The lining of the channel was incombustible. This situation models behaviour of flames under a corridor ceiling and is easier to examine than the unbounded ceiling. The appearance and behaviour of the flames were found to depend strongly on the height of the ceiling above the burner (h in Figure 4.28) and on the gas flowrate: thus if sufficient air was entrained in the vertical portion of the flame, the horizontal flame was of limited extent and burned close to the ceiling (cf. Gross, 1989). Alternatively, with high fuel flowrates or a low ceiling (small h), a burning, fuel-rich layer was found to extend towards the end of the channel with flaming occurring at the lower boundary. The transition between fuel-lean and fuel rich burning was related to a critical value of $(\dot{m}'/\rho_0 g^{1/2})d^{3/2} \approx 0.025$, where $\dot{m}'$ was taken as the rate of burning per unit width of channel (g/m.s) and d is the depth of the layer of hot gas below the ceiling (m). The difference between these two regimes of burning is illustrated clearly in Figure 4.30, which shows vertical temperature distributions below the ceiling at 2.0 m and 5.2 m from the closed end (Figure 4.28). This flame extension will be even greater if the lining material is combustible, as extra fuel vapour will be evolved from the linings and contribute to the flaming process (Hinkley and Wraight, 1969).

It should be noted that some of the processes involved here are common with the development of a layer of fuel-rich gases under the ceiling of a compartment (room) as a fire approaches flashover (Section 9.2.1). A smoke layer forms and descends as the 'reservoir' formed by the walls and the ceiling fills with hot smoke and combustion products. This limits the vertical height through which air can be entrained into the flames, and eventually the flames reaching the ceiling will be very 'fuel rich', with

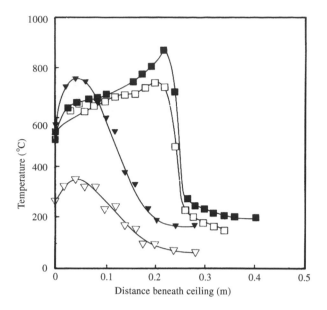

Figure 4.30   Vertical temperature distributions below a corridor ceiling for fuel-lean ($\blacktriangledown$, $\triangledown$) and fuel-rich ($\blacksquare$, $\square$) horizontal flame extensions. Closed and open symbols refer to points 2 m and 5 m from the axis of the vertical fire plume respectively (T_1 and T_2 in Figure 4.28) (Hinkley *et al.*, 1968). (Reproduced by permission of The Controller, HMSO. © Crown copyright)

relatively low levels of oxygen. This fuel-rich layer will eventually burn–a process which is associated with the flashover transition.

4.3.5   The effect of wind on the fire plume

In the open, flame will be deflected by any air movement, the extent of which will depend on the wind velocity. Studies on the effect of wind on LNG pool fires led Raj *et al.* (1979) to propose the relationship (based on earlier work by Thomas (1965))

$$\sin \theta = 1 \quad \text{for } V' < 1 \tag{4.51a}$$

and

$$\sin \theta = (V')^{-1/2} \quad \text{for } V' > 1 \tag{4.51b}$$

where θ is the angle of deflection (Figure 4.31(a)) and

$$V' = V \left(\frac{2c_\mathrm{p} T_\infty \rho_\infty}{\pi \rho_\mathrm{f} \Delta H_\mathrm{c}} \right)^{1/3} \tag{4.51c}$$

in which ρ_∞ and ρ_f are the densities of ambient air and the fuel vapour respectively and ΔH_c is the heat of combustion. V is a dimensionless windspeed, given by v/u^* where v is the actual windspeed, and u^* is a characteristic plume velocity (e.g. Equation (4.69)). A plot of θ versus V for methane (i.e. LNG) is shown in Figure 4.31(b).

On open sites, as encountered in petrochemical plants, deflection of flames by wind can create hazardous situations. In plant layout, this should be taken into account

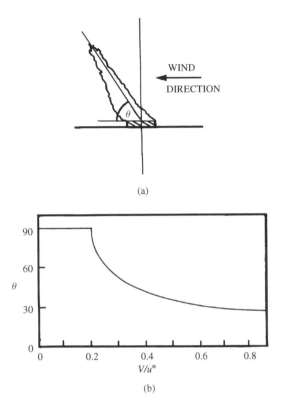

(a)

(b)

Figure 4.31   (a) Deflection of a flame by wind. (b) Relationship between θ and v/u according to Equations (4.51) for LNG (after Quintiere *et al.*, 1981)

when the consequences of fire incidents are being considered. A rule of thumb that is commonly used is that a 2 m/s wind will bend the flame by $\theta = 45°$, and for fires near the ground (e.g. bund fires) the flame will tend to hug the ground downwind of the fuel bed, to a distance of $\sim$0.5 D, where D is the fire diameter (Robertson, 1976; Lees, 1996). This can significantly increase the fire exposure of items downwind, either by causing direct flame impingement, or by increasing the levels of radiant heat flux (Pipkin and Sliepcevich, 1964).

Air movement tends to enhance the rate of entrainment of air into a fire plume. This is likely to promote combustion within the flame and thus reduce its length, although this has not been quantified. However, an investigation has been carried out on entrainment into flames within compartments during the early stages of fire development to determine the influence of the directional flow of air from the ventilation opening. Quintiere *et al.* (1981) have shown that the rate may be increased by a factor of two or three, which could have a significant effect on the rate of fire growth.

4.4   Some Practical Applications

Research in fire dynamics has provided concepts and techniques which may be used by the practising fire protection engineer to predict and quantify the likely effects of fire.

The results of such research have been drawn together as a series of state-of-the-art reviews in the *SFPE Handbook* (Di Nenno *et al.*, 1995) which currently represents the best available source of background information. A 'Guide to the application of fire safety engineering principles' has recently been issued as a Draft for Development by BSI (British Standards Institution, 1997), but in order to be able to translate 'research into practice', a sound understanding of the underlying science is essential. In this section, some of the information which has been presented above is drawn together to illustrate how the knowledge may be applied. However, it should be remembered that this is a developing subject and the reader is encouraged to be continuously on the lookout for new applications of the available knowledge, either to improve existing techniques or to develop new ones. This chapter finishes with a brief review of modelling techniques that are in current use.

4.4.1   Radiation from flames

It was shown in Sections 2.4.2 and 2.4.3 that the radiant heat flux received from a flame depends on a number of factors, including flame temperature and thickness, concentration of emitting species and the geometric relationship between the flame and the 'receiver'. While considerable progress is being made towards developing a reliable method for calculating flame radiation (Tien *et al.*, 1995), a high degree of accuracy is seldom required in 'real world' fire engineering problems, such as estimating what level of radiant flux an item of plant might receive from a nearby fire in order that a water spray system might be designed to keep the item cool (e.g. storage tanks in a petrochemical plant).

Mudan and Croce (1995) discuss the methods available for calculating the radiation levels from large pool fires in detail. Here, two approximate methods described by Lees (1996) are compared, identifying some of the associated problems. Both of these require a knowledge of the flame height (l), which may be obtained from Equation (4.40).

$$l = 0.23\dot{Q}_c^{2/5} - 1.02D \tag{4.40}$$

The total rate of heat release ($\dot{Q}_c$) may be calculated from:

$$\dot{Q}_c = \dot{m}'' \Delta H_c A_f \tag{4.52}$$

where A_f is the surface area of the fuel (m^2). As only a fraction (χ) of the heat of combustion is radiated, it is necessary to incorporate this into the equation, thus:

$$\dot{Q}_r = \chi \dot{m}'' \Delta H_c A_f \tag{4.53}$$

This fraction is sometimes assumed to be 0.3, but studies have shown that it varies not only with the fuel involved, but also with the tank diameter (see Table 4.5, from Mudan and Croce, 1995).

In the first method, it is assumed that $\dot{Q}_r$ originates from a point source on the flame axis at a height $0.5l$ above the fuel surface. The heat flux ($\dot{q}_r''$) at a distance R from the point source (P) is then:

$$\dot{q}_r'' = 0.3\dot{m}'' \Delta H_c A_f / 4\pi R^2 \tag{4.54}$$

Table 4.5 Radiation fraction for a number of hydrocarbon pool fires (Mudan and Croce, 1995)

Hydrocarbon	Pool size (m)	Radiative fraction (χ)
LNG on land	18.0	0.164
	0.3–3.05	0.15–0.34
	1.8–6.1	20.0–25.0
	20.0	0.36
LPG on land	20.0	0.07
Butane	0.3–0.76	19.9–26.9
Gasoline	1.22–3.05	40.0–13.0[a]
	1.0–10.0	60.1–10[a]
Benzene	1.22	0.36–0.38

[a] In these cases, the radiative fraction decreased as the pool size was increased.

as illustrated in Figure 4.32, where $R^2 = (l/2)^2 + d^2$, d being the distance from the plume axis to the receiver, as shown. However, if the surface of the receiver is at an angle θ to the line-of-sight (PT), the flux will be reduced by a factor $\cos\theta$.

$$\dot{q}''_{r,T} = (0.3\dot{m}'' \Delta H_c A_f \cos\theta)/4\pi R^2 \tag{4.55}$$

Consider a fire on a 10 m diameter pool of gasoline. Given that this will burn with a regression rate of 5 mm/min (Section 5.1.1, Figure 5.1) corresponding to a mass flowrate of $\dot{m}'' = 0.058$ kg/m^2.s, then as $\Delta H_c = 45$ kJ/g (Table 1.13), the rate of heat release according to Equation (4.52) will be 206 MW. The heat flux at a distance according to Equation (4.55) is shown in Figure 4.33. It can be seen that it does not apply at short distances for the geometry specified in Figure 4.32.

In the second method, the flame is approximated by a vertical rectangle, $l \times D$, straddling the tank in a plane at right angles to the line of sight. The net emissive power of one face of this rectangle would then be:

$$E = \tfrac{1}{2}(0.3\dot{m}'' \Delta H_c A_f/lD)$$

$$= 151 \text{ kW/m}^2 \tag{4.56}$$

The radiant flux at a distant point can then be obtained from:

$$\dot{q}''_{r,T} = \phi E \tag{4.57}$$

Values of $\dot{q}''_{r,T}$ calculated by this method for the above problems are also shown in Figure 4.33. Higher figures are obtained because the emitter is treated as an extended source: provided that $d > 2D$, values of $\dot{q}''_{r,T}$ are about double those obtained from Equation (4.55). By using Equations (4.56) and (4.57), a very conservative figure is obtained. However, this would result in unnecessary expense as the protection system would then be overdesigned. There are several simplifying assumptions in the above calculations, at least two of which will lead to an overestimate of the heat flux, i.e.

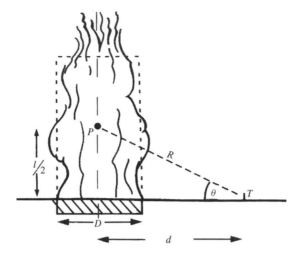

Figure 4.32 Estimating the radiant heat flux received at point T from a pool fire, diameter D. Equivalent point source at P

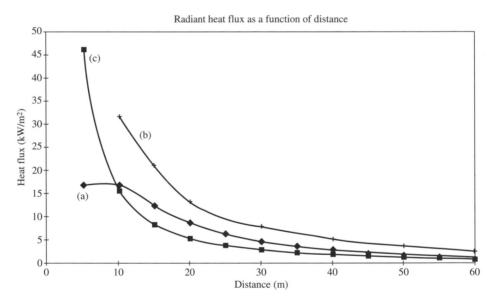

Figure 4.33 Variation of incident radiant heat flux ($\dot{q}''_{\mathrm{r,T}}$) with distance from a 10 m diameter pool of gasoline (see Figure 4.32): (a) assuming point source (◆); (b) assuming that the flame behaves as a vertical rectangle, $l \times D$ (+); and (c) calculated from the correlation by Shokri and Beyler (1989) (■).

that combustion is 100% efficient, and that the flame acts as if it were at a uniform temperature. The former is certainly incorrect, while the latter ignores the fact that there are substantial temperature gradients within such flames and that thermal radiation may be attenuated by soot in the cooler regions of the flame. Shokri and Beyler (1989) have reviewed data on radiant fluxes ($\dot{q}''_{\mathrm{r,T}}$) measured in the vicinity of experimental hydrocarbon pool fires. They estimated the 'effective' emissive power (E) of each by calculating the appropriate configuration factor ϕ (assuming the flame to be cylindrical) and using Equation (4.57). Values of E_{eff} in the range 16–90 kW/m^2 were obtained, with a tendency to decrease with increasing pool diameter: this was attributed to the increasing amounts of black smoke which envelop the flame at larger diameters (see the data for gasoline fires in Table 4.5).

Shokri and Beyler (1989) found that the following empirical expression (in which d is the distance from the pool centreline to the 'target'):

$$\dot{q}''_{r,T} = 15.4 \left(\frac{d}{D} \right)^{-1.59} \text{ kW/m}^2 \tag{4.58}$$

was an excellent fit to the data from over 80 of the pool fire experiments examined in their review which refer to the configuration shown in Figure 4.32 (i.e. vertical, ground-level targets). It can be seen in Figure 4.33 that this predicts lower heat fluxes than the point source method in the far field, but continues to increase monotonically as the distance from the tank decreases. These authors recommend that applying a safety factor of two to this formula will give heat flux values above any of the measurements which were used in its derivation; these values still lie below the heat fluxes obtained using the vertical rectangle approximation.

The above calculations assume that the flames are vertical and are not influenced by wind. If the presence of wind has to be assumed, then the appropriate flame configuration can be deduced from information presented in Section 4.3.5.

4.4.2 The response of ceiling-mounted fire detectors

In Section 4.3.4, it was shown that the temperature under a ceiling could be related to the size of the fire ($\dot{Q}$), the height of the ceiling (H) and the distance from the axis of the fire plume (r) (Figure 4.27) (Alpert, 1972). Equations (4.49) and (4.50) may be used to estimate the response time of ceiling-mounted heat detectors, provided that the heat transfer to the sensing elements can be calculated. Of course, it is easy to identify the minimum size of fire ($\dot{Q}_{min}$ kW) that will activate fixed-temperature heat detectors, as $T_{max} \geq T_L$, where T_L is the temperature rating. Thus, from Equations (4.49) and (4.50), for $r > 0.18H$:

$$\dot{Q}_{min} = r(H(T_L - T_\infty)/5.38)^{3/2} \tag{4.59}$$

and for $r \leq 0.18H$:

$$\dot{Q}_{min} = ((T_L - T_\infty)/16.9)^{3/2} H^{5/2} \tag{4.60}$$

If the detectors are to be spaced at 6 m centres on a flat ceiling in an industrial building, then the maximum distance from the plume axis to any detector head is $(0.5 \times 6^2)^{1/2}$, or $r = 4.24$ m. Thus, for the worst case

$$\dot{Q}_{min} = 4.24(H(T_L - T_\infty)/5.38)^{3/2}$$
$$= 0.34(H(T_L - T_\infty))^{3/2} \tag{4.61}$$

showing that for a given sensor, the minimum size of fire that may be detected is proportional to $H^{3/2}$. Substituting $H = 10$ m and $T_\infty = 20°C$, and assuming $T_L = 60°C$, then the minimum size of fire that can be detected in a 10 m high enclosure is 2.7 MW.

However, rapid activation of the detector will require a high rate of heat transfer ($\dot{q}$) to the sensing element of area A which consequently must be exposed to a temperature significantly in excess of T_L. The rate will be given by (Equation (2.3))

$$\dot{q} = h \cdot A \cdot \Delta T \text{ kW}$$

where h, the heat transfer coefficient for forced convection, will be a function of the Reynolds and Prandtl numbers (Section 2.3).

The response time (t) of the sensing element can be derived from Equations (2.20) and (2.21), setting T_{max} as the steady fire-induced temperature at the head and T (the temperature of the element at time of response) as T_L: thus,

$$t = \frac{Mc}{A}\frac{1}{h} \ln\left(\frac{T_{max} - T_\infty}{T_{max} - T_L}\right) \tag{4.62}$$

i.e.

$$t = -\frac{Mc}{Ah} \ln(1 - \Delta T_L/\Delta T_{max}) \tag{4.63}$$

where Mc is the thermal capacity of the element and A is its surface area (through which heat will be transferred) and ΔT_L and ΔT_{max} are $T_L - T_\infty$ and $T_{max} - T_\infty$, respectively. The quantity

$$\tau = \frac{Mc}{Ah} \tag{4.64}$$

is the time constant of the detector, but while Mc/A is readily calculated, it is very difficult to estimate h from first principles. However, it refers to conditions of forced convection so that if the flow is laminar, then according to Equation (2.39) $h \propto Re^{1/2}$, hence $h \propto u^{1/2}$ and $\tau \propto u^{-1/2}$. This leads to the concept of the Response Time Index (RTI) which was originally introduced by Heskestad and Smith (1976) to characterize the thermal response of sprinkler heads. It is determined experimentally in the 'plunge test': this involves suddenly immersing the sprinkler head in a flow of hot air, the temperature (T_0) and flowrate (u_0) of which are known. The RTI is then defined as:

$$RTI = \tau_0 u_0^{1/2} \tag{4.65}$$

where τ_0 is the value of the time constant determined under the standard conditions from Equation (4.63), in which t becomes the response time t_0. It is assumed* that the product $\tau u^{1/2}$ will be equal to the RTI at any other temperature and flowrate (assumed laminar). Thus, if the value of u under fire conditions can be predicted, the time to

---

*This assumption has been shown to be invalid if conduction losses from the sprinkler head into the associated pipework are significant. This will be the case for conditions in which the sensing element is heated slowly, e.g. as a result of a slowly developing fire. This is discussed by Heskestad and Bill (1988) and Beever (1990).

sprinkler actuation (t) can be calculated from:

$$\frac{t}{t_0} = \left(\frac{u_0}{u}\right)^{1/2} \left(\frac{\ln(1 - \Delta T_L/\Delta T_{max})}{\ln(1 - \Delta T_L/\Delta T_{max,0})}\right) \tag{4.66}$$

This can be used if there is a fire of constant heat output, giving a temperature (T_{max}) and flowrate (u) at the detector head under a large unobstructed ceiling: the temperatures can be calculated using Equations (4.49) and (4.50), while the gas velocities can be calculated from:

$$u_{max} = \frac{0.197\dot{Q}^{1/3}H^{1/2}}{r^{5/6}} \text{ m/s} \tag{4.67}$$

which applies to the ceiling jet ($r > 0.18H$ and $Y \approx 0.01H$, see Figure 4.24), while within the buoyant plume ($r \geq 0.18H$):

$$u_{max} = 0.946 \left(\frac{\dot{Q}}{H}\right)^{1/3} \text{ m/s} \tag{4.68}$$

where $\dot{Q}$ is in kW (Alpert, 1972). This work forms the basis for Appendix B of NFPA 72 (Evans and Stroup, 1986; National Fire Protection Association, 1993a). However, it is necessary to take into account the growth period of a fire when the temperatures and flowrates under the ceiling are increasing. Beyler (1984a) showed how this could be incorporated by assuming that $\dot{Q} \propto t^2$ (Section 9.2.4) and using ceiling jet correlations developed by Heskestad and Delichatsios (1978): this is included in NFPA 72 (Appendix C), but is not discussed further here (see Schifiliti *et al.*, 1995).

It should be remembered that Alpert's equations were derived from steady-state fires burning under horizontal ceilings of effectively unlimited extent. They will not apply to ceilings of significantly different geometries. Obstructions on the ceiling should represent no more than 1% of the height of the compartment. Moreover, higher temperatures and greater velocities would be anticipated if the fire were close to a wall or in a corner (Section 4.3.4).

4.4.3 Interaction between sprinkler sprays and the fire plume

The maximum upward velocity in a fire plume ($u(max)$) is achieved in the intermittent flame, corresponding to $z/\dot{Q}_c^{2/5} = 0.08$ to 0.2 in Table 4.2 (McCaffrey, 1979): thus, McCaffrey's data give

$$u_0(max) = 1.9\dot{Q}_c^{1/5} \text{ m/s} \tag{4.69}$$

where $\dot{Q}_c$ is in kW.

For a sprinkler to function successfully and extinguish a fire, the droplets must be capable of penetrating the plume to reach the burning fuel surface. Rasbash (1962) and Yao (1976) identified two regimes, one in which the total downward momentum of the spray was sufficient to overcome the upward momentum of the plume, while in the other the droplets were falling under gravity. In the gravity regime, the terminal velocity of the water drops will determine whether successful penetration can occur. In Figure 4.34, the terminal velocity for water drops in air at three different temperatures

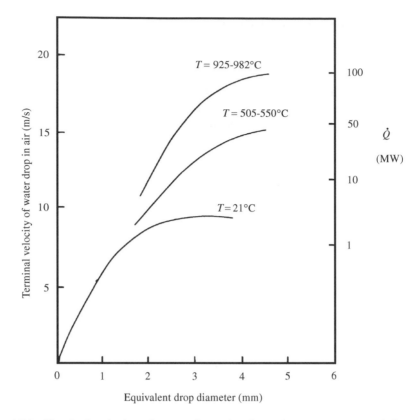

Figure 4.34   Terminal velocity of water drops in air at three temperatures (adapted from Yao, 1980). The right-hand ordinate gives the fire size for which $u_0(\text{max})$ (Equation (4.49)) is equal to the terminal velocity given on the left-hand ordinate

is shown as a function of drop size. For comparison, values of $\dot{Q}$ (MW) (for which $u_0(\text{max})$ correspond to the terminal velocities shown on the left-hand ordinate) are given on the right-hand ordinate, referring to McCaffrey's methane flames. Thus, in the 'gravity regime', drops less than 2 mm in diameter would be unable to penetrate vertically into the fire plume above a 4 MW fire. This can be overcome by generating sufficient momentum at the point of discharge but this will be at the expense of droplet size. Penetration may then be reduced by the evaporative loss of the smallest droplets as they pass through the fire plume. Although this will tend to cool the flame gases, it will contribute little to the control of a fast-growing fire.

It is outside the scope of the present text to explore this subject further, but the above comments indicate some of the problems that must be considered in sprinkler design. Although development of the sprinkler has been largely empirical, there is now a much sounder theoretical base on which to progress (Rasbash, 1962; Yao, 1980). This is being enhanced by recent studies of the interaction of water droplets with sprinkler sprays using CFD modelling (e.g. Hoffmann *et al.*, 1989; Kumar *et al.*, 1997). Moreover, advanced experimental techniques have also been brought to bear on the problem (Jackman *et al.*, 1992b).

4.4.4   The removal of smoke

A large proportion of fire injuries and fatalities can be attributed to the inhalation of smoke and toxic gases (Chapter 11). One technique that may be used in large buildings to protect the occupants from exposure to smoke while they are making way to a place of safety is to provide extract fans in the roof. However, it is necessary to know what rate of extraction will be required to prevent the space 'filling up with smoke', or more specifically, what rate will be sufficient to prevent those escaping from being exposed to untenable conditions due to smoke.

   If we consider a theatre, with a ceiling 20 m high, a large fire at the front of the stalls would produce a buoyant plume which would carry smoke and noxious gases up to the ceiling, there to form a smoke layer which would progressively deepen as the fire continued to burn. This would place the people in the upper balcony at risk, particularly if they had to ascend into the smoke layer to escape. If 'head height' is 5 m from the ceiling of the theatre, then ideally, the smoke layer should not descend below this level. To achieve this, the extraction fans would have to remove smoke at a rate equivalent to the mass flow of smoke into this layer — which is the total mass flowrate at a height 15 m above the floor of the theatre. This can be calculated using Equation (4.48), provided that the rate of heat release from the fire is known. This has to be estimated, identifying likely materials that may become involved, using whatever information is available (see Chapter 9): it is commonly referred to as the 'design fire'. For example, it may be shown that eventually a block of 16 seats at the front of the theatre could be burning simultaneously, each with a maximum rate of heat release of 0.4 MW, giving a total rate of heat release of 6.4 MW. Taking this as the design fire, then the mass flowrate 15 m above the floor will be:

$$
\begin{aligned}
\dot{m} &= 0.071 \dot{Q}_c^{1/3} z^{5/3} [1 + 0.026 \dot{Q}_c^{2/3} z^{-5/3}] \\
&= 0.071(6400)^{1/3} 15^{5/3} [1 + 0.026(6400)^{2/3} 15^{-5/3}] \\
&= 132.1 \text{ kg/s}
\end{aligned}
\tag{4.48}
$$

A correction could be made for the virtual origin (Figure 4.10(b)), but for tall spaces it is a relatively minor correction. Replacing z by $(z - z_0)$, where z_0 is obtained from Equation (4.26), and taking $D = 3$ m (the value for the area of seating involved), $z_0 = -0.3$, and the flowrate becomes 136.1 kg/s.

   The temperature of the layer (T_{layer}) may be estimated from the rate of heat release ($\dot{Q}_{design}$ kW), the mass flowrate ($\dot{m}$ kg/s) and the heat capacity of air (c_p kJ/kg) (Table 2.1), assuming that there are no heat losses (the system is *adiabatic*):*

$$
T_{layer} = T_0 + \Delta T
$$

where T_0 is the ambient temperature. The temperature rise, ΔT, is give by:

$$
\Delta T = \frac{\dot{Q}_{design}}{\dot{m} c_p} = \frac{6400}{136.1 \times 1.0} = 47.0 \text{ K}
$$

---

* The calculation which follows is similar to that used for the calculation of adiabatic flame temperature in Section 1.2.5.

i.e., if the ambient temperature is 20°C, the smoke layer will be at 67°C (340 K).

The volumetric flowrate of the smoke ($\dot{V}_{\text{smoke}}$) can be estimated if the density of air at this temperature is known. Welty *et al.* (1976) give $\rho_{\text{air}} = 1.038$ kg/m^3 at 340 K (see also Table 11.7), thus:

$$\dot{V}_{\text{smoke}} = \frac{\dot{m}}{\rho_{\text{air}}} = 131 \text{ m}^3/\text{s}$$

In practice calculations of this kind could be used to determine the sizes and temperature ratings of smoke extraction fans. In this example a steady fire has been used for illustrative purposes, but real fires grow with time (see Chapter 9). A more general calculation might take this into account coupled with evacuation calculations and analysis of the development of untenable conditions on escape routes to arrive at an appropriate smoke extraction rate.

4.4.5 Modelling

The term 'modelling' has two connotations, physical and mathematical. 'Mathematical' modelling of fire has gained considerable prominence in the last two decades, but it is appropriate to consider 'physical' modelling, and the role it has to play in fire science. Many problems in other branches of engineering have been resolved by the latter approach, applying experimental procedures which permit full-scale behaviour to be predicted from the results of small-scale tests. The prerequisite is that the model is 'similar' to the prototype in the sense that there is a direct correlation between the responses of the two systems to equivalent stimuli or events (Hottel, 1961). Scaling the model is achieved by identifying the important parameters of the system and expressing these in the form of relevant dimensionless groups (Table 4.6). For exact similarity, these must have the same values for the prototype and the model, but in fact it is not possible for all the groups to be preserved. Thus, in ship hull design, small-scale models are used for which the ratio L/V^2 ($L =$ length scale and $V =$ rate of flow of water past the hull) is identical to the full-scale ship, so that the Froude number (V^2/gL) is preserved, although the Reynolds number ($VL\rho/u$) must vary. Corrections based on separate experiments can be made to enable the drag on the full-scale prototype to be calculated from results obtained with the low-Reynolds number model (Friedman, 1971).

There are obvious advantages to be gained if the same approach could be applied to the study of fire. The number of dimensionless groups that should be preserved is quite large as the forces relating to buoyancy, inertia and viscous effects are all involved. However, there are two methods that are available, namely Froude modelling and pressure modelling. Froude modelling is possible for situations in which viscous forces are relatively unimportant and only the group $u_\infty^2 \rho/lg\Delta\rho$ need be preserved. This requires that velocities are scaled with the square root of the principal dimension, i.e. $u/l^{1/2}$ is maintained constant. In natural fires, when turbulent conditions prevail, behaviour is determined by the relative importance of momentum and buoyancy: viscous forces can be ignored. In the introduction to this chapter, it was pointed out that the relevant dimensional group is the Froude number, $\text{Fr} = u^2/gD$ (Table 4.6), which can be expressed in terms of the rate of heat release. The non-dimensional group $\dot{Q}^*$ is the

Table 4.6 Dimensionless groups

	Group	Physical interpretation	References
Biot	$\mathrm{Bi} = \dfrac{hl}{k}$	$\dfrac{\text{internal resistance to heat conduction}}{\text{external resistance to heat conduction}}$	Section 2.2.2
Fourier	$\mathrm{Fo} = \dfrac{\alpha t}{l^2}$	dimensionless time for transient conduction	Section 2.2.2
Froude	$\mathrm{Fr} = \dfrac{u_\infty^2}{lg}$	$\dfrac{\text{inertia forces}}{\text{gravity forces}}$	Section 4.4.4
	$= \dfrac{u_\infty^2 \rho}{lg\Delta\rho}$		
Grashof	$\mathrm{Gr} = \dfrac{gl^3\beta T}{v^2}$	$\dfrac{\text{buoyancy forces} \times \text{inertia forces}}{(\text{viscous forces})^2}$	Equation (2.49)
	$\equiv \dfrac{gl^3\Delta\rho}{\rho v^2}$	$= \mathrm{Re} \cdot \dfrac{\text{buoyancy forces}}{(\text{viscous forces})^2}$	
Lewis	$\mathrm{Le} = \dfrac{D}{\alpha}$	$\dfrac{\text{mass diffusivity}}{\text{thermal diffusivity}}$	Section 5.1.2
Nusselt	$\mathrm{Nu} = \dfrac{hl}{k}$	ratio of temperature gradients (non-dimensionalized heat transfer coefficient)	Section 2.3
Prandtl	$\mathrm{Pr} = \dfrac{\mu c \rho}{k}$	$\dfrac{\text{momentum diffusivity}}{\text{thermal diffusivity}}$	Section 2.3
Reynolds	$\mathrm{Re} = \dfrac{\rho u_\infty l}{\mu}$	$\dfrac{\text{inertia forces}}{\text{viscous forces}}$	Section 2.3

square root of a Froude number:

$$\dot{Q}^* = \frac{\dot{Q}_c}{\rho_\infty c_p T_\infty \sqrt{gD} \cdot D^2} \tag{4.3}$$

The significance of this group has already been demonstrated in correlations of flame height, etc., but it should be noted that early dimensional analysis revealed that the rate of heat release (or rate of burning (g/s)) of a fire had to be scaled with the five-halves power of the principal dimension, i.e. $\dot{Q}_c/l^{5/2}$, or $\dot{Q}_c^2/l^5$ must be preserved. This quotient appears in early correlations of flame height (see Figure 4.17) (Thomas *et al.*, 1961), temperatures and velocities in the fire plume (Figures 4.9(b) and 4.20) (McCaffrey, 1979) and ceiling temperatures arising from a fire at floor level (Alpert, 1972). However, limitations on Froude modelling are found when viscous effects become important, e.g. in laminar flow situations, and when transient processes such as flame spread are being modelled as the response times associated with transient heating of solids follow different scaling laws (de Ris, 1973).

Pressure modelling has the advantage of being able to cope with both laminar and turbulent flow. The Grashof number may be preserved in a small-scale model if the pressure is increased in such a way as to keep the product $\rho^2 l^3$ constant. This can be seen by rearranging the Grashof number thus:

$$\mathrm{Gr} = \frac{gl^3\Delta\rho}{\rho v^2} = \frac{g\rho l^3 \Delta\rho}{\mu^2} = \frac{g}{\mu^2}\left(\frac{\rho}{\Delta\rho}\right)^{-1} \rho^2 l^3 \tag{4.70}$$

where μ (the dynamic viscosity), g and $(\Delta\rho/\rho)$ are all independent of pressure. Thus, an object 1 m high at atmospheric pressure could be modelled by an object 0.1 m high if the pressure was increased to 31.6 atm. In experiments carried out under these conditions, it is also possible to preserve the Reynolds number for any forced flow in the system.

$$Re = \rho u_\infty l / \mu$$

In designing the experiment, l has been scaled with $\rho^{-2/3}$, so that u (the velocity of any imposed air flow) must be scaled with $\rho^{-1/3}$ to maintain a constant Reynolds number. The Froude number is then automatically conserved, as $Fr = Re^2/Gr$. The validity of this modelling technique has been explored by de Ris (1973).

It might appear that Froude modelling or pressure modelling could be used as the basis for designing physical models of fire, but different non-dimensional groups have to be conserved for transient processes such as ignition and flame spread. Perhaps more significant is the fact that it is not possible to scale radiation as it is such a highly non-linear function of temperature. In effect, this means that small-scale physical models of fire cannot be used directly, either as a means of improving our understanding of full-scale fire behaviour, or in assessing the likely performance of combustible materials in fire situations.

Mathematical modelling, on the other hand, has been developed to a stage where it is now possible to gain valuable insight into certain types of fire phenomena, as well as being able to carry out an examination of the consequences of change — answering the 'what if' questions. The approach is based on the fundamental physics of fluid dynamics and heat transfer. Some of the simplest models have already been considered: examples include the ignition of combustible materials exposed to a radiant heat source (Chapter 6) and the transfer of heat to a sprinkler head (Section 4.4.2). Modelling of flame spread has received considerable attention, starting with attempts to derive analytical solutions to the appropriate conservation equations (e.g. de Ris, 1969; Quintiere, 1981). In general, many simplifying assumptions have to be made to achieve this. Latterly, attention has been turned to developing numerical models which can be solved computationally. There has been much impetus to use this type of model to assist in the assessment of the fire hazards associated with combustible wall lining materials, using data obtained from small-scale tests, such as the Cone Calorimeter (Magnusson and Sundstrom, 1985; Karlsson, 1993; Grant and Drysdale, 1995). This is addressed briefly in Chapter 7.

Very significant advances have been made in modelling fires and fire phenomena using 'zone' and 'field' models. Zone models derive from work carried out at Harvard University in the late 1970s and early 1980s by Emmons and Mitler (Mitler and Emmons, 1981; Mitler, 1985). The original model was developed for a fire in a single compartment, or enclosure, which is divided into a small number of control volumes, e.g. the upper smoke layer, the lower layer of clear air, the burning fuel and the fire plume (below the smoke layer). The basic conservation equations (for mass, energy and chemical species) are solved iteratively as the fire develops, entraining air which enters through the lower part of the ventilation opening and expelling hot smoke through the upper part (see Chapter 10). They generally rely on empirical or semi-empirical correlations to enable various features to be incorporated, such as entrainment into the fire plume and the rate of supply of air into the compartment. The effect of the transfer of radiant heat from the smoke layer to the surface of the burning fuel can be included, thus allowing the development of the fire to be modelled. In this text, only the fundamental principles on which the zone models are based can be covered: readers may wish to consult relevant review articles such as those by Quintiere (1989) and Cox (1995) and other papers and texts specifically devoted to the subject.

The term 'field model' is used in the fire community as a synonym for computational fluid dynamics (CFD). The relevant space is divided into a very large number of control volumes — typically 10^4 to 10^6 — and the partial differential conservation equations (the Navier–Stokes equations) are solved iteratively for every control volume, stepping forward in time. This is computationally intensive and only in the last ten years have sufficiently fast computers become available to make this type of modelling feasible. In 1988, one of the first three-dimensional simulations of a major fire spread scenario had to be run on the Cray "supercomputer" at Harwell (Simcox *et al.*, 1992). The advantage of CFD is that it involves solving the fundamental equations of fluid dynamics and can be coupled to flame chemistry and radiation models to give a precise description of a variety of fire phenomena, including smoke movement in large spaces (Cox *et al.*, 1990), flame spread (coupling the gas phase and condensed phases processes (di Blasi *et al.*, 1988)), the structure of the flames of pool fires (Crauford *et al.*, 1985), and the development of a fire in a compartment (Cox, 1983). Cox (1995) provides an excellent review of the application of field models to fire problems.

Finally, it should be emphasized that there are two objectives in developing a fire model: (i) to test that we understand fully the fundamentals of the fire process; and (ii) to provide an alternative to carrying out a large number of expensive, full-scale tests to discover the effect of varying different parameters. For both types of model, it is clearly essential to check the results against experimental data. The term 'validation' is frequently used in this context, but this is midleading. A model can only be said to have been 'validated' if it has been tested successfully against a large number of fire tests, in which case the limits of 'validity' must clearly be stated. One of the major problems in modelling room fires is that experimental data that are available are not sufficiently detailed to enable proper checks to be made, particularly for field models. With large capacity data-logging systems now available, attempts are being made to generate suitable databases for this purpose (Luo and Beck, 1994).

Problems

4.1 Calculate the length of the turbulent diffusion flame formed when pure methane is released at high pressure through a nozzle 0.1 m in diameter. Assume that the flame temperature for methane is 1875°C (Lewis and von Elbe, 1987) and the ambient temperature is 20°C. What would the length be if the nozzle fluid consisted of 50% methane in air?

4.2 Calculate the rate of heat release for the following fires: (a) natural gas (assume methane) released at a flowrate of 20×10^{-3} m³/s (measured at 25°C and normal atmospheric pressure) through a sand bed burner 1.0 m in diameter (assume complete combustion); (b) propane, released under the same conditions as (a).

4.3 Calculate the value of $\dot{Q}^*$ for fires (a) and (b) in Problem 4.2.

4.4 Calculate the heights of the flames (a) and (b) in Problem 4.2. Check your answers to Problem 4.3 by calculating the flame heights from Equation (4.41). Estimate the flame volumes.

4.5   Calculate the frequency of oscillation of the flames (a) and (b) in Problem 4.2.

4.6   Using data from the previous questions, calculate the temperature on the centreline of the fire plume at a height equal to four times the flame height for fires (a) and (b). What difference does it make of allowance is made for the virtual source?

4.7   Given a 1.5 MW fire at floor level in a 4 m high enclosure which has an extensive flat ceiling, calculate the gas temperature under the ceiling (a) directly above a fire; and (b) 4 m and 8 m from the plume axis. Assume an ambient temperature of 20°C and steady state conditions.

4.8   Using the example given in Problem 4.7, calculate the maximum velocity of the gases in the ceiling jet 4 m and 8 m from the plume axis. How long will it take the sensing element of a sprinkler head to activate in these two positions if it is rated at 60°C and the RTI is 100 $m^{1/2}s^{1/2}$? What will be the effect of the RTI being reduced to 25 $m^{1/2}s^{1/2}$? (Assume that the air temperature in the plunge test is 200°C).

4.9   What is the minimum size of fire at floor level capable of activating fixed temperature heat detectors (rated at 70°C) in a large enclosure 8 m high? Assume that the ceiling is flat and that the detectors are spaced at 5 m centres. Ambient temperature is 20°C. Consider three situations in which the fire is (a) at the centre of the enclosure; (b) close to one wall; and (c) in a corner.

4.10 The space described in Problem 4.9 is to be protected by smoke detectors spaced at 5m centres. On the assumption that these will activate when the temperature at the detector head has increased by 15 K, calculate the minimum fire sizes required to activate the heads when the fire is (a) at the centre of the enclosure; (b) close to one wall; and (c) in a corner if the detectors are at 5 m centres.

4.11 If the height of the space described in the previous question is reduced to 2.5 m, calculate the minimum fire sizes that would activate a smoke detector for cases (a), (b) and (c) with the detectors at 5 m centres and 2.5 m centres.

4.12 In Section 4.4.4, the rate of flow of smoke into a smoke layer 15 m above the floor of the front stalls of a theatre was calculated for a design fire of 6.4 MW. Calculate (a) the rate of extraction that must be provided at roof level to prevent smoke descending below 10 m from the floor for the same design fire (6.4 MW); and (b) the rate of extraction required to prevent the smoke layer descending below 15 m if the seats are replaced by ones that individually gave a maximum rate of heat release of 0.25 MW. Assume that the same number of seats (16) are involved in the design fire.

4.13 Calculate the temperatures of the smoke layers formed in parts (a) and (b) of the previous question, assuming no heat losses (adiabatic). Compare these with the value obtained for smoke layer 15 m above floor for the design fire (6.4 MW) in Section 4.4.4.

4.14 A pool fire involving a roughly circular area 25 m in diameter where a substantial spillage of gasoline has accumulated is exposing surrounding structures and materials to radiant heat. Estimate the magnitude of the flux at ground level 25 m from the edge of the pool using the empirical formula derived by Shokri and Beyler (Equation 4.58). On this basis, assuming that the combustion is 80% efficient, calculate the fraction of the heat of combustion that is lost from the fire by radiation.

5

Steady Burning of Liquids and Solids

In the previous chapter, it was shown that the size of a fire as perceived by flame height depends on the diameter of the fuel bed and the rate of heat release due to the combustion of the fuel vapour. The latter may be expressed in terms of the primary variable, the mass flowrate of the fuel vapours ($\dot{m}$), thus:

$$\dot{Q}_c = \dot{m}\chi\Delta H_c \qquad (5.1)$$

where ΔH_c is the heat of combustion of the volatiles and χ is an efficiency factor that takes into account incomplete combustion (Tewarson, 1980, 1995). It was necessary to rely on the mass flowrate, commonly associated with the term 'burning rate', until it became possible to measure $\dot{Q}_c$ directly, either on a small scale with the Cone Calorimeter (Babrauskas, 1995b), or for full-scale items in the furniture calorimeter, the room calorimeter, or one of many variants (Babrauskas, 1992a) using the technique of oxygen consumption calorimetry (Section 1.2.3).

Information on the rate of heat release of liquid fuels and combustible solids is required not only to evaluate flame size (Section 4.3.2) but also to assess likely flame behaviour in practical situations, and estimate the contribution that individual combustible items may make towards fire development in a compartment.

In this chapter, the steady burning of combustible solids and liquids is considered in detail and those parameters which determine how rapidly volatiles are produced under fire conditions are identified.

5.1 Burning of Liquids

In this section, only those liquids that are naturally in the liquid state under normal conditions of temperature and pressure are considered in any detail. Cryogenic liquids such as LNG (liquefied natural gas) and pressurized liquids such as LPG (liquefied petroleum gas) are classified as special hazards and lie outside the scope of this text. However, attention needs to be drawn to their behaviour when released, and when involved in fire. Liquid methane, the main constituent of LNG, boils at $-164°C$, but will behave as a stable liquid at atmospheric pressure if its container, or the surface on which it is lying, is at or below this temperature. A sudden release of LNG on to

Table 5.1   Limiting regression rates for liquid pool
fires (Burgess *et al.*, 1961)

Liquid fuel	Limiting regression rate $(R_\infty \text{ (mm/min)})^a$
LNG	6.6
n-Butane	7.9
n-Hexane	7.3
Xylene	5.8
Methanol	1.7

a From Equation (5.10) and Figure 5.2.

the ground will be followed by vigorous boiling until the surface of the ground has cooled: evaporation will then level off at a relatively high rate (Clancey, 1974). If the pool is ignited, the steady rate of burning is not too dissimilar from that of any other hydrocarbon liquid, such as hexane (Table 5.1) (Burgess *et al.*, 1961).

A pressurized liquid such as propane (one of the main constituents of LPG) remains as liquid at ambient temperature only if held at a pressure corresponding to its vapour pressure at that temperature (over 8 bar at 20°C). If the pressure is released suddenly, a fraction of the liquid will vaporize almost instantaneously throughout its volume, drawing heat from the remaining liquid which will be cooled to the atmospheric boiling point (−42°C). This process is known as 'flashing': if the 'theoretical flashing fraction' is more than *c*.30% (i.e. the energy to vaporize 30% of the liquid is drawn from the remaining 70%), a sudden release of pressure caused by catastrophic failure of the container will produce a 'BLEVE' (Boiling Liquid Expanding Vapour Explosion). This term was first coined for steam boilers which failed at high pressures due to overheating, but is now common parlance in the fire engineering community to describe an event involving a pressurized liquid which is also flammable. It is associated with failure of a pressure vessel (storage or transportation) during fire exposure: ignition occurs and a fireball is formed (Roberts, 1981/82; Dorofeev *et al.*, 1995; Zalosh, 1995; Lees, 1996).

Stable liquids tend to burn as 'pools' with uniform horizontal surfaces. However, in the petrochemical and related industries it is possible to experience a 'running liquid fire' in which a leak of fluid at high level produces a flow of burning liquid over the surfaces below. This type of fire is difficult to extinguish and can cause very substantial damage to plant and structural steelwork. High rates of burning are likely to ensue as the surface area of the burning liquid can become very large. This type of fire is not considered further here, but the burning of liquid droplets is discussed at the end of the section.

5.1.1 Pool fires

The early Russian work on liquid pool fires remains the most extensive single study. Blinov and Khudiakov (1957) (Hottel, 1959; Hall, 1973) studied the rates of burning of pools of hydrocarbon liquids with diameters ranging from 3.7×10^{-3} to 22.9 m. A constant head device was used with all the smaller 'pools' to maintain the liquid surface level with the rim of the container.* They found that the rate of burning expressed as a 'regression rate' R (mm/min) (equivalent to the volumetric loss of liquid per unit surface area of the pool in unit time) was high for small-scale laboratory 'pools' (1 cm diameter and less), and exhibited a minimum at around 0.1 m (Figure 5.1). While the regression rate (R mm/min) is convenient for some purposes, the mass flux (kg/m^2.s) is a more valuable measure of the rate of burning. The conversion is straightforward:

$$\dot{m}'' = \rho_l \cdot R \cdot 10^{-3}/60 \quad \text{kg/m}^2.\text{s} \tag{5.2}$$

where ρ_l is the density of the liquid (kg/m^3). This may be used to estimate a rate of heat release using Equation (1.4).

Three regimes can be distinguished. If the diameter is less than 0.03 m, the flames are laminar, and the rate of burning, R, falls with increase in diameter, while for large diameters ($D > 1$ m), the flames are fully turbulent and R becomes independent of diameter. In the range $0.03 < D < 1.0$ m, 'transitional' behaviour, between laminar and turbulent, is observed. The dependence on pool diameter can be explained in terms of changes in the relative importance of the mechanisms by which heat is transferred

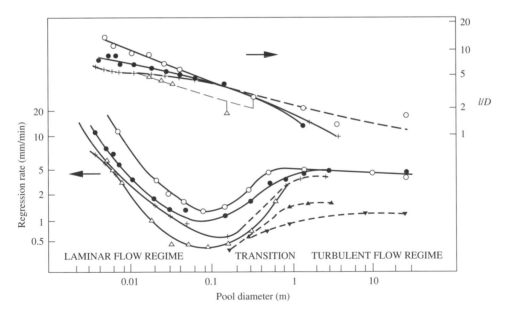

Figure 5.1 Regression rates and flame heights for liquid pool fires with diameters in the range 3.7×10^{-3} to 22.9 m. ○, Gasoline; ●, tractor kerosene; △, solar oil; +, diesel oil; ▲, petroleum oil; ▼, mazut oil (Blinov and Khudiakov, 1957, 1961; Hottel, 1959. By permission)

---

* This point of detail is important in experimental work: if there is an exposed rim above the liquid surface, the characteristics of the flame are altered (Corlett, 1968; Hall, 1973; Orloff and de Ris, 1982; Brosmer and Tien, 1987; Bouhafid *et al.*, 1988) and the rate of burning is affected significantly (de Ris, 1979).

to the fuel surface from the flame. The rate of burning can be expressed as

$$\dot{m}'' = \frac{\dot{Q}_a'' - \dot{Q}_L''}{L_v} \tag{5.3}$$

(Equation (1.3)). In the absence of an independent heat flux $\dot{Q}_a'' = \dot{Q}_F''$ which can be expressed as the sum of three terms, namely:

$$\dot{Q}_F'' = \dot{q}_{conduction}'' + \dot{q}_{convection}'' + \dot{q}_{radiation}'' \tag{5.4}$$

where $\dot{q}_{radiation}''$ takes into account surface reradiation, which would normally be considered as part of $\dot{Q}_L''$ (see Equation (5.7)).

The conduction term refers to heat transfer through the rim of the container, thus

$$\dot{q}_{conduction} = k_1 \pi D (T_F - T_l) \tag{5.5}$$

where T_F and T_l are the flame and liquid temperatures, respectively, and k_1 is a constant which incorporates a number of heat transfer terms. For convection direct to the fuel surface,

$$\dot{q}_{convection} = k_2 \frac{\pi D^2}{4} (T_F - T_l) \tag{5.6}$$

where k_2 is the convective heat transfer coefficient. Finally, the radiation term is given by:

$$\dot{q}_{radiation} = k_3 \frac{\pi D^2}{4} (T_F^4 - T_l^4)(1 - \exp(-k_4 D)) \tag{5.7}$$

where k_3 contains the Stefan-Boltzmann constant (σ) and the configuration factor for heat transfer from the flame to the fuel surface (Section 2.4.1) while $(1 - \exp(-k_4 D))$ is the emissivity of the flame (Equation (2.83)). On inspection it can be seen that k_4 must contain not only some factor of proportionality relating the mean beam length to the pool diameter but also concentrations and emission coefficients of the radiating species in the flame.* Dividing Equations (5.5)–(5.7) by the pool surface area $\pi D^2/4$ and substituting the results into Equation (5.4):

$$\dot{Q}_F'' = \frac{4\Sigma \dot{q}}{\pi D^2} = 4\frac{k_1(T_F - T_l)}{D} + k_2(T_F - T_l) + k_3(T_F^4 - T_l^4)(1 - \exp(-k_4 D)) \tag{5.8}$$

The regression rate would then be given by:

$$R = \frac{\dot{Q}_F'' - \dot{Q}_L''}{\rho L_v} \tag{5.9}$$

where ρ is the density of the liquid, L_v is the latent heat of evaporation and $\dot{Q}_F''$ is given by Equation (5.8). This has the correct mathematical form to account for the shape of the curves in Figure 5.1. When D is very small, conductive heat transfer determines

---

* k_4 must also incorporate a factor for 'radiation blocking' when the layer of fuel vapours above the fuel becomes sufficiently thick to attenuate the flux falling on the surface (see, e.g., de Ris, 1979).

the rate of burning while the radiative term predominates if D is large, provided that k_4 is of sufficient magnitude.

The dominance of radiation in large diameter ($D > 1$ m) hydrocarbon fires is also shown in results obtained by Burgess *et al.* (1961). They correlated their data on the rates of burning of hydrocarbons using the expression:

$$R = R_\infty(1 - \exp(-k_4'D)) \quad (5.10)$$

where R_∞ is the limiting, radiation-dominated regression rate (compare Equations (5.7) and (5.10) and Figures 5.1 and 5.2). Some values of R_∞ obtained by Burgess *et al.* are given in Table 5.1.

Zabetakis and Burgess (1961) recommended that the following expression be used to predict the burning rate (kg/m^2.s) of liquid pools of diameters greater than 0.2 m (cf. Equation (5.10)):

$$\dot{m}'' = \dot{m}''_\infty(1 - \exp(-k\beta D)) \quad (5.11)$$

where the product $k\beta$ is equivalent k_4' in Equation (5.10), and consists of an extinction coefficient (k m^{-1}) and a 'mean beam length corrector' (β). From his survey of the contemporary published data on a range of liquids, Babrauskas (1983b) proposed values of $\dot{m}''_\infty$ and $k\beta$ which are given in Table 5.2. What is surprising is the relatively small range of values of $\dot{m}''_\infty$ that is found: with the exception of the petroleum products quoted, the hydrocarbon fuels lie between 0.06 and 0.10 kg/m^2.s, including two cryogenic liquids, methane and propane (see Section 5.1). Table 5.2 shows that the limiting burning rates for the simple alcohols methanol and ethanol are much less than that of the hydrocarbons. This is partly due to the greater values of L_v (see Equation (1.2)) for these liquids (Table 5.3), but is also a result of the much lower emissivity of the alcohol flames (see Section 2.4.2). The significance of the latter was first demonstrated by Rasbash *et al.* (1956) who made a detailed study of the flames above 30 cm diameter pools of alcohol, benzene, kerosene, and petrol. Their apparatus is shown in Figure 5.3.

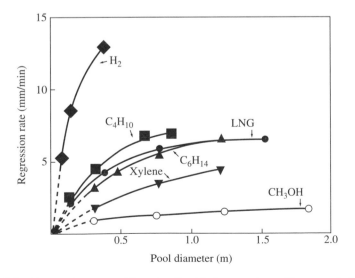

Figure 5.2 Regression rates of burning liquids in open trays (Burgess *et al.*, 1961)

Table 5.2 Data for estimating the burning rate of large pools (Babrauskas, 1983b)

Liquid	Density (kg/m^3)	$\dot{m}''_\infty$ (kg/m^2.s)	$k\beta$ (m^{-1})
Cryogenics			
Liquid methane	415	0.078	1.1
Liquid propane	585	0.099	1.4
Alcohols			
Methanol	796	0.017	—
Ethanol	794	0.015	—
Simple organic fuels			
Butane	573	0.078	2.7
Benzene	874	0.085	2.7
Hexane	650	0.074	1.9
Heptane	675	0.101	1.1
Acetone	791	0.041	1.9
Petroleum products			
Gasoline	740	0.055	2.1
Kerosene	820	0.039	3.5
Crude oil	830–880	0.022–0.045	2.8

Table 5.3 Latent heats of evaporation of some liquids (Lide, 1993/94)

Liquid	Boiling point (°C)	L_v (kJ/g)[a]
Water	100	2.258
Methanol	64.6	1.100
Ethanol	78.3	0.838
Methane	−161.5	0.512
Propane	−42.1	0.433
Butane	−0.5	0.387
Benzene	80.09	0.394
Hexane	66.73	0.335
Heptane	98.5	0.318
Decane	174.15	0.273

[a] The latent heat of evaporation refers to the boiling point at atmospheric pressure.

They measured the burning rates of these liquids and estimated the emission coefficients of the flames (K) from measurements of flame shape, temperature and radiant heat loss, assuming that emissivity can be expressed as:

$$\varepsilon = 1 - \exp(-KL) \tag{5.12}$$

where L is the mean beam length (Equation (2.83)). Their results are summarized in Table 5.4. One significant feature of these data is that the temperature of the non-luminous alcohol flame is much higher than that of the hydrocarbon flames which

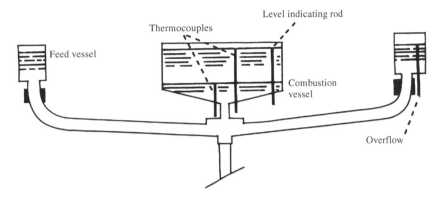

Figure 5.3   Details of the apparatus used by Rasbash *et al.* (1956) to study liquid pool fires

Table 5.4   Radiation properties of flames above 0.3 m diameter pool fires (Rasbash *et al.*, 1956). The liquid level was maintained 20 mm below the rim of the vessel

	Flame temperature (°C)[a]	Flame width (m)	K (m^{-1})	Emissivity (ε)
Alcohol	1218	0.18	0.37	0.066
Petrol	1026	0.22	2.0	0.36
Kerosene	990	0.18	2.6	0.37
Benzene	921			
after 2 min		0.22	3.9	0.59
after 5 min		0.29	4.1	0.70
after 8 min		0.30	4.2	0.72

[a] Time-averaged flame temperatures measured by the Schmidt method (Gaydon and Wolfhard, 1979).

Table 5.5   Radiative heat transfer rates to the surface of burning liquids compared with the net heat transfer rates (Rasbash *et al.*, 1956)

Liquid	Heat required to maintain steady burning rate (kW)	Estimated radiant heat transfer from flames to surface (kW)
Alcohol	1.22	0.21
Benzene	2.23	2.51
Petrol	0.94	1.50
Kerosene	1.05	1.08

lose a considerable proportion of heat by radiation from the soot particles within the flame (Section 2.4.3). The amount of heat radiated to the surface of the pool was calculated using Equation (5.12) with the appropriate configuration factor and compared with the rate of heat transfer required to produce the observed rate of burning. The results of these calculations are given in Table 5.5 and show that the

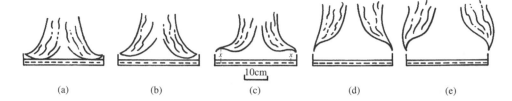

(a)   (b)   (c)   (d)   (e)

Figure 5.4   Shapes of flames immediately above the surface of burning liquids (Rasbash *et al.*, 1956. Reproduced from *Fuel*, 1956, **31**, 94–107, by permission of the publishers, Butterworth & Co. (Publishers) Ltd. ©

radiative flux to the surface in the case of the alcohol falls far short of that required to maintain the flow of volatiles. The difference must be supplied by convection: observations of the flame shape are compatible with this conclusion (Figure 5.4). The pale blue alcohol flame burns very close to the surface, apparently touching it as shown in Figure 5.4(a), while with the other fuels there is a discernible vapour zone immediately above the liquid. This was particularly apparent for benzene which after some time adopted the shape illustrated by Figure 5.4(d), thereafter oscillating occasionally between 5.4(d) and 5.4(e). The vapour zone above the hydrocarbon liquids would be expected to attenuate the radiation reaching the surface. This is consistent with the positive discrepancy that was observed between the estimated radiant heat flux and the heat flux required to produce the flow of vapours. However, it is not possible to draw any inferences regarding the relative magnitudes of the effect in comparison to the thicknesses for the vapour zones (at 'x' in Figure 5.4(c): these were 50 mm, 40–50 mm and 25–30 mm for benzene, petrol and kerosene respectively.

The blocking effect has been studied by a number of authors. Brosmer and Tien (1987) calculated the radiant heat feedback to the surface of PMMA 'pool fires' ($\dot{Q}''_a$ in Equation (5.3)), comparing their results with measurements taken by Modak and Croce (1977) on PMMA fires of diameter 0.31, 0.61 and 1.22 m. They made two important observations: (i) it was necessary to model the flame shape accurately to allow a realistic mean beam length to be calculated; and (ii) the predicted feedback was too high unless absorption of radiation by the cool vapour layer above the surface was included in the model. Moreover, for pool diameters greater than c.0.5 m, it was predicted that the maximum radiant intensity would not lie at the centre of the PMMA 'pool', but some way towards the perimeter as a consequence of the 'blockage' effect.

The rate of burning is unlikely to be constant across any horizontal fuel surface. In their study of *small* liquid pool fires in concentric vessels, 10–30 mm in diameter, Akita and Yumoto (1965) found that the rate of evaporation was greater near the perimeter than at the centre, an effect that was most pronounced with methanol. This is consistent with the fact that flames above small pools are of low luminosity and that convection will dominate the heat transfer process (see Figure 5.4(a)). As the size of the pool is increased, radiative heat transfer will come to dominate, and will result in more rapid burning towards the centre, but moderated by absorption by the cool vapours (de Ris, 1979; Brosmer and Tien, 1987).

Only the surface layers of a deep pool of a pure liquid fuel will be heated during steady burning. A temperature distribution similar to that shown in Figure 5.5 will become established below the surface, although this may take some time (Rasbash

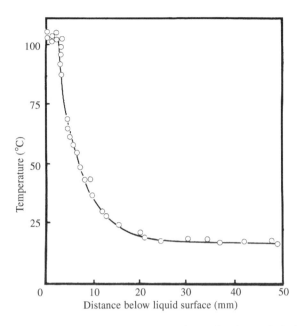

Figure 5.5 Temperature distribution below the surface of *n*-butanol during steady burning ('pool' diameter 36 mm) (Blinov and Khudiakov, 1961, by permission)

et al., 1956). Khudiakov showed that the distribution in Figure 5.5 can be described by the empirical equation:

$$\frac{T - T_\infty}{T_S - T_\infty} = \exp(-\kappa x) \tag{5.13}$$

where x is the depth and T_S and T_∞ are the temperatures at $x = 0$ and $x = \infty$ respectively. Equation (5.13) is the solution of Equation (2.15), in which t is replaced by x/R_∞, where R_∞ is the regression rate, i.e.

$$\frac{\mathrm{d}^2 T}{\mathrm{d}x^2} = \frac{R_\infty}{\alpha} \left(\frac{\mathrm{d}T}{\mathrm{d}x} \right) \tag{5.14}$$

Accordingly, the constant κ in Equation (5.13) should be equal to $R_\infty \rho c / k$, but the agreement is poor for the results shown in Figure 5.5, probably as a result of the finite area of the pool and heat loss through the container walls. The shallow surface layer at constant temperature is certainly not consistent with Equation (5.14) and may indicate in-depth absorption of radiation (e.g. Inamura *et al.*, 1992).

As a pool of liquid burns away and its depth decreases, there will come a time when heat losses to the base of the container will become increasingly important. If this represents a substantial heat sink, the rate will diminish and burning will eventually cease if the heat losses are sufficient to lower the surface temperature to below the firepoint (Section 6.2). This effect is encountered when attempts are made to burn oil slicks floating on water. Once the slick has 'weathered' or otherwise lost its light ends, e.g. by burning, the residue, although combustible, cannot burn if its thickness is less than a few millimetres (e.g. Petty, 1983).

The surface temperature of a freely-burning liquid is close to, but slightly below, its boiling point. Liquid mixtures, such as petrol, kerosene and fuel oil, do not have a fixed boiling point and the lighter volatiles will tend to burn off first. The surface temperature will therefore increase with time as the residual liquid becomes less volatile. A hazard associated with some hydrocarbon liquid blends (particularly crude oils) is that of hot zone formation (Burgoyne and Katan, 1947). In such cases, a steady-state temperature distribution similar to that shown in Figure 5.5 does not form. Instead, a 'hot-zone' propagates into the fuel at a rate significantly greater than the surface regression rate. This is illustrated in Figure 5.6. The danger arises with fires involving large storage tanks containing these liquids if the temperature of the hot zone is significantly greater than 100°C. If the hot zone reaches the foot of the tank and encounters a layer of water (which is commonly present) then explosive vaporization of the water can occur when the vapour pressure becomes sufficient to overcome the head of liquid above, thus ejecting hot, burning oil. This is known as 'boilover': the likely consequences need not be elaborated (Vervalin, 1973; Koseki, 1993/94; Lees, 1996). Table 5.6 compares the regression rates with the rate of descent of the hot zone (or 'heat wave') for a number of crude oils.

The precise mechanism of hot zone formation has not been established but as it is a phenomenon associated exclusively with fuel mixtures, it is likely to involve selective evaporation of the light ends. In his original review, Hall (1973) was unable to choose between the mechanism proposed by Hall (1925) involving the continuous migration of light ends to the surface, followed by distillation, and that of Burgoyne and Katan (1947), who suggested that light ends volatilize at the interface of the hot oil and the cool liquid below, and then rise to the surface. Certainly, bubbles are produced within the hot zone, and rise to the surface (Hasegawa, 1989), apparently enhancing the mixing process which may account for the uniform properties of the hot zone. The mechanism of downward propagation (descent) of the hot zone is unclear, but may involve slow vertical oscillations at the interface which have been observed by Hasegawa (1989) and others. The onset of the boilover process is impossible to predict, but the time to boilover increases with depth of liquid (Koseki, 1993/94), and some

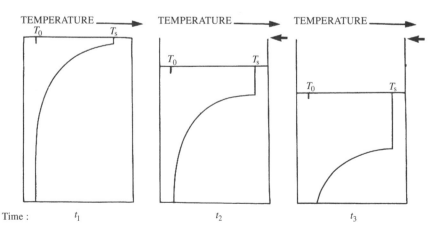

Figure 5.6   Three stages in the formation and propagation of a hot zone during a storage tank fire ($t_1 < t_2 < t_3$). The position of the original surface is marked by arrows

Table 5.6 Comparison between rates of propagation of hot zones and regression rates of liquid fuels (Burgoyne and Katan, 1947)[a]

Oil type	Rate of descent of hot zone (mm/min)	Regression rate (mm/min)
Light crude oil		
<0.3% water	7–15	1.7–7.5
>0.3% water	7.5–20	1.7–7.5
Heavy crude and fuel oils		
<0.3% water	up to 8	1.3–2.2
>0.3% water	3–20	1.3–2.3
Tops (light fraction of crude oil)	4.2–5.8	2.5–4.2

[a] Discussed by Hall (1973).

warning of its onset may be obtained by monitoring the noise emitted at the oil/water interface when the water first starts to vaporize (Fan *et al.*, 1995).

5.1.2 Burning of liquid droplets

If a combustible liquid is dispersed as a suspension of droplets in air, ignition can result in very rapid burning, even if the liquid is below its firepoint temperature (see Section 6.2). This is an extremely efficient method by which liquid fuels may be burnt and is used widely in industrial furnaces and other devices (e.g. the combustion chamber of the diesel engine). Similarly, accidental formation of a flammable mist or spray — for example, as a result of a small leak in a high pressure hydraulic system — will present a significant fire and/or explosion hazard. Such is the concern about spray fires that much effort has been expended in developing test methods for assessing this particular hazard for various hydraulic fluids (Holmstedt and Persson, 1985; Yule and Moodie, 1992). A pinhole leak in a high-pressure hydraulic system can produce a discharge which, if ignited, will sustain a very large flame. The rate of discharge ($\dot{m}$ kg/s) is given by the expression (see, e.g., Wells, 1997):

$$\dot{m} = C_D A \sqrt{2\rho(p - p_0)} \tag{5.15}$$

where C_D is a discharge coefficient (normally taken to be 0.61), A (m^2) is the cross-sectional area of the 'pinhole', ρ (kg/m^3) is the density of the liquid, and p (Pa) is the pressure in the system (p_0 is the ambient pressure, 101 300 Pa). Thus, with a system operating at 10 bar (1013 kPa), the discharge rate through a 1 mm hole for a typical hydrocarbon fluid (750 kg/m^3) will be 0.0177 kg/s. Taking the heat of combustion as 45 MJ/kg (Table 1.13), this will correspond to a rate of heat release of 0.8 MW.

As indicated earlier (Section 3.1.3) flammability limits of mists exist and can be measured: for hydrocarbon liquids, the lower limit corresponds to 45–50 g/m^3 although this figure apparently decreases as the droplet size is increased (Figure 3.7: Burgoyne and Cohen, 1954; Cook *et al.*, 1977). Explosions involving flammable mists are recognized as a serious risk in certain well-characterized situations, e.g. crankcases of marine engines (Burgoyne *et al.*, 1954).

A significant amount of research has been carried out into the burning of mists and sprays (see Zabetakis, 1965; Kanury, 1975; Holmstedt and Persson, 1985). Part of this overall activity has encompassed a study of the burning of single droplets (Williams, 1973), a problem that has highlighted the interrelationship between heat and mass transfer at the burning surface. It is appropriate to discuss this work here as it provides an introduction to Spalding's mass transfer number, commonly referred to as the 'B-number' (Spalding, 1955; Kanury, 1975). It was derived originally in an analysis of droplet vaporization in which the latent heat of evaporation is supplied to a droplet at uniform temperature by convection from the surrounding free gas stream. The derivation, which is discussed in full by Spalding (1955) and Kanury (1975), hinges on the fact that the mass flux ($\dot{m}_s''$) from the droplet surface can be expressed in two ways, either in terms of heat transfer

$$\dot{m}_s'' \cdot L_v = k_g \left(\frac{dT}{dr} \right)_s \tag{5.16}$$

(where L_v is the latent heat of vaporization, k_g is the thermal conductivity of air, $(dT/dr)_s$ is the gas phase temperature gradient at the surface and r is the radial distance from the centre of the droplet (using spherical coordinates)); or in terms of mass transfer:

$$\dot{m}_s'' \cdot Y_{fR} = \dot{m}_s'' Y_{fs} + \left(-\rho_g D_f \left(\frac{dY_f}{dr} \right)_s \right) \tag{5.17}$$

(where Y_f is the mass fraction of the fuel within the droplet (Y_{fR}) and at the surface (Y_{fs}), ρ_g is the density of air and D_f is the diffusivity of the fuel vapour in air). These equations can be cast in the following forms:

$$\dot{m}_s'' = \rho_g \alpha_g \frac{d}{dr} \left(\frac{c_g(T - T_\infty)}{L_v} \right)_s = \rho_g \alpha_g \left(\frac{db_T}{dr} \right)_s \tag{5.18}$$

$$\dot{m}_s'' = \rho_g D_f \frac{d}{dr} \left(\frac{Y_f - Y_{f\infty}}{Y_{fs} - Y_{fR}} \right)_s = \rho_g D_f \left(\frac{db_D}{dr} \right)_s \tag{5.19}$$

(where α_g is the thermal diffusivity of the gas, and T_∞ and $Y_{f\infty}$ are the values of T and Y_f at $r = \infty$). These equations are identical if the ratio $Le = \alpha_g/D_f = 1$, where Le is the Lewis number: this is a common approximation in combustion problems (Lewis and von Elbe, 1987). Equations (5.18) and (5.19) indicate that b_T and b_D are conserved variables which determine the direction and magnitude of the mass flux. The mass transfer number is defined as the difference between b_∞ (i.e. at $r = \infty$) and b_s, the value at $r = R$, where R is the droplet radius, thus,

$$B = b_\infty - b_s \equiv \frac{c_g(T_\infty - T_s)}{L_v} \equiv \frac{(Y_{f\infty} - Y_{fs})}{(Y_{fs} - Y_{fR})} \tag{5.20}$$

Spalding showed that the rate of mass loss from the surface of the droplet is given by

$$\dot{m}''_s = \frac{h}{c_g} \ln(1 + B) \tag{5.21}$$

where h is the convective heat transfer coefficient averaged over the entire surface of the droplet (Spalding, 1955; Kanury, 1975) and c_g is the thermal capacity of air. As $h = \text{Nu} \cdot k/2R$ (Section 2.3), it can be seen that the rate of evaporation is inversely proportional to droplet diameter, a factor of significance when rapid evaporation is required, as in a diesel engine.

If evaporation is accompanied by combustion of the vapour, some of the heat released in the flame will contribute to the volatilization process. Analysis of the conservation equations (energy, fuel, oxygen and products) permits identification of a series of conserved variables, similar to 'b' above, which are equivalent provided that $\text{Le} = 1$ and it is assumed that the diffusion flame is of the Burke-Schuman type (Section 4.1), i.e. the reaction rate is infinite and burning is stoichiometric in the flame zone. (This implies that there is no oxygen on the fuel side of the flame.) The resulting mass transfer number is normally quoted as

$$B \approx \frac{\Delta H_c(Y_{O_2\infty}/r_{ox}) + c_g(T_\infty - T_s)}{L_v} \tag{5.22}$$

where r_{ox} is the mass stoichiometric ratio (gm oxygen/gm fuel). The first term in the numerator is the heat of combustion per unit mass of air consumed, i.e. $\sim$3000 J/g (Section 1.2.3). The second term is small and may be neglected so that B reduces to:

$$B \approx \frac{3000}{L_v} \tag{5.23}$$

From Equation (5.21), it is seen that combustible liquids with low heats of evaporation will tend to burn more rapidly. This is illustrated in Table 5.7 which compares the

Table 5.7   B-numbers of various fuels in air at 20°C (Friedman, 1971)

Fuel	B^a
n-Pentane	8.1
n-Hexane	6.7
n-Heptane	5.8
n-Octane	5.2
n-Decane	4.3
Benzene	6.1
Toluene	6.1
Xylene	5.8
Methanol	2.7
Ethanol	3.3
Acetone	5.1
Kerosene	3.9
Diesel oil	3.9

[a] These refer to evaporation at ambient temperature.

B-numbers of a range of fuels and highlights the difference between methanol and the alkanes which has already been discussed (Figure 5.2).

However, it must be emphasized that in the derivation the mode of heat transfer from the flame to the liquid surface is assumed to be convection. Consequently, the B-number concept as originally developed by Spalding (1955) cannot be applied directly to situations in which radiative heat transfer is significant. In the case of a single droplet, the associated flame, being small, is non-luminous and has such a very low emissivity (Spalding, 1955) that convective heat transfer will predominate. Thus Equation (5.21) may be used to calculate the rate of burning, which will vary inversely with droplet size. These equations cannot be used to calculate the burning rate of droplets in mists and sprays as radiative heat transfer is important.

5.2   Burning of Solids

It was pointed out in Chapter 1 that the burning of a solid fuel almost invariably requires chemical decomposition to produce fuel vapours ('the volatiles') which can escape from the surface to burn in the flame. A discussion of these chemical processes is beyond the scope of the present text but it is important to emphasize the complexity involved, and the wide variety of products that are formed in polymer degradation (Madorsky, 1964; Cullis and Hirschler, 1981; Beyler and Hirschler 1995). While the composition of the fuel vapours has direct relevance to the combustion process and product formation, the fire safety engineer normally bypasses this complexity by relying on the results of small-scale tests to provide relatively simple data which may be used in the assessment of the fire hazard of a given material. The best known example is perhaps the Cone Calorimeter (Section 1.2.3; Babrauskas, 1992a, 1995b), but the results (i.e. the performance of the material in the test) must be interpreted correctly if they are to be of any value. This requires a thorough understanding of the fire process and careful analysis of the test results–whether obtained using 'new generation' test procedures, or the older, more 'traditional' tests such as BS 476 Part 7 (British Standards Institution, 1987a) and ASTM E84 (ASTM, 1995a). In the remainder of this chapter, some of the fundamental issues relating to the steady burning of combustible solids will be addressed as these are central to our understanding of many fire processes, including ignition (Chapter 6) and flame spread (Chapter 7), and are essential for the interpretation of fire test data.

The behaviour of synthetic polymers will be considered separately from that of wood, which merits special treatment (Section 5.2.2). The fire behaviour of finely divided solids is discussed briefly in Section 5.2.3.

5.2.1   Burning of synthetic polymers

Unlike liquids, solids can be burnt in any orientation although thermoplastics will tend to melt and flow under fire conditions (Section 1.1.2). The important factors which determine rate of burning have already been identified in the equation:

$$\dot{m}'' = \frac{\dot{Q}_a'' - \dot{Q}_L''}{L_v} \tag{5.3}$$

Table 5.8 'Flammability parameters' determined by Tewarson and Pion (1976)

Combustibles[a]	L_v (kJ/g)	$\dot{Q}_F''$ (kW/m²)	$\dot{Q}_L''$ (kW/m²)	$\dot{m}_{ideal}''$ (g/m².s)
FR phenolic foam (rigid)	3.74	25.1	98.7	11[b]
FR polyisocyanurate foam (rigid, with glass fibres)	3.67	33.1	28.4	9[b]
Polyoxymethylene (solid)	2.43	38.5	13.8	16
Polyethylene (solid)	2.32	32.6	26.3	14
Polycarbonate (solid)	2.07	51.9	74.1	25
Polypropylene (solid)	2.03	28.0	18.8	14
Wood (Douglas fir)	1.82	23.8	23.8	13[b]
Polystyrene (solid)	1.76	61.5	50.2	35
FR polyester (glass-fibre reinforced)	1.75	29.3	21.3	17
Phenolic (solid)	1.64	21.8	16.3	13
Polymethylmethacrylate (solid)	1.62	38.5	21.3	24
FR polyisocyanurate foam (rigid)	1.52	50.2	58.5	33
Polyurethane foam (rigid)	1.52	68.1	57.7	45
Polyester (glass fibre reinforced)	1.39	24.7	16.3	18
FR polystyrene foam (rigid)	1.36	34.3	23.4	25
Polyurethane foam (flexible)	1.22	51.2	24.3	32
Methyl alcohol (liquid)	1.20[a]	38.1	22.2	32
FR polyurethane foam (rigid)	1.19	31.4	21.3	26
Ethyl alcohol (liquid)	0.97	38.9	24.7	40
FR plywood	0.95	9.6	18.4	10[b]
Styrene (liquid)	0.64[a]	72.8	43.5	114
Methylmethacrylate (liquid)	0.52	20.9	25.5	76
Benzene (liquid)	0.49[a]	72.8	42.2	149
Heptane (liquid)	0.48[a]	44.3	30.5	93

[a] Weast, 1974/75.
[b] Charring materials. $\dot{m}_{ideal}''$ taken as the peak burning rate.

Surface temperatures of burning solids tend to be high (typically >350°C) so that radiative heat loss from the surface is significant. The heat required to produce the volatiles or 'heat of gasification' (L_v) is considerably greater for solids than for liquids as chemical decomposition is involved (compare $L_v = 1.76$ kJ/g for solid polystyrene with 0.64 kJ/g for liquid styrene monomer, Table 5.8 (Tewarson and Pion, 1976)). Materials which char on heating (e.g. wood (Section 5.2.2), polyvinyl chloride, certain thermosetting resins, etc. (Table 1.2)) build up a layer of char on the surface which will tend to shield the unaffected fuel beneath. Even higher surface temperatures are achieved and the burning behaviour is modified accordingly.

The apparatus developed at the Factory Mutual Research Corporation to examine parameters that determined 'flammability' (Tewarson and Pion, 1976) is illustrated in Figure 5.7. It permits a small sample of solid material (~0.007 m² in area) to be weighed continuously as it burns in a horizontal configuration: the oxygen concentration in the surrounding atmosphere and the intensity of an external radiant heat flux can be varied as required. This allows various quantities implicit in Equation (5.3) to be determined. Writing $\dot{Q}_a''$ in terms of its components, namely $\dot{Q}_F''$ and $\dot{Q}_E''$ which refer to the heat fluxes to the surface from the flame and from the external radiant heaters

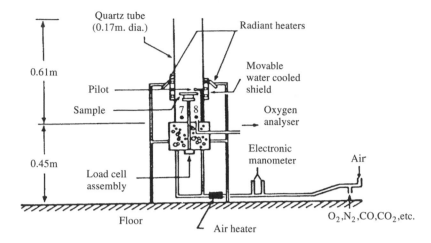

Figure 5.7   The apparatus developed at the Factory Mutual Research Corporation for determining 'flammability' parameters (Tewarson and Pion, 1976). By permission

respectively, $\dot{m}''$ becomes:

$$\dot{m}'' = (\dot{Q}''_F + \dot{Q}''_E - \dot{Q}''_L)/L_v \tag{5.24}$$

As the rate of burning is strongly dependent on the oxygen concentration, it was assumed that $\dot{Q}''_F = \xi\eta_{O_2}^{\alpha'}$, where ξ and α' are constants, and the relationship between $\dot{m}''$ and η_{O_2}, the mole fraction of oxygen in the surrounding atmosphere, examined. It was found that when $\dot{Q}''_E$ was held constant, $\dot{m}''$ is a linear function of η_{O_2} (i.e. $\alpha' = 1$) over the range of oxygen concentrations studied (Figure 5.8). The slope of the line in Figure 5.8 gives a value for ξ/L_v provided that $(\dot{Q}''_E - \dot{Q}''_L)/L_v$ is constant: this appears to be the case.

$$\dot{m}'' = \frac{\xi\eta_{O_2}}{L_v} + \frac{\dot{Q}''_E - \dot{Q}''_L}{L_v} \tag{5.25}$$

Similarly, if η_{O_2} is held constant, then a plot of $\dot{m}''$ against $\dot{Q}''_E$ (which is also an experimental variable) will give a straight line of slope $1/L_v$ (Figure 5.9). Values of L_v, the heat required to produce the volatiles, for a number of polymeric materials are given in Table 5.8. These compare favourably with values obtained by other methods, such as Differential Scanning Calorimetry (Tewarson and Pion, 1976). As both ξ/L_v and L_v can be derived by this method, the constant ξ is known so that the product $\xi\eta_{O_2}$ can be calculated for air ($\eta_{O_2} = 0.21$). This is the heat transferred from the flame to the surface of the fuel, i.e. $\dot{Q}''_F$. In Table 5.8, values of $\dot{Q}''_F$ are compared with those of $\dot{Q}''_L$ which have been calculated directly from Equation (5.24). This identifies clearly materials which will not burn unless an external heat flux is applied to render the numerator of Equation (5.24) positive (e.g. fire retarded phenolic foam).

Tewarson proposed that the quantity:

$$\dot{m}''_{ideal} = \frac{\dot{Q}''_F}{L_v} \tag{5.26}$$

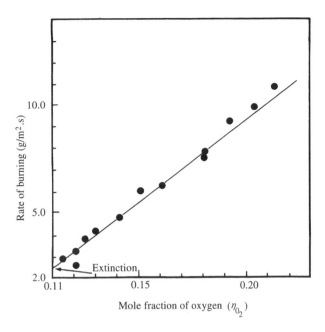

Figure 5.8   Mass burning rate of polyoxymethylene as a function of mole fraction of oxygen
(η_{O_2}) with no external heat flux ($\dot{Q}_E'' = 0$). (Adapted from Tewarson and Pion, 1976,
by permission of the Combustion Institute)

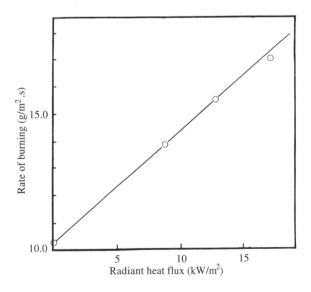

Figure 5.9   Mass burning rate of polyoxymethylene as a function of external heat flux ($\dot{Q}_E''$)
in air ($\eta_{O_2} = 0.21$) (reproduced by permission of the Combustion Institute from
Tewarson and Pion, 1976)

be used as a measure of the 'burning intensity' of a material (Tewarson and Pion, 1976) (Table 5.8), i.e. the maximum burning rate that a material could achieve if all heat losses were reduced to zero or exactly compensated by an imposed heat flux $\dot{Q}_E'' = \dot{Q}_L''$ (Equation (5.24)). While this gives results that appear to correlate reasonably well with data from existing fire tests, it would be more logical if heat loss by surface re-radiation was included in $\dot{m}_{ideal}''$. This would seem to provide a means of calculating burning rates under different heat gain and loss regimes, thus:

$$\dot{m}'' = \dot{m}_{ideal}'' + (\dot{Q}_E'' - \dot{Q}_L'')/L_v \tag{5.27}$$

However, the values of $\dot{m}_{ideal}''$ shown in Table 5.8 were obtained with a small-scale apparatus in which the sample area was no more than 0.007 m^2 (0.047 m in diameter). It has already been shown (Section 5.1.1) that as the diameter of a burning pool of liquid is increased to 0.3 m and beyond, radiation becomes the dominant mode of heat transfer, except for fuels which burn with non-luminous flames, such as methanol (Figure 5.2). The same is true for solids. Regardless of the nature of fuels in Tewarson's original experiments, the transfer of heat from the flame to the surface ($\dot{Q}_F''$) would have been predominantly by convection. Such data cannot be used directly to predict large-scale behaviour, or to make a hazard assessment. The problem of enhancing the radiation component at the small scale was addressed by Tewarson and his co-workers by carrying out the experiments at increased oxygen concentrations (Tewarson *et al.*, 1981). They showed that radiation becomes the dominant mode of heat transfer even on this scale if the oxygen concentration in the surrounding atmosphere is increased (Figure 5.7). This effect is achieved because elevated oxygen concentrations produce hotter, more emissive flames which radiate a greater proportion of the net heat of combustion back to the surface. This is consistent with the observation that the fraction of the heat of combustion that is radiated increases asymptotically as the oxygen concentration is increased. The increase in rate is self-limiting partly because the flow of volatiles will absorb a significant proportion of the radiative flux as well as tending to block convective transfer to the surface. (Section 5.1.1; Brosner and Tien, 1987).

There is strong evidence that radiation is the dominant mode of heat transfer for large fuel beds. Markstein (1979) observed that the emissivity of the flames above polymethylmethacrylate (PMMA) increased approximately three-fold as the diameter of the fuel bed was increased from 0.31 to 0.73 m: this was accompanied by an increase in $\dot{m}''$ from 10 to 20 g/m^2.s. Also, Modak and Croce (1977) determined that over 80% of the heat transferred to the surface of a burning PMMA slab, 1.22 m square, was by radiation. Approximately the same figure was obtained by Tewarson *et al.* (1981) by extrapolating their data for small samples to high oxygen concentrations. They used a version of Spalding's B-number (Equation (5.22)), corrected for radiation (Equation (5.28)), to deduce the contributions to the heat flux to the surface by radiation and convection, and showed convincingly for a number of fuels that as the oxygen concentration was increased, radiation became predominant.

$$B_R = \frac{(Y_{og}\Delta H_c/r)(1 - (\chi_R/\chi_A)) + c(T_g - T_s)}{L_v(1 - E)} \tag{5.28}$$

where χ_R/χ_A is the fraction of heat released in the flame that is radiated, and

$$E = (\dot{Q}''_E + \dot{Q}''_{FR} - \dot{Q}''_L)/\dot{m}'' \cdot L_v \qquad (5.29)$$

The comparison is given in Table 5.9 for polymethylmethacrylate (PMMA) and polypropylene (PP). Both fuels in Table 5.9 show the increasing importance of radiative heat transfer to the surface. Indeed, $\dot{Q}''_{F,r}$ increases while $\dot{Q}''_{F,c}$ decreases. The latter is a consequence of the so-called 'blowing effect' brought about by the fact that the increasing flow of fuel vapours *from* the surface inhibits convective heat transfer in the opposite direction (*to* the surface). This accounts for the increase of $\dot{Q}''_{FR}/\dot{Q}''_{FC}$ from less than 1.0 in air to more than 3.0 in mixtures consisting of over 50% oxygen. Iqbal and Quintiere (1994) have applied a one-dimensional analytical model, also using the modified Spalding B-number (Equation (5.28)) which showed the same pattern of behaviour (i.e. increasing $\dot{Q}''_{FR}/\dot{Q}''_{FC}$) with increased size of fire. This is entirely consistent with experimental work described above.

Clearly, in steady burning of an isolated fuel bed, flame emissivity and the heat required to produce the volatiles are important properties which can be assigned to the material itself, rather than to interactions with its environment. Markstein (1979) has compared the radiative output of flames above 0.31 m square slabs of polymethylmethacrylate (PMMA), polypropylene (PP), polystyrene (PS), polyoxymethylene (POM) and polyurethane foam (PUF) and found the emissivities to decrease in the order

$$PS > PP > PMMA > PUF > POM$$

Table 5.9 Convective and radiative components of $\dot{Q}''_F$ (Tewarson *et al.*, 1981)

Fuel[a]	$m_{O_2}^b$	$\dot{Q}''_{F,c}$ (kW/m^2)	$\dot{Q}''_{F,r}$ (kW/m^2)	$\dot{Q}''_{F,c} + \dot{Q}''_{F,r}$ (kW/m^2)	$\dot{Q}''_{F,r}/\dot{Q}''_{F,c}$
PMMA	0.183	17	4	21	0.23
	0.195	16	7	23	0.44
	0.207	17	7	24	0.41
	0.233	15	15	30	1.0
	0.318	13	26	39	2.0
	0.404	12	38	50	3.2
	0.490	13	43	56	3.3
	0.513	12	44	56	3.7
PP	0.196	20	3	23	0.15
	0.208	15	14	28	0.93
	0.233	17	14	31	0.82
	0.266	15	23	38	1.5
	0.310	12	37	49	3.1
	0.370	20	41	61	2.1
	0.427	18	44	62	2.4
	0.507	13	53	66	4.1

[a] Fuel bed area = 0.0068 m^2.
[b] Mass fraction of oxygen in atmosphere ($m_{O_2} = 0.232$ in air).

Table 5.10   Burning rates of plastics fires (Markstein, 1979)

Fuel[a]	Emissivity[b]	$\dot{m}''$ (g/m².s)
Polystyrene	0.83	14.1 ± 0.8
Polypropylene	0.4	8.4 ± 0.6
Polymethylmethacrylate	0.25	10.0 ± 0.7^c
Polyurethane foam	0.17	8.2 ± 1.8
Polyoxymethylene	0.05	6.4 ± 0.5

[a] Except for polyurethane foam, the fuels were burnt as pools, 0.31 × 0.31 m. Data for PUF deduced from a spreading fire.
[b] As measured 0.051 m above the fuel bed.
[c] 0.73 m diameter pool of PMMA gave $\dot{m}'' = 20.0 \pm 1.4$ g/m².s.

This agrees closely but not exactly with the ranking of these plastics according to their rates of burning (Table 5.10), namely:

$$PS > PMMA > PP > PUF > POM$$

Only PP and PMMA are out of sequence, which can be explained at least in part by the differences in the heats required to produce the volatiles: that for PP is 25% larger than that for PMMA (Table 5.8).

If the material is burning in an enclosure fire, in which the heat flux to the surface comes from general burning within the space (Chapter 10), the rate at which it will contribute heat to the compartment will be calculated from Equation (5.1), i.e.

$$\dot{Q}_c = \dot{m}'' \chi \Delta H_c A_F \tag{5.1a}$$

where A_F is the fuel surface area. Writing $\dot{Q}''_{net}$ as the net heat flux entering the surface, Equation (5.1a) may be rewritten:

$$\dot{Q}_c = \frac{\dot{Q}''_{net}}{L_v} \chi \Delta H_c A_F \tag{5.1b}$$

or

$$\frac{\dot{Q}_c}{A_F} = \dot{Q}''_{net} \chi \left(\frac{\Delta H_c}{L_v} \right) \tag{5.1c}$$

Given that χ lies within a relatively narrow range (0.4–0.7, according to Tewarson (1980)), it can be seen that the rate of heat release from a burning material is strongly dependent on $\Delta H_c/L_v$, which Rasbash (1976) refers to as the 'combustibility ratio'. Values calculated from Tewarson's data (but using the heat of combustion of the solid) are given in Table 5.11 (Tewarson, 1980). This shows that combustible solids have values in the range 3 (for red oak) to 30 (for a particular rigid polystyrene foam) and places materials in a ranking order which in its broad outline matches the consensus based on common knowledge of the steady burning behaviour of these materials. Liquid fuels tend to have much larger values of $\Delta H_c/L_v$, ranging up to 93 for heptane, with methanol having a low value in line with its high latent heat of evaporation and relatively low ΔH_c (Table 1.13) (see Section 5.1). As hydrocarbon

Table 5.11 $\Delta H_c / L_v$ values for fuels (Tewarson, 1980)

Fuel[a]	$\Delta H_c / L_v^b$
Red oak (solid)	2.96
Rigid PU foam (43)	5.14
Polyoxymethylene (granular)	6.37
Rigid PU foam (37)	6.54
Flexible PU foam (1-A)	6.63
PVC (granular)	6.66
Polyethylene 48% Cl (granular)	6.72
Rigid PU foam (29)	8.37
Flexible PU foam (27)	12.26
Nylon (granular)	13.10
Flexible PU foam (21)	13.34
Epoxy/FR/glass-fibre (solid)	13.38
PMMA (granular)	15.46
Methanol (liquid)	16.50
Flexible PU foam (25)	20.03
Rigid polystyrene foam (47)	20.51
Polypropylene (granular)	21.37
Polystyrene (granular)	23.04
Polyethylene (granular)	24.84
Rigid polyethylene foam (4)	27.23
Rigid polystyrene foam (53)	30.02
Styrene (liquid)	63.30
Heptane (liquid)	92.83

[a] Numbers in parentheses are PRC sample numbers (Products Research Committee, 1980).

[b] ΔH_c measured in an oxygen bomb calorimeter and corrected for water as a vapour for fuels for which data are not available: L_v is obtained by measuring the mass loss rate of the fuel in pyrolysis in N_2 environment as a function of external heat flux for fuels for which data are not available. *Note:* If ΔH_c is replaced by the heat of combustion of the volatiles, $(\Delta H_c + L_v)$, then all the 'combustibility ratios' are increased by 1.00 and the ranking order is unchanged.

polymers (e.g. polyethylene) tend to have much higher heats of combustion than their oxygenated derivatives (e.g. polymethylmethacrylate), their 'combustibility ratios' tend to be greater. However, while these figures are likely to give a reasonable indication of the ranking order for different materials, logically they should be calculated from the heat of combustion of the volatiles, rather than the net heat of combustion of the solid. The latter is determined by oxygen bomb calorimetry and, for char forming materials (e.g. wood), will include the energy released in oxidation of the char which would normally burn very slowly in a real fire, much of it after flaming combustion has ceased. The result of this is that the combustibility ratio for charring fuels will be slightly overestimated (see Table 5.11).

Fire retardants can influence the 'combustibility ratio' by altering ΔH_c and/or L_v. This can be achieved by changing the pyrolysis mechanism (see Section 5.2.2) or effectively 'diluting' the fuel by means of an inert filler such as alumina trihydrate (Lyons, 1970). However, the rate of heat release (Equation (5.1a)) is influenced by χ, the combustion 'efficiency', which for some fire-retarded species may be as low as 0.4. Tewarson (1980) suggests that χ may vary from 0.7 to 0.4, decreasing in the following order:

Aliphatic > Aliphatic/Aromatic > Aromatic > Highly halogenated species

Some values obtained using the Factory Mutual Flammability Apparatus are given in Table 5.12 (Tewarson, 1982).

If the surface of a combustible solid is vertical, the interaction between the flame and the fuel is quite different. The flame clings to the surface, entraining air from one side only (Figure 5.10(a)), effectively filling the boundary layer and providing convective heating as the stream of burning gas flows over the surface. The surface 'sees' a flame whose thickness is a minimum at the base of the vertical surface where the flow is laminar, but increases with height as fresh volatiles mix with the rising plume to yield turbulent flaming above $\sim$0.2 m. Measurements on thick, vertical slabs of PMMA, 1.57 and 3.56 m high, have shown that the local steady burning rate exhibits a minimum at approximately 0.2 m from the lower edge, thereafter increasing with height and reaching a maximum at the top (Figure 5.10(b)) (Orloff *et al.*, 1974, 1976). Calculations based on measurements of the emissive power of the flame as a function of height indicate that this can be attributed to radiation. It was estimated that 75–87% of the total heat transferred to the surface was by radiation (Orloff *et al.*, 1976). While these results refer specifically to PMMA, it is likely that this conclusion will apply generally to steady burning of vertical surfaces. However, it should be borne in mind that many synthetic materials (i.e. most thermoplastics) will melt and flow while burning. Not only will this lead to the establishment of a pool fire at the base of the wall, but it will also

Table 5.12 Fraction of heat of combustion released during burning in the Factory Mutual Flammability Apparatus (Figure 5.7) (Tewarson, 1982)

Fuel	$\dot{Q}''_E$ (kW/m^2)	χ	χ_{conv}	χ_{rad}
Methanol(*l*)	0	0.993	0.853	0.141
Heptane (*l*)	0	0.690	0.374	0.316
Cellulose	52.4	0.716	0.351	0.365
Polyoxymethylene	0	0.755	0.607	0.148
Polymethylmethacrylate	0	0.867	0.622	0.245
	39.7	0.710	0.340	0.360
	52.4	0.710	0.410	0.300
Polypropylene	0	0.752	0.548	0.204
	39.7	0.593	0.233	0.360
	52.4	0.679	0.267	0.413
Styrene (*l*)	0	0.550	0.180	0.370
Polystyrene	0	0.607	0.385	0.222
	32.5	0.392	0.090	0.302
	39.7	0.464	0.130	0.334
Polyvinylchloride	52.4	0.357	0.148	0.209

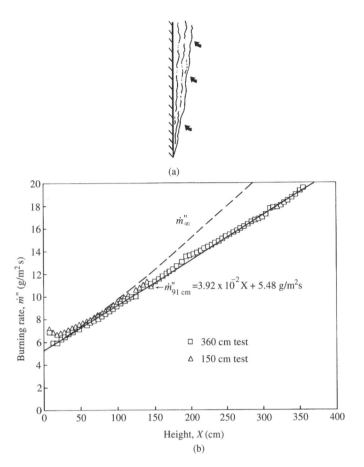

Figure 5.10   (a) Illustration of burning at a vertical surface. (b) Variation of local steady burning rate per unit area with distance from the bottom of vertical PMMA slabs 0.91 m wide, 3.6 m high (□) and 1.5 m high (△): – – – – Predicted burning rate for an infinitely wide slab. (Reproduced by permission of the Combustion Institute from Orloff *et al.*, 1976)

affect the burning behaviour of the vertical surface. The combined effect of the flames from the horizontal and vertical surfaces will inevitably produce vigorous burning which will create special problems in confined spaces and enclosures (Chapter 10). Indeed, the use of large-scale tests as a means of assessing the fire hazard of wall lining materials reflects the awareness of potentially dangerous situations of this type (e.g. ISO, 1993b).

Burning of horizontal, downward-facing combustible surfaces tends not to occur in isolation and consequently has received limited attention. Combustible ceiling linings may become involved during the growth, or pre-flashover, period of a compartment fire (Section 9.2) and will contribute to the extension of flames under the ceiling (Hinkley and Wraight, 1969) (Section 4.3.4), but will rarely ignite and burn without a substantial input of heat from the primary fire at floor level or elsewhere. It has been shown that flames on the underside of small slabs of polymethylmethacrylate tend to be very thin and weak, providing a relatively low heat flux to the surface $(\dot{Q}_F'')$ compared with burning in the pool configuration. Thus, Ohtani *et al.* (1981) estimated $\dot{Q}_F''$ to be 8 kW/m^2 and 22 kW/m^2 for bottom and top burning, respectively, from data obtained with 50 mm square slabs of PMMA. The appearance and behaviour of flames in this configuration have been investigated by Orloff and de Ris (1972) using downward-facing porous gas burners to allow fuel flowrate to be independent of the rate of heat transfer. Flames with a cellular structure are produced, their size and behaviour depending on the flowrate of gaseous fuel. The 'cells' were small and quite distinctive, growing in size with increasing fuel flowrate, but always present even at the highest flows that could be achieved in their apparatus. Cellular-like flame structures have been observed occasionally on the underside of combustible ceilings during the later stages of compartment fires.

The discussion so far has referred to plane surfaces burning in isolation or in an experimental situation with an imposed heat flux. In 'real fires', isolated burning will only occur in the early stages before fire spreads beyond the item first ignited. Once the area of fire has increased, cross-radiation from flames and between different burning surfaces will enhance both the rate of burning and the rate of spread (Section 7.2.5). Indeed, wherever there is opportunity for heat to build up in one location, increased rates of burning will result (Section 9.1). This can be expected in any confined space in which combustible surfaces are in close proximity (Section 2.4.1). The most hazardous configurations in buildings are ducts, voids and cavities which, if lined with combustible materials, provide optimal configurations for rapid fire spread and intense burning. Such conditions must be avoided or adequately protected.

5.2.2 Burning of wood

Unlike synthetic polymers, wood is an inhomogeneous material which is also non-isotropic, i.e. many of its properties vary with the direction in which the measurement is made. It is a complex mixture of natural polymers of high molecular weight, the most important of which are cellulose ($\sim$50%), hemicellulose ($\sim$25%) and lignin ($\sim$25%) (Madorsky, 1964), although these proportions vary from species to species. Moreover, it normally contains absorbed moisture, the amount of which will vary according to the relative humidity and the conditions of exposure. Cellulose, which is the principal constituent of all higher plants, is a condensation polymer of the hexose sugar, D-glucose (Figure 5.11(a)), and adopts the linear structure shown in Figure 5.11(b). This configuration allows the molecules to align themselves into bundles (microfibrils) which provide the structural strength and rigidity of the cell wall. The microfibrils are bound together during the process of lignification when the hemicellulose and lignin are laid down in the growing plant.

The structure of hemicellulose is similar, based on pentose sugars, but that of lignin is vastly more complex (Greenwood and Milne, 1968). Thermogravimetric analysis of the degradation of wood, cellulose and lignin (Figure 5.12) shows that the constituents decompose to release volatiles at different temperatures, typically:

$$\begin{array}{ll}
\text{Hemicellulose} & 200 - 260°\text{C} \\
\text{Cellulose} & 240 - 350°\text{C} \\
\text{Lignin} & 280 - 500°\text{C}
\end{array}$$

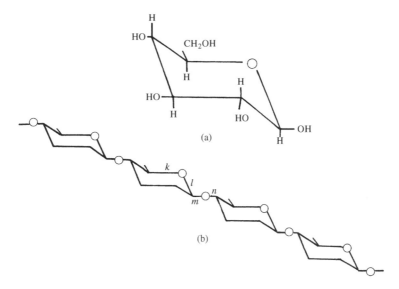

(a)

(b)

Figure 5.11   (a) β-D-Glucopyranose (the stable configuration of D-glucose); (b) part of a cellulose molecule

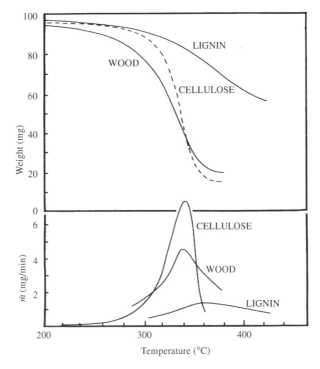

Figure 5.12   (a) Thermogravimetric analysis of 90–100 mg samples of wood (Ponderosa pine), cellulose powder (Whatman) and lignin, heated under vacuum at 3° rise in temperature per minute. (b) Derivative TGA curves from (a). (Browne and Brenden, 1964. Reproduced by permission of Forest Products Laboratory, Forest Service, USDA, Madison, Wisconsin)

(Roberts, 1970). If lignin is heated to temperatures in excess of 400–450°C, only about 50% volatilizes, the balance of the mass remaining as a char residue. On the other hand, pure 'α-cellulose' — the material extracted from cotton and washed thoroughly to leach out any soluble inorganic impurities — leaves only 5% char after prolonged heating at 300°C. However, if inorganic impurities (e.g. sodium salts, etc.) are present, much higher yields are found: for example, viscose rayon (a fibre consisting of regenerated cellulose and having a relatively high residual inorganic content) can give over 40% char (Madorsky, 1964). When wood is burnt, or heated above 450°C, 15–25% normally remains as char, much of this coming from the lignin content (up to 10–12% of the original weight of the wood). However, the yield from the cellulose (and presumably the hemicellulose) is variable, depending on the temperature or rate of burning, and more importantly on the nature and concentration of any inorganic salts that are present. The significance of this is that the nature of the volatiles must change if there is a change in the yield of char, and consequently the fire behaviour will be altered.

Several authors have reported on the effects on the decomposition of wood of temperature (Madorsky, 1964) and inorganic impurities (Lyons, 1970). Brenden (1967) illustrates the latter most clearly by comparing the yields of 'char', 'tar', water and 'gas' (mainly CO and CO_2) from samples of Ponderosa pine which had been treated with a number of salts capable of imparting some degree of fire retardancy (Table 5.13). The fraction designated 'tar' contains the combustible volatiles and consists of products of low volatility, the most important of which is believed to be levoglucosan:

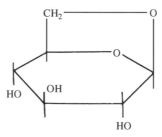

It appears that there are two competing mechanisms of cellulose degradation. Referring to Figure 5.11(b) (Madorsky, 1964), if any of the bonds of the type marked k or l break, a six-membered ring will open but the continuity of the polymer chain remains

Table 5.13   Pyrolysis of Ponderosa pine (Brenden, 1967)

	Concentration of applied solution	Treatment level	Char	Tar	Water	Gas[a]
Untreated wood	—	—	19.8	54.9	20.9	4.4
$+Na_2B_4O_7$	5%	4.28%	48.4	11.8	30.4	9.4
$+(NH_4)_2.HPO_4$	5%	6.69%	45.5	16.8	32.0	5.7
+ ammonium polyphosphate	5%	5.0%	43.8	19.0	34.6	2.6
$+H_3BO_3$	5%	3.9%	46.2	10.7	33.9	9.2
+ ammonium sulphamate[b]	5%	6.3%	49.8	2.6	33.4	14.2
$+H_3PO_4$	5%	6.8%	54.1	2.5	37.3	6.1

[a] 'Non-condensable gases': CO, CO_2, H_2, CH_4.
[b] $NH_4.NH_2.SO_3$.

intact. It is suggested that under these circumstances the products are char, with CO, CO_2 and H_2O as the principal volatiles. If, on the other hand, bonds *m* or *n* break, the polymer chain 'backbone' is broken, leaving exposed reactive ends from which levoglucosan molecules can break away and volatilize from the high temperature zone. Low rates of heating, or relatively low temperatures, appear to favour the char-forming reaction. Similarly, the range of fire retardants commonly used to improve the response of wood to fire (e.g. phosphates and borates) act by promoting the char-forming process at the expense of 'tar' formation. Table 5.13 shows how the char yields can be more than doubled by treating pine with phosphates and borates, while at the same time the composition of the volatiles changes in favour of a lower proportion of the flammable 'tar' constituent (Brenden, 1967). As a result, the heat of combustion of the volatiles is decreased which will lower the amount of heat that can be transferred to the surface from the flame ($\dot{Q}_F''$ in Equation (5.24)) so that $\dot{m}_{ideal}''$ will be reduced (Equation (5.26)). Consequently a higher imposed heat flux ($\dot{Q}_E''$ in Equation (5.24)) may be necessary to allow sustained burning (see Table 5.8). As it accumulates, the layer of char will protect the unaffected wood below and higher surface temperatures will be required to provide the necessary heat flow to produce the flow of volatiles. The higher surface temperatures will mean greater radiative heat losses (included in $\dot{Q}_L''$) but against this there will be some surface oxidation of the char which will contribute positively to the heat balance.

As a slab of wood burns away and a layer of char accumulates, it seems likely that the composition of the volatiles will change. Roberts (1964a,b) found no evidence for this although his samples may have been too small to show the effect. He carried out combustion bomb calorimetry on small samples of wood (dry European beech), partially decomposed wood samples, and 'char', which enabled him to deduce the 'heat of combustion of the volatiles'. His results are given in Table 5.14. In these experiments the char yield was 16–17% of the original wood, indicating on the basis of data in Table 5.14 that it accounted for ~30% of the total heat production and consumed ~33% of the total air requirement of the wood. Browne and Brenden (1964) carried out similar experiments with dry Ponderosa pine and found evidence that the composition of the volatiles did change but, perhaps contrary to expectation, became *more* combustible as the degradation proceeded: their results were as follows:

at 10% weight loss ΔH_c (volatiles) $= 11.0$ kJ/g

at 60% weight loss ΔH_c (volatiles) $= 14.2$ kJ/g

Parent wood ΔH_c (wood) $= 19.4$ kJ/g

Table 5.14   Combustion of wood and its degradation products (Roberts, 1964a)

	Wood[a]	Volatiles	Char
Gross heat of combustion (kJ/g)[b]	19.5	16.6	34.3
Mean molecular formula	$CH_{1.5}O_{0.7}$	CH_2O	$CH_{0.2}O_{0.02}$
Theoretical air requirements (g air/g fuel)[c]	5.7	4.6	11.2

[a] European beech.
[b] By combustion bomb calorimetry.
[c] Section 1.2.3.

This apparent difference remains to be resolved. Browne and Brendan (1964) also showed that ΔH_c (the heat of combustion of the volatiles) was less for a fire retarded wood than for the parent wood, in agreement with the observations made in Table 5.13.

(a) *Burning of wooden slabs and sticks.*   The complexity of wood makes it difficult to interpret the burning behaviour in terms of Equation (5.3). Because of the grain structure, properties vary with direction: thus the thermal conductivity parallel to the grain is about twice that perpendicular to the grain, and there is an even greater difference in gas permeability (of the order of 10^3 (Roberts, 1971a)). Volatiles generated just below the surface of the unaffected wood can escape more easily along the grain than at right angles towards the surface. The appearance of jets of volatiles and flame from the end of a burning stick or log, or from a knot, is evidence for this.

Wood discolours and chars at temperatures above 200–250°C, although prolonged heating at lower temperatures ($\gtrsim 120°C$) will have the same effect. The physical struc-ture begins to break down rapidly at temperatures above 300°C. This is first apparent on the surface when small cracks appear in the char, perpendicular to the direction of the grain. This permits volatiles to escape easily through the surface from the affected layer (Figure 5.13) (Roberts, 1971a). The cracks will gradually widen as the depth of char increases, leading to the characteristic 'crazed' pattern which is frequently referred to as 'crocodiling' or 'alligatoring'. The appearance of such patterns in a fire-damaged building is widely believed to give an indication of the rate of fire development (e.g. Brannigan, 1980), but there has been no systematic investigation of this and the method must be regarded as highly questionable, as illustrated by De Haan (1997) (see also Cooke and Ide, 1985).

Clearly, the burning of wood is a much more complex process than that of synthetic polymers, charring or non-charring. Any theoretical analysis must take into account not only the terms in Equation (5.3) which will be complicated by the presence of the layer of char, but also the interactions within the hot char. The latter may be undergoing oxidation as small quantities of oxygen diffuse to the surface. This, being an exothermic process, would contribute heat to the decomposing wood and reduce the apparent heat of gasification.

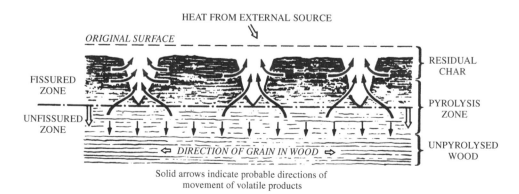

Figure 5.13   Representation of a cross-section through a slab of burning, or pyrolysing, wood (reproduced by permission of the Combustion Institute from Roberts, 1971a)

There has been a lack of consensus in the literature regarding the value of L_v for wood, with values from 1.8 (and less) to 7 kJ/g having been reported for a range of species, including both hardwoods and softwoods (Thomas *et al.*, 1967b; Tewarson and Pion, 1976; Petrella, 1979). There is, of course, a wide variation in the composition and structure between woods of different species. Thomas *et al.* (1967b) proposed that there might be a correlation between L_v and permeability, while Hadvig and Paulsen (1976) suggested a link with the lignin content. Janssens (1993) carried out a very thorough study of six solid woods using the Cone Calorimeter to obtain the experimental data, which was then analysed by means of an integral heat transfer model. He found that L_v was not constant, but varied as the depth of char increased: for example, Victorian ash showed an initial value of about 3 kJ/g, increasing slightly (to 3.5–4 kJ/g) then decreasing slowly to about 1 kJ/g when the char depth was *c*.14 mm. The average value was 2.57 kJ/g. The averages for six species are shown in Table 5.15. There is an apparent difference between the softwoods ($L_v \simeq 3.2$ kJ/g) and the hardwoods ($L_v \simeq 2.6$ kJ/g), with one exception: L_v for Douglas Fir, a softwood with a high resin content, is 2.64 kJ/g.

Results such as these still require detailed interpretation. Janssens' work shows where some of the variability in reported values of L_v may lie. Thus, Thomas *et al.* (1967b) found the values of L_v to increase with mass loss (comparing measurements at 10% and 30% loss), but the absolute values which they reported are much higher than Janssens' (e.g. 5.1 kJ/g at 10% mass loss for Douglas fir). (It should be noted that Tewarson and Pion (1976) and Petrella (1979) studied horizontal samples, while Thomas *et al.* (1967b) and Janssens (1993) held their samples vertically.)

It is known that the rate of decomposition of wood, or cellulose in particular, is very sensitive to the presence of inorganic impurities, such as fire retardants (Figure 5.14). Thus the difference between L_v for Douglas fir and fire-retarded plywood reported by Tewarson and Pion (1976) (1800 J/g and 950 J/g respectively) is at least consistent with the catalytic action of the retardant chemicals on the char-forming reaction referred to above (see also Table 5.8). While the variation in the relative proportions of the three main constituents of wood from one species to another is likely to have an effect, variations in the content of inorganic constituents may predominate. This has not been investigated on a systematic basis.

It is common experience that a thick slab of wood will not burn unless supported by radiation (or convection) from another source (e.g. flames from a nearby fire or burning surface). This is in agreement with Tewarson's observation that $\dot{Q}_F'' \approx \dot{Q}_L''$ for

Table 5.15 Average values of L_v for various woods (Janssens, 1993)

Material	L_v (kJ/g)
Western red cedar	3.27
Redwood	3.14
Radiata pine	3.22
Douglas fir	2.64
Victorian ash	2.57
Blackbutt	2.54

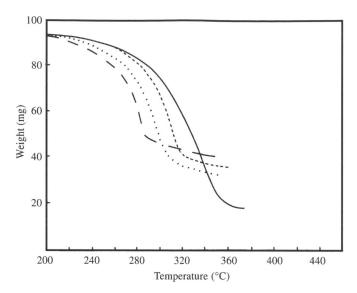

Figure 5.14   Thermogravimetric analysis of samples of wood (Ponderosa pine), untreated
————, and treated with various inorganic salts: - - - -, 2% $Na_2B_4O_7.10H_2O$;
$\cdots\cdots$, 2% NaCl; — — — —, 2% $NH_4.H_2.PO_4$ (Brenden, 1967. Reproduced by
permission of Forest Products Laboratory, Forest Service, USDA, Madison,
Wisconsin)

Douglas fir (Table 5.8) (Tewarson and Pion, 1976), i.e. the heat transfer from the
flame was theoretically just sufficient to match the heat losses from the sample under
burning conditions. Results obtained by Petrella (1979) with an apparatus similar to
that of Tewarson indicate that generally $\dot{Q}_F'' < \dot{Q}_L''$ for several species of wood. Clearly,
the ability of wood to burn depends on an imposed heat flux. Nevertheless, in the
literature — particularly that relating to fire investigation — the burning rate of wood
is quoted as 1/40 inch per minute (0.6 mm/min). This comes from observations of the
depth of char on wooden beams and columns which have been exposed to a standard
fire test (e.g. BS 476 Part 21 (British Standards Institution, 1987b)). It may be compared
with Tewarson's $\dot{m}_{ideal}''$ for Douglas fir (13 g/m^{2}s) (Tewarson and Pion, 1976) which
corresponds to an 'ideal burning rate' of $\sim$2 mm/min. Of course the rate of burning
R_W is *not* constant, as can be seen in Figure 5.15, but varies substantially with heat
flux according to

$$R_W = 2.2 \times 10^{-2} \ I \ \text{mm/min} \tag{5.30}$$

where I is in kW/m^2 (Butler, 1971). In a compartment fire. localized temperatures as
high as 1100°C may be achieved: the corresponding black body radiation is 200 kW/m^2,
which could result in 'rates of burning of wood' as high as 4.4 mm/min. Figure 5.15
refers to slabs of wood which are sufficiently thick as to be regarded as semi-infinite
solids for the duration of burning, remembering that only a relatively thin layer below
the regressing surface of the unaffected fuel is heated (cf. Figure 5.5). Higher rates
of burning will be observed for samples that are thermally thin, unless the heat
losses from the rear face (included with $\dot{Q}_L''$ of Equation (5.3)) are high.

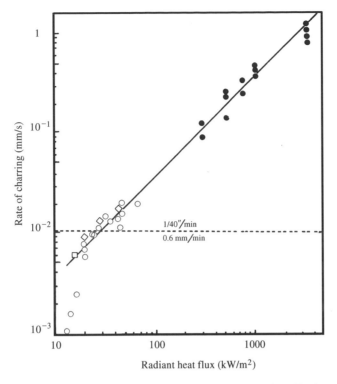

Figure 5.15   Variation of charring rate of wood with radiant heat flux (Butler, 1971). (Reproduced by permission of The Controller, HMSO. © Crown copyright)

While a thick slab of wood cannot burn in isolation, 'kindling' and thin pieces (e.g. wood shavings and matchsticks) can be ignited relatively easily and will continue to burn, although flaming may have to be established on all sides. This is possible because thin samples will behave as systems with low Biot number (Section 2.2.2) so that once ignition has occurred and the ignition source has been removed, the rate of heat loss from the surface into the body of the sample will be minimal (Section 6.3.2). It is possible to estimate the maximum thickness of wood that can still be regarded as 'thin' from the point of view of ignition. It will depend mainly on the duration of contact with the ignition source (assumed to be a flame). The depth of the heated layer is of the order $(\alpha t)^{1/2}$, where t is the time duration in seconds (Section 2.2.2). Thus, with a 10 s application time, the maximum thickness of a splint of oak ($\alpha = 8.9 \times 10^{-8}$ m^2/s) that may be ignited will be of the order of $2 \times (\alpha t)^{1/2} \approx 2 \times 10^{-3}$ m (assuming immersion of the splint in the flame). In principle, thicker samples could be ignited after longer exposure but then other factors, such as depletion of volatiles from the surface layers and direction of subsequent flame spread (Section 7.2.1), become important. The duration of burning of a 'thin' stick of wood varies roughly in proportion to D^n, where D is the diameter and $n \approx 1.6 \pm 0.2$ (e.g. Thomas, 1974a).

(b) *Burning of wood cribs*   In the earlier sections, burning at plane surfaces has been discussed as this is directly relevant to materials burning in isolation, although in real fires complex geometries and configurations can cause interactions which will influence behaviour strongly (Chapters 9 and 10). One type of fuel bed in which such complexities dominate behaviour is the wood crib, which comprises crossed layers of sticks. The confinement of heat within the crib and cross-radiation between burning surfaces allows sticks of substantial cross-section to burn efficiently (cf. log fires). They are still used as a means of generating reproducible fire sources for research and testing purposes, although they are being superseded by the sand-bed gas burner, particularly as the ignition source in large-scale fire tests (Babrauskas, 1992b). Several parameters may be controlled independently to produce a fuel bed which will burn at a known rate and for a specific duration: these include stick thickness (b), the number of layers (N), the separation of sticks in each layer (s) and the length of the sticks (Gross, 1962; Block, 1971). Moisture content is also controlled.

Gross (1962) identified two regimes of burning corresponding to 'under-ventilated' and 'well-ventilated' cribs. In the former, corresponding to densely packed cribs, the rate of burning is dependent on the ratio A_v/A_s, where A_s is the total exposed surface area of the sticks and A_v is the open area of the vertical shafts. He scaled the rate of burning with stick thickness, as $Rb^{1.6}$, where R is the rate in per cent per second, and compared it with a 'porosity factor' $\Phi = N^{0.5}b^{1.1}(A_v/A_s)$. This parameter is derived from the ratio $\dot{m}_{ac}/\dot{m}$ where $\dot{m}_{ac}$ is the mass flowrate of air through the vertical shafts and $\dot{m}$ is the total rate of production of volatiles. As $\dot{m}'' \propto b^{-0.6}$ (Gross, 1962) and assuming that $\dot{m}_{ac} \propto h_c^{1/2} \cdot A_v$, where $h_c(= Nb)$ is the height of the crib, then:

$$\frac{\dot{m}_{ac}}{\dot{m}} \propto \frac{(Nb)^{1/2} \cdot A_v}{A_s \cdot b^{-0.6}} = N^{1/2}b^{1.1}\frac{A_v}{A_s} \qquad (5.31)$$

Gross's plot of $Rb^{1.6}$ versus Φ is shown in Figure 5.16. For $\Phi < 0.08$, a linear relationship exists between $Rb^{1.6}$ and Φ, but $Rb^{1.6}$ is approximately constant when $\Phi > 0.1$. The latter case corresponds to good ventilation and substantial flaming within the crib, the rate of burning being controlled by the thickness of the individual sticks. (Sustained burning is not possible if $\Phi > 0.4$.)

5.2.3 Burning of dusts and powders

While finely divided combustible materials can behave in fire as simple fuel beds as described in the previous sections, two additional modes of behaviour must be considered, namely the ability to give rise to smouldering combustion and the potential to create an explosible dust cloud. Smouldering can only occur with porous char-forming materials such as sawdust or wood flour. The mechanism requires that surface oxidation of the rigid char generates enough heat to cause the unaffected fuel adjacent to the combustion zone to char which, in turn begins to oxidize. Flaming combustion on a pile of sawdust (as an example) may self-extinguish (if $\dot{Q}_E''$ is insufficient) or be extinguished, only to leave behind a smouldering fire which will progress inwards. A non-charring thermoplastic powder when exposed to heat will simply melt and form a liquid pool. The phenomenon of smouldering will be discussed at greater length in Section 8.2.

Many, but not all, combustible dusts are capable of burning rapidly in air if they are thrown into suspension as a dust cloud. This is equivalent to the burning of a flammable

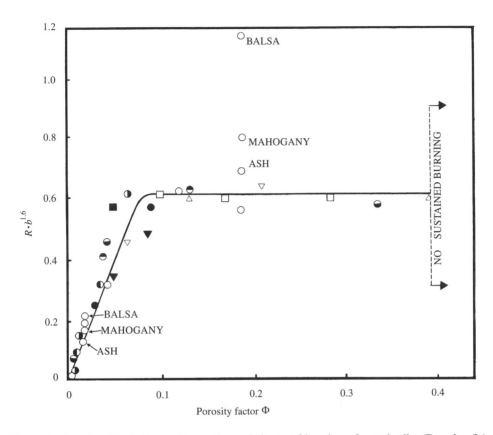

Figure 5.16   The effect of porosity on the scaled rate of burning of wood cribs (Douglas fir). Symbols refer to a range of stick thicknesses (*b*) and spacings (*s*) (Gross, 1962)

mist or droplet suspension to which reference has already been made (Section 5.1.2). The hazard associated with the formation and ignition of dust clouds in industrial and agricultural environments has long been recognized: extremely violent and damaging explosions are possible (e.g. coal dust explosions in mines, dust explosions in grain elevators, etc.). As with flammable gases and vapours, it is possible to identify limits of flammability (or more correctly *explosibility*), minimum ignition energies, auto-ignition temperatures, etc. The subject has been studied in great depth but is considered to be outside the scope of the present text. There are several excellent reviews and monographs which cover the subject at all levels (Palmer, 1973; Bartknecht, 1981; Field, 1982; Lees, 1996; Eckhoff, 1997).

Problems

5.1   Calculate the rate of heat release from a fire involving a circular pool of *n*-hexane of diameter 2 m, assuming that the efficiency of combustion is 0.85. Use the limiting regression rate given in Table 5.1. Compare your result with that given for *n*-heptane in Table 4.3.

5.2 Carry out the same calculation as in Problem 5.1 for a 2 m diameter pool of methanol, but assume 100% combustion efficiency. Consider what information you would need to calculate the radiant flux at a point distant from this flame.

5.3 An explosion occurs in a crude oil tank (15 m in diameter), leaving behind a fully developed pool fire. Massive attack with firefighting foam at 4 hours has the fire extinguished within 15 minutes. By this time, the hydrocarbon liquid level has fallen by 0.5 m. What was the average rate of burning and the average rate of heat release, if the density of the oil is 800 kg/m^3? Assume 85% combustion efficiency.

5.4 Considering the fire described in Problem 5.3, use the formula by Shokri and Beyler to calculate the maximum radiant heat flux falling on adjacent, identical tanks within the same bund. The minimum distance between tanks is 10 m.

5.5 Compare Tewarson's 'ideal burning rates' (Table 5.8) for n-heptane and methanol with the limiting burning rates for large pools of these liquids (see Tables 5.1 and 5.2). (Take R_∞(n-heptane) $= 7.3$ mm/min.) What can you deduce from the results?

5.6 Using the data in Table 5.2, calculate the rates of burning of 0.5 m diameter trays of the following liquid fuels: (a) Ethanol; (b) Hexane; and (c) Benzene. (Note the differences between these and the respective limiting values.)

5.7 You are required to 'design' a fire which will burn for about 10 minutes with a heat output of approximately 750 kW. What diameter of tray, or pan, would you require to contain (a) ethanol; and (b) heptane to achieve a fire with these characteristics? Assume that combustion is 100% efficient, and take the heat of combustion of heptane to be 45 MJ/kg. (Hint: first try estimating the tray diameter assuming the limiting rates of burning.)

5.8 A horizontal sheet of black polymethylmethacrylate is allowed to burn at its upper surface while its lower surface is maintained at 20°C. Treating the sheet as an infinite slab, calculate the rate of burning if the burning surface is at 350°C, and the sheet is (a) 4 mm thick and (b) 2 mm thick. (Assume $\varepsilon = 1$ and the thermal conductivity is constant.)

6
Ignition: The Initiation of Flaming Combustion

Ignition may be defined as that process by which a rapid, exothermic reaction is initiated, which then propagates and causes the material involved to undergo change, producing temperatures greatly in excess of ambient. Thus, ignition of a stoichiometric propane/air mixture triggers the oxidation reaction which propagates as a flame through the mixture, converting the hydrocarbon to carbon dioxide and water vapour at temperatures typically in the range 2000–2500 K (Chapter 1). It is convenient to distinguish two types of ignition, namely piloted — in which flaming is initiated in a flammable vapour/air mixture by a 'pilot', such as an electrical spark or an independent flame — and spontaneous — in which flaming develops spontaneously within the mixture. To achieve flaming combustion of liquids and solids, external heating is required, except in the case of pilot ignition of flammable liquids which have firepoints below ambient temperature (see Section 6.2.1). The phenomenon of spontaneous ignition within bulk solids, which leads to smouldering combustion, will be discussed separately in Chapter 8.

The objectives of this chapter are to gain an understanding of the processes involved in ignition and to examine ways in which the 'ignitability' or 'ease of ignition' of combustible solids might be quantified. The phenomenon of extinction has many features in common with ignition and is discussed briefly in Section 6.6. However, as initiation of flaming necessarily involves reactions of the volatiles in air, it is appropriate to start with a review of ignition of flammable vapour/air mixtures.

6.1   Ignition of Flammable Vapour/Air Mixtures

It has been shown elsewhere (Section 1.2.3) that the reaction between a flammable vapour and air is capable of releasing a substantial amount of energy, but it is the rate of energy release that will determine whether or not the reaction will be self-sustaining and propagate as a flame through a flammable mixture (Section 3.3). To illustrate this point, it can be assumed that the rate of the oxidation processes obeys an Arrhenius-type temperature dependence (Equation (1.2)). The rate of heat release within a small volume would then be given by:

$$\dot{Q}_c = \Delta H_c V C_i^n A \exp(-E_A/RT) \tag{6.1}$$

where A is the 'pre-exponential factor' whose units will depend on the order of the reaction, n, C_i is the concentration (mole/m^3) and ΔH_c is the heat of combustion (kJ/mole). There is no temperature limit below which $\dot{Q}_c = 0$: oxidation occurs even at ordinary ambient temperatures, although in most cases at a negligible rate. The heat generated is lost to the surroundings and consequently there is no significant rise in temperature and $\dot{Q}_c$ remains negligible. This is illustrated schematically in Figure 6.1 as a plot of $\dot{Q}_c$ against temperature, superimposed on a similar plot of the rate of heat loss, $\mathscr{L}$. The latter is assumed to be directly proportional to the temperature difference, ΔT, between the reaction volume and the surroundings, i.e.

$$\dot{\mathscr{L}}_1 = hS\Delta T \tag{6.2}$$

where h is a heat transfer coefficient and S is the surface area of the reaction volume through which heat is lost. The intersection at p_1 represents a point of equilibrium $(\dot{Q}_c = \dot{\mathscr{L}})$, limiting the temperature rise to $\Delta T = (T_{p_1} - T_{a_1})$. (This is exaggerated in Figure 6.1 for clarity.) Slight perturbations about this point are stable and the system will return to its equilibrium position. This cannot be said for p_2: although $\dot{Q}_c = \dot{\mathscr{L}}$ at this point, perturbations lead to instability. For example, if the temperature is reduced infinitesimally, then $\dot{\mathscr{L}} > \dot{Q}_c$ and the system will cool and move to p_1. Alternatively, at a temperature slightly higher than T_{p_2}, $\dot{Q}_c > \dot{\mathscr{L}}$, and the system will rapidly increase in temperature to a new point of stability at p_3. This corresponds to a stable, high temperature combustion reaction which can propagate as a premixed flame. Although Figure 6.1 is schematic, arguments based on it are valid in a qualitative sense. However,

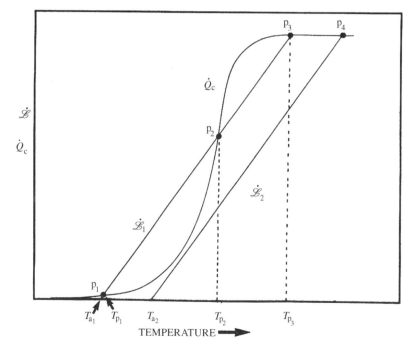

Figure 6.1   Rates of heat production ($\dot{Q}_c$) and heat loss ($\dot{\mathscr{L}}$) as functions of temperature

it should be noted that it does not indicate that there is a limit to the temperature that the reacting mixture can achieve because of the thermal capacity of the products (Section 1.2.3), nor that the heat loss function will change at higher temperatures, especially when radiative losses become significant (Section 2.4.2).

Referring to Figure 6.1, it can be seen that to ignite a flammable vapour/air mixture at an ambient temperature T_{a_1}, sufficient energy must be available to transfer the system from its stable state (p_1) at a low temperature (T_{p_1}) to an unstable condition at a temperature greater than T_{p_2}. The concept of a minimum ignition energy for a given flammable vapour/air mixture (Figure 3.3) is quite consistent with this, although when the ignition source is an electrical spark, Figure 6.1 is not entirely satisfactory. Given that an electrical discharge generates a transient plasma, rich in atoms, free radicals and ions, free radical initiation must contribute significantly in spark ignition. The energy dissipated in the weakest spark capable of igniting a stoichiometric propane/air mixture (0.3 mJ) is capable of raising the temperature of a spherical volume of diameter equal to the quenching distance (2 mm) by only a few tens of degrees. Without free radical initiation, a rise of several hundred degrees is necessary to promote rapid ignition (see below).

The minimum ignition energies quoted in Table 3.1 refer to electrical sparks generated between two electrodes whose separation cannot be less than the minimum quenching distance (Table 3.1), otherwise heat losses to the electrodes will cause the reaction zone to cool and prevent flame becoming established (Section 3.3). If the electrodes are free, as defined in Figure 6.2, then ignition can be achieved when their separation is less than the quenching distance (d_q) simply by increasing the spark energy to overcome heat losses to the electrodes. However, if the electrodes are flush mounted through glass discs (Figure 6.2) and their separation is less than d_q, ignition

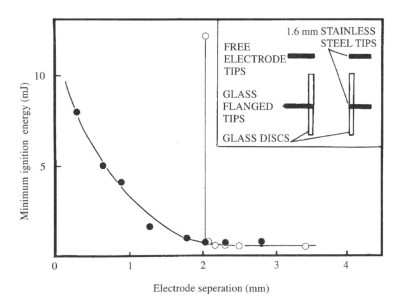

Figure 6.2   Minimum ignition energies for free (●) and glass-flanged (○) electrode tips as a function of electrode distance (stoichiometric natural gas/air mixture). (Reproduced by permission of Academic Press from Lewis and von Elbe, 1987)

is not possible because the flame will be quenched as it propagates away from the spark.

Other ignition sources include mechanical sparks, hot surfaces and glowing wires.* These involve convective heat transfer from the solid surface to the gas, and ignition occurs spontaneously in the hot boundary layer. Figure 6.1 may be used qualitatively to illustrate the mechanism. Imagine a small volume of flammable vapour/air mixture within the boundary layer which, for simplicity, is assumed to be at a uniform temperature T_{a_2}. Under these conditions, $\dot{Q}_c > \dot{\mathscr{L}}$, and the temperature of the element of volume will rise rapidly. In this particular case, the rate of heat loss is unable to prevent a runaway reaction, and ignition will occur as the system transfers to the intersection p_4, corresponding to the high temperature combustion process (see above). In reality, the temperature in the boundary layer is not uniform (Figure 2.15) and the rate of heat loss will be influenced strongly by any air movement or turbulence. Consequently, whether or not flame will develop will depend on the extent of the surface, its geometry and temperature, as well as the ambient conditions. The minimum temperatures for ignition of stoichiometric vapour/air mixtures which are quoted in the literature (British Standards Institution, 1977; National Fire Protection Association, 1997) refer to uniform heating of a substantial volume of mixture (>0.2 litre) enclosed in a spherical glass vessel. Under these conditions, the mixture is static and there will be a measurable delay (τ) before ignition occurs, particularly at temperatures close to the minimum 'auto-ignition temperature' (Figure 6.3) when τ may be found to be of the order of 1 s or more.

The existence of a critical ignition temperature for flammable mixtures led to the development of thermal explosion theory, based on Equations (6.1) and (6.2) (Semenov, 1928). Semenov assumed that the temperature within the reacting gas

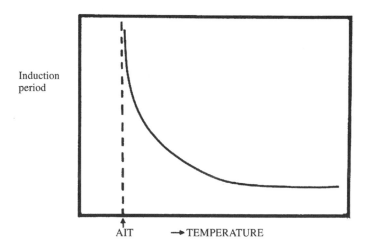

Figure 6.3   Variation of induction period with temperature for stoichiometric fuel/air mixtures (schematic) (AIT = Autoignition Temperature)

---

* Smouldering cigarettes cannot ignite common flammable gases and vapours such as methane, propane, and petrol (gasoline) vapour. However, there is evidence that they can cause ignition of hydrogen, carbon disulphide, diethyl ether and other highly reactive species (Holleyhead, 1996).

mixture remained uniform (Figure 6.4(a)) and that heat losses were described accurately by Equation (6.2), where ΔT is the temperature difference between the gas and the walls of the enclosing vessel. It was also assumed that reactant consumption was negligible and that the rate followed the Arrhenius temperature dependence (Equation (6.1)). Figure 6.5 shows $\dot{Q}_c$ and $\dot{\mathscr{L}}$ plotted against temperature for three values of the ambient (i.e. wall) temperature.

The critical ambient temperature ($T_1 = T_{a,cr}$) is identified as that giving a heat loss curve which intersects the heat production curve, $\dot{Q}_c$, tangentially. This may be expressed mathematically as:

$$\dot{Q}_c = \dot{\mathscr{L}} \tag{6.3a}$$

and

$$\frac{d\dot{Q}_c}{dT} = \frac{d\dot{\mathscr{L}}}{dT} \tag{6.3b}$$

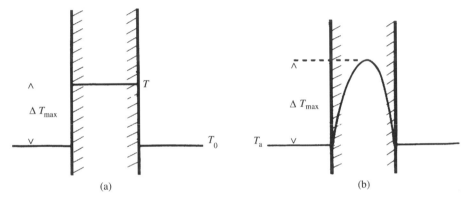

Figure 6.4 Temperature profiles inside spontaneously-heating systems according to the models of (a) Semenov and (b) Frank-Kamenetskii (schematic)

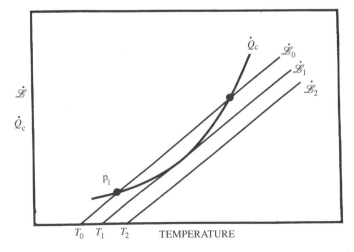

Figure 6.5 Illustrating Semenov's model for spontaneous ignition ('thermal explosion theory') (Equations (6.3)–(6.7))

which, stated in full, give

$$\Delta H_c V C_i^n A \exp(-E_A/RT) = hS(T - T_{a,cr}) \qquad (6.4)$$

and

$$\frac{E_A}{RT^2} \Delta H_c V C_i^n A \exp(-E_A/RT) = hS \qquad (6.5)$$

Dividing Equation (6.4) by Equation (6.5) gives:

$$\frac{RT^2}{E_A} = T - T_{a,cr} \qquad (6.6)$$

where $T_{a,cr}$ is the critical ambient temperature and T is the corresponding (equilibrium) gas temperature. The difference $(T - T_{a,cr})$ is the maximum temperature rise that can occur spontaneously within this system without ignition taking place. Provided that $E_A \gg RT$, Equation (6.6) can be solved by binomial expansion to give

$$\Delta T_{crit} = T - T_{a,cr} \approx \frac{RT_{a,cr}^2}{E_A} \qquad (6.7)$$

For typical values of $T_{a,cr}$ and E_A (700 K and 200 kJ/mol respectively) $\Delta T_{crit} \approx 20$ K.

However, the Semenov model is unrealistic in that it ignores temperature differences within the reacting system: it is a 'lumped thermal capacity', or low Biot number model, as defined in Section 2.2.2. Frank-Kamenetskii (1939) developed a high Biot number model (Bi > 10: see Figure 6.4(b)) based on Equation (2.14), i.e.

$$\nabla^2 T + \frac{\dot{Q}_c'''}{k} = \frac{1}{\alpha} \frac{\partial T}{\partial t} \qquad (6.8)$$

To simplify the solution, this can be reduced to the one-dimensional case with uniform, symmetrical heating, thus:

$$\frac{\partial^2 T}{\partial r^2} + \frac{\kappa}{r} \frac{\partial T}{\partial r} + \frac{\dot{Q}_c'''}{k} = \frac{1}{\alpha} \frac{\partial T}{\partial t} \qquad (6.9)$$

where κ takes values of 0, 1 and 2 for an infinite slab (thickness $2r_0$), an infinite cylinder (radius r_0) and a sphere (radius r_0) respectively. Further simplification was achieved by assuming that (i) the reaction rate can be described by a single Arrhenius expression (Equation (6.1)); (ii) there is no reactant consumption (cf. Semenov, 1928); (iii) the Biot number is sufficiently large for conduction within the reacting volume to determine the rate of heat loss; and (iv) the thermal properties of the system are constant (independent of temperature). Thus, the boundary conditions for Equation (6.9) are:

$$T_{r_0} = T_0 \text{ at } t \geq 0 \text{ (surface)} \qquad (6.9a)$$

$$\frac{\partial T}{\partial r} = 0 \quad \text{at } r = 0 \text{ and } t \geq 0 \text{ (centre)} \qquad (6.9b)$$

$$\text{Rate of heat flow at surface} = k \left(\frac{\partial T}{\partial r} \right)_{r=r_0} \qquad (6.9c)$$

(see Figure 6.4(b)). If the reacting system is capable of achieving a stable steady state, analogous to the intersection p_1 in Figures 6.1 and 6.5, then Equation (6.9) will have a solution when $(\partial T/\partial t) = 0$.

Conventionally (e.g. Gray and Lee, 1967), the following dimensionless variables are introduced:

$$\theta = \frac{T - T_a}{RT_a/E_A}$$

$$z = \frac{r}{r_0}$$

which allow Equation (6.9) to be rewritten, with $\partial T/\partial t = 0$,

$$\frac{k}{r_0^2}\left(\frac{RT_a^2}{E_A}\right)\left(\frac{\partial^2\theta}{\partial z^2} + \frac{\kappa}{z}\frac{\partial\theta}{\partial z}\right) = -\Delta H_c C_i^n A \exp\left[-\left(\frac{E_A}{RT_a}\cdot\frac{\theta}{1-\varepsilon\theta}\right)\right] \qquad (6.10)$$

where $\varepsilon = RT_a/E_A$. Provided that $\varepsilon \ll 1$, then Equation (6.10) can be approximated by

$$\frac{\partial^2\theta}{\partial z^2} + \frac{\kappa}{z}\frac{\partial\theta}{\partial z} = -\frac{r_0^2 E_A \Delta H_c A C_i^n}{kRT_a^2}\exp(-E_A/RT_a)\cdot\exp(\theta) \qquad (6.11)$$

or

$$\nabla^2\theta = -\delta\exp(\theta) \qquad (6.12)$$

where

$$\delta = \frac{r_0^2 E_A \Delta H_c A C_i^n}{kRT_a^2}\exp(-E_A/RT_a) \qquad (6.13)$$

Solutions to Equation (6.12) exist only for a certain range of values of δ corresponding to various degrees of self-heating. It may be assumed that conditions lying outside this range, i.e. when $\delta > \delta_{cr}$, correspond to ignition. The phenomenon of criticality is illustrated very well by results of Fine *et al.* (1969) on the thermal decomposition of diethyl peroxide (Figure 6.6) although in this case the exothermic reaction is the decomposition of an unstable compound rather than an oxidation process. Mathematically it is possible to identify values for δ_{cr} for a number of different shapes (Table 6.1) (Gray and Lee, 1967; Boddington *et al.*, 1971). Equation (6.13) may then be used to investigate the relationship between the characteristic dimension of the system (r_0) and the critical ambient temperature ($T_{a,cr}$) above which it will ignite. Rearranging Equation (6.13), and taking Napierian logarithms:

$$\ln\left(\frac{\delta_{cr}T_{a,cr}^2}{r_0^2}\right) = \ln\left(\frac{E_A \Delta H_c A C_i^n}{kR}\right) - \frac{E_A}{RT_{a,cr}} \qquad (6.14)$$

shows that provided the assumptions are valid (i.e. the first term on the right-hand side of Equation (6.14) is constant), $\ln(\delta_{cr}T_{a,cr}^2/r_0^2)$ should be a linear function of $1/T_{a,cr}$. This has been found to hold for many systems and is widely used to investigate the spontaneous combustion characteristics of bulk solids (Section 8.1 and Figure 8.2). It indicates the strong inverse relationship that exists between r_0 and the critical ambient temperature $T_{a,cr}$, which is apparent when auto-ignition temperatures are measured in

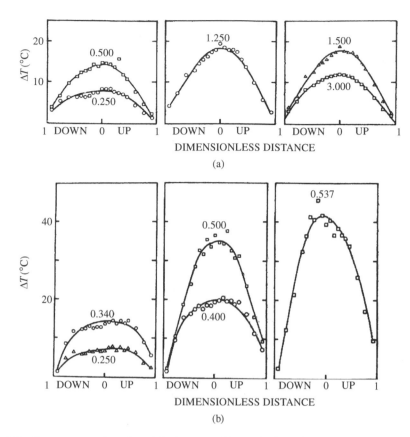

Figure 6.6 Spontaneous (exothermic) decomposition of diethyl peroxide. (a) Instantaneous temperature distributions across the diameter of a reaction vessel for subcritical system (1.1 torr of peroxide at 203.7°C); (b) as (a), but supercritical (1.4 torr at 203.7°C). Times given in seconds. (Reproduced by permission of the Combustion Institute from Fine *et al.*, 1969)

Table 6.1 Critical values of Frank-Kamenetskii's δ (Equation (6.13))

Shape	κ	δ_{cr}
Slab, thickness $2r_0$	0	0.88
Cylinder, radius r_0	1	2.00
Sphere, radius r_0	2	3.32
Cube, side $2r_0$	3.28	2.52

reaction vessels of different sizes (Table 6.2). Such figures, many of which are quoted in the literature without qualification, should be regarded as indicative as they refer to particular experimental conditions. Typical values are given in Table 6.3.

For spontaneous ignition to occur in the boundary layer close to a heated surface, the surface must be hot enough to produce temperatures sufficient for auto-ignition at

Table 6.2 Comparison of the minimum auto-ignition temperatures (°C) of combustible liquids in spherical vessels of different sizes (Setchkin, 1954)[a]

	Volume of vessel ($m^3 \times 10^6$)				
	8	35	200	1000	1200
Diethylether	212	197	180	170	160
Kerosene	283	248	233	227	210
Benzene	668	619	579	559	—
Methanol	498	473	441	428	386
n-Pentane	295	273	—	258	—
n-Heptane	255	248	—	223	—

[a] The test involves introducing a small sample of liquid into the vessel. It is, of course, the vapour/air mixture that ignites.

Table 6.3 Typical values of the minimum auto-ignition temperature for flammable gases and vapours

	Minimum auto-ignition temperature (°C)
Hydrogen	400
Carbon disulphide	90
Carbon monoxide	609
Methane	601[a]
Propane	450
n-Butane	288[b]
iso-Butane	460[b]
n-Octane	206[b]
iso-Octane (2,2,4-trimethylpentane)	415[b]
Ethene	450
Acetylene (ethyne)	305
Methanol	385
Ethanol	363
Acetone	465
Benzene	560

[a] Data taken from the *NFPA Handbook* (National Fire Protection Association, 1997) except for that for methane which was redetermined (Robinson and Smith, 1984).
[b] Note that branched alkanes have much higher auto-ignition temperatures than their straight-chain isomers.

a distance greater than the quenching distance, d_q. For a surface of limited extent, the temperature necessary for ignition increases as the surface area is decreased (Powell, 1969): this is shown clearly in Figure 6.7 (Rae *et al.*, 1964; Laurendeau, 1982). With mechanical sparks which comprise very small incandescent particles (<0.1 mm) generated by frictional impact between two solid surfaces, even higher temperatures must be achieved. The temperature of impact sparks is limited by the melting points of

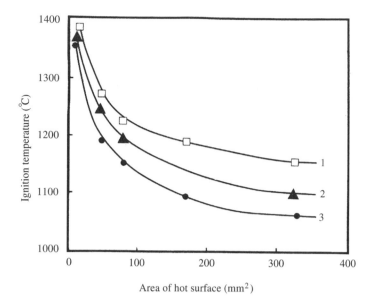

Figure 6.7   Minimum temperatures for ignition of 6% methane/air mixtures by hot surfaces of different areas and locations within an explosion chamber: □, hot surface on wall; ▲, on ceiling; ● on floor. (Rae *et al.* (1964). quoted by Laurendeau (1982). Reproduced by permission of the Combustion Institute)

the materials involved (Powell, 1969; Laurendeau, 1982) and ignition must be rapid as the particles will cool quickly. Pyrophoric sparks — in which the particles (e.g. aluminium and magnesium) oxidize vigorously in air — are capable of achieving very high temperatures (>2000°C) and can ignite the most stubborn mixtures. The thermite reaction between aluminium and ferric oxide (rust) can be initiated by impact, e.g. aluminium paint on rusty iron struck by any rigid object: the resultant shower of sparks is highly incendive.

6.2   Ignition of Liquids

Combustible liquids are classified according to their flashpoints, i.e. the lowest temperature at which a flammable vapour/air mixture exists at the surface (Burgoyne and Williams-Leir, 1949). This is normally determined using the Pensky–Martens Closed Cup Apparatus (ASTM, 1994b). The liquid is heated slowly (5–6°C per minute) in an enclosed vessel (Figure 6.8(a)) and a small, non-luminous pilot flame is introduced into the vapour space at frequent intervals through a port which is opened and closed automatically by a shutter. The flashpoint is taken as the lowest temperature of the liquid at which the vapour/air mixture ignites: some examples are given in Table 6.4. The proportion of vapour in air at the flashpoint can be calculated from the equilibrium vapour pressure of the liquid (Equation (1.14)): for pure liquids, the calculated vapour concentration is in good agreement with data on the lower flammability limit. Taking again the example of *n*-decane (Section 3.1.3), its vapour pressure at the flashpoint (44°C) is 5.33 mm Hg (Equation (1.14) and Table 1.12) or $5.33/760 = 7.0 \times 10^{-3}$ atm, i.e. 0.7 per cent by volume at normal atmospheric pressure. This should be compared with the accepted figure of 0.75% for the lower flammability limit of *n*-decane vapour at 25°C (Table 3.1). This agreement is satisfactory, given the uncertainty in the measured value of the limit: the small difference in temperature would not be expected to have a significant effect (Equation (3.3d)).

Flashpoints of mixtures of flammable liquids can be estimated if the vapour pressures of the components can be calculated. For 'ideal solutions', to which hydrocarbon mixtures approximate, Raoult's law can be used (Equation (1.15)). As an example, consider the problem of deciding whether or not *n*-undecane ($C_{11}H_{24}$) containing 3% of *n*-hexane (by volume) should be classified as a 'highly flammable liquid' as defined

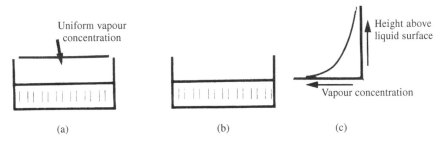

Figure 6.8 Determination of flashpoint: (a) closed cup, (b) open cup, (c) partial vapour pressure gradient above the surface of the fuel in the open cup

Table 6.4 Flashpoints and firepoints of liquids[a]

	Closed cup flashpoint (°C)	Open cup flashpoint (°C)	Firepoint (°C)
n-Hexane	−22	—	—
Cyclohexane	−20	—	—
n-Octane	13	—	—
iso-Octane	−12	—	—
n-Decane	44	52[b]	61.5[b]
n-Dodecane	72	—	103[b]
Benzene	−11	—	—
Toluene	4	7	—
p-Xylene	25	31[b]	44[b]
Methanol	12	1[b] (13.5)[c]	1[b] (13.5)[c]
Ethanol	13	6[b] (18.0)[c]	6[b] (18.0)[c]
n-Propanol	15	16.5[b] (26.0)[c]	16.5[b] (26.0)[c]
n-Butanol	29	36[b] (40.0)[c]	36[b] (40.0)[c]
n-Hexanol	45	74	—
Acetone	−14	−9	—

[a] Unless otherwise indicated, the data are taken from Kanury (1995a).
[b] Glassman and Dryer (1980/81).
[c] Figures in brackets refer to ignition by a spark (see the last part of Section 6.2) (also Glassman and Dryer, 1980/81).

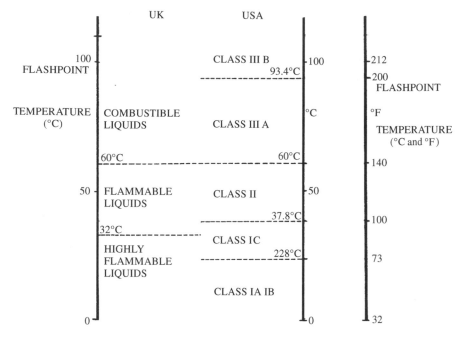

Figure 6.9   A comparison of the UK and US classifications of flammable and combustible liquids

in the UK Regulations (Figure 6.9). This can be reduced to determining if the mixture has a flammable vapour/air mixture above its surface at 32°C. Molar concentrations can be calculated from the formula $V\rho/M_w$ where V is the vol% of the component in the liquid mixture and ρ and M_w are the density and molecular weight respectively. Thus, as $\rho_{n-\text{undec}} = 740$ kg/m³ and $\rho_{n-\text{hex}} = 660$ kg/m³, the respective mole fractions (Equation (1.16)) become:

$$n_{n-\text{undec}} = 0.952 \text{ and } n_{n-\text{hex}} = 0.048$$

The vapour pressures of the pure liquids at 32°C are calculated (Equation (1.14) and Table 1.12) to be

$$p^o_{n-\text{undec}} = 1.08 \text{ mm Hg and } p^o_{n-\text{hex}} = 179.08 \text{ mm Hg}$$

and their partial pressures above the liquid mixture are obtained by applying Raoult's law (Equation (1.15)), thus:

$$p_{n-\text{undec}} = 1.03 \text{ mm Hg and } p_{n-\text{hex}} = 8.60 \text{ mm Hg}$$

Whether or not this mixture would be flammable in a normal atmosphere can be established by applying Le Chatelier's law in the form given in Equation (3.2b). The lower flammability limits of n-undecane and n-hexane are 0.68% and 1.2% respectively (at 25°C), thus Equation (3.2b) gives:

$$\frac{100 \times 1.03/760}{0.68} + \frac{100 \times 8.60/760}{1.2} = 1.14 > 1$$

indicating the lower limit has been exceeded and therefore the liquid mixture has a flashpoint below 32°C.

The reliability of this calculation depends on a number of assumptions which are known to be approximate, in particular that the liquid mixture behaves according to Raoult's law. For non-ideal mixtures, the activities of the components of the mixture must be known and applied as described briefly in Section 1.2.2 (Equation (1.15) *et seq.*).

Flashpoint measurements can be made in an open cup (Figure 6.8(b)) but in this case vapour is free to diffuse away from the surface, producing a vapour concentration gradient which decreases monotonically with height (Figure 6.8(c)). As ignition can only occur when the vapour/air mixture is above the lower flammability limit at the location of the pilot flame, it is found that the open cup flashpoint is dependent on the height of the ignition source above the liquid surface (Burgoyne *et al.*, 1967): this is shown in Figure 6.10 (Glassman and Dryer, 1980/81). In the standard open cup test (ASTM, 1990a) the height and size of the pilot flame are strictly specified. In general, open cup flashpoints are greater than those measured in the closed cup. However, ignition of the vapour in the open cup test will lead to sustained burning of the liquid only if its temperature is greater than the firepoint. This is found to be significantly greater than the flashpoint for hydrocarbons, although there are surprisingly few data quoted in the literature (see Table 6.4). Values that do exist indicate that the concentration of vapour at the surface must be greater than stoichiometric for a diffusion flame to become established (Roberts and Quince, 1973). (Alcohols appear to behave differently and will be discussed below.) Ignition of the vapour above the liquid gives a transient premixed flame which will consume all of the vapour/air mixture lying within the flammability limits. A diffusion flame will then remain only if the rate of supply of vapour is sufficient to support it. If the rate is too low (i.e. the liquid is below its firepoint), such a flame cannot survive as heat losses to the surface cause

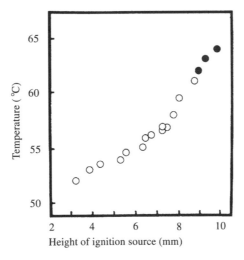

Figure 6.10 Variation of the measured open cup flashpoint with height of ignition source above the liquid surface (*n*-decane). ○ flash only; ● flash followed by sustained burning (from Glassman and Dryer, 1980/81, by permission)

extinction (Section 3.3). Roberts and Quince (1973), Rasbash (1975) and others, have argued that the nascent flame will self-extinguish if it loses more than ~30% of the heat of combustion to the surface. However, at and above the firepoint, the flame will stabilize and the continuing heat loss to the surface becomes available to heat the surface layer and increase the rate of supply of vapour. Consequently, the diffusion flame will strengthen and grow in size, culminating in steady burning when the surface temperature achieves a value close to the boiling point and a stable temperature profile has been established below the surface (e.g. Figure 5.5).

The firepoint of a liquid is defined as the lowest temperature at which ignition of the vapours in an open cup is followed by sustained burning. As the temperature is maintained by external heating, the rate of evaporation under steady-state (non-fire) conditions is given by:

$$\dot{m}''_{\text{evap}} = \frac{\dot{Q}''_E - \dot{Q}''_L}{L_v} \tag{6.15}$$

where $\dot{m}''_{\text{evap}}$ is the mass flux leaving the surface, L_v is the latent heat of evaporation of the liquid, and $\dot{Q}''_E$ and $\dot{Q}''_L$ are the rates of external heating and heat loss, respectively, per unit surface area. If flame is established at the surface, then the rate of evaporation (now the 'rate of burning') is increased:

$$\dot{m}''_{\text{burn}} = \frac{f \Delta H_c \dot{m}''_{\text{burn}} + \dot{Q}''_E - \dot{Q}''_L}{L_v} \tag{6.16}$$

where f is the fraction of the heat of combustion of the vapour (ΔH_c) that is transferred back to the surface and is made up of radiative and convective components, indicated by f_r and f_c, respectively. This may be rewritten:

$$((f_r + f_c)\Delta H_c - L_v)\dot{m}''_{\text{burn}} + \dot{Q}''_E - \dot{Q}''_L = S \tag{6.17}$$

where $S = 0$ for 'steady burning'. The proportions f_r and f_c vary as the size of the fire. At the firepoint, f_r is small as the flame is non-luminous and may be assumed to be zero for the present argument. However, f_c (which becomes small for steady burning of large fuel beds (Section 5.1)) approaches a maximum value (ϕ) at the firepoint corresponding to the stability limit of flame at the surface. Setting $f_r = 0$ and $f_c = \phi$ and writing $\dot{m}''_{\text{burn}} = \dot{m}''_{\text{cr}}$, the limiting or critical flowrate of the volatiles at the firepoint (Section 6.3.2), Equation (6.17) can be used to identify the conditions under which ignition of the volatiles will lead to sustained burning:

$$(\phi\Delta H_c - L_v)\dot{m}''_{\text{cr}} + \dot{Q}''_E - \dot{Q}''_L = S \tag{6.18}$$

This equation, first proposed by Rasbash (1975), will be discussed in more detail in relation to the ignition of solids (Section 6.3.2). Steady burning will develop after ignition of the volatiles, only if $S > 0$, i.e. sufficient excess heat is available to cause the surface temperature to increase, thereby increasing the rate of volatilization and strengthening the flame. Equation (6.18) identifies formally those factors which determine whether or not sustained burning can develop.

Glassman and Dryer (1980/81) found an apparent anomaly in the relationship between open and closed cup flashpoints and firepoints for alcohols. With a pilot flame ignition source, it is possible to ignite an alcohol in an open cup at a temperature significantly lower than the closed cup flashpoint (Table 6.4). This may be explained if the pilot flame is contributing heat to the surface, raising the temperature locally to the firepoint; the anomaly disappears if a spark ignition source is used instead, although the flashpoint and firepoint remain coincident. This behaviour is quite different from that of hydrocarbon liquids but while the explanation may lie in the different infrared absorption characteristics of these fuels, the effect has not been fully investigated. However, this is not simply an academic point, as it suggests that the closed cup flashpoint may not be a valid method of assessing the fire risks associated with all liquid fuels.

In a container of flammable liquid, it is possible for the vapour/air mixture in the headspace to be flammable. This is the case with methanol and ethanol at ambient temperatures (15–20°C): there have been several serious incidents in restaurants where an attempt has been made to refill flambé lamps without first ensuring that the flame had been extinguished. Flame propagates into the headspace, causing an explosion which can be sufficient to rupture the container and cause burning liquid to be widely scattered. This will not occur with petrol (gasoline) tanks as the vapour pressure of the gasoline lies well above the upper flammability limit at normal temperatures and the mixture in the headspace is too rich to burn. However, under extreme winter conditions in some inhabited parts of the world, temperatures as low as −30°C may be experienced, at which the vapour pressure of gasoline may be reduced sufficiently to create a flammable vapour/air above the liquid surface. Potentially, this is extremely hazardous. (See also Section 3.1.3.)

6.2.1 Ignition of low flashpoint liquids

Bearing in mind the above caveat, classifying combustible liquids according to their flashpoints is a convenient way of indicating their relative fire hazards. Liquids with 'low' flashpoints present a risk at ambient temperatures as their vapours may be ignited by a spark or flame. Explosive forces may be generated if such a vapour/air mixture is confined (Section 1.2.5), although fire will only result if the liquid is above its firepoint. The closed cup flashpoint is always used to indicate the hazard as this will err on the side of safety if the risk is only that of fire. In the United Kingdom, liquids with flashpoints less than 32°C as measured in the Abel Closed Cup apparatus (British Standards Institution, 1982) are classified as 'highly flammable' and come under specific legislation (The Highly Flammable Liquids and Liquefied Petroleum Gases Regulations, 1972). Liquids with flashpoints between 32°C and 60°C are termed 'flammable', while those with flashpoints above 60°C are classified as 'combustible'. A similar scheme in use in the USA is summarized in Figure 6.9, where it can be seen that the low flashpoint liquids are also divided into two groups. It is worth noting that the upper bound of the 'Group I' type flammable liquid is higher in the USA than in the UK, reflecting the higher ambient temperatures that are encountered in the States. In still warmer climates, it would be appropriate to set an even higher upper boundary.

In the open a large pool of a highly flammable liquid (e.g. petrol or gasoline) will produce a significant volume of flammable vapour/air mixture which can extend beyond the limits of the pool boundary. Introduction of an ignition source to this volume will cause flame to propagate back to the pool, burning any vapour near the surface which was initially above the upper limit, producing a large, transient diffusion flame before steady burning is established. This is sometimes referred to as a 'flash fire'. If there is a wind blowing, the flammable region will extend downwind of the pool, its extent depending on the vapour pressure of the liquid and the windspeed and degree of turbulence in the atmosphere. These factors determine how rapidly the vapour will be dispersed: however, further discussion of this topic is beyond the scope of this text (see Wade, 1942; Sutton, 1953; Clancey, 1974).

6.2.2 Ignition of high flashpoint liquids

A high flashpoint liquid can only be ignited if it is heated to above its firepoint. In the standard test (American Society for Testing and Materials, 1990a) the bulk of the liquid is heated uniformly, although for sustained ignition to occur, only the surface layers need be heated. While this may be achieved by applying a heat flux to the entire surface (e.g. a pool of liquid exposed to radiation from a nearby fire) it is more common to encounter local application of heat, such as a flame impinging on, or burning close to the surface. Some time will elapse before flame spreads to cover the entire surface, as heat is dissipated rapidly from the affected area by a convective mechanism which is maintained by surface tension variations at the surface (Sirignano and Glassman, 1970). Surface tension — defined as the force per unit length acting at the surface of a liquid — is temperature-dependent, decreasing significantly with increase in temperature. The result of this is that there is a net force acting at the surface, drawing hot liquid away from the heated area, causing fresh, cool liquid to take its place from below the surface: the convection currents so induced are illustrated in Figure 6.11. In a pool of limited extent, ignition will eventually be achieved following flame spread across the surface, but only after a substantial amount of heat has been transferred convectively to the bulk of the liquid and the temperature at the surface has increased to the firepoint (Burgoyne and Roberts, 1968).

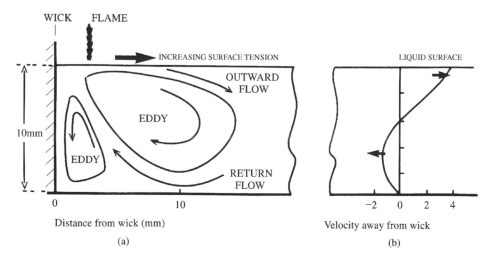

Figure 6.11   (a) Surface-tension driven flows and convective motion in a liquid subjected to a localized ignition source. (b) Velocity profile 10 mm from wick. (After Burgoyne *et al.*, 1968 by permission)

However, a high firepoint liquid can be ignited very easily if it is absorbed on to a wick, i.e. a porous medium of low thermal conductivity such as is used in kerosene lamps and candles. Application of a flame to the fuel-soaked wick causes a rapid local increase in temperature, not only because the layer of liquid is too thin for convective dissipation of heat to occur, but also because the wick is an effective thermal insulator (low $k\rho c$). Ignition can be achieved at a point and will be followed by flame spread over the wick surface (Section 7.1). One mechanism by which pools of high firepoint liquids may be ignited without recourse to bulk heating is to ignite the liquid absorbed on a wick (a cloth or any other porous material) which is lying in the pool. Flame will establish itself in a position in which it can begin to heat the surface layers of the free liquid. This type of ignition was studied by Burgoyne *et al.* (1968). The liquid was contained in a long trough and ignited on a wick located at one end (Figure 6.12). By monitoring the temperature at various points within the liquid (Figure 6.12), it was found that flame began to spread away from the wick only when the temperature at the surface exceeded the flashpoint. The associated induction period is illustrated in Figure 6.13(a). It was shown that the amount of heat transferred from the flame at the wick to the liquid was in agreement with the amount of heat stored in the liquid at the end of the 'induction period'.

In view of this, Burgoyne *et al.* (1968) examined the effect of reducing the depth of the liquid on the induction period. They found that the latter exhibited a minimum at about 2 mm, no ignition being possible for the fuels used at depths less than 1 mm (Figure 6.13(b)). The minimum exists because less heat needs to be transferred from the flame on the wick to raise the surface temperature of a shallow pool to the fire-point. However, if the pool is too shallow, then in addition to convective flow being restricted, heat losses to the supporting surface become too great and the firepoint will not be attained. The same observation applies to hydrocarbon liquids floating on water: reference was made to this effect in Section 5.1.1 and there will be further discussion in Section 7.1 in relation to flame spread over liquids (Mackinven *et al.*, 1970).

With wick ignition, the source of heat is also acting as the means by which the volatiles are ignited, but other situations can be envisaged in which the heat source is quite independent of the agency which initiates flaming (as in the determination of firepoint using the Cleveland Open Cup (American Society for Testing and Materials,

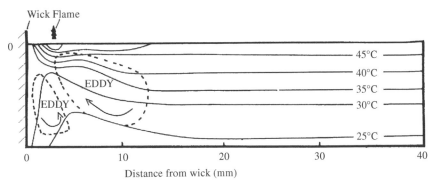

Figure 6.12   Wick ignition of a pool of high flashpoint liquid. Temperature distribution immediately before flame spread over the surface (Burgoyne *et al.*, 1968, by permission)

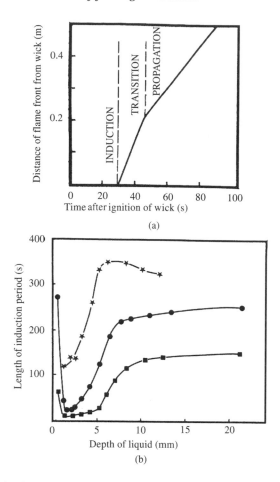

Figure 6.13 (a) Induction period for wick ignition of high firepoint fuels. (b) Effect of liquid depth on the duration of the induction period. ★ Hexanol; ● isopentanol; ■ butanol (Burgoyne *et al.*, 1968, by permission)

1990a)). If the temperature of the environment is increased to above the firepoint of the liquid, then it will eventually behave as a highly flammable liquid as it becomes heated (Section 6.2.1). If heating is confined to the liquid and its container, then additional evaporation will take place, giving a localized flammable zone, although the vapour may condense on nearby cold surfaces. Over a long period of time the liquid may evaporate completely if sufficiently volatile. On the other hand, a relatively involatile combustible liquid may be heated sufficiently for spontaneous ignition to occur. This can happen with most cooking oils and fats and can be demonstrated easily using the Cleveland Open Cup. Careful observation shows that the flame starts in the plume of hot volatile products, well clear of the liquid surface, then flashes back to give immediately an intense fire, as the liquid is already close to its boiling point (see Figure 6.25). The liquid temperature at which auto-ignition occurs will depend on the surface area of the liquid and be very sensitive to any air movement which would tend to disturb and cool the plume (Section 6.1).

6.3 Pilot Ignition of Solids

The phenomena of 'flashpoint' and 'firepoint' can be observed with solids under conditions of surface heating (e.g. Deepak and Drysdale, 1983; Thomson and Drysdale, 1987) but cannot be defined in terms of a bulk temperature. The generation of flammable volatiles involves chemical decomposition of the solid which is an irreversible process: there is no equivalent to the equilibrium vapour pressure which may be used to calculate the flashpoint of a liquid fuel.

However, it is reasonable to assume that the same principles apply, namely that the flashpoint is associated with the minimum conditions under which pyrolysis products achieve the lower flammability limit close to the surface, and that the firepoint corresponds to a near-stoichiometric mixture at the surface. As the system is open (cf. the open cup flashpoint test), these concentrations will be associated with specific rates of pyrolysis, or mass fluxes (Rasbash *et al.*, 1986; Janssens, 1991a), which should be capable of measurement. A number have been measured by Tewarson (1995) and Drysdale and Thomson (1989), but under significantly different conditions: these will

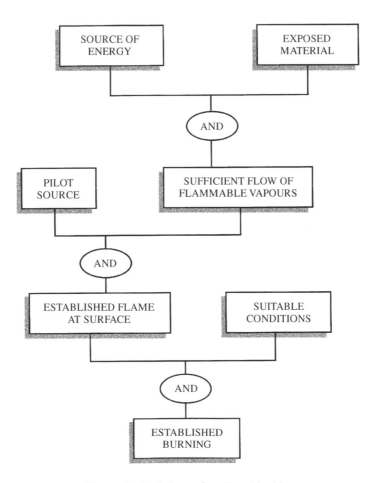

Figure 6.14 Scheme for piloted ignition

be discussed in Section 6.3.2. If it is assumed that sustained, piloted ignition can only occur if a critical mass flux of fuel vapours is exceeded, then the process of piloted ignition can be described in terms of Figure 6.14 (Drysdale, 1985a), where 'sufficient flow of volatiles' corresponds to a mass flux greater than the critical value, and 'suitable conditions' implies that the environmental conditions remain favourable for the flame to become established.

This critical condition may be identified as the firepoint of the solid. It might be supposed that it can be defined more simply in terms of a surface temperature (which can be measured, albeit with difficulty (Atreya *et al.*, 1986; Thomson and Drysdale, 1987)), but this is not valid in all circumstances. Stabilization of the flame has to be interpreted in terms of a detailed heat balance at the surface using Equation (6.18). Thus, it depends whether the heat flux responsible for raising the surface to the 'firepoint' condition remains approximately constant after the volatiles have been ignited or whether it is in the form of a pulse, decaying to a lower value (perhaps zero) after ignition. In the latter case the heat losses from the flame to the surface and the rate of heat loss from the surface of the material will determine whether or not flaming will be sustained.

For the sake of clarity, the following sections deal with the piloted ignition scenario which is illustrated in Figure 6.14, i.e. the source of energy and the 'pilot' are distinct. However, when there is direct flame impingement, the flame acts both as the source of energy and as the ignition source for the vapours. This is more difficult to analyse from first principles as the heat flux imposed by the flame depends on the size of the flame, its radiation characteristics, and the geometry of the adjacent surfaces, which can influence the flow dynamics, and thus the magnitude of the convective component (Hasemi, 1984; Kokkala, 1993; Back *et al.*, 1994; Foley and Drysdale, 1995).

6.3.1   Ignition during a continuous heat flux

If the heat flux is continuous, the firepoint condition may be characterized by the minimum surface temperature at which the flow of volatiles is sufficient to allow flame to persist at the surface. It is possible to identify a number of factors that are likely to contribute to the attainment of the firepoint, including the effects of chemical reaction at and below the surface and the movement of the volatiles through the surface layers (e.g. Steward, 1974a), but if it is assumed that the material is completely inert, the problem can be reduced to one of heat transfer to a surface (Simms, 1963; Kanury, 1972; Mikkola and Wichman, 1989). This was originally reviewed by Kanury (1972) who considered various solutions to the one-dimensional heat conduction equation (Equation (2.15)) in which the boundary conditions were chosen to represent a number of configurations, including both the 'infinite slab' and the 'semi-infinite solid' discussed in Section 2.2.2. In all cases, the solid is assumed to be opaque and inert, with uniform thermal properties which are independent of temperature. Chemical decomposition and the associated energy changes in the solid are neglected although clearly this is questionable. However, their effect can be relegated to second order status to allow the underlying principles which influence the ignition behaviour of solids to be studied. Most theoretical and experimental investigations have concentrated on ignition brought about by radiative heat transfer. The original impetus for this followed the realization that thermal radiation levels from a nuclear explosion would be sufficient to ignite combustible materials at great distances from the blast centre. However, it has become apparent that radiation is of fundamental importance in the growth and spread of fire in many situations, such as open fuel beds (cribs, etc.) and compartments (Chapter 9). The high level of interest in radiative ignition has been maintained although ignition by convection cannot be neglected. In the following sections, relevant solutions for the one-dimensional heat conduction equation

$$\frac{\partial^2 T}{\partial x^2} = \frac{1}{\alpha}\frac{\partial t}{\partial t} \tag{6.19}$$

are discussed.

(a) *The thin slab*   It was shown in Section 2.2.2 that the temperature (T) of a thin 'slab' exposed to convective heating at both faces varies with time according to the expression (Equation (2.21)):

$$\frac{T_\infty - T}{T_\infty - T_0} = \exp(-2ht/\tau\rho c) \tag{6.20}$$

where T_0 and T_∞ refer to the initial and final (i.e. gas stream) temperatures respectively, and τ is the thickness. For example, this would be relevant to a curtain fabric exposed to a rising current of hot air. If the 'firepoint' of the material can be identified as a specific temperature T_i, then assuming ignition of the volatiles, the time to ignition (t_i) would be given by:

$$t_i = \frac{\tau\rho c}{2h}\ln\left(\frac{T_\infty - T_0}{T_\infty - T_i}\right) \tag{6.21}$$

This indicates that the time to ignition is directly proportional to weight per unit area ($\tau\rho$), assuming that both T_i and c are constant, and inversely proportional to h, the convective heat transfer coefficient. The same general conclusions may be deduced if the positive heat transfer is on one side only. Note that for a single material (e.g. paper), t_i is directly proportional to the thickness of the sample, at least up to the thin (low Biot number) limit. This is best illustrated in terms of flame *spread* over thin fuels (see Figure 7.10).

The problem of an infinite slab exposed to radiant heating on one face ($x = +l$) with convective cooling at both faces ($x = \pm l$) (Figure 6.15) can be solved with the following boundary conditions applied to Equation (6.19):

$$T = T_0 \text{ for all } x \text{ at } t = 0 \tag{6.22a}$$

$$a\dot{Q}''_R = h\theta - k\frac{d\theta}{dx} \text{ for } x = +l \text{ and } t > 0 \tag{6.22b}$$

and

$$h\theta = -k\frac{d\theta}{dx} \text{ at } x = -l \text{ and } t > 0 \tag{6.22c}$$

where $\theta = T - T_0$, a is the absorptivity (assumed constant) and $\dot{Q}''_r$ is the radiant heat flux falling on the surface. Radiative heat loss is neglected here, although it will become significant at temperatures approaching T_i when it cannot be ignored (Mikkola and Wichman, 1989). The full analytical solution is available (Carslaw and Jaeger, 1959)

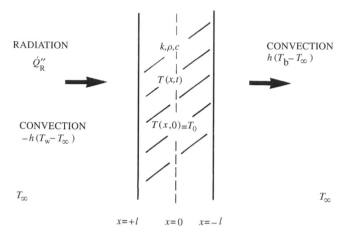

Figure 6.15   Radiative heat transfer to one face of a vertical, infinite slab, with convective heat losses at $x = \pm l$

but Simms (1963) adopted the 'lumped thermal capacity' approach (Section 2.2.2) to obtain the following equation for a thermally thin material:

$$a\dot{Q}_R'' = \tau \rho c \frac{d\theta}{dt} + 2h\theta \qquad (6.23)$$

where τ, the thickness of the material, is equivalent to $2l$ (Figure 6.15). As θ is now independent of x, Equation (6.23) may be integrated to give

$$\theta = T - T_0 = \frac{a\dot{Q}_R''}{2h}(1 - \exp(-2ht/\rho c\tau)) \qquad (6.24)$$

If $T = T_i$, this can be rearranged to give the time to ignition as:

$$t_i = \frac{\tau \rho c}{2h} \ln\left(\frac{a\dot{Q}_R''}{a\dot{Q}_R'' - 2h(T_i - T_0)}\right) \qquad (6.25)$$

Equations (6.21) and (6.25) are similar in form, showing that regardless of the mode of heat transfer, the time to ignition for thin materials is directly proportional to the thermal capacity per unit area ($\tau\rho c$). They also define the limiting conditions for ignition, i.e. $T_\infty > T_i$ for convective heating and $a\dot{Q}_R'' > 2h(T_i - T_0)$ for radiative heating. (Recall that radiative losses are neglected in this derivation.)

Analysis of ignition of sheets or slabs which cannot be regarded as thermally thin (e.g. Bi > 0.1) requires more complex solutions (e.g. Equation (2.18) for convective heat transfer to an infinite slab). Time to ignition increases asymptotically with τ to a limiting value corresponding to the ignition of a semi-infinite solid for which the mathematics is simpler.

(b) *The semi-infinite solid*   The surface temperature of a semi-infinite solid exposed to convective heating (or cooling) varies with time according to Equation (2.26), i.e.

$$\frac{\theta_s}{\theta_\infty} = \frac{T_s - T_0}{T_\infty - T_0} = 1 - \exp(\beta^2) \cdot \mathrm{erfc}(\beta) \qquad (6.26)$$

where T_s is the surface temperature and

$$\beta = (h(\alpha t)^{1/2}/k) = (ht^{1/2}/(k\rho c)^{1/2}) = \text{Bi} \cdot \text{Fo}^{1/2}.$$

If sustained piloted ignition is possible for $T_s \geq T_i$, then an estimate of the minimum time to ignition under convective heating may be deduced from the value of β corresponding to θ_i/θ_∞ (from Figure 6.16), if the thermal inertia $(k\rho c)$ and the convective heat transfer coefficient (h) are known. (As stated in Section 2.2.2, a thick slab will approximate to 'semi-infinite' behaviour provided that $\tau > 2(\alpha t)^{1/2}$ where t is the duration of heating.)

For radiative heating, the appropriate boundary conditions for Equation (6.19) are:

$$a\dot{Q}_R'' - h\theta = -k\frac{\mathrm{d}\theta}{\mathrm{d}x} \text{ at } x = 0 \text{ for } t > 0 \qquad (6.27a)$$

$$T = T_0 \text{ at } t = 0 \text{ for all } x \qquad (6.27b)$$

However, the solution may be obtained from Equation (6.26), by substituting $(T_\infty - T_0) = a\dot{Q}_R''/h$, which refers to the steady state $(t = \infty)$ *if heat loss by radiation is ignored*, i.e.

$$a\dot{Q}_R'' = h(T_\infty - T_0) \qquad (6.28)$$

Thus Equation (6.26) becomes

$$\theta_s = \frac{a\dot{Q}_R''}{h}(1 - \exp(\beta^2) \cdot \mathrm{erfc}(\beta)) \qquad (6.29)$$

Simms (1963) referred to β as the 'cooling modulus' and cast Equation (6.29) in a different form by rearranging and multiplying both sides by β, giving:

$$\gamma = \beta(1 - \exp(\beta^2) \cdot \mathrm{erfc}(\beta))^{-1} \qquad (6.30)$$

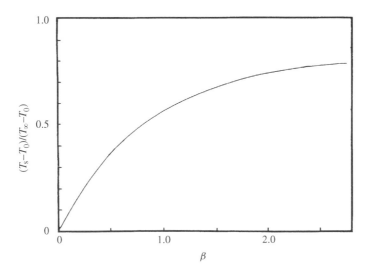

Figure 6.16   Variation of θ_s/θ_∞ with β (Equation (6.26))

where

$$\gamma = a\dot{Q}''_R t / \theta_s \rho c (\alpha t)^{1/2}$$

and may be called the 'energy modulus'. Simms (1963) used Equation (6.30) to correlate data of Lawson and Simms (1952) on the piloted ignition of wood, in which vertical samples (5 cm square) of several species were exposed to radiant heat fluxes in the range 6.3–63 kW/m². A pilot flame was held in the plume of fuel vapours rising from the surface, as shown in Figure 6.17, and the time to ignition recorded. Simms plotted γ vs β, and selected a single value of $\theta_s = \theta_i$ that give the most satisfactory correlation for all the data. This is shown in Figure 6.18, and while there is a non-random scatter which reflects density differences, the correlation is reasonable for a value of $\theta_i = 340°C$. The exception is fibre insulating board: this could be due to a number of factors, including the onset of smouldering after prolonged exposure times.

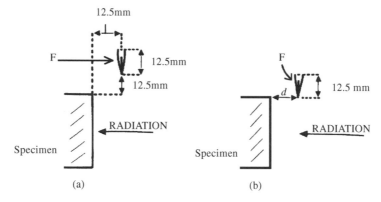

Figure 6.17   Pilot ignition of wood exposed to a radiant heat flux by means of a flame burning downwards (F) (Simms, 1963). (Reproduced by permission of The Controller, HMSO. © Crown Copyright)

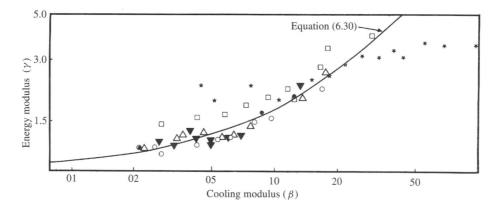

Figure 6.18   Correlation between the energy modulus (γ) and the cooling modulus (β) for woods of different density (Equation (6.30)). Data obtained from configuration shown in Figure 6.17(a). $\theta_i = 340°C$, $h = 33$ W/m².K.; ★, fibre insulation board; □, cedar; △, freijo; ○, mahogany; ●, oak; ▼, iroko. (Simms, 1963.) (Reproduced by permission of The Controller, HMSO. © Crown Copyright)

It was found that if the location of the pilot flame was changed the value of θ_i had to be altered to maintain a correlation. Thus, with the tip of the pilot flame level with the top edge of the sample, as shown in Figure 6.17(b), θ_i was found to be 300°C, 380°C and 410°C for $d = 6.2$, 12.5 and 19 mm respectively, the time to ignition also increasing in this order (Figure 6.19). This observation is similar to that made earlier regarding the measurement of the firepoint of a liquid fuel (see Figure 6.10): the pilot source must be within the zone of flammability of the volatiles. If $d > 20$ mm (see Figure 6.17(b)) then no ignition was possible, even at the highest heat fluxes, as the ignition source was outside the stream of volatiles. While care must be taken in interpreting results of this type of experiment, these in no way negate the initial concept of a critical surface temperature acting as a limiting ignition criterion, at least under a continuous heat flux (see Section 6.3.2).

A critical radiant heat flux is sometimes quoted as the limiting criterion for pilot ignition (e.g. Figure 6.19) although this will be sensitive to changes in heat loss from the surface and hence the orientation and geometry of that surface. Lawson and Simms (1952) obtained an estimate of the limiting flux for vertical samples of wood by extrapolating a plot of $\dot{Q}_R''$ versus $\dot{Q}_R''/t_i^{1/2}$ to $t_i = \infty$, where t_i is the time to ignition under a radiant heat flux $\dot{Q}_R''$ (Figure 6.20): however, it should be remembered that in the above model, radiative heat losses are neglected. From these and other data, a minimum flux for pilot ignition of wood was deduced as approximately 12 kW/m² (0.3 cal/cm².s). This figure was incorporated into the Scottish Building Regulations in 1971 as a basis for determining building separation (Section 2.4.1). It is believed that this was the first application of quantitative 'Fire Safety Engineering'.

Care must be taken in the interpretation of the critical heat flux. A considerable amount of data has been gathered on the 'time to ignition' (t_{ig}) as a function of the imposed (incident) heat flux in the Cone Calorimeter (e.g. Babrauskas and Parker, 1987), in the ISO Ignitability Test (e.g. Bluhme, 1987), and in other experimental apparatuses (Simms, 1963; Thomson *et al.*, 1988), including the FMRC Flammability

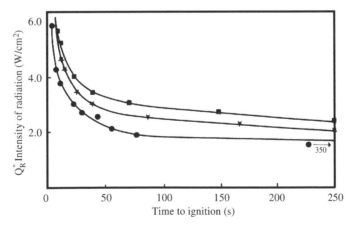

Figure 6.19 Effect of position of the pilot flame on the time to ignition of Columbian pine (Figure 6.17(b)). ●, $d = 6.2$ mm; ★, $d = 12.5$ mm; ■, $d = 19.0$ mm. (Simms, 1963.) (Reproduced by permission of The Controller, HMSO. © Crown Copyright)

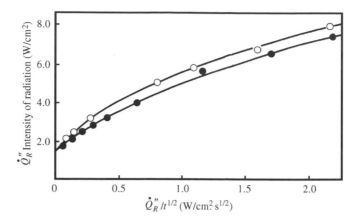

Figure 6.20   Determination of the critical radiant heat flux for pilot ignition of oak (Figure 6.17(b)). ●, $d = 6.2$ mm: ○, $d = 19.0$ mm. (Simms, 1963.) (Reproduced by permission of The Controller, HMSO. © Crown Copyright)

Apparatus (e.g. Tewarson and Ogden, 1992). Such data may be used to estimate a critical heat flux either as the minimum heat flux capable of producing the conditions for piloted ignition after some arbitrary (but experimentally valid) time interval (e.g. 15 minutes (ISO, 1997a)) or, as in Figure 6.20, by carrying out an extrapolation to obtain the minimum value of the imposed heat flux at which ignition is theoretically possible (i.e. $t_{ig} = \infty$). This may also be done by plotting $\dot{Q}_R''$ vs t_{ig} and estimating the asymptotic value of $\dot{Q}_R''$ (Simms, 1963 (Figure 6.19); Quintiere, 1981) or using a correlation based on a simplification of the heat transfer models discussed in Chapter 2. Thus, taking Equation (2.30) (in which $\dot{Q}_R''$ is constant and *heat losses are ignored*) and setting $\theta_{ig} = T_{ig} - T_0$, we obtain the time to ignition for a thermally thick solid:

$$t_{ig} = \frac{\pi}{4} k\rho c \frac{(T_{ig} - T_0)^2}{\dot{Q}_R''^2} \tag{6.31}$$

For thin fuels:

$$t_{ig} = \rho c \tau \frac{(T_{ig} - T_0)}{\dot{Q}_R''} \tag{6.32}$$

which may be derived from Equation (6.23) by ignoring the heat loss term ($h\theta = 0$) and setting $a = 1.0$.

These suggest that data for thick fuels should be examined by plotting $1/\sqrt{t_{ig}}$ vs $\dot{Q}_R''$, and for thin fuels, $1/t_{ig}$ vs $\dot{Q}_R''$. These have been widely used, but sometimes with scant regard to their origins (Mikkola and Wichman, 1989). Deceptively simple correlations can be obtained which lend themselves to linear extrapolation, yielding an apparent value for the critical radiant flux, and a straight line whose slope (for a thermally thick sample) is related to the thermal inertia ($k\rho c$) of the material (Equation (6.31)). However, non-linearities are found if data sets are extended to low heat fluxes, corresponding to long ignition times (>5–10 min) when it is impossible to ignore the effects of radiative and convective heat losses (Mikkola and Wichman, 1989). It was shown in Section 2.2.2 that the characteristic thermal conduction length ($\sqrt{\alpha t}$) could be used as an indicator of the depth of the heated layer of a thick material, and that heat losses from the rear face of a material would be negligible if $L > 4 \times \sqrt{\alpha t}$, indicating 'semi-infinite behaviour' (Figure 2.9). A thermally thin material could be defined as one with $L < \sqrt{\alpha t}$. 'Thermal thickness' increases with $\sqrt{t}$, and for a sufficiently long exposure times a physically thick material will no longer behave as a semi-infinite solid, and will begin to show behaviour which is neither 'thick' nor 'thin'. The consequences of this are normally lost in the scatter unless the data set includes ignition times significantly greater than c.5 min. This is clearly shown in data presented by Toal *et al.* (1989) and Tewarson and Ogden (1992) (Figure 6.21). A related observation can be made when the insulation of the rear face of a physically thick sample is altered: the material will behave as a semi-infinite solid ('thermally thick') while undergoing piloted ignition under a high radiative heat flux and the results are independent of the conditions at the rear face. However, at low heat fluxes, the differences in the heat losses through the rear face show up as differences in the times to ignition (Figure 6.22) (Thomson *et al.*, 1988).

Delichatsios *et al.* (1991) have examined the interpretation of such data and have shown how a linear extrapolation of the plot of $1/\sqrt{t_{ig}}$ vs $\dot{Q}''_R$ is likely to give a critical heat flux which is only 70% of the true value (cf. Figure 6.21). Their work emphasizes the importance of matching the correlation to the physical characteristics of the fuel (i.e. in the limits, thick or thin).

Clearly, the critical value is not a material property *per se* as it is influenced — sometimes strongly — by the heat transfer boundary conditions associated

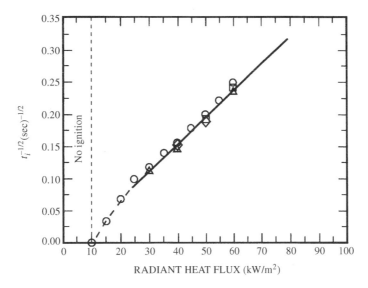

Figure 6.21 Time to ignition data of Tewarson and Ogden (1992) for 25 mm thick black PMMA, obtained in the FMRC Flammability Apparatus and plotted according to Equation (6.31) ($t_i^{-1/2}$ vs $\dot{Q}''_R$). The surface was coated with a thin layer of fine graphite powder to ensure a high, reproducible absorptivity. Symbols refer to different flows of air past the sample (0–0.18 m/s) (By permission of the Combustion Institute.)

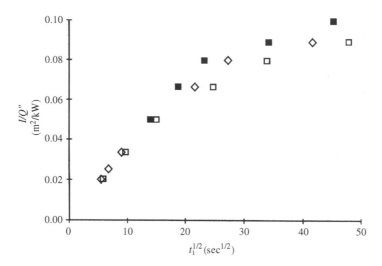

Figure 6.22   Time to ignition data of Thomson *et al.* (1988) for 6 mm thick PMMA measured in the ISO Ignitability Apparatus (ISO, 1997a) and plotted according to Equation (6.31) ($1/\dot{Q}_R''$ vs $t_i^{1/2}$); ($\diamond$), clear PMMA; ($\square$), black PMMA; ($\blacksquare$), black PMMA backed with the insulating material Kaowool rather than the standard ISO backing material. Reprinted from Thomson *et al.* (1988), by permission

with the configuration of the sample. Thus, some composite materials may be ignited at relatively low heat fluxes, e.g. a thin cellulosic material such as a fabric over an insulating substrate such as cotton batting or foam cushioning (Alvarez, 1975), or where delamination occurs, effectively isolating the delaminated surface layer from the bulk of the material and turning it into a 'thin fuel' (Rasbash and Drysdale, 1983). In the test based on the LIFT Apparatus (Quintiere, 1981), figures for the minimum heat flux for piloted ignition are deduced from measurements of time to ignition, but these are consistently higher than values obtained by other authors (e.g. Simms, and Tewarson (see Table 6.5)) and may reflect significantly higher heat losses in this apparatus.

Tewarson and Ogden (1992) and Tewarson, (1995) have introduced the concept of 'thermal response parameter' (TRP), given by an expression which has its origin in Equation (6.31):

$$\text{TRP} = (T_{ig} - T_0)\sqrt{k\rho c} \qquad (6.33)$$

It is derived from the linear part of the plot of $1/\sqrt{t_{ig}}$ vs $\dot{Q}_R''$, and has been suggested as a means of assessing the ignition resistance of materials. However, it is valid only while the material is behaving as a thermally thick solid. Tewarson (1995) tabulates values obtained in the FMRC flammability apparatus and compares them with a few data obtained in the Cone Calorimeter, but it is clear that it is apparatus dependent. Its value is primarily for ranking different materials: applying the TRP to 'real' fire scenarios would require a detailed knowledge of the likely heat transfer boundary conditions in the fire.

Despite these caveats, useful information has been gleaned from such data sets. One that is historically interesting, and still perfectly valid, is the original work by Lawson and Simms (1952) on the ignition of a range of samples of wood in which they found

Table 6.5 Criteria for ignition (various sources)

Material	Critical radiant heat flux (kW/m^2)		Critical surface temperature (°C)	
	Pilot	Spontaneous	Pilot	Spontaneous
'Wood'	12[a]	28[a]	350[b]	600[c]
Polymethylmethacrylate	21[d]	—	—	—
Polymethylmethacrylate	11[e]	—	310 ± 3[g]	—
Polyoxymethylene	13[f]	—	281 ± 5[g]	—
Polyethylene	15[f]	—	363 ± 3[g]	—
Polypropylene	15[f]	—	334 ± 4[g]	—
Polystyrene	13[f]	—	366 ± 4[g]	—

[a] General value for wood, vertical samples, Lawson and Simms (1952). The pilot value is consistent with the range of values found by Mikkola and Wichman (1989).
[b] Deduced from flame spread under conditions of radiant heating (Harvard University). Value for wood compatible with Simms (1963).
[c] Deduced by Simms (1963) for radiative heating. Lower value observed for convective heating (Section 6.4).
[d] Quintiere (1981). Comparatively, these values are very high (see Section 7.2.5(c)).
[e] Thomson *et al.* (1988). Horizontal samples.
[f] Tewarson (1995). Horizontal samples, Factory Mutual Flammability Apparatus.
[g] Thomson and Drysdale (1987). Horizontal samples.

a relationship between time to ignition and the 'thermal inertia' of the solid, which is revealed in an excellent correlation between $(\dot{Q}_R'' - \dot{Q}_{R,0}'')t_i^{2/3}$ and $k\rho c$, where $\dot{Q}_{R,0}''$ is the minimum radiant intensity for pilot ignition of a vertical sample and t_i is the time to ignition under an imposed heat flux $\dot{Q}_R''$. This is shown in Figure 6.23 in which the straight line is given by

$$(\dot{Q}_R'' - \dot{Q}_{R,0}'')t_i^{2/3} = 0.6(k\rho c + 11.9 \times 10^4) \quad (6.34)$$

As wood is the most common combustible material in general use, it is not surprising that there have been numerous studies relating to its ignition (Janssens, 1991a). Invariably, its complex chemical and physical nature (see Section 5.2.2) influences its behaviour, and in particular the ease with which it can be ignited. Thus, its surface absorptivity ('*a*' in Equation (6.27a) *et seq.*) is unlikely to remain constant when exposed to radiant heating as a layer of char forms as the surface temperature exceeds 150–200°C. This behaviour is common to all materials which char on heating, but unlike synthetic char-forming materials (such as the polyisocyanurates), wood exhibits unique behaviour due to its anisotropy and its ability to absorb water.

Regarding anisotropy, it takes longer to ignite a piece of wood at a cut end than on its surface, as the thermal conductivity (hence $k\rho c$) is greater along the grain than across it (Vytenis and Welker, 1975). Exposed knots are difficult to ignite for the same reason although their greater density also has some influence. There is considerable resistance to the flow of volatiles perpendicular to the grain. Only when the structure of the wood begins to break down at temperatures in excess of 300°C will volatiles tend to move directly to the surface, across the grain (cf. Figure 5.13). Even so, they can be observed to issue from a cut end or from around a knot where the resistance to movement is much less.

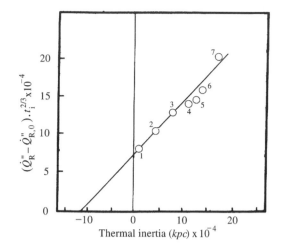

Figure 6.23   Correlation between time to ignition and thermal inertia for a number of woods, plus fibre insulation board (Equation (6.34)) (Lawson and Simms, 1952). 1. Fibre insulation; 2. Cedar; 3. Whitewood; 4. Mahogany; 5. Freija; 6. Oak; 7. Iroko. (Reproduced by permission of The Controller, HMSO. © Crown Copyright)

Table 6.6   Effect of moisture on time to ignition (horizontal samples, Douglas fir) (from Atreya and Abu-Zaid, 1991)

Moisture content (%)	Time to ignition (s)
0	55
11	100
17	145
27	215

   Moisture contained in the wood affects the ignition process in two ways: physically, by increasing the effective thermal capacity of a sample, and chemically, as the water vapour dilutes the pyrolysis products, effectively reducing the heat of combustion of the evolved vapours (see Section 6.3.2). Thus, the ignition time is found to increase with increasing moisture content (Table 6.6) (Atreya and Abu-Zaid, 1991). This is considered in more detail by Mikkola (1992). However, the 'critical heat flux', measured asymptotically as in Figure 6.19, is unaffected by the initial moisture content for the simple reason that for long ignition times, exposed samples will have had time to dry out.

6.3.2   *Ignition during a discontinuous heat flux*

While a critical surface temperature may be a satisfactory method of characterizing the firepoint of a solid exposed to a constant heat flux and may be suitable for engineering calculations (Thomson *et al.*, 1988), it is not always suitable if the flux is discontinuous.

Bamford *et al.* (1946) studied the ignition of slabs of wood (deal) by subjecting both sides to the flames from a pair of 'batswing' burners and determining how long it took to reach a stage where flaming would persist when the burners were removed. By comparing the results of a numerical analysis of transient heat conduction within the slabs, they concluded that a critical volatile flowrate from the surface, $\dot{m}_{cr}'' \geq 2.5$ g/m^{2}s was necessary for sustained ignition to occur. However, this is not a sufficient condition. It remained for Martin (1965), Weatherford and Sheppard (1965) and others (see Kanury, 1972) to emphasize the significance of the heating history and temperature gradient within the solid (at the moment of 'ignition'). This can be illustrated directly using the firepoint equation (Equation (6.18)):

$$(\phi\Delta H_c - L_v)\dot{m}_{cr}'' + \dot{Q}_E'' - \dot{Q}_L'' = S \tag{6.18}$$

This describes 'ignition' if $S \geq 0$ but 'extinction' if $S < 0$. It is possible for self-extinction to occur following ignition if the imposed heat flux $\dot{Q}_E''$, which was initially responsible for the surface achieving the firepoint temperature, is reduced or removed completely. Consider Equation (6.18): values of L_v are available from the literature (Table 5.6), and while there is some disagreement over measured values of $\dot{m}_{cr}''$, it is clear that it is a meaningful concept (see below). Rasbash (1975) has argued that the value of ϕ may be taken as 0.3 for many combustible materials, although it will be less than 0.2 for those that are fire retarded, and may be as high as 0.4 for certain oxygenated polymers (Table 6.7). Only the terms $\dot{Q}_E''$ and $\dot{Q}_L''$ need to be calculated for the relevant fuel configuration before Equation (6.18) can be applied.

As an example, consider a thick slab of PMMA (assumed to act as a semi-infinite solid) exposed to a radiant heat flux of 50 kW/m^2 in the presence of a pilot ignition source. Given that the firepoint temperature of PMMA is 310°C (Thomson and Drysdale, 1987) and the thermal inertia ($k\rho c$) is 3.2×10^5 W^{2}s/m^{4}K^2 (Table 2.1),

Table 6.7 Parameters in the firepoint equation (Equation (6.18))

Sample	Forced convection[a]		Natural convection[b]		Thomson and Drysdale (1988)[c]	
	$\dot{m}_{cr}''$ (g/m^{2}s)	ϕ	$\dot{m}_{cr}''$ (g/m^{2}s)	ϕ	T_{ig} (°C)	$\dot{m}_{cr}''$ (g/m^{2}s)
Polyoxymethylene	4.4	0.43	3.9	0.45	281	1.8
Polymethylmethacrylate	4.4	0.28	3.2	0.27	310	2.0
Polyethylene (LD)	2.5	0.27	1.9	0.27	363	1.31
Polyethylene (HD)	2.5	0.27	1.9	0.27	—	—
Polypropylene	2.7	0.24	2.2	0.26	334	1.1
Polystyrene	4.0	0.21	3.0	0.21	366	1.0

[a] Tewarson and Pion (1978). $\dot{m}_{cr}''$ was determined experimentally and ϕ calculated from Equation (6.36), using $h/c_p = 13$ g/m^{2}s for forced convection.

[b] As[a], but using $h/c_p = 10$ g/m^{2}s for natural convection.

[c] These data were obtained in an apparatus of design based on the ISO Ignitability Test (ISO 5657), quite different from the Factory Mutual Flammability Apparatus of Tewarson. The values of $\dot{m}_{cr}''$ are about one-half of those determined by Tewarson and Pion. This has not been resolved, but it seems more likely that it reveals a strong sensitivity to the value of the heat transfer coefficient h rather than a fundamental flaw in the concept of a critical mass flux at the firepoint (see Drysdale and Thomson, 1989). This is discussed further in the text.

Equation (6.31) can be used to show that it will take approximately 8.5 s to reach the firepoint. After this period of exposure, the depth of the heated layer will be approximately $(\alpha t)^{1/2} = 10^{-3}$ m (see Section 2.2.2, following Equation (2.24b)). $\dot{Q}''_L$ can then be estimated as:

$$\dot{Q}''_L = \varepsilon \sigma T_i^4 + h(T_i - T_0) + k \left(\frac{dT}{dx} \right)_{surface} \tag{6.35}$$

where $T_i = 583$ K, $T_0 = 293$ K and $k = 0.19$ W/mK. For this argument, ε and h are taken as 0.8 (dimensionless) and 15 (W/m^{2}K) respectively, and it is assumed that the temperature gradient at the surface $(dT/dx)_{surface}$ can be approximated by $(583 - 293)/(\alpha t)^{1/2}$.

Converting from W to kW, the heat loss term in Equation (6.18) is estimated as:

$$\dot{Q}''_L = 5.2 + 4.4 + 55.1 = 64.7 \text{ kW/m}^2$$

Taking Equation (6.18) with $\phi = 0.3$, $\Delta H_c = 26$ kJ/g, $L_v = 1.62$ kJ/g (Tewarson and Pion, 1976) and $\dot{m}''_{cr} = 2.5$ g/m^{2}s (Thomson and Drysdale, 1988), it is possible to test whether or not a flame will remain stabilized at the surface if the supporting radiation is removed immediately piloted ignition takes place (i.e. $\dot{Q}''_R \to 0$ at $t = 8.5$ s):

$$S = (0.3 \times 26.0 - 1.62) \times 2.5 - 64.7$$

$$= -49.25 \text{ kW/m}^2 < 0$$

i.e., flaming will not be sustained.

If exactly the same calculation is carried out for a slab of low thermal inertia material, such as polyurethane foam (PUF), the same conclusion is drawn. However, while theoretically correct, the exercise is unrealistic, as the firepoint temperature (assumed for convenience to be 310°C (Drysdale and Thomson, 1990)) is achieved after only 0.025 s, when the depth of the heated layer will be only 0.17 mm. (The conductive heat loss then will be 57 kW/m^2.) Other than under strictly controlled experimental conditions, such a short exposure time is physically impossible: 0.5 s would be more realistic, after which the heated layer would be 0.8 mm thick and the surface temperature would be much higher than 310°C. The latter has a major effect on subsequent behaviour. The rate of production of fuel vapours will now greatly exceed the critical value ($\dot{m}'' > \dot{m}''_{cr}$), and the associated flame will be capable of providing a much greater heat flux to the surface (by radiation and convection), thus sustaining the burning process.

Similarly, to achieve sustained burning on a thick slab of PMMA, the surface temperature must be increased to well above 310°C before the imposed heat flux is removed. A relevant example of this effect is to be found in the original British Standard Test for the ignitability of materials (BS 476 Part 4, now withdrawn): this involved exposing a vertical sheet of material to a small impinging diffusion flame for 10 s. Most dense materials greater than 6 mm thick will 'pass' the test, as flaming will not be sustained when the igniting flame has been removed. Typically, sustained flaming requires an exposure duration of 30 s or more.

Equation (6.18) can only be used at the firepoint, when $\dot{m}'' = \dot{m}''_{cr}$. When $\dot{m}''$ is greater than $\dot{m}''_{cr}$, the *proportion* of the heat of combustion that can be transferred from the flame to the surface (by convection and radiation) reduces, but the actual heat flux increases as the flame strengthens. Values of f_c and f_r (see Equations (6.17) and (6.18)) can no longer be assumed. These factors must be borne in mind when attempting to interpret the ignition process in terms of this equation.

An alternative to this detailed argument provides a more general overview which can also be used to explain directly why low density materials, once ignited, progress to give an intense fire so quickly. As burning continues, the depth of the heated layer ($\sim \sqrt{(\alpha t)}$) increases as heat is conducted into the body of the solid. It is possible to calculate the effective 'thermal capacity' of this layer as a function of time for different materials. This is shown in Table 6.8 as the product $\rho c \sqrt{(\alpha t)}$ ($= \sqrt{(k \rho c t)}$) which has the units J/m^2K, i.e. the amount of energy required to raise the *average* temperature of unit area of the heated layer of material by one degree. The amount required for the polyurethane foam is only 6% of that for polymethylmethacrylate. Thus, even if the flames above these two materials had the same heat transfer properties with respect to the surface, the polyurethane foam would achieve fully developed burning in only a fraction of the time taken by the PMMA. These results show clearly that thermal inertia ($k\rho c$) is an important factor in determining rate of fire development as well as ease of ignition.

Of the other factors that influence the ignition process (T_i, ΔH_c, $\dot{m}''_{cr}$ and ϕ in Equation (6.18)), values are becoming available for the firepoint temperature and the critical mass flux. However, there is a lack of agreement between different sources, particularly regarding $\dot{m}''_{cr}$. For example, Thomson's values are approximately half of those of Tewarson (Table 6.7). It seems likely that this is due to differences in the flow conditions which exist at the surface of the samples in the two experimental arrangements, but this has still to be resolved (Drysdale and Thomson, 1989). What is encouraging is that within each data set, there are clear trends: for example, the critical mass fluxes for the oxygenated polymers (POM and PMMA) are very roughly twice those for the hydrocarbon polymers (PE, PP and PS), qualitatively consistent with the lower heats of combustion of the oxygenated polymers (cf. Equation (6.18)).

The critical mass flux ($\dot{m}''_{cr}$) can be determined experimentally by monitoring the weight of a sample of material continuously while it is exposed to a constant radiant heat flux and subjected at regular intervals to a small pilot flame (cf. the Open Cup Flashpoint Test (American Society for Testing and Materials, 1990a)). Unfortunately, the rate of mass loss at the firepoint is too small to be measured routinely in the standard Cone Calorimeter: a load cell of very high resolution is required.

Table 6.8 Effective 'thermal capacities' of surfaces

	Time of heating (s)	Thermal diffusivity[a] (m²/s)	Depth of heated layer[b] (m)	'Effective thermal capacity' (J/m²K)
PMMA	10	1.1×10^{-7}	1×10^{-3}	1690
Polypropylene	10	1.3×10^{-7}	1.1×10^{-3}	1965
Polystyrene	10	8.3×10^{-8}	0.9×10^{-3}	1188
Polyurethane foam	10	1.2×10^{-6}	3.5×10^{-3}	98

[a] Data taken from Tables 1.2 and 2.1.
[b] Assumed to be equal to $(\alpha t)^{1/2}$.

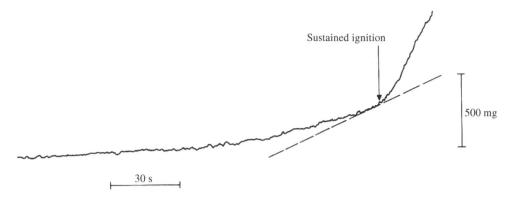

Figure 6.24   Determination of the critical mass flowrate ($\dot{m}_{cr}$) at the firepoint. (Deepak and
Drysdale, 1983, by permission)

Figure 6.24 shows a typical record of sample weight as a function of time, the
gradient changing as the surface ignites and sustains flame (Deepak and Drysdale,
1983). 'Flashing' is normally observed before this point is reached. $\dot{m}''_{cr}$ is then taken
from the gradient of the first section of the curve at the discontinuity, as shown. It has
been suggested (Rasbash, 1976) that $\dot{m}''_{cr}$ can be related to a critical value of Spalding's
mass transfer number (Spalding, 1955), thus:

$$\dot{m}''_{cr} = \frac{h}{c_p} \ln{(1 + B_{cr})} \qquad (6.36)$$

where h is the coefficient which applies to convective heat transfer between the
flame and surface, c_p is the thermal capacity of air, and $B_{cr} = A/\phi \Delta H_c$ where $A \approx$
3000 kJ/g (Section 5.1.2). Equation (6.36) applies to the 'steady state' firepoint condition (Equation (6.18), with $S = 0$) and assumes that $L_v = \phi \Delta H_c$: it offers the only
means by which ϕ may be estimated, and has been used by both Rasbash (1975) and
Tewarson (1980).

If burning was stoichiometric, then a value of $\phi = 0.45$ would be anticipated on the
basis that the flame must be quenched to a temperature of $\sim$1600 K (Section 3.1.2).
The fact that it is found to be $\sim$0.3 for many materials could be due to a number of
reasons, including non-stoichiometric burning in the limiting flame. The reactivity of
the volatiles will certainly influence the magnitude of this factor. Fire retardants which
reduce reactivity by inhibiting the flame reactions increase the limiting flame temperature and consequently reduce ϕ. The effect on the ignition properties of a material
can be seen by examining Equation (6.18). However, the value of h/c must be known
accurately. Tewarson selected $h = 13$ W/m^{2}K and 10 W/m^{2}K for 'forced convection'
and 'natural convection', respectively, and obtained a reasonable range of values of
ϕ (Table 6.7). However, the lower values of $\dot{m}''_{cr}$ obtained by Thomson give impos-
sibly high values (i.e. $\phi > 0.45$) with the same heat transfer coefficients (Thomson and
Drysdale, 1988). Yet, as before, each data set gives an internally consistent group of
values, suggesting that further research would prove fruitful to our understanding of
this aspect of the ignition process.

Indeed, from the above discussion, it is possible to identify several material properties which influence ease of ignition. A material will be difficult to ignite if L_v is large and ϕ and/or ΔH_c are small—or if $\dot{Q}_L''$ is large. Materials may be selected on the basis of these properties, or treated with fire retardants to alter these properties in an appropriate way. For example, retardants containing bromine and chlorine release the halogen into the gas phase along with the volatiles, rendering the latter less reactive, thus decreasing ϕ (Section 3.5.4). (However, they can be driven off under a sustained low-level heat flux, thus causing the polymer to lose its fire retardant properties (Drysdale and Thomson, 1989).) The use of alumina trihydrate as a filler for polyesters increases the thermal inertia ($k\rho c$) of the solid and effectively lowers ΔH_c as water vapour is released with the fuel volatiles. Phosphates and borates, when added to cellulosic materials, promote a degradation reaction which leads to a greater yield of char, and an increase in the proportion of CO_2 and H_2O in the volatiles, which reduces ΔH_c (Table 5.10). Thermally stable materials which have high degradation temperatures will exhibit greater radiative heat losses at the firepoint (increased $\dot{Q}_L''$). Similarly, the formation of a layer of char insulates the fuel beneath and higher temperatures will be required at the surface of the char to maintain the flow of volatiles. However, the effect of thermal inertia ($k\rho c$) can have a dominating influence on the ignition characteristics of a thick solid material, as has been demonstrated above.

6.4   Spontaneous Ignition of Combustible Solids

If a combustible solid is exposed to a sufficiently high heat flux in the absence of a pilot source, the fuel vapours may ignite spontaneously if somewhere within the plume the volatile/air mixture is at a sufficiently high temperature (Section 6.1). This is summarized in Figure 6.25 but Kanury (1995b) describes the processes in detail. Spontaneous ignition requires a higher heat flux than piloted ignition because a higher surface temperature is required: a value is quoted in Table 6.5 for 'wood', but it must be recognized that this is highly apparatus-dependent, and should not accepted as a general value (see below).

The mechanism for spontaneous ignition of a uniformly heated flammable vapour/air mixture has already been discussed, but in this case the mixture is neither homogeneous nor uniformly heated. Nevertheless, the same thermal mechanism is responsible for the instability that leads to the appearance of flame. Under an imposed radiative heat flux it is possible that absorption of radiation by the volatiles may contribute towards the onset of reaction. Kashiwagi (1979) showed that the volatiles can attenuate the radiation reaching the surface quite strongly. Volatiles can be ignited by subjecting them to intense radiation from a laser, but it is not known if the effect is significant in real fire situations.

Kanury (1972) has made an interesting observation on the surface temperatures required for pilot (PI) and spontaneous ignition (SI) of wood under radiative and convective heating, thus:

Mode of heat transfer	*Surface temperature of wood for:*	
	SI	PI
Radiation	600°C	300–410°C
Convection	490°C	450°C

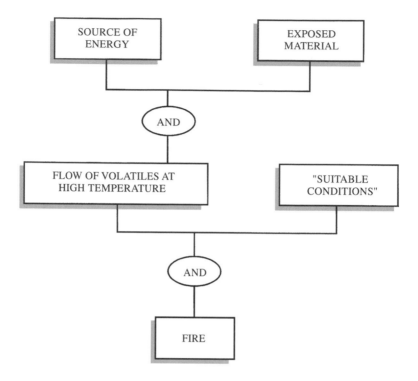

Figure 6.25   Scheme for spontaneous ignition

These results were obtained in an experimental arrangement similar to that shown in Figure 6.17. They can easily be explained if it is remembered that a forced convective flow will dilute the volatiles and consequently a higher surface temperature will be required to produce a mixture which is above the lower flammability limit at the source of ignition. On the other hand, for spontaneous ignition to occur as a result of radiative heat transfer, the volatiles released from the surface must be hot enough to produce a flammable mixture above its auto-ignition temperature when it mixes with unheated air. With convective heating, as the volatiles are entering a stream of air that is already at a high temperature, they need not be so hot.

6.5   Surface Ignition by Flame Impingement

This mode of ignition has been discussed in Section 6.3.2 and refers to the situation in which a pilot flame impinges on the surface of the material, with or without an imposed radiant flux. Ignition will occur after a delay, the duration of which will depend on $k\rho c$ and the total heat flux to the surface, comprising heat transfer from the flame and any external source: thereafter, flame may spread across the surface. This can occur at heat fluxes much lower than those required for piloted ignition, in which the only source of heat is from the external source. Simms and Hird (1958) found that the minimum heat flux for igniting pinewood in this way was only 4 kW/m^2, compared with 12 kW/m^2 for the mode of ignition illustrated in Figure 6.17. It is essentially a flame spread property which has been investigated in detail by Quintiere (1981) (Section 7.2.5(c)).

It should be noted that the size and characteristics of the impinging flame are extremely important. A small, laminar flame of low emissivity will transfer heat mainly by convection, but with increasing size and (in particular) thickness, radiation from the flame will come to dominate and the flow may become turbulent. Under these circumstances, the effective heat transfer to the combustible surface can be dramatically increased, causing the surface in contact with the flame to burn vigorously, contributing vapours which will in turn enhance the size of the flame. Taking as an example a vertical surface, such as a combustible wall lining, exposed to a large flame (such as from a burning item of furniture (Williamson *et al.*, 1991)), the heat transfer may be sufficiently great to overcome any ignition resistance that the material may be judged to have on the basis of results from small-scale tests, such as the ISO Ignitability Test, or the Cone Calorimeter. Such data must be interpreted very carefully, bearing in mind the end-use scenario in which the material is to be used. The importance of flame impingement as a means (or source) of ignition has been recognized by the International Organization for Standardization which has examined the characteristics of a wide range of flame sources (ISO, 1997b). (A set of standard ignition sources were developed for testing upholstered furniture in the UK (BSI, 1979).)

6.6   Extinction of Flame

Conceptually, extinction can be regarded as the obverse of pilot ignition and may be treated in a similar fashion, as a limiting condition or criticality. As with pilot ignition, there are two principal aspects to the phenomenon, namely: (i) extinction of the flame, (ii) reducing the supply of flammable vapours to below a critical value ($\dot{m}'' < \dot{m}''_{cr}$).

While the latter will cause the flame to go out, it is possible to achieve (i) without (ii), leaving the risk of re-ignition, which may occur spontaneously with combustible solids and liquids of high flashpoint (e.g. cooking oil) which have been burning for some time. The risk of re-ignition from a pilot will remain until the fuel cools to below its firepoint. However, for gas leaks and low flashpoint liquids, suppression of the flame will leave a continuing release of gaseous fuel which in an enclosed space could lead to the formation of a flammable atmosphere. Under these circumstances, the rate of supply of fuel vapour must be stopped, or at the very least reduced to a non-hazardous level.

6.6.1   Extinction of premixed flames

The stability of premixed flames was discussed in Section 3.3 in relation to the existence of flammability limits. In a confined space, an explosion following the release of a flammable gas can be prevented by creating and maintaining an atmosphere which will not support flame propagation even under the most severe conditions (Section 3.1 and Figure 3.12). This is 'inerting' rather than extinction and must exist before the ignition event occurs in order that flame does not become established. A premixed flame can be extinguished if a suitable chemical suppressant is released very rapidly ahead of the flame front. This is achieved in explosion suppression systems by early detection of the existence of flame, usually by monitoring a small pressure rise within the compartment, and rapid activation of the discharge of the chemical (Bartknecht, 1981; Gagnon,

1997). Typical agents include the halons CF_2Br_2 and CF_2BrCl, as well as certain dry powders (see Section 1.2.4). The halons have been phased out (Section 3.5.4) over the last few years and are being replaced by 'ozone-friendly' alternatives. As none of the gaseous substitutes are as effective, there has been much interest in the development of water mist systems (see, e.g., see Brenton *et al.*, 1994; Di Nenno, 1995). Nitrogen and carbon dioxide are suitable only for pre-emptive inerting as the amounts required are too large to be released at the rates necessary for rapid suppression.

Premixed flames may also be extinguished by direct physical quenching. This involves (*inter alia*) cooling the reaction zone and is believed to be the principal mechanism by which a flame arrester operates. This device consists of a multitude of narrow channels, each with an effective internal diameter less than the quenching distance, through which flame cannot propagate. The mechanism is described at some length in Section 3.3(a). Flame arresters are normally installed to prevent flame propagation into vent pipes and ducts in which flammable vapour/air mixtures may form (Health and Safety Executive, 1980; National Fire Protection Association, 1997).

6.6.2   Extinction of diffusion flames

In addition to cutting off the supply of fuel vapours (e.g. closing a valve to stop a gas leak or blanketing the surface of a flammable liquid with a suitable firefighting foam), diffusion flames may be extinguished by the same agents that are used for premixed flames. However, as there are already considerable heat losses from a diffusion flame, theoretically, less agent is required than for premixed flames: in practice, this distinction becomes less significant as the fire size is increased. It is understood that the mechanism by which extinction occurs is essentially the same as in premixed flames. Thus, 'inert' diluents (e.g. N_2 and CO_2) cool the reaction zone by increasing the effective thermal capacity of the atmosphere (per mole of oxygen) (Sections 1.2.5 and 3.5.4) and chemical suppressants such as the halons inhibit the flame reactions (Sections 1.2.4 and 3.5.4).

These agents may be applied locally from hand-held appliances directed at the flame. Small, developing fires are easily extinguished in this way as the local concentration of agent can greatly exceed the minimum requirement. Greater skill is required as the fire size increases, particularly if the supply of agent is limited. All flame must be extinguished before the supply runs out otherwise the fire will simply re-establish itself. This problem can be overcome by 'total flooding', provided that the compartment in which the fire has occurred can be effectively sealed to maintain the necessary concentration of the agent. This is economic only in special circumstances, e.g. when the possibility of water damage by sprinklers is unacceptable, such as in the protection of works of art and valuable documents, and of marine engine rooms and ships' holds. The advantage of chemical suppressants in this role is that the protection system can be activated while personnel are still within the compartment, while prior evacuation is necessary in the case of carbon dioxide (and nitrogen) as the resultant atmosphere is non-habitable. In principle, a halon system can be activated sooner than a CO_2 system, but in addition to the environmental problem, the agent is much more expensive and can generate harmful and corrosive degradation products at unacceptable levels if the fire is already too large when the agent is released. It should be noted that while total flooding may be used to hold in check a deep-seated smouldering fire, it is unlikely to extinguish it completely as such fires can continue at very low oxygen concentrations, particularly if they are well established. Cooling of the smouldering mass by water is the ultimate means of control.

The role of cooling in fire control must not be overlooked as this is the predominant method by which fires are extinguished. Water is particularly effective as it has a high latent heat of evaporation (2.4 kJ/g at 25°C). Indeed, it can extinguish a diffusion flame *per se* if it can be introduced into the flame in the form of a fine mist* or as steam. The suppression of fires by water, including the use of sprays, has been reviewed in detail by Grant *et al.* (1998). The most commonly used mode of suppression by water is in cooling the fuel surface: in terms of Equation (6.17), an additional heat loss term, $\dot{Q}''_w$ is introduced:

$$((f_r + f_c) \cdot \Delta H_c - L_v)\dot{m}''_{burn} + \dot{Q}''_E - \dot{Q}''_L - \dot{Q}''_w = S \tag{6.37}$$

When S becomes negative, the surface of the fuel will cool until ultimately

$$\dot{m}''_{burn} < \dot{m}''_{cr}$$

and flame can no longer exist at the surface. These concepts are considered in some detail by Beyler (1992).

Water is ideal for fires involving solids and can be effective with high flashpoint hydrocarbon liquid fires provided it is introduced at the surface as a high velocity spray which causes penetration of the droplets and cooling of the surface layers. If this is not effective, the water will sink to the bottom and may eventually displace burning liquid from its containment.

Diffusion flames may also be extinguished by the mechanism of 'blowout', familiar with the small flames of matches and candles. It is also the main method by which oilwell fires are tackled. The mechanism involves distortion of the reaction zone within the flame in such a way as to reduce its thickness, so that the fuel vapours have a much shorter period of time in which to react. If the reaction zone is too thin, then combustion will be incomplete and the flame is effectively cooled, ultimately to a level at which it can no longer be sustained ($T_f < 1600$ K). This can be interpreted in terms of a dimensionless group known as the Damkohler number (D).

$$D = \frac{\tau_r}{\tau_{ch}} \tag{6.38}$$

where τ_r is the 'residence time', which refers to the length of time the fuel vapours remain in the reaction zone, and τ_{ch} is the chemical reaction time, i.e. the effective duration of the reaction at the temperature of the flame. A critical value of D may be identified, below which the flame will be extinguished. The residence time will depend on the fluid dynamics of the flame, but as the reaction time, τ_{ch}, is inversely proportional to the rate of the flame reaction, we can write

$$D \propto \tau_r \exp(-E_A/RT_f) \tag{6.39}$$

---

* In Sweden, it has been shown that firefighters can influence the development of flashover in a compartment (Section 9.2) by application of water spray into the hot ceiling layer (e.g. see Schnell, 1996).

Blowout will occur if sufficient airflow can be achieved to reduce τ_r and T_f, thus reducing D ultimately to below the critical value. This approach is quite compatible with the concept of a limiting flame temperature and has been explored by Williams (1974, 1982) as a means of interpreting many fire extinction problems. It can account for chemical suppression which acts by increasing the effective chemical time by reducing the reaction rate (Section 1.2.4). As with local application of a limited quantity of extinguishant, blowout must be totally successful. This is particularly true with an oilwell fire where the flow of fuel will continue unabated after extinction has been attempted.

Problems

6.1 Using data given in Tables 1.12 and 3.1 calculate the closed cup flashpoint of *n*-octane. Compare your results with the value given in Table 6.4.

6.2 Would a mixture of 15% iso-octane + 85% *n*-dodecane by volume be classified as a 'highly flammable liquid' in the UK? Assume that the mixture behaves ideally, and that the densities of iso-octane and *n*-dodecane are 692 and 749 kg/m³ respectively. Take the lower flammability limit of *n*-dodecane to be 0.6%.

6.3 Calculate the temperature at which the vapour pressure of *n*-decane corresponds to a stoichiometric vapour–air mixture. Compare your result with the value quoted for the firepoint of *n*-decane in Table 6.4.

6.4 *n*-Dodecane has a closed cup flashpoint of 74°C. What percentage by volume of *n*-hexane would be sufficient to give a mixture with a flashpoint of 32°C?

6.5 A vertical strip of cotton fabric, 1 m long, 0.2 m wide and 0.6 mm thick is suspended by one short edge and exposed uniformly on one side to a radiant heat flux of 20 kW/m². Is it possible for the fabric to achieve its pilot ignition temperature of 300°C and if so, approximately how long will this take? Assume that the convective heat transfer coefficient $h = 12$ W/m²K, $\rho = 300$ kg/m³ and $c = 1400$ J/kg K and that the fabric surface has an emissivity of 0.9. The initial temperature is 20°C.

6.6 What difference would there be to the result of Problem 5 if the unexposed face of the fabric was insulated (cf. fabric over cushioning material)? (Assume the insulation to be perfect.)

6.7 A vertical slab of a 50 mm thick combustible solid is exposed to convective heating which takes the form of a plume of hot air flowing over the surface. If the firepoint of the solid is 320°C, how long will the surface take to reach this temperature if the air is at (a) 600°C; and (b) 800°C? (See Equation (2.26).) Assume that the convective heat transfer coefficient is 25 W/m²K, and that the material has the same properties as Yellow Pine (Table 2.1). Ignore radiative heat losses and assume that the properties of the wood remain unchanged during the heating process.

6.8 With reference to Problem 6.7, estimate the conductive heat losses from the surface into the body of the slab of wood at the moment the firepoint is reached. How does this compare with the radiative losses from the exposed face (at 320°C, assuming $\varepsilon = 1$)?

7

Spread of Flame

The rate at which a fire will develop will depend on how rapidly flame can spread from the point of ignition to involve an increasingly large area of combustible material. In an enclosure, the attainment of fully developed burning requires growth of the fire beyond a certain critical size (Section 9.1) capable of producing high temperatures (typically >600°C) at ceiling level. Although enhanced radiation levels will increase the local rate of burning (Section 5.2), it is the increasing area of the fire that has the greater effect on flame size and rate of burning (Thomas, 1981). Thus the characteristics of flame spread over combustible materials must be examined as a basic component of fire growth.

Flame spread can be considered as an advancing ignition front in which the leading edge of the flame acts both as the source of heat (to raise the fuel ahead of the flame front to the firepoint) and as the source of pilot ignition. It involves non-steady state heat transfer problems similar if not identical to those discussed in the context of pilot ignition of solids (Section 6.3). Consequently the rate of spread can depend as much on the physical properties of a material as on its chemical composition. The various factors which are known to be significant in determining the rate of spread over combustible solids are listed in Table 7.1 (Friedman, 1977). Several reviews of flame spread have been published in the intervening years (Fernandez-Pello and Hirano, 1983; Fernandez-Pello, 1984, 1995; Wichman, 1992).

Following the precedent set in earlier chapters, the behaviour of liquids will be reviewed before that of solids. Spread of flame through flammable vapour/air mixtures has already been discussed (Section 3.2).

7.1 Flame Spread Over Liquids

Flame spread over the surface of a pool of combustible liquid at temperatures below its firepoint involves surface tension-driven flows, first identified by Sirignano and Glassman (1970) (Section 6.2.2). The mechanism depends on the fact that surface tension decreases as the temperature is raised. Consequently, at the surface of the liquid, the temperature decrease ahead of the flame front is directly responsible for a net force which causes hot fuel to be expelled from beneath the flame, thereby displacing the cooler surface layer (McKinven *et al.*, 1970; Akita, 1972). This movement of hot liquid is accompanied by advancement of the flame. Some of the observations made by

Table 7.1 Factors affecting rate of flame spread over combustible solids (after Friedman, 1977)

Material factors		Environmental factors
Chemical	Physical	
Composition of fuel	Initial temperature	Composition of atmosphere
Presence of retardants	Surface orientation	Pressure of atmosphere
	Direction of propagation	Temperature
	Thickness	Imposed heat flux
	Thermal capacity	Air velocity
	Thermal conductivity	
	Density	
	Geometry	
	Continuity	

McKinven *et al.* (1970) on hydrocarbon fuels contained in trays or channels (1.2–3.0 m in length) merit discussion as they have relevance to spread of flame on solids. In these experiments, preheating of the fuel — which occurs with wick ignition (Burgoyne *et al.*, 1968) — was avoided by partitioning a short section at one end of the channel with a removable barrier and igniting only the enclosed liquid surface. The barrier was then taken away and flame allowed to spread over liquid whose surface was still at ambient temperature.

The diagram shown in Figure 7.1 represents flame spread over the surface of a liquid which is initially below its flashpoint. Behind the flame front, steady pool burning will develop (Section 5.1). Under quiescent conditions, a flow of air will be set up against the direction of spread as a direct consequence of entrainment into the base of the

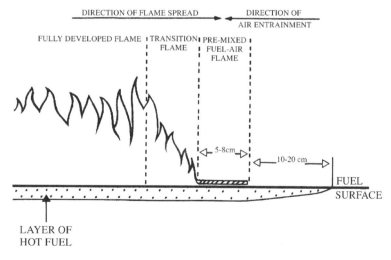

Figure 7.1 Spread of flame across a combustible liquid, initially at a temperature significantly less than its flashpoint, showing the net effect of the surface tension-driven flow. (Reproduced by permission of Gordon and Breach from McKinven *et al.* 1970.) The extent to which the flow and the flashes of premixed flame precede the main flame decrease if the initial temperature is increased towards the flashpoint (Akita, 1972)

developing fire. This spread mechanism is described as 'counter-current' (or 'opposed flow'). (Concurrent spread may be created by an imposed airflow parallel to and in the same direction as the spread (Section 7.2.5(e)).)

The leading edge of the spreading flame is blue, similar in appearance to a pre-mixed flame, and pulsates or 'flashes' ahead of the main flame (Glassman and Dryer, 1980/81). This is the behaviour one would expect if the temperature of the region of surface just ahead of the main flame lay between the flashpoint and the firepoint. A flash of premixed flame would occur periodically whenever a flammable concentration of vapour in air had formed (cf. Section 6.2.2) (Ito *et al.*, 1991). Raising the bulk temperature of the liquid has the effect of reducing the pulsation period and increasing the rate of spread. If the temperature of the liquid is below its flashpoint, then it is found that for shallow pools, the rate will decrease as the depth is reduced McKinven *et al.*, 1970; Miller and Ross, 1992). This is due mainly to restriction of the internal convection currents which accompany the surface tension-driven flow (Figure 6.11). In the limit, these will be completely suppressed, as with a liquid absorbed on to a wick (Section 6.2.2): if the heat losses to the supporting material are too great, then flame spread (and ignition, Section 6.2.2) will be impossible (Figure 7.2).*

McKinven *et al.* (1970) also found that the rate of spread is independent of the width of the channel containing the liquid for widths between 15 and 20 cm (Figure 7.3). For narrower channels, heat losses to the sides are important, while for wider ones the established flame behind the advancing front becomes so large that radiative heat transfer to the unaffected fuel becomes significant.

If the liquid is above its firepoint, then the rate of flame spread is determined by propagation through the flammable vapour/air mixture above the surface. Burgoyne and

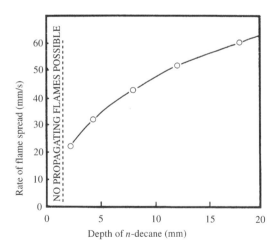

Figure 7.2   Variation of the rate of flame spread over *n*-decane floating on water as a function of depth of fuel. Container dimensions: 1.8 m × 0.195 m wide and 25 mm deep. Total depth of liquid (*n*-decane + water) 18 mm, initial temperature 23°C; flashpoint of *n*-decane 46°C. (Reproduced by permission of Gordon and Breach from McKinven *et al.*, 1970)

---

* Materials that are used for wicks are invariably good insulators and will permit flame spread, although the mechanism by which heat is transferred ahead of the flame will be conduction, as for solids (Section 7.3).

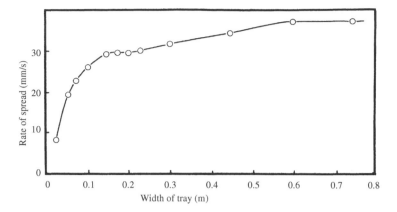

Figure 7.3   Rate of flame spread over *n*-decane as a function of tray width (4 mm of *n*-decane floating on water, other conditions as in Figure 7.2). (Reproduced by permission of Gordon and Breach from McKinven *et al.*, 1970)

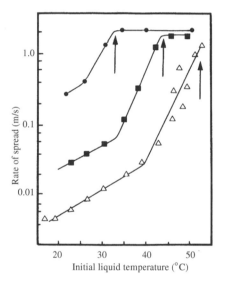

Figure 7.4   Dependence of the rate of spread of flame over flammable liquids on initial temperature. ●, propanol; ■, butanol; △, isopentanol. Container 33 mm wide, liquid depth 2.5 mm (Burgoyne and Roberts, 1968, by permission). The arrows indicate the temperatures at which a stoichiometric vapour/air mixture exists at the surface

Roberts (1968) and Akita (1972) have noted that the apparent rate of spread increases to a limiting value at around the temperature at which the vapour pressure corresponds to a stoichiometric mixture at the surface (Figure 7.4), the limiting rate being four or five times the fundamental burning velocity (as defined in Section 3.4). This suggests that what is being observed is flame propagation in a partially confined system in which the unburnt gas is pushed ahead of the flame front (cf. Figure 3.21). A similar effect has been reported in flame spread through layers of methane/air mixtures trapped beneath the ceiling of an experimental mine gallery (e.g. Phillips, 1965).

7.2 Flame Spread Over Solids

Attention will now be focused on the surface spread of flame over combustible solids, examining the factors listed in Table 7.1 systematically. Unlike liquid pools, the surface of a solid can be at any orientation, which can have a dominating effect on fire behaviour. This is particularly true for the phenomenon of flame spread as it is controlled by the mechanism by which heat is transferred ahead of the burning zone: this is strongly influenced by surface geometry and inclination.

As defined above (Figure 7.1), the term 'counter-current spread' describes the situation in which there is a flow of air opposed to the direction of spread, while concurrent spread is said to exist when the flow of air and the direction of spread are in the same direction. While these are clearly relevant to cases in which there is an imposed airflow, it also applies to flame spread where the air movement is generated naturally, by the dynamics of the flame. Naturally induced counter-current flow is developed in an otherwise quiescent atmosphere by a flame spreading along a horizontal surface (Figure 7.1), while naturally induced concurrent flow is observed when a flame is spreading upwards on a vertical surface (Section 7.2.1). The consequences of these effects will be discussed below.

7.2.1 *Surface orientation and direction of propagation*

In general, solid surfaces can burn in any orientation, but flame spread is most rapid if it is directed upwards on a vertical surface. This can be illustrated with results of Magee and McAlevy (1971) on the upward propagation of flame over strips of filter paper at inclinations between the horizontal and the vertical (Table 7.2). Propagation of flame was monitored by following the leading edge of the burning, or pyrolysis, zone. Downward propagation is much slower, and the rate less sensitive to change in orientation. With computer cards as fuel, Hirano *et al.* (1974) found the rate of spread to be approximately constant (*c*.1.3 mm/s) as the angle of orientation was changed from $-90°$ (vertically downwards) to $-30°$, while increasing more than threefold when the angle was changed from $-30°$ to $0°$ (horizontal) (Figure 7.5a). These combined results suggest at least a 50-fold increase in the rate of spread between $-90°$ and $+90°$ for thin fuels. The situation for thick fuels will be discussed below.

The reason for this behaviour lies in the way in which the physical interaction between the flame and the unburnt fuel changes as the orientation is varied (Figure 7.6). For downward and horizontal spread, air entrainment into the flame leads to 'counter-current spread' (i.e. spread against the induced flow of air), but with upward spread on a vertical surface, the natural buoyancy of the flame generates 'concurrent spread'.

Table 7.2   Rate of flame spread over strips of filter paper (Magee and McAlevy, 1971)

Orientation	Rate of flame spread (mm/s)
0° (horizontal)	3.6
+22.5°	6.3
+45°	11.2
+75°	29.2
+90° (vertically upwards)	46–74 (erratic)

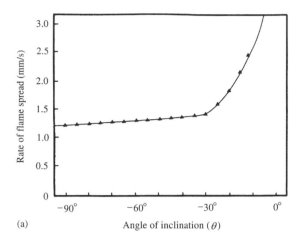

(a)

Angle of inclination (θ)

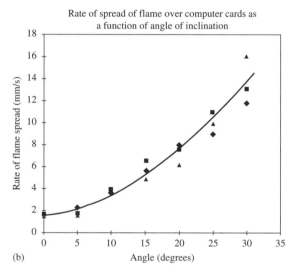

(b)

Figure 7.5   Variation of rate of flame spread over a thin fuel (computer card) as a function of angle of inclination (θ). (a) $\theta = -90°$ (vertically downwards) to $\theta = 0°$ (horizontal) (Reproduced by permission of the Combustion Institute from Hirano *et al.*, 1974); (b) $\theta = 0°$ to $\theta = 30°$. (Reproduced by permission of Elsevier Science from Drysdale and Macmillan, 1992)

This produces greatly enhanced rates of spread as the flame and hot gases rise in the same direction, filling the boundary layer and creating high rates of heat transfer ahead of the burning zone. With physically thin fuels, such as paper or card, burning can occur simultaneously on both sides. This must be taken into account when interpreting flame spread behaviour. The enhanced rate for downward spread on computer cards at inclinations between $-30°$ and $0°$ reported by Hirano *et al.* (1974) (Figure 7.5a) is due to the flame on the underside of the card contributing to the forward heat transfer process. Kashiwagi and Newman (1976) have linked this to the onset of instability of the flame on the underside at low angles of inclination. Upward flame spread at inclinations greater than $0°$ increases monotonically (Figure 7.5b), for the same reason:

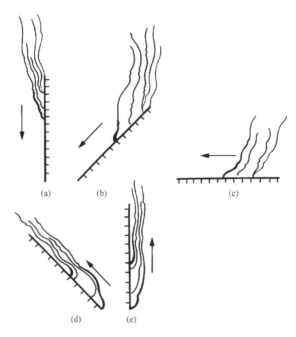

Figure 7.6 Interaction between a spreading flame and the surface of a (thick) combustible solid for different angles of inclination: (a) $-90°$; (b) $-45°$; (c) O°; (d) $+45°$; (e) $+90°$. (a)–(c) are counter-current spread, while (d) and (e) are co-current spread. The switch from counter-current to co-current spread takes place at an angle of $c15$–$20°$ (see Figure 7.7)

The upward inclination allows the flame and hot gases on the underside of the card to flow ahead of the burning zone (Drysdale and Macmillan, 1992).

Quite different behaviour is observed with physically thick fuels. At intermediate orientations, enhancement of upward spread is observed only when the inclination of the surface is increased above 15–20°. This possibility was first noted by Markstein and de Ris (1972), but not demonstrated experimentally until 1988, following a fatal fire involving a wooden escalator at King's Cross Underground Station in London (Fennel, 1988; Drysdale and Macmillan, 1992). The effect is quite pronounced if the flame is spreading up a plane incline (Figure 7.7 curve a), but is greatly enhanced if entrainment of air from the side is prevented — as occurred in the escalator fire (Figure 7.7 curve b). This has become known as the 'trench effect', and has received much attention, both experimental (Smith, 1992; Atkinson *et al.*, 1995) and through CFD modelling (Simcox *et al.*, 1992; Cox *et al.*, 1989; Woodburn and Drysdale, 1997). There is a switch from counter-current spread at inclinations below about 15°, to concurrent spread which is certainly operating at inclinations of 25° and above (Figure 7.8): the critical angle depends on the geometry of the fire and the trench (Woodburn and Drysdale, 1997). With thin fuels, this behaviour is completely masked by the flow of hot gases which occurs on the underside when there is the slightest upward inclination (Figure 7.6). Macmillan found that if this flow is inhibited, flame may not be able to propagate: e.g. with computer cards held above a metal baseplate, the flame self-extinguished if the gap was 4 mm or less.

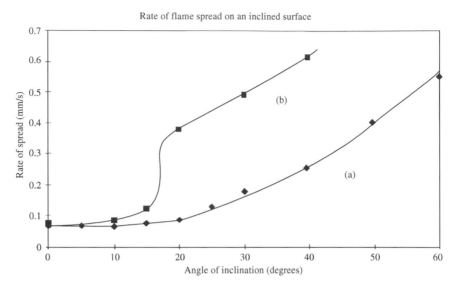

Figure 7.7 Rate of spread of flame on an inclined surface. 60 mm wide samples of PMMA with (■) and without (◆) 'sidewalls' (Drysdale and Macmillan, 1992)

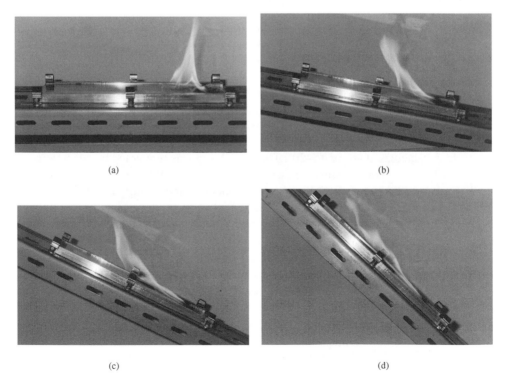

Figure 7.8 Flame spreading over the surface of PMMA slabs, 60 mm wide, 150 mm long and 6 mm thick (a) $\theta = 0°$ (horizontal); (b) $\theta = 10°$; (c) $\theta = 20°$; and (d) $\theta = 45°$. (photographs by Stuart Kennedy and David Oliphant)

Downward spread ($-90°$) achieves a slow, steady rate of propagation almost immediately, but upward spread ($+90°$) accelerates toward a quasi-steady state. This has been observed for vertical PMMA slabs (Orloff *et al.*, 1974) and for freely suspended strips of fabric (Thomas and Webster, 1960; Markstein and de Ris, 1972). Markstein and de Ris (1972) observed that following ignition at the bottom edge there is a short period of laminar burning which quickly develops turbulence as the flame size increases. In their experiments which involved 1.5 m lengths of fabric (maximum width 0.6 m), they showed that the instantaneous rate of flame spread was dependent on the length of the pyrolysing zone, i.e. the zone from which the volatiles were being released. As the rate of flow of volatiles determines the height of the flame (Thomas and Webster, 1960) it also determines the extent of preheating of the unaffected fabric which in turn determines how quickly fresh fuel is brought to the firepoint. Markstein and de Ris (1972) showed that:

$$V_p \propto l_p^n$$

where V_p is the rate of vertical spread, l_p is the length of the pyrolysis zone (Figure 7.9) and n is a constant, approximately equal to 0.5, and developed a simple mathematical model to describe the behaviour. Information regarding forward heat transfer was obtained experimentally using a combination of flat sintered metal gas burners and water-cooled heat transfer plates. Their model shows that the growth of the fire depends *only* on the total forward heat transfer which was limited by the onset of turbulence and the burnout time of the fabric. With a heavier fabric the burnout time would be longer (i.e. l_p would be longer) and a more rapid rate of upward spread would be expected: indeed, it follows that upward flame spread on a semi-infinite solid can never achieve a steady state. Alpert and Ward (1984) propose that the development of such a fire can be approximated by exponential growth (Section 7.5).

For vertically upward propagation over 'thick' PMMA slabs, radiation from the flame can be responsible for over 75% of the total heat transfer ahead of the flame

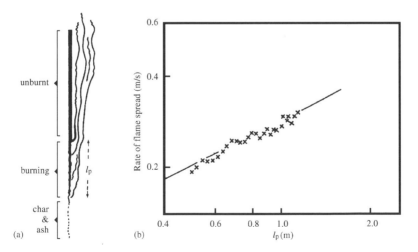

Figure 7.9 Upward spread of flame on a vertical strip of fabric (a) showing the burning (pyrolysis) zone of length l_p; (b) increase of rate of upward spread with increasing length of the pyrolysis zone. Cotton broadcloth (103 g/m²) of width 0.457 m and length 1.524 m (Markstein and de Ris, 1972)

front (Orloff *et al.*, 1974). A smaller proportion would be anticipated in the case of thin materials (e.g. fabrics) as the flames are unlikely to be so large, and consequently will be optically thinner (Section 2.4.3).

Substantial lengths of fabric strips are required to observe the development of steady upward burning (Thomas and Webster, 1960). Markstein and de Ris (1972) extrapolated their data to give the limiting rates of spread and obtained values of up to 0.45 m/s (cf. Table 7.2), which is more than two orders of magnitude greater than the limiting rate of spread on computer cards found by Hirano *et al.* (1974) for vertically downward propagation. This provides another example of the difference between concurrent and counter-current spread. In vertically downward spread, the flame gases flow away from the unburnt material so that convective transfer cannot occur and radiation is likely to be unimportant as the relevant configuration factor (i.e. the flame to the unburnt fuel) will be very small. Conduction through the gas phase is believed to be the dominant mechanism for thin fuel beds (Parker, 1972; Hirano *et al.*, 1974) and through the solid for 'thick' fuels (Fernandez-Pello and Williams, 1974; Fernandez-Pello and Santoro, 1980) (Section 7.3).

The remaining orientation to be discussed is the horizontal ceiling. A combustible ceiling will burn and spread flame, but the process generally has to be driven by a 'ceiling jet' created by a fire at floor level, or by a wall fire (see Chapter 9). Without this imposed flow, the rate of spread will be slow as the hot gases will tend to stagnate underneath the horizontal surface. It can only spread effectively as a co-current flow. Not only will the ceiling jet create a flow of hot combustion products which will preheat the surface, it will also carry flame over unburnt fuel ahead of the burning area (cf. upward spread on a vertical surface, Figure 7.9) (Fernandez-Pello, 1995). (It is anticipated that upward spread on a sloping ceiling would not require an imposed flow.)

7.2.2   Thickness of the fuel

While flame spread can be treated theoretically as a quasi-steady state problem (e.g. de Ris, 1969) it involves transient heat transfer processes identical to those encountered in the previous chapter on ignition (Section 6.3). The flame front represents a formal boundary, referred to by Williams (1977) as the 'surface of fire inception', which lies between the two extreme states of unburnt and burning fuel. Movement of this boundary over the fuel can be regarded as the propagation of an ignition front: just as with ignition, the rate of heat transfer by conduction from the surface to the interior of the fuel influences the process significantly. Thus, if the fuel is very thin and can be treated by the 'lumped thermal capacity model' in which there is no temperature gradient between the faces of the sample (Section 2.2.2), it can be shown theoretically that the rate of spread will be inversely proportional to the thickness (τ) of the material (Sections 7.2.3 and 7.3). There is ample evidence in the literature to support this conclusion. Magee and McAlevy (1971) quote data obtained by Royal (1970) which show that for downward spread on vertical specimens of thin cellulosic fuels, $V \propto \tau^{-1}$, for thicknesses less than 1.5 mm (Figure 7.10a). This has been confirmed by results of Suzuki *et al.* (1994) with samples prepared from filter paper, and similar relationships have been demonstrated for fabrics where the rate has been shown to be inversely proportional to the 'weight' of the fabric. Air permeability of the material has no significant effect (Moussa *et al.*, 1973).

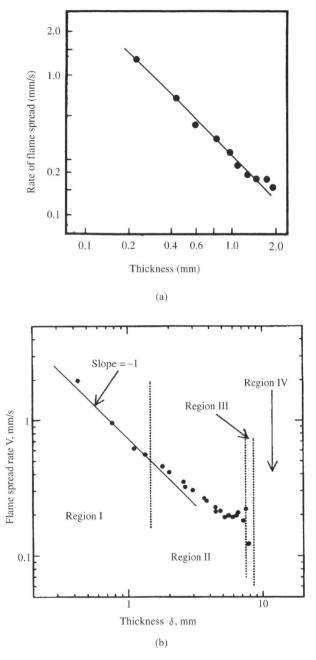

(a)

(b)

Figure 7.10 (a) Variation of downward flame spread rate with thickness for thin cellulosic specimens (Royal (1970), quoted by Magee and McAlevy, 1971). Reproduced by permission of Technomic Publishing Co. Inc., Lancaster, PA 17604, USA (b) Dependence of rate of downward flame spread on thickness (paper samples). Regions I and II: stable spread; Region III: unstable spread; Region IV: no spread (thickness > 8.4 mm) (Suzuki *et al.*, 1994 by permission of the Combustion Institute)

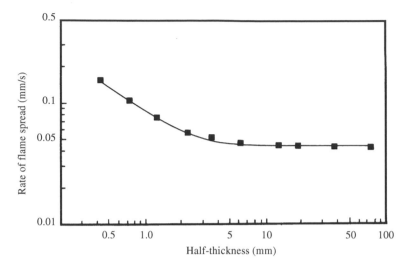

Figure 7.11   Variation of the rate of downward flame spread on vertical sheets of PMMA of different thickness (reproduced by permission of the Combustion Institute from Fernandez-Pello and Williams, 1974)

As the thickness is increased the rate of spread ultimately becomes independent of thickness. This is seen in Figure 7.10b. Suzuki *et al.* (1994) extended the range of thickness of paper to 10 mm and found that the dependence on thickness became less pronounced when $\tau > 1.5$ mm, becoming approximately constant in the range *c*.5.0–7.5 mm. For thicker samples, the rate became erratic, and if $\tau > 8.4$ mm, continuous spread was not possible. However, with PMMA (Fernandez-Pello and Williams, 1974), which is a non-charring material, downward spread achieves a constant rate when the sample becomes 'thermally thick' ($\tau > c.5$ mm) in the sense described earlier (Section 2.2.2) (Figure 7.11). In going from thermally thin to thermally thick fuels, there is a change in the dominant mode by which heat is transferred ahead of the flame front, from heat conduction through the gas phase for thin fuels (see Parker, 1972, Section 7.3) to heat conduction through the solid for thick fuels (Fernandez-Pello and Williams, 1974; Fernandez-Pello and Santoro, 1980).

It should be borne in mind that the above discussion relates to materials that do not change their physical form during the burning process. Processes such as melting and delamination should be considered. Delamination can turn a composite material which initially has the hallmarks of a 'thermally thick' fuel into one in which the surface lamina detaches and becomes a thin fuel, more easily ignited and with the propensity to spread flame rapidly. Another example is the delamination of multiple layers of gloss paint on a non-combustible substrate such as plaster. This is recognized as a serious fire hazard in old buildings such as hospitals which have been regularly maintained (Murrell and Rawlins, 1996).

7.2.3   *Density, thermal capacity and thermal conductivity*

The distinction between 'thermally thick' and 'thermally thin' materials has already been made and the concept developed in Section 6.3.1. The 'depth of heating' is given approximately by $(\alpha t)^{1/2}$, where α is the thermal diffusivity $(k/\rho c)$ and t is the time in seconds during which the surface of the solid is exposed to a heat flux. For an advancing flame front, the exposure time for the unburnt fuel is l/V, where V is the rate of spread and l is the 'heating length', i.e. the length of sample perpendicular to the advancing flame over which the temperature rises from T_0 (ambient) to the temperature corresponding to the firepoint. (This is marked clearly in Figure 7.18 (Section 7.3).) A critical thickness τ_{cr} for flame spread is then estimated from the expression:

$$\tau_{cr} = (\alpha l/V)^{0.5} \tag{7.1}$$

Thin fuels ($\tau < \tau_{cr}$) may be treated by the lumped thermal capacity model for which the time to achieve the firepoint (t_i) under a given heat flux is directly proportional to the product $\rho c \tau$ (Equation (6.30)). As the rate of spread will be inversely proportional to t_i, then

$$V \propto (\rho c \tau)^{-1} \tag{7.2}$$

which gives the dependence on τ which was noted above (Figure 7.10a, Section 7.2.2). To discover how the thermal properties of a 'thick' fuel ($\tau > \tau_{cr}$) influence the rate of spread, τ must be replaced by an expression for the depth of the heated layer at the surface of the material (δ): thus, following Equation (7.1),

$$\delta = (\alpha l/V)^{0.5} \tag{7.3}$$

If this is substituted in place of τ in Equation (7.2), rearrangement gives

$$V \propto \frac{1}{k\rho c} \tag{7.4}$$

provided that l is constant. The appearance of thermal inertia ($k\rho c$) is consistent with the fact that ease of ignition is strongly dependent on this factor. Examining it in more detail, as the thermal conductivity (k) of a solid is roughly proportional to its density (ρ), it can be seen from Equation (7.4) that the rate of flame spread is extremely sensitive to the density of the fuel bed (approximately, $V \propto \rho^{-2}$). This is why foamed plastics and other combustible materials of low density spread flame and develop fire so rapidly: only a very small mass of material at the surface needs to be heated to allow the flame to spread (Table 6.8). Some authors have attributed this behaviour to surface roughness and porosity (e.g. Magee and McAlevy, 1971), but these have at most only second-order effects (Moussa *et al.*, 1973).

7.2.4 Geometry of the sample

(a) *Width*   The width of a sample has little or no effect on the rate of vertically downward spread provided that edge effects do not dominate the behaviour (see (b) below). The situation is different for vertically upward burning. Thomas and Webster (1960) found that for freely suspended strips of cotton fabric:

$$V \propto (\text{width})^{0.5} \tag{7.5}$$

For widths ranging from 6–100 mm this result can be correlated with the increase in flame height arising from the increased area of burning (Section 4.3.2). This illustrates one of the difficulties encountered in devising a small-scale test for assessing the fire behaviour of fabrics. In addition, the propensity of some synthetic materials to melt and drip while burning can lead to results that are difficult to interpret. Consequently, fabric flammability tests that have been developed more recently call for substantial widths of material (Holmes, 1975; British Standards Institution, 1976) rather than narrow strips as specified in BS 3119 (British Standards Institution, 1959) and DOC FF-3-71 (US Department of Commerce, 1971a).

While the effect of width on the rate of flame spread over thick materials has not been studied, one would anticipate that the same general features would be observed, but modified by the fact that the flame behind the leading edge will be substantially greater as the area of burning (the 'pyrolysis zone' of Section 7.2.1) will be larger. This will affect the rate of horizontal spread if the width of the fuel bed is great enough. Thus, McKinven *et al.* (1970) found that radiative heat transfer ahead of the flame front became significant in the spread of flame over the surface of *n*-decane contained in trays more than 0.2 m wide (Figure 7.3). A similar effect would be expected for solids. Indeed, for fires spreading on fuel beds of substantial width, radiative heat transfer ahead of the flame front can dominate the fire spread mechanism, simply because of the size of the flames. Kashiwagi (1976) noted in other experiments (Section 7.2.5c) that this became significant once the flame height exceed 0.15 m.

(b) *Presence of edges*   Flame spreads much more rapidly along an edge or in a corner than over a flat surface. This has been studied systematically by Markstein and de Ris (1972) using 'wedges' of PMMA as shown in Figure 7.12(a). The rate of downward propagation at the edge was measured as a function of the angle (θ), the following dependence being found for $20° \leq \theta \leq 180°$:

$$V \propto \theta^{4/3}$$

(Figure 7.12(b)). The narrower the angle θ, the closer the edge of the solid approaches thin-fuel behaviour, with flame spreading down both sides. Thus it is partly a thermal capacity effect and partly due to the fact that heat is being transferred to the fuel on both sides of the edge. The rate of downward spread is a minimum for $\theta = 180°$. If θ is greater than 180° (i.e. the angle is re-entrant) the rate of downward spread is enhanced by cross-radiation near the junction of the walls (Figure 7.12(c)). Markstein and de Ris (1972) estimated that $V \approx 0.56$ mm/s for $\theta = 270°$, this referring to the maximum, close to the corner. When $\theta > 270°$, cross-radiation effects will increasingly dominate, approaching the case of two inward-facing burning surfaces (see Section 2.4.1).

7.2.5   *Environmental effects*

(a) *Composition of the atmosphere*   Combustible materials will ignite more readily, spread flame more rapidly and burn more vigorously if the oxygen concentration is increased. This is of practical significance as there are many locations where oxygen enriched atmospheres (OEAs) may be produced accidentally (e.g. leakage from the oxygen supply system in a hospital or leakage from oxygen cylinders used for oxyacetylene welding) and several where OEAs are produced deliberately (e.g. 'oxygen tents' in

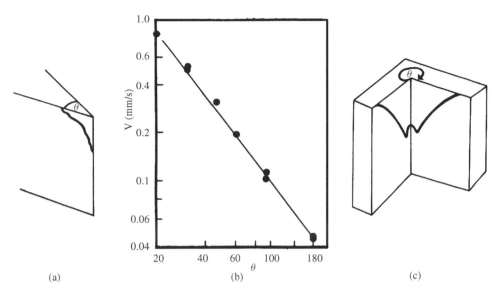

Figure 7.12   Rate of downward spread of flame on edges and in corners. (a) Definition of the angle (θ) at an edge. (b) Variation of V for $20° < \theta \le 180°$. (c) Definition of θ for a corner. Figures (a) and (c) show typical shapes of the leading edge of the advancing flame (Markstein and de Ris, 1975). Reproduced by permission of Technomic Publishing Co. Inc., Lancaster, PA 17604, USA

intensive care units). An OEA is one in which the partial pressure of oxygen is greater than that of the normal atmosphere (160 mm Hg) (National Fire Protection Association, 1994), so that any environment in which the air pressure has been increased artificially (e.g. in a diving bell or for tunnel boring operations) must be considered potentially hazardous (see Section 7.2.5(d)).

Any increase in oxygen concentration in the air is accompanied by an increase in the rate of flame spread (see Figure 7.16). This is because the flame is hotter and can lose more heat to the fuel (i.e. ϕ is greater, see Section 6.3.2). It will lie closer to the fuel surface, thus increasing the rate of heat transfer. Magee and McAlevy (1971) record that the dependence of propagation rate (V) on oxygen concentration is greater for 'thick' fuels than for 'thin', at least for vertically downward spread. Huggett *et al.* (1966) found that the rate of flame spread over wooden surfaces in O_2/N_2 and O_2/He mixtures depended on the ratio $[O_2]/c_g$, where $[O_2]$ is the concentration of oxygen and c_g is the heat capacity of the atmosphere. Varying c_g also affects the flame temperature (Table 1.16), allowing the flame to move closer to the surface if c_g is lower than that for air.

A corollary to the work of Huggett is that it is possible to create an atmosphere which will support life but not flame. If the thermal capacity of the atmosphere is increased to more than 50 J/g.K (corresponding to 12–13% O_2 in N_2) flame cannot exist under normal ambient temperatures. This level of oxygen is too low for normal human activity but if this atmosphere is pressurized until the oxygen partial pressure equals 160 mm Hg, it becomes perfectly habitable although it will still be unable to support burning (Huggett, 1973).

(b) *Fuel temperature*   Increasing the temperature of the fuel increases the rate of flame spread. Intuitively this is to be expected as the higher the initial fuel temperature the less heat is required to raise the unaffected fuel to the firepoint ahead of the flame. Magee and McAlevy (1971) report that for thin fuels:

$$V \propto \frac{1}{(T_p - T_0)} \text{ (approximately)} \tag{7.6}$$

while for thick fuels:

$$V \propto \frac{1}{(T_p - T_0)^2} \text{ (approximately)} \tag{7.7}$$

where T_p is the minimum temperature at which decomposition occurs and T_0 is the initial fuel temperature. These experimental results are consistent with the theoretical analysis carried out by de Ris (1969) (see Section 7.3).

(c) *Imposed radiant heat flux*   An imposed radiant heat flux will cause an increase in the rate of flame spread, primarily by preheating the fuel ahead of the flame front (Fernandez-Pello, 1977a,b; Hasemi *et al.*, 1991) (cf. Equations (7.6) and (7.7)). However, the increased rate of burning behind the flame front will give stronger flames which will provide additional forward heat transfer and thus enhance the process (Kashiwagi, 1976; Hirano and Tazawa, 1978). These observations are quite general, although the relative importance of the effects will depend on orientation: thus the effect is particularly marked if the spread is vertically upwards (Saito *et al.*, 1989; Hasemi *et al.*, 1991).

    A relatively low level is capable of producing a measurable effect. Alvares (1975) reports that a radiant heat flux equivalent to three to four times that of the summer sun in the UK will increase the rate of flame spread over inclined upholstered panels by 70%. It is even more pronounced with thin fuels such as paper (Hirano and Tazawa, 1978). Such an effect is significant during the early stages of a compartment fire when the levels of radiant heat flux from the compartment boundaries and the layer of hot smoky gases trapped below the ceiling are increasing (Chapter 9).

    The response of a surface to the imposed flux is not instantaneous and the effect of transient heating must be considered if the flame begins to spread over the surface before thermal equilibrium has been reached (Figure 7.13) (Kashiwagi, 1976; Fernandez-Pello, 1977a; Quintiere, 1981). This is illustrated by the results of Kashiwagi which are shown in Figure 7.13(a): he found the rate of flame spread over the surface of a thermally thick material to increase with the duration of exposure to a constant heat flux. The initial response to thermal radiation will depend on the thermal inertia ($k\rho c$) of the material (see Figure 2.10), while the steady state velocity (corresponding to a long preheating time) will be determined by the surface temperature achieved at equilibrium. This will depend on the heat losses from the surface. Kashiwagi's experiments, in which the 'fuel bed' was a carpet, showed that the steady-state flame spread was faster if the carpet had an underlay to reduce heat losses to the floor (Figure 7.13(a)). It would be anticipated that if the equilibrated surface temperature is greater than the firepoint, the rate of spread would be very high as the flame would be propagating through a pre-mixed vapour/air mixture (compare with Figure 7.4).

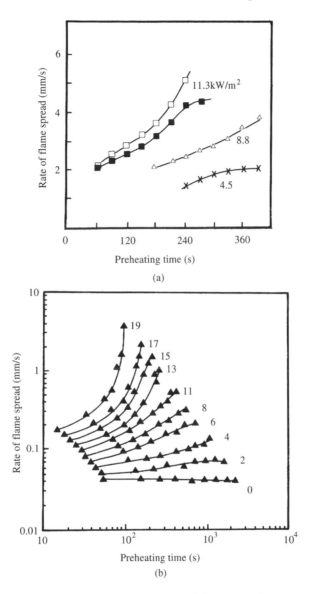

Figure 7.13 Effect of preheating time on the rate of flame spread over surfaces exposed to radiant heat. (a) Horizontal acrylic carpet; ×, △ and □ with underlay; ■ without underlay. Reproduced by permission of the Combustion Institute from Kashiwagi (1975). (b) Vertical PMMA sheets (downward spread). Figures refer to the imposed radiant heat flux (kW/m²). Reproduced by permission of Gordon and Breach from Fernandez-Pello (1977a)

This behaviour will not necessarily be observed for thermally thin materials. For example, although an imposed heat flux will promote flame spread over paper and similar materials, high levels will cause charring and rapid degradation which can deplete the fuel bed of most of its 'volatiles' before the arrival of the flame front (Fernandez-Pello, 1977a). Ultimately, this will place a limit on the maximum rate of spread. (Note that thin thermoplastic films and sheets will soften and melt if exposed to the same conditions.)

When it was realized that radiation had such a profound influence on the rate of flame spread (and thus rate of fire growth), laboratory tests involving exposure to radiant heat were developed to assess the flame spread potential of wall lining materials (e.g. British Standards Institution, 1987a; ASTM, 1994c; ISO, 1980). These were developed on an *ad hoc* basis to provide a means by which materials could be 'ranked' for regulatory purposes. However, the results of these tests are strongly apparatus-dependent, and cannot yield quantitative data on material performance: indeed, different tests designed for the same purpose in different European countries give widely different ranking orders for a given set of materials (Emmons, 1974).

This was seen as a barrier to international trade which would become increasingly unacceptable, particularly within Europe after the formation of the European Union. A programme of work was initiated to develop a new generation of tests which would surmount such problems. These were described collectively as 'reaction to fire' tests, and were designed and developed with the benefit of an improved understanding of fire science. These test methods are listed in Table 7.3.

Quintiere (1981) analysed the ISO Surface Spread of Flame test in which flame is allowed to spread horizontally along a vertical sheet of material 155 mm high and 800 mm long which is set with its long axis at an angle of 60° to a radiant panel as shown in Figure 7.14. Thus the radiant heat flux falling on the surface diminishes with distance from the panel, from 50 kW/m^2 at one end to 2 kW/m^2 at the other. The sample is ignited by a pilot flame close to the radiant heater and the subsequent flame spread is observed.

Quintiere (1981) carried out a detailed mathematical analysis of the heat transfer at the flame boundary, based on that developed earlier to analyse flame spread over carpets (Quintiere, 1975b; NFPA, 1995). Several simplifying assumptions were required to enable the equations to be solved, including the following: (i) the solid is thermally thick; (ii) heat is transferred forward from the flame through the gas phase; and (iii) flame moves forward when the unburnt fuel reaches an 'ignition temperature', T_{ig} (equivalent to the firepoint) and below which the material is regarded as inert. Although there is some doubt about the compatibility of (i) and (ii), the equations

Table 7.3   ISO Reaction to Fire Tests (Vandevelde *et al.*, 1996)

Test method	Reference
Ignitability of building products	ISO 5657 (1997)
Lateral ignition and flame spread of building products	ISO 5658-3
Rate of heat release from building products — Cone Calorimeter	ISO 5660 (1993)[a]
Smoke generated by building products — dual chamber test	ISO 5924 (1990)
Full-scale room test for surface products	ISO 9705 (1993)

[a] See also British Standards Institution, 1993.

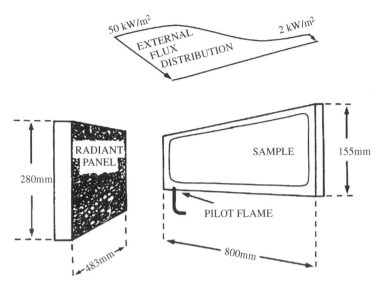

Figure 7.14 Details of the ISO Surface Spread of Flame Apparatus (Quintiere, 1981). Reproduced with permission

reduce to

$$V^{-1/2} = \left[\frac{(\pi k\rho c)^{1/2}}{2hl^{1/2}\dot{q}_F''}\right]\left[h(T_{ig} - T_0) - \dot{q}_E''\right] \tag{7.8}$$

where $\dot{q}_F''$ is the heat flux from the flame to the surface across a distance l, $\dot{q}_E''$ is the externally imposed heat flux, T_0 is the initial temperature and h is an effective surface heat transfer coefficient. This equation can be rewritten:

$$V^{-1/2} = C(\dot{q}_{0,i}'' - \dot{q}_E'') \tag{7.9}$$

where C is referred to as a 'rate coefficient' and $\dot{q}_{0,i}''$ is the limiting heat flux necessary for pilot ignition (Section 6.3.1). These quantities contain the basic properties of the material and could reasonably be expected to be 'constants' in a given flame spread situation. If so, a plot of $V^{-1/2}$ versus $\dot{q}_E''$ would be a straight line of gradient $-C$ and intercept $C \cdot \dot{q}_{0,i}''$. Quintiere (1981) examined this hypothesis by using data on the variation of V and $\dot{q}_E''$ as a function of distance from the radiant panel in the ISO Surface Spread of Flame Apparatus. However, he found that $V^{-1/2}$ was a linear function of $\dot{q}_E''$ only for rapid rates of spread and only if the sample had been allowed a substantial preheat time before it was ignited (Figure 7.15), i.e. thermal equilibrium must be achieved. Quintiere's analysis provides a means by which C and $\dot{q}_{0,i}''$ can be evaluated, but in addition identifies those materials for which there is a minimum external heat flux necessary to permit flame spread, $\dot{q}_{0,s}''$. At low values of $\dot{q}_E''$, Equation (7.9) does not fit the data: $V^{-1/2}$ is found to increase asymptotically (corresponding to $V \to 0$) as $\dot{q}_E''$ is reduced. The value of $\dot{q}_E''$ at the asymptote was taken to be $\dot{q}_{0,s}''$. Some of these figures are quoted in Table 7.4 although it must be emphasized that these relate to the heat transfer conditions pertaining to the configuration under study (cf. Section 6.3.1). Thus, these figures should not be accepted blindly for flame spread over any other

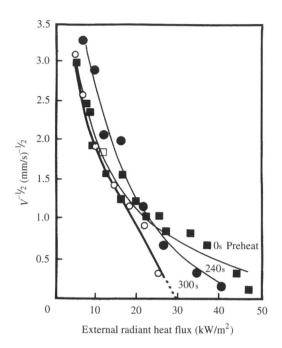

Figure 7.15   Flame spread correlation for hardboard according to Equation (7.9). ■, no preheat; ●, 240 s preheat; ○, 300 s preheat (Quintiere, 1981). Reproduced with permission

Table 7.4   Values of C, $\dot{q}_{0,s}''$ and $\dot{q}_{0,i}''$ from the Surface Spread of Flame Apparatus (Quintiere, 1981)

Material	$\dot{q}_{0,s}''$ (kW/m^2)	$\dot{q}_{0,i}''$ (kW/m^2)	C (s/mm)$^{1/2}$ (m^2/kW)
Coated fibreboard	12	19	0.30
Uncoated fibreboard	≤ 2	19	0.057
Chipboard	7	28	0.12
Hardboard	4	27	0.13
PMMA	≤ 2	21	0.16
Plywood panelling (vinyl side)	15	29	0.17
Plywood panelling (rear face)	8	29	0.08
Polyester	≤ 2	28	0.11

orientation of surface, although in principle the *model* may be used for any orientation (e.g. Quintiere, 1975a). The concept of a limiting heat flux for flame spread has already been incorporated into a standard test for assessing the fire properties of carpets, etc. (NFPA, 1995). Some examples are given in Table 7.5: these show clearly the insulation effect of an underlay or pad. (Note that if Equation (7.9) is rearranged in the form

$$V = \left\{ C(\dot{q}_{0,i}'' - \dot{q}_E'') \right\}^{-2} \tag{7.10}$$

then, as $C \propto (k\rho c)^{1/2}$, this predicts $V \propto 1/k\rho c$ as deduced in Section 7.2.3 for a semi-infinite solid.)

Table 7.5　Limiting radiant heat fluxes for flame spread over carpets

Carpet	Underlay	Critical radiant flux (kW/m^2)	Reference
Nylon	—	2.4	Alderson and Breden, 1976[a]
Nylon	Hair jute	1.2	Alderson and Breden, 1976[a]
Wool	—	10.0	Alderson and Breden, 1976[a]
Wool	Hair jute	6.6	Alderson and Breden, 1976[a]
Acrylic	—	4.1	Alderson and Breden, 1976[a]
Acrylic	Hair jute	2.5	Alderson and Breden, 1976[a]
Polyester	—	3.2	Alderson and Breden, 1976[a]
Polyester	Hair jute	<1.0	Alderson and Breden, 1976[a]
100% wool	Hair jute	11	Moulen and Grubits, 1976[b]
80% wool, 20% nylon	Hair jute	12	Moulen and Grubits, 1979[b]

[a] Critical flux determined using the flooring radiant panel test (NFPA, 1995).
[b] Critical flux determined using a modified ISO ignitability apparatus with central ignition (methenamine tablet). Underlay was hessian, 240 g/m^2.

(d) *Atmospheric pressure*　Higher rates of flame spread are observed at increased atmospheric pressure on account of the effective oxygen enrichment (Section 7.2.5(a)) which enhances flame stability at the surface. This dependence is illustrated in Figure 7.16 (McAlevy and Magee, 1969) although it should be noted that the dependence is much less for thin than for thick materials.

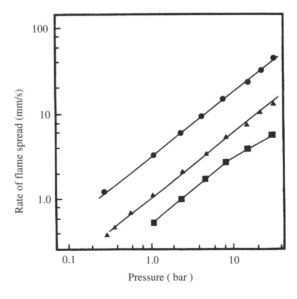

Figure 7.16　Variation of rate of flame spread over horizontal samples of PMMA with pressure of mixtures of oxygen and nitrogen. ■, 46% O_2; ▲, 62% O_2; ●, 100% O_2. (Reproduced by permission of the Combustion Institute from McAlevy and Magee, 1969)

(e) *Imposed air movement (wind)*   In general, confluent air movement will enhance the rate of spread of flame over a combustible surface (cf. co-current spread). Friedman (1968) reports that the rate will increase quasi-exponentially up to a critical level at which extinction, or blow-off, will occur (Section 6.6.2), but there has been no fundamental study of this dependence. The mechanism is understood to involve flame deflection which, combined with enhanced burning behind the flame front, will increase the rate of forward heat transfer. If the direction of air flow is opposed to the spread of flame, the net effect depends on the air velocity. Sufficiently large velocities will cause the rate of spread to decrease (and ultimately the flame to extinguish). but low velocities promote flame spread. Magee and McAlevy (1971) describe some small-scale studies by Lastrina (1970) involving flame spread along horizontal samples of PMMA (10 mm wide by 100 mm long) against an imposed flow of mixtures of oxygen and nitrogen. The results are summarized in Figure 7.17. However, as it is not known to what extent 'edge effects' can account for the response at the lower flowrates, the result cannot be generalized for wider fuel beds.

It should be observed that this corresponds to 'counter-current spread' which occurs naturally by entrainment if a flame is spreading horizontally (Figure 7.1), or downwards. An imposed counter-current flow will have two effects, (i) promoting mixing and combustion at the leading edge of the flame (tending to increase the spread rate) and (ii) cooling the fuel ahead of the flame front (tending to decrease the rate). The latter will become increasingly important as the air velocity is increased.

Air movement is more common out of doors, and can have a dramatic effect on the development of forest and bush fires, i.e. in 'open' fuel beds. Experimental studies have made use of wood cribs and other open fuel bed configurations. This work is discussed briefly in Section 7.4, but is reviewed in depth by Pitts (1991).

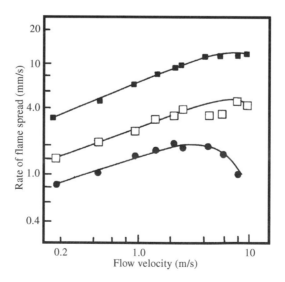

Figure 7.17   Variation of the rate of flame spread on horizontal slabs of PMMA (10 cm × 1 cm) with velocity of a counterflow of O_2/N_2 mixtures. ●, 46% O_2; □, 62% O_2; ■, 100% O_2 (Lastrina (1970), reported by Magee and McAlevy, 1971). Reproduced by permission of Technomic Publishing Co. Inc., Lancaster, PA 17604, USA

7.3   Flame Spread Modelling

Early attempts at modelling flame spread were confined to analytical studies, casting the problem in terms of the heat transfer processes and then seeking solutions of the appropriate differential equations (de Ris, 1969; Quintiere, 1981). This gave valuable insight into the processes involved, but because the problem is so complex, some sweeping assumptions had to be made to allow the equations to be solved. However, these studies laid the groundwork for numerical simulations which became increasingly sophisticated as computational facilities became readily available. As an additional bonus, these numerical techniques are now providing the means by which data from tests such as the Cone Calorimeter may be used directly to model the fire spread process.

Williams (1977) provided an excellent review of contemporary knowledge of flame spread in which he attempted to unify all types of fire spread (premixed flames, smouldering, forest fires, etc.) by using the concept of the 'surface of fire inception'. This was first used by Fons (1946), Emmons (1965), Thomas *et al.* (1964) and others, to describe mass fires and defines the boundary between burning and non-burning fuel. The rate of heat transfer across this surface then determines the rate of fire spread. The 'fundamental equation of fire spread' is a simple energy conservation equation (Williams, 1977):

$$\rho V \Delta h = \dot{q} \tag{7.11}$$

where $\dot{q}$ is the rate of heat transfer across the surface. ρ is the fuel density, V is the rate of spread, and Δh is the change in enthalpy as unit mass of fuel is raised from its initial temperature (T_0) to the temperature (T_i) corresponding to the 'firepoint'. If it is possible to identify $\dot{q}$ in a given fire spread situation, some insight can be gained into the factors that affect the rate of spread. Just such a model was developed by Parker (1972) to test the hypothesis that the rate of downward flame spread over a vertical piece of card (as a thermally thin, typical cellulosic fuel) was determined by the rate of conductive heat transfer through the gas phase from the leading edge of the flame to the unaffected fuel. He monitored the temperature at a point at the mid-plane of the card as the flame progressed downwards: the record of one temperature/time history is shown in Figure 7.18. Just below the advance edge of the visible flame the temperature was

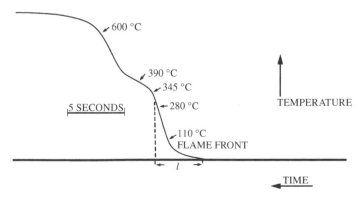

Figure 7.18   Temperature history of a point on a vertical card during downward flame propagation (Parker, 1972). Reproduced by permission of Technomic Publishing Co. Inc., Lancaster, PA 17604, USA

found to be only 110°C, thereafter rising steeply to 300°C, above which the cellulose begins to decompose, producing volatiles. From the rate of temperature rise between 100° and 300°C, Parker calculated that the rate of heat transfer ahead of the flame was 20 kW/m².

Thus the flame appears to advance ahead of the pyrolysis zone: its leading edge is blue (cf. Figure 7.1) and stands about 1 mm off the surface. Parker (1972) derived the following simple expression for the rate of downward propagation (V) over white index cards, assuming that heat losses by radiation are negligible and that heat transfer to the fuel ahead of the burning zone is by gas phase conduction from the leading edge of the flame:

$$V = \frac{l}{t_p} = l.\frac{k_g(T_f - T_0)}{d} . \frac{2}{\rho c \tau (T_p - T_0)} \tag{7.12}$$

where:

t_p = time required to heat the surface to the pyrolysis temperature T_p;
l = length of the preheat zone (measured as 1.5 mm);
τ = thickness of material ($\tau = 0.12$ mm);
ρ = density of the material ($\rho\tau = 182$ g/m²);
c = heat capacity of the material ($c = 1.26$ J/g.K);
T_p = threshold temperature for active pyrolysis (280°C);
T_f = temperature of the leading edge of the flame (1300°C);
T_0 = initial temperature (20°C);
k_g = thermal conductivity of air (2.5×10^{-2} W/m.K at 20°C); and
d = 'standoff distance' (measured as 1 mm).

Although approached in a slightly different way, this equation has essentially the same form as Equation (7.11), where

$$\rho\Delta h = \rho c \tau (T_p - T_0) \tag{7.13}$$

and

$$\dot{q} = \frac{k_g(T_f - T_0)}{d}.l \tag{7.14}$$

where l is effectively the area per unit width of the material to which heat is transferred from the flame: the factor of two (in Equation (7.12)) takes account of the fact that flame spreads down both sides of the card simultaneously. Using the data given, the expression gives $V = 1.6$ mm/s, close to the observed value of 1.5 mm/s (Parker, 1972). Although this remarkable agreement must be regarded as fortuitous as several factors (k_g, ρ and c) are temperature-dependent, this result was taken as vindication of the original assumption regarding the mechanism of forward heat transfer. It has since been confirmed experimentally for downward spread over thin fuels generally (Hirano *et al.*, 1974; Fernandez-Pello and Santoro, 1980). Note that Equation (7.12) predicts that $V \propto \tau^{-1}$, and $V \propto (T_p - T_0)^{-1}$, as found experimentally (Figure 7.10 and Equation (7.6)).

More sophisticated mathematical models of flame spread over horizontal surfaces have been developed (e.g. de Ris, 1969). Generally these give solutions for the rate of spread that are consistent with observation, but many assumptions need to be made to make the mathematics tractable and capable of yielding workable solutions. Thus, in his analysis of thermally thick fuels, de Ris (1969) found it necessary to ignore conduction through the solid as a means by which heat is transferred ahead of the flame front, although it has since been shown that this is the principal mechanism (Fernandez-Pello and Williams, 1974; Fernandez-Pello and Santoro, 1980). The difficulty arises because many of the factors which are important in determining the rate of spread are interdependent and cannot be uncoupled. If the gas phase flame kinetics are fast compared with the rate at which the surface responds to heating, then the flame spread mechanism is dominated by heat transfer from the flame to the surface ahead of the burning zone. However, if the flame reactions are show, e.g. at low oxygen concentrations, the flame kinetics have to be taken into account. This has been shown clearly by Fernandez-Pello and co-workers for counter-current spread (Fernandez-Pello, 1995). Accurate prediction of flame spread rates cannot be obtained by analytical solutions to the governing equations, but considerable success has been achieved through numerical modelling. However, in all cases, significant assumptions have to be made.

Involvement of combustible wall lining materials in a fire is recognized as a particularly hazardous scenario. Upward flame spread on a vertical surface is known to be rapid, and for this reason there has been great interest in developing a model which could be used to calculate the rate of fire development. Theoretical treatments have been explored by Markstein and de Ris (1972), Orloff *et al.* (1974) and Williams (1982), but while they assist in our understanding of the processes involved, they cannot be used directly in practical applications, *per se*. There are too many unknown factors, and it is necessary to develop a semi-empirical approach which can incorporate relevant data relating to fire performance. A suitable numerical model involves analysing flame spread in terms of a series of discrete steps which are defined in terms of the rate at which material ahead of the burning zone is heated to a temperature at which it will support flame (i.e. the firepoint). A simplified illustration of the model used by Karlsson and others (Karlsson, 1993) is given in Figure 7.19 It has been shown that for vertical flame spread, the material above the burning zone is exposed to a heat flux of $c.25$ kW/m^2 (Quintiere *et al.*, 1985), at least in the early stages of fire development when the flames are relatively thin. The flame height ($(x_{fo} - x_{po})$ in Figure 7.19) is taken as a function of the rate of heat release per unit width ($\dot{Q}_c'$) over the height x_{po}:

$$x_{fo}(t) - x_{po}(t) = K(\dot{Q}_c'(t))^n \tag{7.15}$$

where K and n are empirical constants (Tu and Quintiere (1991) give $K = 0.067$ and $n = 2/3$ for x in m and $\dot{Q}_c'$ in kW/m). By incorporating data on time to ignition and rate of heat release from measurements in the Cone Calorimeter, the rate of development of a wall fire may be calculated (e.g. Magnusson and Sundstrom, 1985; Saito *et al.*, 1985; Hasemi *et al.*, 1991; Karlsson, 1993). The inclusion of burnout of the fuel is more difficult to incorporate (Figure 7.20), but has been addressed by Grant and Drysdale (1995) to model early fire growth on a vertical sheet of cardboard using data files from the Cone Calorimeter directly.

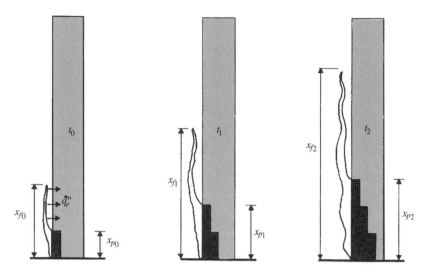

Figure 7.19 Simple model for upward flame spread, without burnout. The heat flux from the flame is assumed to be constant in the region $x_f < x < x_p$ (Karlsson, 1993) (Grant and Drysdale, 1995, by permission)

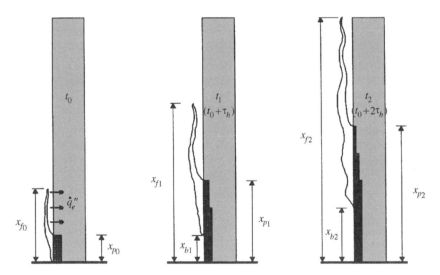

Figure 7.20 Simple model for upward flame spread, with burnout, where x_b is the burnout height (cf. Figure 7.19) (Grant and Drysdale, 1995, by permission)

The value of this type of model is in providing a means by which the results of small-scale tests can be used to assess full-scale fire behaviour. At the present time, it is considered necessary to subject wall lining materials to large-scale 'standard' tests such as ISO 9705 (1993), which are much more expensive than existing 'bench-scale' tests. For this reason, the models are being developed with a view to modelling large-scale test behaviour, particularly the rate of growth of rate of heat release (e.g. Magnusson and Sundstrom, 1985; Karlsson, 1993). This approach is discussed in general terms by Cleary and Quintiere (1991) (see also Section 9.2.2).

7.4   Spread of Flame through Open Fuel Beds

Previous sections have dealt specifically with the spread of flame over the surface of a continuous fuel bed, but equally important is the behaviour of the 'open' or 'discontinuous' fuel. This is relevant to the spread of fire through bush and forest and to the growth of fire in an enclosure (Section 9.2.2).

Most of the research into this topic has been carried out by the forest fire research community (e.g. see Pitts, 1991) although experimental studies using wood cribs (Section 5.2.2(b)) have provided most of the underlying principles. The mechanism of propagation has been shown to involve radiative heat transfer ahead of the flame front (Thomas *et al.*, 1964, 1965). The advancing front adopts a vertical orientation under still air conditions and, in agreement with Equation (7.11), the rate of spread is inversely proportional to the bulk density of the fuel bed:

$$V = \frac{\dot{q}}{\rho \Delta h} \qquad (7.11)$$

where $\dot{q} = \varepsilon \sigma (T_F^4 - T_0^4)$. Although this equation does not incorporate all the factors that affect the rate of spread of flame (e.g. moisture content), it may be applied directly if the individual fuel element (sticks) are thermally thin: otherwise ρ must be replaced by an effective density which takes into account the fact that only the surface layer of each element (Equation (7.3)) will be heated before the arrival of the flame front (Thomas, 1974a, b; Williams, 1982) (cf. Table 6.7). General conclusions from these studies are compatible with many field observations. Thus, in a forest, fire will spread most rapidly through the brush at ground level, or through the 'crown' where the bulk density is low and the fuel elements are thin (Emmons, 1965). Very high rates are possible under drought conditions when moisture content will be low.

7.5   Applications

As fire growth proceeds by flame spread from the point of ignition, it would be valuable to be able to analyse any potential fire spread situation and quantify the rate of development. In this context, much of the information in this chapter is essentially qualitative but two aspects merit closer examination, namely spread under the influence of radiation and upward spread on a vertical surface. A limited amount of quantification is possible with each.

7.5.1   *Radiation-enhanced flame spread*

The enhancement of flame spread by radiant heat is very significant and is shown clearly in Figure 7.13(b). Its relevance to fire development in compartments is discussed in Section 9.2 where it is argued that when the heat flux at floor level reaches 20 kW/m^2 there is an onset of a very rapid change in the character of the fire, from localized burning to full room involvement, when all combustible items are burning. This transition is known as flashover (Section 9.1) and requires that the fire has grown beyond a certain minimum size, mainly by flame spread (Thomas, 1981).

Table 7.6 Rate of spread of flame

Material	Orientation and direction	Rate of spread (mm/s)	Reference
Cards	−90°	1.3	Hirano *et al.* (1974)
	0°	∼4	
PMMA (thick)	−90°	0.04	Fernandez-Pello
	0°	∼0.07	and Williams (1974)
Polyurethane[a]			
$\rho = 15$ kg/m³	0°	3.7	Paul (1979)
$\rho = 22$ kg/m³	0°	2.5	
$\rho = 32$ kg/m³	0°	1.6	

[a] Formulation unknown.

Unfortunately, there is little quantitative information available on the dependence of flame spread rate on radiant heat flux. A few data on the rate of spread over thermally thick materials in the absence of supporting radiation are given in Table 7.6 for horizontal fuel beds. Some materials will spread flame only if supported by a radiant flux: the values of the minimum flux necessary for flame spread ($\dot{q}''_{0.s}$) obtained by Quintiere (1981) using the ISO Surface Spread of Flame Apparatus (Figure 7.14) are recorded in Table 7.4. These refer to horizontal spread along a vertical surface and would not apply to thicknesses and orientations where different heat loss conditions hold. Similar data on the spread of flame over carpets obtained using NFPA 253 (National Fire Protection Association, 1995) are also given in Table 7.5. These show clearly that an underlay reduces $\dot{q}''_{0.s}$, which can be attributed to the reduced heat losses through the fuel bed. Information of this kind can be used to assess how far fire would spread away from a source of radiant heat over a combustible surface (e.g. carpets (Quintiere, 1975b) and wall lining materials in corridors).

On exposure to radiant (or convective) heat, the surface temperature of a combustible solid will increase, eventually to reach a steady-state (equilibrium) value. This can be estimated on the basis of simple heat transfer calculations, assuming the material to be inert, and taking heat losses into account. Equations (7.6) and (7.7) could then be used to estimate the limiting rate of flame spread (V_2) under a given heat flux, provided that we know T_p (the pyrolysis temperature) and the flame spread rate V_1 at a single fuel temperature T_1. Thus for a thick fuel

$$V_2 = V_1 \frac{(T_p - T_1)^2}{(T_p - T_2)^2} \tag{7.16}$$

where T_2 — the equilibrium surface temperature reached when exposed to the heat flux — must first be calculated. The limits of applicability of this method are not known but it will break down unless the fuel surface temperature is considerably less than T_p.

7.5.2   Rate of vertical spread

Alpert and Ward (1984) have pointed out that as the spread of flame up a vertical surface accelerates exponentially, the rate of spread from an initial (small) fire can

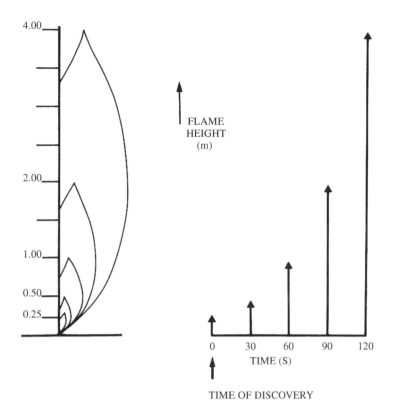

Figure 7.21   Illustration of the exponential growth in flame height for vertical spread. Doubling time 30 s (after Alpert and Ward, 1984, with permission)

be estimated if the time it takes for the fire to double in size is known. This can be obtained simply from relatively small-scale laboratory tests, and provides a measure of the hazard of a vertical surface. If it is found that the doubling time for a particular material is 30 s and a fire is discovered at the base of a 4 m high vertical wall of the same material when the flame is 0.25 m high, the flame will be 0.50 m high at 1 min and will reach the top of the wall at $2\frac{1}{2}$ min (Figure 7.21). This type of information may be used to assess the risks associated with storage of goods in high bay warehouses and to develop fire protection strategies based on early detection of an outbreak of fire.

Problems

7.1   The rate of downward spread of flame on a vertical piece of card 0.4 mm thick is found to be 1.0 mm/s. What rate of spread would be expected for a card 0.7 mm thick if it is made from the same material and held in the same configuration?

7.2   Flame was found to spread down one face of a large vertical sheet of PMMA, 6 mm thick, at a rate of 0.05 mm/s when the initial temperature was 20°C. What would the equilibrium spread rate be if that face was exposed to a uniform radiant flux of 4 kW/m^2? Assume that the surface has an emissivity of 0.5 and that the convective heat transfer coefficient is $h = 12$ W/m^{2}K. Ignore radiative heat loss for the calculation, but check subsequently if this is justified. Take the ambient air temperature to be 20°C and the 'firepoint' temperature to be 280°C.

7.3 Taking the example in Problem 7.2, what would the rate of downward spread be if the rear face was maintained constant at 20°C during exposure to the same radiant flux?

7.4 Timber flooring will not burn and spread flame unless it is exposed to (radiant) heating from elsewhere. The minimum level is of the order of 4 kW/m^2. Figure 9.4 shows the results of a series of experiments carried out at Fire Research Station, Borehamwood, which show the radiant heat flux at floor level produced by a crib fire at the end of a corridor — with and without the ceiling in place. Estimate roughly how far flame would spread along a wooden floor for the three situations illustrated in the diagram.

8

Spontaneous Ignition within Solids and Smouldering Combustion

The phenomenon of spontaneous ignition in vapour/air mixtures was discussed in some detail in Section 6.1 in relation to gaseous, liquid and solid fuels. However, certain combustible solids can ignite as a result of internal heating which arises spontaneously if there is an exothermic process liberating heat faster than it can be lost to the surroundings. There appear to be two main factors necessary for this type of spontaneous combustion to occur. First, the material must be sufficiently porous to allow air (oxygen) to permeate throughout the mass, and secondly it must yield a rigid char when undergoing thermal decomposition. The phenomenon is normally associated with relatively large masses of material and its principal characteristic is that combustion begins deep inside the material where the effects of self-heating are greatest (Figure 6.4(b)). It will give rise to a smouldering wave which will slowly propagate outwards.

Smouldering involves surface oxidation of the char which provides the heat necessary to cause further thermal degradation of the neighbouring layer of combustible material. Successful propagation requires that volatiles are progressively driven out ahead of the zone of active combustion to expose fresh char which will then begin to burn. This is described in more detail in Section 8.2. However, the chapter begins with a brief discussion of spontaneous combustion and how it may be quantified by the application of thermal explosion theory. The subject is covered in depth by Gray and Lee (1967) and Bowes (1984) and is reviewed by Beever (1995).

8.1 Spontaneous Ignition in Bulk Solids

The principal characteristic of spontaneous ignition in a bulk solid is that combustion starts as a smouldering reaction within the material and propagates slowly outwards. The fire is initially deep-seated, and although it may lead to flaming combustion when it breaks through to the surface, it leaves behind evidence of prolonged burning at depth. The phenomenon can be described by thermal explosion theory in the form developed by Frank-Kamenetskii (1939) (Section 6.1). Heat must be generated within the mass by an exothermic process, in this context a heterogeneous surface oxidation within the interstices of a porous combustible material, although other sources are observed (e.g. decomposition of endothermic compounds (cf. Figure 6.6) and microbiological heating (Section 8.1.4)). If the energy is not removed as rapidly as it is released, then the material will self-heat, perhaps leading to smouldering ignition if the conditions are right. One prerequisite is that the mass must have a low surface to volume ratio. A list of materials which are known to be subject to spontaneous heating is given in the NFPA Handbook (1997), from which Table 8.1 has been derived. However, no information is given on the conditions under which spontaneous ignition might occur. The list includes those oils, such as linseed, which present a hazard when absorbed on to fibrous combustible materials such as cotton rags and waste. The oil is dispersed on a rigid porous structure, thereby presenting a large surface area at which oxidation can take place. A similar situation is found in chemical plant where thermal insulation, or lagging, is applied to pipes and vessels carrying hot process fluids. Leakage from a faulty flange or valve into the lagging can lead to a 'lagging fire', even if the insulation material is non-combustible. This phenomenon has been examined in some detail by Bowes (1974) and Gugan (1976).

The main difference between spontaneous ignition within a combustible solid and that of a flammable vapour/air mixture is that the latter is a premixed system. With a porous solid, oxygen must diffuse into the material to maintain the oxidation reactions on the surface of the individual particles or fibres wherever oxidation is occurring. The rate of development of self-heating which precedes ignition is slow and will be significant only if the insulation provided by the surrounding mass of material is sufficient. Consequently, the delays or 'induction periods' for spontaneous ignition of bulk solids tend to be many orders of magnitude greater than those associated with the auto-ignition of gases (Section 6.1 and Figure 6.3). On the other hand, in practical situations the risk of spontaneous ignition in bulk solids exists at temperatures much lower than typical auto-ignition temperatures quoted for gases and vapours (Table 3.1). If the accumulation of material is large enough, spontaneous combustion can occur at relatively low temperatures, perhaps as low as ambient (Bowes, 1984) (see Section 8.1.1). There can be substantial losses of stocks of coal by this process if very large accumulations are permitted to lie undisturbed for long periods: the induction period is likely to be of the order of weeks (National Fire Protection Association, 1997).

8.1.1 Application of the Frank-Kamenetskii model

It is possible to use the result obtained from the Frank-Kamenetskii model (Equation (6.14)) to investigate the spontaneous heating characteristics of different materials:

$$\ln\left(\frac{\delta_{cr} T_{a,cr}^2}{r_0^2}\right) = \ln\left(\frac{E_A \Delta H_c A C_i^n}{k \cdot R}\right) - \frac{E_A}{R T_{a,cr}} \tag{8.1}$$

The relationship between $T_{a,cr}$ (the critical ignition temperature) and r_0 (the characteristic dimension of the sample) for a given geometry of material may be determined experimentally. Thus, cubical samples of the material can be prepared and exposed to constant elevated temperatures in a thermostatically-controlled oven, monitoring the temperature at the centre of the sample by means of a thermocouple. In this way, the extent to which a sample of given size will tend to self-heat or ignite at different

Table 8.1 Some materials subject to spontaneous heating[a]

Name	Tendency to spontaneous heating	Usual shipping container or storage method	Precautions against spontaneous heating	Remarks
Charcoal	High	Bulk, bags	Keep dry. Supply ventilation.	Hardwood charcoal must be carefully prepared and aged. Avoid wetting and subsequent drying.
Distillers' dried grains with oil content (brewing grains)	Moderate	Bulk	Maintain moisture 7–10%. Cool below 38°C before storage.	Very dangerous if moisture content is 5% or lower.
Fish meal	High	Bags, bulk	Keep moisture 6–12%. Avoid exposure to heat.	Dangerous if overdried or packaged over 38°C.
Foam rubber in consumer products	Moderate		Where possible remove foam rubber pads, etc. from garments to be dried in dryers or over heaters. If garments containing foam rubber parts have been artificially dried, they should be thoroughly cooled before being piled, bundled or put away. Keep heating pads, hair dryers, other heat sources from contact with foam rubber pillows, etc.	Foam rubber may continue to heat spontaneously after being subjected to forced drying as in home or commercial dryers and after contact with heating pads and other heat sources. Natural drying does not cause spontaneous heating.
Grain (various kinds)	Very slight	Bulk, bags	Avoid moisture extremes.	Ground grains may heat if wet and warm.
Hay	Moderate	Bulk, bags	Keep dry and cool.	Wet or improperly cured hay is almost certain to heat in hot weather. Baled hay seldom heats dangerously.

(*continued overleaf*)

Table 8.1 (*Continued*)

Name	Tendency to spontaneous heating	Usual shipping container or storage method	Precautions against spontaneous heating	Remarks
Linseed oil	High	Tank cars, drums, cans, glass	Avoid contact of leakage from containers with rags, cotton or other fibrous combustible materials.	Rags or fabric impregnated with this oil are extremely dangerous. Avoid piles, etc. Store in closed containers, preferably metal.
Manure	Moderate	Bulk	Avoid extremes of low or high moisture contents. Ventilate the piles.	Avoid storing or loading uncooled manures.
Oiled rags	High	Bales	Avoid storing in bulk in open.	Dangerous if wet with drying oil.
Rags	Variable	Bales	Avoid contamination with drying oils. Avoid charring. Keep cool and dry.	Tendency depends on previous use of rags. Partially burned or charred rags are dangerous.
Sawdust	Possible	Bulk	Avoid contact with drying oils. Avoid hot, humid storage.	Partially burned or charred sawdust may be dangerous.
Wool wastes	Moderate	Bulk, bales, etc.	Keep cool and ventilated or store in closed containers. Avoid high moisture.	Most wool wastes contain oil, etc. from the weaving and spinning and are liable to heat in storage. Wet wool wastes are very liable to spontaneous heating and possible ignition.

temperatures can be determined. The value of $T_{a,cr}$ is obtained for each cube size (side $= 2r_0$) by a process of trial-and-error and 'bracketing', an example of which is shown in Figure 8.1. Once $T_{a,cr}$ has been determined for several cube sizes, then using $\delta_{cr} = 2.52$ (Table 6.1), the data may be plotted in the form $\ln(\delta_{cr} T_{a,cr}^2 / r_0^2)$ versus $1/T_{a,cr}$, as suggested by Equation (8.1). This has been done in Figure 8.2 with data from various sources for samples of wood fibre insulation board (FIB) in the form of cubes ($\delta_{cr} = 2.52$), slabs ($\delta_{cr} = 0.88$) and octagonal piles ($\delta_{cr} = 2.65$) (Thomas and Bowes, 1961a; Thomas, 1972a).

For experiments of this type, corrections to δ_{cr} may be necessary if the assumptions used in the original derivation of Equation (8.1) do not apply, for example, if the Biot number (Bi $= hr_0/k$) is less than about 10 (see Section 8.1.2). This will not be necessary for FIB unless $r_0 < 0.05$ m as the thermal conductivity (k) of this material is very low (0.041 W/m.K; Table 2.1). The linear correlation shown in Figure 8.2 suggests that the Frank-Kamenetskii model is a good approximation for this material within the range of temperatures covered. It may be used tentatively to predict spontaneous ignition behaviour outside this range, provided that the extrapolation is not too long. However, in addition to the assumptions introduced into the derivation of Equation (8.1) (Section 6.1) it is implicit that there is no change in the mechanism of heat production within the temperature range studied. If an extrapolation outside this range is to be attempted, it must be tacitly assumed that the same conditions apply.

Spontaneous ignition is not instantaneous, but exhibits a delay, or induction period, τ' (Figure 6.3). However, whereas the induction period for auto-ignition of a flammable vapour/air mixture is seldom longer than 1 s, that for a bulk solid may be hours, days or even weeks, depending on the mass of the stored material and the ambient temperature. The larger the mass, the lower the temperature which will lead to spontaneous ignition. This is illustrated by data of Bowes and Cameron (1971) on the spontaneous ignition of chemically active carbon (Table 8.2). This work was part of an investigation of fires

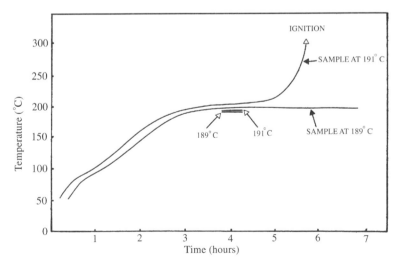

Figure 8.1   Determination of the critical temperature for 50 mm cubes of ground grain. Ignition occurred after 5 h at $T_a = 191°C$, but there was no ignition at $T_a = 189°C$. Therefore $T_{a,cr} = 190°C$. (Curves displaced for clarity)

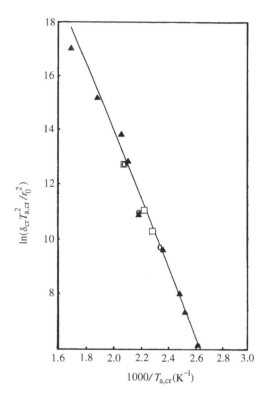

Figure 8.2 Correlation of spontaneous ignition data for wood fibre insulating board according to Equation (8.1) (Thomas and Bowes, 1961a). ▲, Mitchell, 1951; ○ (cube) and □ (slab), Thomas and Bowes, 1961a. (Reproduced by permission of The Controller, HMSO. © Copyright)

Table 8.2 Ignition of cubes of chemically activated carbon (Bowes and Cameron, 1971)

Linear dimension $(2r_0/\text{mm})$	Critical temperature $(T_{a,cr}/°C)$	Time to ignition (τ/h)
51	125	1.3
76	113	2.7
102	110	5.6
152	99	14
204	90	24
601	60	68

involving substantial cargoes of active carbon (4–14 tonnes) stored in the holds of ships passing through the tropics. They established that spontaneous ignition was indeed the cause, and that the delay before the discovery of the fire (3–4 weeks) was compatible with, although slightly less than, an extrapolation of the small scale experimental data to the ambient temperature relevant to the cargo spaces (29–38°C) (Figure 8.3). It

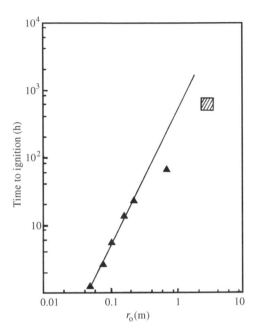

Figure 8.3 Time to spontaneous ignition of cubes of chemically activated carbon. ▲, Data from Table 8.2. Hatched area represents the times to discovery of fires on board ship involving this material (Bowes and Cameron, 1971). (Reproduced by permission of The Controller, HMSO. © Crown Copyright)

was subsequently recommended that this material be baled for transportation, with a 400 gauge polyethylene liner incorporated into the sacking to prevent the ingress of oxygen (Board of Trade, 1966; Bowes and Cameron, 1971).

8.1.2 The Thomas model

The Frank-Kamenetskii model assumes that heat loss from the bulk of the material is restricted only by conduction through the solid (Section 6.1), i.e. that the temperature of the surface and the ambient atmosphere are equal, as illustrated in Figure 8.4(a) (see also Figure 6.4(b)). This will hold provided that the Biot number is large, i.e. Bi > 10. For very small values of Bi (i.e. Bi → 0), the Semenov model will be applicable (Section 6.1), but between these limits conduction through the solid and convective heat loss from the surface must both be considered. Thus the surface boundary conditions, Equations (6.9a) and (6.9c) are replaced by:

$$h(T_s - T_a) = k(\mathrm{d}T/\mathrm{d}r) \tag{8.2}$$

This is illustrated schematically in Figure 8.4(b). Provided that the heat of combustion (ΔH_c) is large, δ_{cr} becomes a function of Bi, as shown in Figure 8.5 (Thomas, 1960, 1972a). In small scale experiments, Bi may be sufficiently low to make it necessary to use corrected values of δ_{cr}, but otherwise application of the method is identical to that described in the previous section (Thomas and Bowes, 1961a).

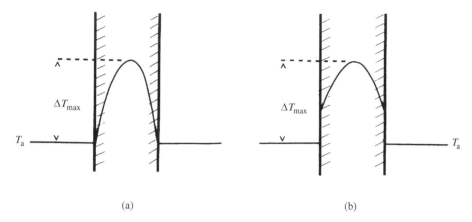

(a) (b)

Figure 8.4 Temperature profiles inside spontaneously heating materials according to the models of (a) Frank-Kamenetskii. 1939 and (b) Thomas, 1960

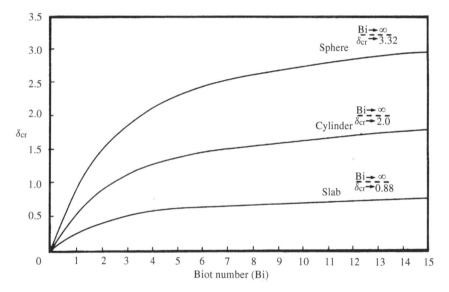

Figure 8.5 δ_{cr} as a function of Biot number for a sphere, a cylinder and a slab (Thomas, 1958). (Reproduced by permission of The Royal Society of Chemistry)

Closely related to the above is the concept of 'hot spot criticality' which has already been encountered in Section 6.1: given some mechanism by which a local temperature rise occurs, under what circumstances will ignition follow? In the case of a flammable vapour/air mixture, the energy source could be a mechanical spark, although for liquid explosives adiabatic heating of minute trapped air bubbles may be sufficient to cause ignition if the liquid is subjected to a shock. The 'hot spot' will cool by conductive losses to the surrounding material but if the rate of heat production within the affected volume exceeds the losses then runaway reaction will occur. Thomas (1973) has developed an approximate theory of 'hot spot criticality' which may be applied to the problem of storage of large quantities of hot material in a cool environment

(Anthony and Greaney, 1979), e.g. hot, dry sheets of fibre insulation board stored directly from the production line (Back, 1981/82) or hot linen removed from industrial tumble driers and placed in large baskets, as is common practice in hospital laundries (Perdue, 1956). Fires resulting from this type of incident can be avoided by cooling the material before it is stored in bulk quantities.

8.1.3   Ignition of dust deposits and oil-soaked lagging

It is known that a layer of a combustible dust lying on a hot surface has the potential to ignite 'spontaneously'. Bowes and Townsend (1962) have examined this in detail and have shown that the Frank-Kamenetskii model may be modified for this situation. They carried out experiments in which layers of dust of various thicknesses were heated on a hot plate (Figure 8.6(a)) and determined the minimum plate temperatures which would cause ignition of each thickness by the process of 'bracketing'. Figure 8.6(b) shows the result of one of their tests (cf. Figure 8.1). The temperature of the dust (beech sawdust) was monitored by means of a thermocouple located 10 mm above the hot surface. They found that the tendency to ignite was insensitive to particle size in the

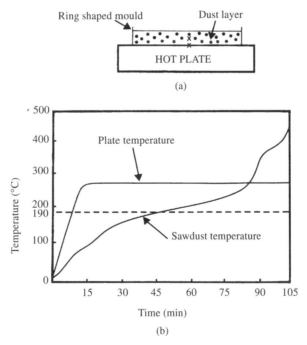

Figure 8.6   (a) Schematic diagram of the apparatus used to determine critical ignition temperatures for dust layers. Thermocouples located at ×: the mould was removed before heating the plate to a preselected temperature. (b) Ignition of a 25 mm thick layer of beech sawdust on a surface at 275°C. The sawdust temperature refers to a point 10 mm above the centre of the plate. If the sawdust was chemically inert, the equilibrium temperature at this location would be 190°C (Bowes and Townsend, 1962). (Reproduced by permission of The Controller, HMSO. © Crown Copyright)

range of sieve fractions 18–120 BS (~853–124 μm particle diameter) and that packing density had a significant effect only with the thinnest layers (~5 mm). (A more recent study of the ignition of dust layers of coal, oil shale, corn starch, grain and brass powder is reported by Miron and Lazzara (1988).)

The layer may be treated as an infinite slab, of thickness $2r_0$, resting on a hot surface, temperature T_{pl} at $r = 0$. The problem can be reduced to the same form as Equation (6.12), i.e.

$$\frac{d^2\theta}{dz^2} = -\delta \exp(\theta) \qquad (8.3)$$

where $\theta = (T - T_{pl})E_A/RT_{pl}^2$ and $z = r/r_0$, but the boundary conditions are more complex as a result of non-symmetrical heating. These are:

$$\theta = 0 \text{ at } z = 0 \qquad (8.3a)$$

and

$$-\frac{d\theta}{dz} = \frac{hr_0}{k}(\theta_s - \theta_0) \text{ at } z = 2 \qquad (8.3b)$$

where $hr_0/k = \text{Bi}$. At the upper surface of the layer (i.e. at $r = 2r_0$) the temperature is $T = T_0$. As in the case of symmetrical heating, solutions to Equation (8.3) exist only for $\delta < \delta_{cr}$. Consequently δ_{cr} must be determined to enable critical states to be identified.

Thomas and Bowes (1961b) showed δ_{cr} to be a function of both Bi and θ_0, as indicated in Figure 8.7, both of which must be evaluated. Bi can be obtained from the properties of the dust layer and the heat transfer conditions at the surface; θ_0 can only be

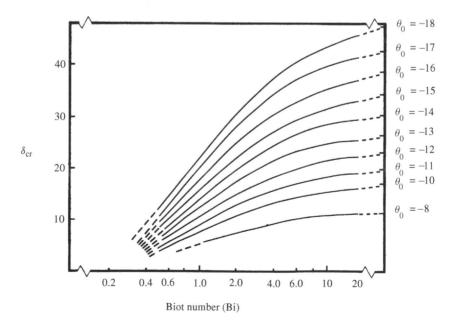

Figure 8.7 δ_{cr} as a function of Biot number and θ_0 (Thomas and Bowes, 1961b). (Reproduced by permission of The Royal Society of Chemistry)

calculated if E_A (the activation energy) is known. This may be derived from the gradient of a plot of $\log(\delta_{cr} \cdot T_{pl}^2/r_0^2)$ versus $1/T_{pl}$ (cf. Figure 8.2), but initially an assumed, trial value must be used. The procedure can then be repeated using the 'improved' value of E_A obtained from the plot based on the trial value. The success of this approach may be judged by the data shown in Figure 8.8. Bowes and Townsend (1962) use this method to extrapolate their small-scale data on wood sawdust to substantial thicknesses (Figure 8.9).

A similar approach is required to interpret spontaneous ignition of oil-soaked lagging (Bowes, 1974). This type of fire involves the insulation ('lagging') which is used to reduce heat losses from pipes carrying hot process fluids in chemical plants and refineries. The model is based on the geometry of the infinite cylinder. The lagging material is normally non-combustible, but a slow leak of combustible liquid will be absorbed by the insulation and spontaneous ignition will occur under certain conditions, which include the following:

(i) the liquid is insufficiently volatile to evaporate too rapidly;

(ii) the lagging is sufficiently porous to allow oxygen to diffuse to the surface of the absorbed liquid;

(iii) the leak is not so rapid as to fill the pores of the lagging material and thereby exclude oxygen from the high temperature region.

These fires can develop undetected and remain so for a considerable period of time, perhaps becoming noticeable only when the leak increases and the whole mass bursts

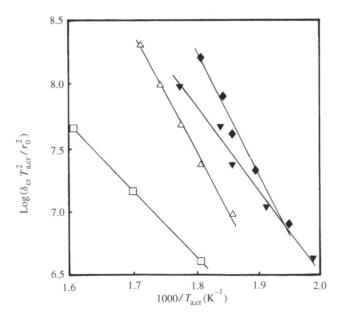

Figure 8.8 Correlation of spontaneous ignition data for dust layers on a hot surface according to Equation (8.1). □, Cork; △, provender; ♦, cocoa; ▼, cotton seed hull 'pepper' (Bowes and Townsend, 1962). (Reproduced by permission of The Controller, HMSO. © Crown Copyright)

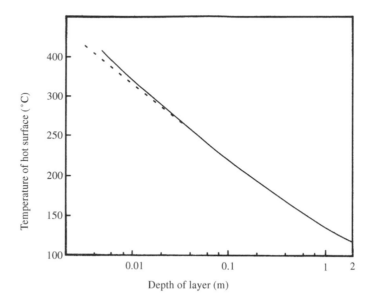

Figure 8.9   Extrapolation of results on spontaneous ignition of layers of sawdust on hot surfaces. ———, Mixed hardwood; - - - - -, beech (Bowes and Townsend, 1962). (Reproduced by permission of The Controller, HMSO. © Crown Copyright)

into flame. Even if detected before this stage they are difficult to deal with since when the lagging is stripped away, extensive flaming can develop spontaneously. One precaution that can be taken to avoid this type of incident is to exclude oxygen by using non-porous lagging (e.g. foamed glass) or to seal the surface of the insulating material. One can also identify those parts of the system that are particularly susceptible to leaks (e.g. flanges, valves, etc.) and isolate these from the lagging material. Although it is technically possible to identify the maximum allowable thickness of insulation to avoid spontaneous ignition, this is not an acceptable method of prevention as the required thicknesses are generally insufficient to provide a satisfactory reduction in heat loss.

8.1.4   Spontaneous ignition in haystacks

Spontaneous ignition of large piles of hay is one of the best known manifestations of this phenomenon. However, the mechanism is complex with at least three stages, the first of which involves microbiological activity. This is capable of raising the temperature at a location within the stack to over 70°C. Chemical oxidation then takes over, although the reactions involved at these relatively low temperatures appear to be catalysed by moisture. The initial heating stage requires a relatively high moisture content for vigorous bacterial growth (∼63–92% by weight), but higher water contents will enhance the effective thermal conductivity, increase the heat losses and thus limit the extent of self-heating (Rothbaum, 1963).

There is no information on the critical size for a haystack, although Bowes (1971) has estimated theoretically that it will be of the order of $2r_0 = 2$ m at ambient temperatures. This is not unreasonable and is consistent with the fact that there have been no reports of spontaneous ignition of rolled hay bales. The normal practice adopted to avoid self-heating and ignition of hay (and similar materials) is to make certain the material is not damp when it is stored (Walker, 1967).

8.2   Smouldering Combustion*

Only porous materials which form a solid carbonaceous char when heated can undergo self-sustained smouldering combustion. Included are a wide range of materials of vegetable origin such as paper, cellulosic fabrics, sawdust, fibreboard and latex rubber, as well as some thermosetting plastics in expanded form. Materials which can melt and shrink away from a heat source will not exhibit this mode of combustion. The reason for this will be clear once the mechanism of smouldering has been discussed.

Moussa *et al.* (1977) have examined the propagation of smouldering along horizontal, cylindrical cellulosic elements and have shown that the combustion wave has three distinct regions as shown in Figure 8.10, namely:

Zone 1   Pyrolysis zone in which there is a steep temperature rise and an outflow of visible airborne products from the parent material.

Zone 2   A charred zone where the temperature reaches a maximum, the evolution of the visible products stops and glowing occurs.

Zone 3   A zone of very porous residual char and/or ash which is no longer glowing and whose temperature is falling slowly.

Heat is released in Zone 2 where surface oxidation of the char occurs and it is here that the temperature reaches a maximum, typically in the region 600–750°C for smouldering

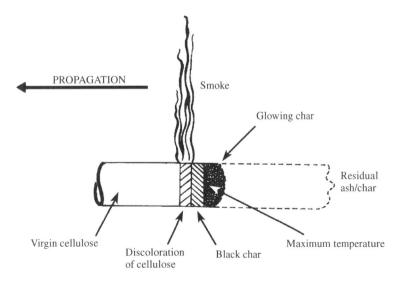

Figure 8.10   Representation of steady smouldering along a horizontal cellulose rod. (Reproduced by permission of the Combustion Institute from Moussa *et al.*, 1977)

* This section was originally based on an article by the author (Drysdale, 1980).

of cellulosic materials in still air. Heat is conducted from Zone 2 into the virgin fuel (across the 'surface of fire inception', Section 7.3) and is responsible for the temperature rise observed in Zone 1. This causes thermal decomposition of the fuel, yielding the outflow of pyrolysis products and a residue of carbonaceous char. Temperatures greater than 250–300°C are required to accomplish this change in most organic materials. The three zones discussed by Moussa *et al.* (1977) appear to be common to all self-smouldering processes, although the reactions occurring in Zone 1 are complex (see Section 8.2.1). A distinction needs to be made between two types of propagation: forward smoulder, in which the flow of air is in the same direction as the propagation process, and reverse smoulder, when the airflow opposes the direction of smoulder. These are discussed in the next section.

The volatile degradation products which are driven out from Zone 1 are not oxidized significantly. They represent the gaseous fuel which, in flaming combustion, would burn as a flame above the surface of the fuel. They are released ahead of the zone of active combustion and comprise a very complex mixture of products including high boiling point liquids and tars which condense to form an aerosol, quite different from smoke produced in flaming combustion (Section 11.1). These products are combustible and if allowed to accumulate in an enclosed space can, in principle, give rise to a flammable atmosphere. A smoke explosion arising from a smouldering fire involving latex rubber mattresses was reported in England in 1974 (Anon., 1975; Woolley and Ames, 1975).

From the model of smouldering described above (Figure 8.10), it can be seen that if there is any tendency for the unaffected material to shrink away from the source of heat (Zone 2), there will be an effective reduction in the forward heat transfer. If this is sufficiently large, propagation of smouldering will not be possible. An interesting corollary is that it is possible for combustible liquids to exhibit smouldering provided that they are absorbed on a rigid, but porous substrate (cf. lagging fires, Section 8.1.3).

8.2.1   Factors affecting the propagation of smouldering

The mechanism by which smouldering propagates through a porous fuel bed has received increasing attention in recent years (Ohlemiller and Lucca, 1983; Dosanjh *et al.*, 1987; Torero and Fernandez-Pello, 1995, 1996). The process is modelled most simply by using the concept of the 'surface of fire inception' introduced in Section 7.3. This emphasizes the role of heat transfer ahead of the zone of active combustion, and the need to identify the rate controlling process. The basic equation is:

$$V = \dot{q}''/\rho\Delta h \tag{8.4}$$

(Equation (7.11)), where V is the rate of spread, $\dot{q}''$ is the net energy flux across the surface of fire inception, ρ is the bulk density of the fuel, and Δh is the thermal enthalpy change in raising unit mass of fuel from ambient temperature to the ignition temperature (Williams, 1977). This can be used to obtain an order of magnitude estimate of the rate of propagation of a smouldering combustion wave. If it is assumed that the ignition temperature is not too different from the maximum temperature in Zone 2, i.e. T_{max}, then:

$$\Delta h = c(T_{max} - T_0) \tag{8.5}$$

where c is the heat capacity of the fuel and T_0 is the ambient temperature. If it is also assumed that heat transfer from Zone 2 to Zone 1 is by conduction and that a quasi-steady state exists, then

$$V \approx \frac{k(T_{max} - T_0)}{x} \cdot \frac{1}{\rho c(T_{max} - T_0)} \tag{8.6}$$

i.e.

$$V \approx \frac{k}{\rho c} \cdot \frac{1}{x} = \frac{\alpha}{x} \tag{8.7}$$

where k and α are the thermal conductivity and the thermal diffusivity of the fuel respectively; x is the distance over which heat is being transferred (Figure 8.11) and is found to be of the order of 0.01 m (Palmer, 1957). Taking fibre insulation board, for which $\alpha \approx 10^{-7}$ m²/s (Table 2.1) as an example, the rate of propagation will be approximated by:

$$V = O(10^{-5}) \text{ m/s}$$

or

$$= O(10^{-2}) \text{ mm/s}$$

i.e. the rate of propagation is of the order of 10^{-2} mm/s, which is as observed. However, although this model gives the right order of magnitude for V, it is too crude: the calculations predict that the rate of propagation will be independent of the maximum temperature in Zone 2 although this is known to be incorrect. At elevated oxygen concentrations, Moussa *et al.* (1977) found a significant increase in the rate of spread which correlated with an increase in the temperature of Zone 2 (Figure 8.12). They examined a model in which the rate of heat generation in Zone 2 is determined only by the rate of diffusion of oxygen through a free convective boundary layer around the char zone. The steady rate of propagation and the maximum temperature (in Zone 2) are then determined for a given ambient condition by combining (i) the heat flux required to degrade the cellulose in Zone 1; (ii) the heat generated by the oxidation of the char in Zone 2; and (iii) the heat losses from the system. The agreement between two solutions of this model and the experimental data are shown in Figure 8.12. In view of the uncertainties regarding the heat required to degrade the cellulose, the agreement is remarkably good and provides support for the assumption that the rate of oxygen transport (by convection and diffusion) determines the rate of heat release. (It should

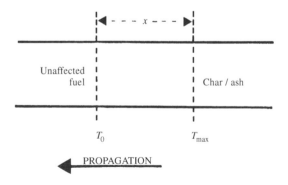

Figure 8.11 Simple heat transfer model for propagation of smouldering

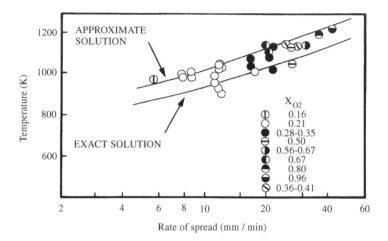

Figure 8.12   Data of Moussa *et al.* (1977) showing the correlation between rate of spread and the maximum temperature in Zone 2 for smouldering along horizontal cellulose rods. The various symbols refer to different proportions of oxygen in nitrogen (X_{O_2}), except for the symbol $\ominus$ which refers to oxygen/helium mixtures. Reproduced by permission of the Combustion Institute from Moussa *et al.* (1977)

be noted that much more sophisticated models for smouldering have been developed, e.g. Dosanjh *et al.*, 1987; Schult *et al.*, 1995.)

When a smoker draws on a cigarette, he is creating a flow of air moving in the same direction as the smoulder 'wave'. This mode of spread has come to be known as 'forward smoulder' (Ohlemiller and Lucca, 1983). Reverse smoulder would be created by blowing through the cigarette — an action unlikely to satisfy the smoker. However, the concepts of forward and reverse smoulder are relevant to our understanding of the mechanism by which a smoulder wave propagates upwards or downwards, e.g. through a large accumulation of sawdust, under conditions of natural convection. Buoyancy will induce a flow of air into the smoulder wave: in the case of downward propagation (e.g. if a cigarette initiates smouldering on the top of a mattress or at the upper surface of a deep layer of sawdust), the spread will be against the induced airflow, and hence will be 'reverse smoulder'. 'Forward smoulder' occurs for upward propagation.

These two modes of spread have been studied experimentally as one-dimensional spread with an imposed airflow (usually of the order of 1 mm/s, but higher flows have been examined) (Ohlemiller and Lucca, 1983; Torero and Fernandez-Pello, 1996). There is a significant difference between the two modes. For reverse smoulder, the front edge of the smoulder wave (Moussa's Zone 1) encounters fresh, cool air with a full complement of oxygen. With forward smoulder, the flow of air reaching Zone 1 will have been significantly (if not completely) depleted of oxygen, and will be hot, and contain combustion products. There is good evidence that two char-forming reactions will be in competition in this zone, one of which requires oxygen (Torero and Fernandez-Pello (1995) refer to this as the 'smoulder'), while the other is a true pyrolysis reaction and will proceed without oxygen, requiring only a sufficiently high temperature. In their experiments, Ohlemiller and Lucca (1983) observed that the velocity of reverse smoulder through cellulosic insulation was significantly greater than that of forward smoulder.

Experimental studies   Very little research into smouldering combustion was carried out until there was an upsurge of interest in the fire properties of polyurethane foams. Early work by Palmer (1957) dealt mainly with the propagation of smouldering in layers of dust. Until the late 1970s, only isolated studies of smouldering in cellulosic materials (Kinbara *et al.*, 1963; Moussa *et al.*, 1977), string (Heskestad, 1976) and cigarettes (Egerton *et al.*, 1963; Baker, 1977), had been published, but there is now interest in the behaviour of synthetic materials, including polyurethane foam, and cellulose products used for insulation purposes. Only a brief review of this work can be given here.

Table 8.3   Minimum depth for sustained smouldering of cork dusts (Palmer, 1957)

Mean particle diameter (mm)	Minimum depth (mm)
0.5	~12
1.0	~36
2.0	~47
3.6	~36[a]

[a] Accompanied by glowing.

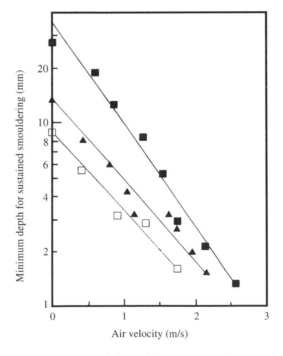

Figure 8.13   Variation with airflow of the minimum depth for sustained smouldering. ■, Deal sawdust (mean particle diameter, 1.0 mm); ▲, beech sawdust (0.48 mm); □, beech sawdust (0.19 mm). (Reproduced by permission of the Combustion Institute from Palmer, 1957)

(a) *Dust layers*   The effect of the depth of a layer of dust on its ability to propagate smouldering horizontally through the fuel bed was studied by Palmer (1957) using wedge-shaped dust deposits in which smouldering was allowed to propagate from the thick towards the thin end of the wedge. The point at which propagation ceased was taken as a measure of the limiting depth. The minimum depths for sustained smouldering of cork dusts is shown in Table 8.3 as a function of particle size. For sawdust, the minimum depth was reduced substantially if the sample was subjected to an imposed airflow. This is shown in Figure 8.13. The effect of an imposed airflow on the rate of propagation of forward smouldering in layers of cellulosic material has been studied quantitatively by Sato and Sega (1988) and Ohlemiller (1990). They used cellulosic powder (thickness 4–10 mm and bulk density 463 kg/m^3) and loose cellu-losic insulation (thickness 107 mm and bulk density 80–100 kg/m^3) respectively. The rate of spread in the higher density material increased from 0.02 mm/s to 0.07 mm/s as the free-stream air velocity (u_0) was increased to 3 m/s (see Figure 8.14a), but it then levelled out, with self-extinction occurring when $u_0 > 6$ m/s. It was observed that this was partly due to instability of the burning edge of the layer which became increasingly steep as the airflow was increased. Ohlemiller also observed an increase in the rate of spread, but at values of u_0 above c.2.5 m/s, transition to flaming occurred (Figure 8.14b). In a limited set of experiments with reverse smoulder, Ohlemiller found the rate of spread to increase linearly up to 5 m/s (the limit of the apparatus), without any transition to flaming occurring (Figure 8.14b). He records the fact that localized blue flames could be seen in cracks and holes in the char at the higher flowrates (Ohlemiller, 1990).

There appears to be no upper limit to the depth of layer that can support smouldering. Palmer (1957) found that if smouldering is initiated at the bottom of a deep layer, the time taken for the combustion wave to propagate vertically to the top surface (forward smoulder) is proportional to the square of the depth. Increasing the moisture content and the density of packing also increase the time of penetration. Packing density has this effect because the rate of propagation is dependent on the penetration of oxygen to the surface of the reacting char. On the other hand, heat losses are low for deep-seated smouldering as the surrounding material provides insulation.

Upward propagation tends to be more rapid than horizontal as hot combustion gases and degradation products rise into the unaffected fuel. The overlying layers absorb the degradation products and there may be no indication that smouldering is in progress within a heap of dust until the combustion zone approaches the surface. The first visible sign is dampness at the surface. As this dries out a musty smell develops, to be followed by the appearance of charring. During those later stages, substantial quantities of steam and acrid smoke are released. Palmer (1957) found that smouldering initiated at the foot of a pile of mixed wood sawdust 0.85 m deep took 10 days to penetrate to the surface.

Most of Palmer's work dealt with the horizontal propagation of smouldering through dust layers: some of the many factors which affect the rate are summarized in Table 8.4 (Palmer, 1957). He observed the onset of flaming at high imposed airflows: comments have already been made about Ohlemiller's observations of the transition to flaming in low density cellulosic insulation undergoing forward smouldering at relatively low airflows (c.2.5 m/s) (Ohlemiller, 1990).

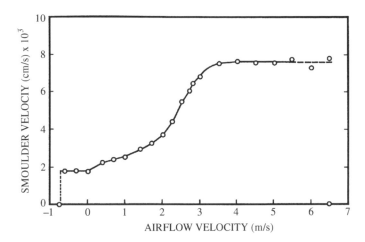

Figure 8.14a Effect of airflow on the rate of spread of smouldering in layers. Dependence of the rate of spread of smoulder along a bed of cellulose powder (density 463 kg/m^3, thickness 8 mm) on the free-stream air velocity u_0 (Sato and Sega, 1988). Forward smoulder. Propagation was not stable for $u_0 > c.6$ m/s: the smouldering self-extinguished. Transition to flaming was not observed. (Reproduced with permission)

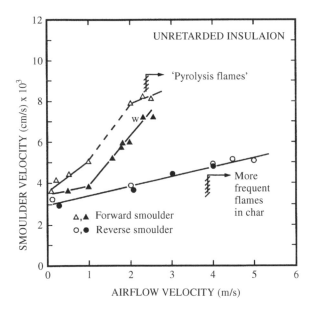

Figure 8.14b Effect of airflow on the rate of spread of smouldering in layers. Dependence of the rate of spread of smoulder along a bed of non-retarded loose cellulosic insulation (density 80–100 kg/m^3, thickness 107 mm) on the free-stream air velocity u_0 (Ohlemiller, 1990). Forward and reverse smoulder, as indicated. Open and closed symbols represent different batches of material. Forward smoulder gave transition to flaming with $u_0 > c.2$ m/s. There was no such transition for reverse smoulder (see text) (By permission of the Combustion Institute)

Table 8.4 Factors affecting the horizontal rate of propagation of smouldering in dust layers (Drysdale, 1980)[a]. Reproduced by permission of the Fire Protection Association

Factor	Dust	Effect on rate of propagation	Interpretation
Particle size	Beech	Rate increases very slightly (~15%) as mean particle diameter decreased from 0.5 to 0.1 mm (16.5 mm deep layer).	Not possible to resolve. Several conflicting factors including changes in ease of access of O_2 to surface, thermal diffusivity and thermal conductivity (improved insulation). Behaviour suggests latter is slightly dominant
Moisture content	Beech	Decreases slightly (~20%) as moisture content increased from 0.5% to 18.8%.	Thermal capacity of unaffected fuel increased; more heat required to drive volatiles from Zone 1.
Depth of layer	Cork, grass	No effect on coarse dusts. With fine dusts (mean particle diameter <0.065 mm) rate decreases with depth.	Restricted access of oxygen to combustion zone.
Imposed air velocity	Beech, deal, cork, grass	Increases; effect much more pronounced if airflow in same direction as propagation. Effect greater for coarse dusts. Glowing occurs at higher airflows (may lead to flaming).	Improved transfer of oxygen to combustion zone. When airflow in same direction as propagation, heat transfer to Zone 1 also improved (forced convection).

[a] Taken mostly from Palmer (1957).

(b) *Cellulose strips and boards*   The various studies of smouldering in cellulose rods and fibreboard tend to confirm the mechanism of propagation described above. Moussa *et al.* (1977) found the rate of propagation correlated well with the maximum temperature in the combustion zone at different oxygen concentrations and partial pressures as shown in Figure 8.12. Palmer's results show a similar correlation for dust layers when there was an imposed airflow blowing in the same direction as the smoulder. This will enhance the rate of supply of oxygen to Zone 2.

Propagation is more rapid in an upward direction (Palmer, 1957; Heskestad, 1976) as convection will contribute to the heat transfer, yet not oppose access of oxygen to Zone 2. Most work has been carried out with horizontal and vertical, downward-burning samples, to simplify the heat transfer problem, but insufficient data have been obtained to generalize about the effect of sample size and shape.

(c) *Rubber latex foams*   Although rubber latex can be made to smoulder easily (Section 8.2.3), no fundamental study of this material has been published. The incident mentioned earlier regarding the production of a flammable smoke involved smouldering latex rubber mattresses stored in a large warehouse (Anon., 1975).

(d) *Leather*   Only a few commercially available leathers have been found to smoulder. Those which have been prepared by traditional vegetable tanning will not burn in this way, but there is evidence that chrome-tanned leathers will smoulder if they contain a high level of chrome (Prime, 1981, 1982).

(e) *Flexible polyurethane foams*   There is now a large variety of flexible 'polyurethane foams' available commercially. The original 'standard' PUF which was widely used in upholstered furniture in the 1970s has virtually disappeared from the market, at least in the UK. This was in response to widespread concern about its behaviour in fire (Joint Fire Prevention Committee, 1978). Certain standard-grade foams are capable of smouldering in isolation, while others, e.g. some high-resilience foams, require contact with a substantial smouldering source, such as a heavy fabric, for sustained smouldering to be initiated (piloted smoulder (Rogers and Ohlemiller, 1978)). This is illustrated in Figure 8.15 (McCarter, 1976). The rate of propagation is likely to be slow as the maximum temperatures achieved in Zone 2 are not significantly greater than 400°C under conditions of natural convection (Ohlemiller and Rogers, 1978; Torero *et al.*, 1994), although this will depend on the thermal insulation afforded by the bulk of the foam. It is substantially less than the temperatures associated with smouldering cellulosics ($\geq 600°C$), which may be one reason why most flexible polyurethane foams will only continue to smoulder (if at all) while there is a supporting heat flux (e.g. from a smouldering fabric cover). Ohlemiller and Rogers (1978) suggest that the propagation mechanism in these foams may involve radiative heat transfer across the open cell structure. Nevertheless, transport of oxygen is the dominant factor in determining the rate of spread under conditions of natural convection (e.g. Torero *et al.*, 1994).

There appears to have been no systematic study of the smouldering potential of the 'new generation' combustion modified flexible foams, of which there are many varieties. If there is any tendency for the foam to soften or melt on exposure to heat (e.g. those formulated with melamine), it is unlikely to smoulder. Otherwise, it is not possible to make any prediction in the absence of some carefully controlled experiments.

(f) *Rigid polyurethane and polyisocyanurate foams*   A limited amount of information regarding the behaviour of rigid foams is reported by Ohlemiller and Rogers (1978). These are used mainly for insulation and tend to have a closed cell structure, hence a very low, if not zero, air permeability. Although the cell walls rupture at relatively low temperatures (less than 275°C) the resulting permeability remains low until pyrolysis and charring commence. Consequently, smouldering of a rigid foam is a surface phenomenon and cannot occur at depth unless some mechanical breakdown has occurred to permit air access. Configuration is important as this will determine the heat loss from the reaction zone. Thus, smouldering combustion may be initiated and will propagate at an inner face of a pair of closely spaced parallel plane slabs.

By studying the smouldering of dusts prepared from these rigid foams, Ohlemiller and Rogers (1978) found that the peak smouldering temperatures in rigid polyurethanes

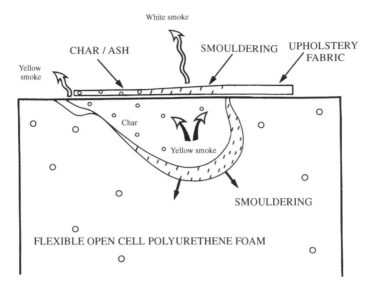

Figure 8.15   Representation of the interaction in a smouldering fabric-foam assembly (after McCarter, 1976). Reproduced by permission of Technomic Publishing Co. Inc., Lancaster, PA 17604, USA

were similar to those observed in flexible polyurethanes (~400°C), but those in rigid polyisocyanurate foams were significantly higher (~500°C) (see also Ohlemiller and Lucca, 1983).

(g) *Phenolformaldehyde foams*   Certain phenolformaldehyde foams can undergo a smouldering process known as 'punking'. Ohlemiller and Rogers (1978) showed that in a configuration which was conducive to minimal heat loss, smouldering in a particular foam could be initiated at a much lower temperature than necessary for a rigid polyurethane foam, although the peak smoulder temperatures of these materials (in powder form) were very similar. Once initiated, the smouldering may continue until the sample has been completely consumed. No smoke is produced but some gaseous volatiles are released which are reported to have a pleasant antiseptic smell. No systematic study of the phenomenon has been made, although it appears that a highly rigid, almost brittle char is required; moreover, there is evidence that the open cell foams are more likely to 'punk' than the closed cell ones. Concern over the possibility of smouldering in this type of material which is used extensively for insulation led to the investigation of a test suitable for assessing 'punking tendency' (British Standards Institution, 1980).

8.2.2   Transition from smouldering to flaming combustion

Flaming combustion can only be established at the surface of a combustible solid if the flowrate of volatiles exceeds a certain critical flux (Section 6.2.3). Thus the application of a small flame to the visible products rising from a smouldering source in still air is unlikely to lead to flaming combustion. Yet transition from smouldering to flaming is known to occur spontaneously even under still air conditions. In these circumstances the flux of volatiles must have increased to at least the critical value and the volatiles must either be ignited by some external means (pilot ignition) or ignite spontaneously. Observation would appear to favour the latter mechanism (Palmer, 1957).

Increasing the temperature of Zone 2 will increase the rate of the production of the volatiles in Zone 1. This is observed in smouldering of coarse dusts (see Table 8.4) and loose cellulosic insulation (Ohlemiller, 1990) if the velocity of the ambient air is increased (Figure 8.14b). Ohlemiller and Rogers (1978) have reported spontaneous flaming to develop from smouldering initiated on the inside faces of closely spaced parallel slabs of rigid polyurethane and polyisocyanurate foams. Under these conditions, the conservation of heat, combined with an increased convective flow of air, enhance the temperature and rate of combustion in Zone 2. Transition to flaming in an item of upholstered furniture occurs only after smouldering has been underway for a substantial period of time (e.g. Babrauskas, 1979). The limited studies that have been carried out seem to suggest that propagation of the smouldering wave down through the material is a prerequisite of the transition. As the smouldering wave breaks through the underside of the cushioning material, a chimney is created through which air can be drawn as a buoyancy-induced flow. If the cushion is resting on a solid base, the transition is inhibited as this flow is prevented (Salig, 1981).

Torero and Fernandez-Pello (1995) observed the transition to flaming following upward (forward) smoulder through a polyurethane foam under conditions of natural convection. The upward velocity was found to increase as the smoulder wave approached the upper surface. They attributed this to an increasing buoyancy-driven flow through the fuel bed as the resistance to flow decreased with progressive conversion of virgin foam to char which has a much greater porosity. The transition was observed only for the deepest fuel beds used (300 mm) (Figure 8.16). Under these circumstances of forward smoulder, the vapours driven off 'Zone 1' would be highly flammable pyrolysis products, and would ignite spontaneously when they achieved a sufficient temperature and concentration in air above the free surface. (In these carefully controlled experiments, air was free to enter through the lower face of the fuel bed. This is unlikely to be the case in practical situations, e.g. smouldering initiated at the foot of a deep layer of sawdust (Palmer, 1957), but more vigorous smouldering will occur as the smoulder wave approaches the surface, thus gaining freer access to fresh air. This has not been studied quantitatively.)

8.2.3 Initiation of smouldering combustion

There are virtually no experimental data on the initiation of this mode of combustion. It is well known that smouldering can follow flaming combustion of many cellulosic materials if the flame is extinguished (Section 8.3) and that a smouldering source such as a burning cigarette can act as a pilot (e.g. British Standards Institution, 1979), but there is no readily available information on other modes of ignition. The basic requirement is the provision of a source of heat which will produce some char and initiate its oxidation, but it would appear that the temperature of the source and/or the rate of supply of heat (e.g. from a radiant heater) are important, more so with some materials than others. For example, a rubber latex foam pillow can be ignited

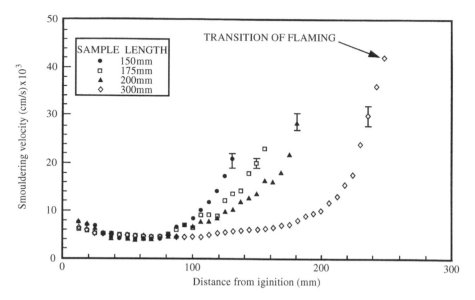

Figure 8.16   Rate of smoulder in polyurethane foam under conditions of natural convection. Upward spread following ignition at the lower face of a cylindrical sample (forward smoulder). The data represent the variation of smouldering velocity with distance from the ignition source for four samples of different height (150, 175, 200 and 300 mm). Transition to flaming was observed in the 300 mm sample only (Torero and Fernandez-Pello, 1995). By permission

within a few minutes by direct contact with a burning 40 W light bulb. Smouldering can commence within the latex foam with little more than a smoke stain visible on the inner and outer cotton covers (Drysdale, 1981). Fibreboard cannot be ignited in this way.

Initiation of smouldering must be interpreted in terms of spontaneous combustion. Indeed, it was emphasized in Section 8.1 that within bulk solids, smouldering is the consequence of sustained self-heating. External heating, whether by means of a light bulb or an electric radiator, will generate a high temperature at the exposed face, a situation similar to the layer of dust on a hot surface (Section 8.1.3). In this case, the simple boundary condition given by Equation (8.3c) would have to be replaced by a heat transfer relationship which would determine the surface temperature. In principle, 'ease of smouldering ignition' could be defined for a given material in terms of a heat flux, although as with flaming ignition this would be very sensitive to configuration and heat losses (Section 6.3). It may be that smouldering will not develop if the heat flux is too great. This would account for the observation that with some materials smouldering can only be initiated by a smouldering source.

By way of summary, the following situations may give rise to smouldering combustion in porous materials:

(i)   contact with a smouldering material;

(ii)   uniform heating (spontaneous ignition within the material);

(iii) unsymmetrical heating (dust on a hot surface, material exposed to a heat flux on one face); and

(iv) the development of a hot spot within the mass of material, e.g. as the result of an electrical fault.

It should be noted that smouldering — or a smoulder-like process — can occur on the surface of a non-porous combustible material exposed to an external heat flux, or under conditions where heat is conserved. Thus, Ohlemiller (1991) observed the propagation of a surface smoulder in wooden channels (white pine and red oak) 51 mm wide and 114 mm deep, but only under forced flow conditions. (Transition to flaming occurred at higher flows.) It was concluded that cross-radiation from the smouldering surfaces was an essential factor in this process.

8.2.4 The chemical requirements for smouldering

The principal requirement for smouldering is that the material must form a rigid char when heated. Materials yielding non-rigid chars or fluid tarry products will tend not to smoulder. This suggests that the molecular structure and mode of degradation of the parent material is of fundamental importance in determining its behaviour. For example, McCarter (1976) has noted that flexible foams made from polyols grafted with acrylonitrile/styrene produce very rigid chars in high yield and have a high propensity to smoulder; similar observations have been made by Rogers and Ohlemiller (1978). The implication of work of this kind is that flexible foams with low propensity to smoulder could be 'designed' chemically or simply selected from materials now available for applications in which a low smoulder tendency is desirable. The importance of the mechanism of degradation can best be illustrated by referring to the smouldering of cellulosic materials. Although smouldering is most often associated with cellulosic materials, pure α-cellulose has little or no tendency to burn in this way. This is compatible with the fact that pure cellulose yields very little char on heating (Section 5.2.2), compared with cotton and rayon, both of which are known to smoulder. The difference can be explained by the catalytic activity of the inorganic impurities which promote the char-forming reaction, some of which are listed in Table 8.5. This has been examined in detail by McCarter (1977, 1978).

Table 8.5 Inorganic compounds[a] which enhance smouldering in cellulosics (McCarter, 1977)

Lithium	Rubidium chloride
chloride	Caesium chloride
hydroxide	
Sodium	
chloride	Silver nitrate
bromide	
iodide	
hydroxide	Ferrous chloride
nitride	Ferrous sulphate
nitrate	
carbonate	Ferric chloride
sulphate	
acetate	Lead nitrate

[a] Applied to the material from solutions.

Table 8.6   Powders which inhibit smouldering in cellulosics (McCarter, 1977)

Sulphur	Copper (I) chloride
Aluminium chloride 6H$_2$O	Iron (II) chloride 4H$_2$O
Antimony chloride	Iron (III) chloride 6H$_2$O
	Magnesium chloride 6H$_2$O
Calcium	Nickel chloride 6H$_2$O
chloride	Stannous chloride 2H$_2$O
chloride 2H$_2$O	Zinc chloride
Chromium (III) chloride 6H$_2$O	

The reduction of smoulder potential can be achieved by selection of materials either which do not form rigid chars or whose char yield can be depressed. An alternative is to use a smoulder suppressant. Boron- and phosphorus-containing compounds are known to inhibit smouldering in cellulosic materials: thus, Ohlemiller (1990) found that 15% boric acid reduced the smoulder velocity through a layer of cellulosic insulation, and increased the airflow velocity necessary to induce flaming combustion (see Section 8.2.1). However, the mechanism of action is not clear as many of these promote the formation of char. McCarter (1978) suggests that the reactive sites on the char are blocked but more work is required not only to elucidate the mechanism but also to improve our understanding of the surface oxidation reaction. McCarter discovered a range of compounds which were capable of suppressing smoulder potential in cellulosic materials if they were dispersed as fine powders: one of the best of these is elemental sulphur (see Table 8.6), although the mechanism is unclear (see Gann *et al.*, 1981).

8.3   Glowing Combustion

Glowing combustion is associated with the surface oxidation of carbonaceous materials or char. The spontaneous ignition of activated charcoal which was referred to in Section 8.1.1 results in glowing combustion within the mass: this differs from smouldering only in that thermal degradation of the parent fuel does not occur, nor is it required. An analysis of such a combustion wave would reveal that the pyrolysis zone (Zone 1) is replaced by a simple pre-heat zone (cf. Section 3.2).

While almost all char-forming fuels–including coal, anthracites, etc.–will undergo flaming combustion, the involatile char will remain, but continue to burn slowly, even after cessation of flaming. A substantial amount of the total heat of combustion of the material is associated with the char, approximately 30% in the case of wood (Section 5.2.2). However, in general, glowing combustion will only continue if heat is conserved at the reacting surface. If a heap of glowing charcoal is opened out to expose the burning surfaces and eliminate cross-radiation, the radiative and convective heat losses may overwhelm the rate of heat production if the individual pieces are large enough and have not become uniformly heated (cf. Section 6.3.2). Under these circumstances, the burning surfaces will cool and the residual fire will self-extinguish.

The term 'afterglow' describes the residual glowing combustion of the char that is left after burnout of a piece of fabric, splint of wood (e.g. a match) or similar item. It is normally of short duration and can easily be eliminated by means of a suitable fire retardant treatment (Lyons, 1970).

Problems

8.1 Given the following data on the ignition of cubes of activated charcoal, calculate by extrapolation the depth of a 'semi-infinite slab' (i.e. a layer) of this material that would be a spontaneous ignition risk at 40°C.

r_0 (half side of cube) (mm)	T_a (crit) (K)
25.40	408
18.60	418
16.00	426
12.50	432
9.53	441

If the thermal conductivity of the activated charcoal was 0.05 W/m K, is the Frank-Kamenetskii model strictly applicable for this range of sample sizes?

8.2 A serious fire broke out in the cable trays in the machine hall of a large factory producing tissue paper. The trays, 60 cm wide, were located below the ceiling at a height of 15 m, and had accumulated a fibrous fluffy deposit over a substantial period of time: the deposit was thought to be 25 cm deep. The ambient temperature below the ceiling was known to be about 40°C during operation of the plant below. The cables (covered by the dust) were working well below their design capacity and there was no evidence of an electrical fault having caused the fire. Spontaneous ignition was suggested. A series of tests was carried out which gave the following data on the critical temperatures of four sizes of sample in the form of cubes.

Cube size $(2r_0, \text{mm})$	Critical temperature (°C)
25	188
50	166
100	142
250	119

Use these data to estimate whether or not spontaneous ignition of the dust layer (at 40°C) could have caused the fire. Assume as a first approximation that the layer can be treated as an infinite slab.

9

The Pre-Flashover Compartment Fire

The term 'compartment fire' is used to describe a fire which is confined within a room or similar enclosure within a building. The overall dimensions are important, but for the most part we shall be directing our attention to room-like volumes of the order of 100 m^3. Fire behaviour in elongated chambers, or in very large spaces ($>$1000 m^3) will depend very much upon the geometry of the enclosure: comments will be made as appropriate.

Following ignition, while still small, the fire will be burning freely: following the definition of Walton and Thomas (1995), the pyrolysis rate and the energy release rate are affected only by the burning of the fuel itself and not by the presence of the boundaries of the compartment. If it can grow in size, either as a result of flame spread over the item first ignited, or by spreading to nearby objects, a stage will be reached when confinement begins to influence its development. If there is sufficient ventilation to allow the growth to continue, then its course can be described in terms of the rate of heat release (power) of the fire as a function of time (Figure 9.1). (This diagram is often presented as a plot of the *average* compartment temperature against time (e.g. Drysdale, 1985b; Walton and Thomas, 1995): the form is similar.)

Although purely schematic, Figure 9.1 illustrates that the compartment fire can be divided into three stages:

(i) the growth or pre-flashover stage in which the average compartment temperature is relatively low and the fire is localized in the vicinity of its origin;

(ii) the fully developed or post-flashover fire, during which all combustible items in the compartment are involved and flames appear to fill the entire volume; and

(iii) the decay period, often identified as that stage of the fire after the average temperature has fallen to 80% of its peak value.

While the average temperature in stage (i) is low, high local temperatures exist in and around the burning zone. During the growth period, the fire increases in size, to and beyond the point at which interaction with the compartment boundaries becomes significant. The transition to the fully developed fire (stage (ii)) is referred to as 'flashover' and involves a rapid spread from the area of localized burning to all combustible

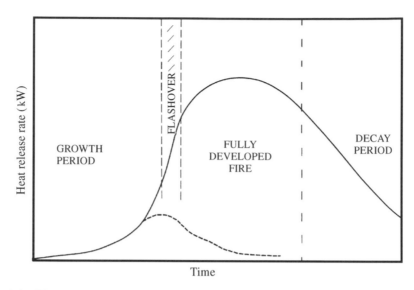

Figure 9.1 The course of a well-ventilated compartment fire expressed as the rate of heat release as a function of time. The broken line represents depletion of fuel before flashover has occurred

surfaces within the room. The transition is normally short in comparison to the duration of the main stages of the fire and is sometimes considered to be a well-defined physical event, in the same way that ignition is an event. This is not the case (Section 9.2), although in many circumstances it is tempting to treat it as such, particularly as it marks the beginning of the fully developed fire (Chapter 10). Anyone who has not escaped from a compartment before flashover is unlikely to survive. Rasbash (1991) has drawn attention to the role that flashover has played in a number of major fire disasters.

During the fully developed stage of a fire, the rate of heat release reaches a maximum and the threat to neighbouring compartments — and perhaps adjacent buildings (Section 2.4.1) — is greatest. Flames may emerge from any ventilation opening, spreading fire to the rest of the building, either internally (through open doorways) or externally (through windows). In addition to the obvious threat to occupants remaining within the building, it is during this stage that structural damage may occur, perhaps leading to partial or total collapse of the building. During the cooling period (stage (iii)), the rate of burning diminishes as the fuel is depleted of its volatiles. Flaming will eventually cease, leaving behind a mass of glowing embers which will continue to burn, albeit slowly, for some time, maintaining high local temperatures (Section 8.3).

An understanding of the pre-flashover stage of the fire is directly relevant to life safety within the building. Once flashover has occurred in one compartment, the occupants of the rest of the building can be threatened directly. The significance of the various events in the sequence may be seen by considering the inequality discussed by Marchant (1976):

$$t_p + t_a + t_{rs} \leq t_u \tag{9.1}$$

where t_p is the time elapsed from ignition to the moment at which the fire is detected; t_a is the delay between detection and the beginning of the 'escape activity'; t_{rs} is the time to move to a place of relative safety; and t_u is the time (from ignition) for the fire to produce untenable conditions at the location under consideration. While automatic detection can reduce t_p, in some cases very significantly, successful escape will ultimately depend on how rapidly the fire grows, i.e. on t_u. Consequently, the time to flashover is a significant factor in determining the fire hazard associated with a particular compartment. The greater its duration, the more opportunity there will be for detection and suppression (manual or automatic) and for safe evacuation of personnel. Smoke detectors would be expected to operate during the earlier stages of the pre-flashover fire before it has grown much beyond ∼10 kW, which is large enough to carry sufficient combustion products to ceiling level in a 3 m high enclosure (Section 4.3.1). Operation of fixed temperature heat detectors and conventional sprinkler heads depends on how quickly the sensing element under the ceiling can be raised to a specific operating temperature (e.g. 68°C). This is determined by the response time index (RTI) (see Section 2.2.2 and 4.4.2). A relatively large fire is required to produce a flow of sufficiently hot gas (Section 4.4.2) and, given the additional delay due to the relatively high thermal mass of the sensing elements used in conventional sprinklers, the fire may be fast approaching the end of the growth period before activation occurs (e.g. 0.5–1.0 MW in an average sized room). The advantage of fast response sprinklers, with sensing elements of low thermal mass, is clear.

This chapter is devoted to the early stages of the compartment fire, up to and including the onset of flashover. The criteria necessary for flashover to occur and the factors which determine the duration of the growth period will be identified and discussed.

9.1   The Growth Period and the Definition of Flashover

After localized burning has become established, one of three things may happen:

(i)   the fire may burn itself out without involving other items of combustible material, particularly if the item first ignited is in an isolated position (see Figure 9.1);

(ii)   if there is inadequate ventilation, the fire may self-extinguish or continue to burn at a very slow rate dictated by the availability of oxygen (the scenario which may lead to 'backdraught' (Section 9.2.1)); or

(iii)   if there is sufficient fuel and ventilation, the fire may progress to full room involvement in which all exposed combustible surfaces are burning.

The following discussion will concentrate on the latter scenario. A number of issues need to be reviewed, but these may be grouped into two broad areas: the conditions associated with the onset of flashover, and the factors which determine the duration of the growth period. The latter is extremely important as it has a major impact on life safety: if the 'time to flashover' is short, the time available for escape (t_u in Equation (9.1)) may be inadequate.

The conditions that are necessary and sufficient for the fire to develop to the fully developed stage need to be understood and quantified. The term 'flashover' was identified in Figure 9.1 and described in the accompanying text. Its duration will depend on a number of factors, including the nature and distribution of the combustible contents, as well as the size and shape of the compartment. In a typical living room, with upholstered furniture containing standard polyurethane foam, the growth period may be short and the flashover process may last only 15–30 s (Building Research Establishment, 1989). However, in large spaces, such as a warehouse, it is likely to take longer, and indeed it is *conceivable* that the area first involved may burn out before the most distant area becomes involved.

Flashover is generally associated with enclosed spaces; however, the Bradford Football Stadium fire which occurred in May 1985 (Popplewell, 1986; Building Research Establishment, 1987) revealed that the term is also applicable to relatively open situations where confinement is provided by little more than the roof of the structure. (This is illustrated clearly in Figure 9.2 below.)

The basic features of the growth period and the onset of flashover may be understood qualitatively in terms of the elements of fire behaviour which have been discussed in earlier chapters. In the open, an isolated fuel bed will burn at a rate determined by the heat flux from the flame to the surface, $\dot{Q}''_F$, i.e.

$$\dot{m}'' = \frac{\dot{Q}''_F - \dot{Q}''_L}{L_v} \tag{9.2}$$

where $\dot{Q}''_F$ will be dominated by radiation if the fuel bed is greater than about 0.3 m in diameter (Section 5.2.1). For most fuels, approximately 30% of the heat liberated in the flame is radiated to the environment and the rest is dispersed convectively in the buoyant plume. If the item is burning in a compartment, this heat is not totally lost from the environs of the fuel as fire gases will be deflected and trapped below the ceiling (Section 4.3.4), which will become heated. If the size of the fire is such that the natural flame height is greater than the height of the room, then the flame may extend

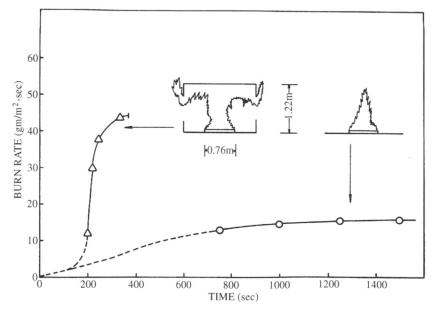

Figure 9.2   The effect of enclosure on the rate of burning of a slab of polymethylmethacrylate (0.76 m × 0.76 m) (Friedman, 1975)

as a ceiling jet (Section 4.3.4) and thus contribute significantly to the heat transfer to the ceiling. This in turn will provide an increasing radiant heat flux back to the fuel as the temperature of the ceiling rises. However, from an early stage a layer of hot smoky gases will be accumulating below the ceiling and will radiate to the lower parts of the room with increasing intensity as the smoke concentration, the layer thickness and the temperature all increase. The effect on the burning fuel will be to increase the rate of burning, according to the equation:

$$\dot{m}'' = \frac{\dot{Q}_F'' + \dot{Q}_E'' - \dot{Q}_L''}{L_v} \tag{9.3}$$

However, more significantly, it will promote flame spread over the item first ignited (Section 7.2.5) and to contiguous surfaces and adjacent items (see Section 4.3.4), thereby increasing the area of burning (A_F) and consequently the rate of generation of fuel vapours (Equation (9.4)):

$$\dot{m} = \left[\frac{\dot{Q}_F'' + \dot{Q}_E'' - \dot{Q}_L''}{L_v}\right] A_F \tag{9.4}$$

Thomas (1981) has argued that the rate of increase in the area of burning is more important than the rate of increase of $\dot{m}''$ (Equation (9.3)) in determining the rate of fire development to the flashover stage. Any scenario which leads to fast fire spread — and hence a rapidly increasing area of burning — will promote the onset of flashover. An extreme example of this was encountered in the fire involving a wooden escalator at the King's Cross Underground Station, London, in November 1987 (Fennel, 1988; Rasbash, 1991), but there are many other examples.

Some of these features are well illustrated in Figure 9.2 which shows the effect of an enclosure on the burning of a slab of polymethylmethacrylate (Friedman, 1975). A hood (acting as a ceiling and the upper parts of the walls) above the burning slab deflects the flames and increases the radiative heat feedback to the fuel surface: air had unrestricted access to all four sides in these experiments. The maximum rate of burning is three times greater than that in the open and appears to be achieved in about one-third of the time. This effect is quite general although its magnitude depends on the nature of the fuel and the size of the compartment. For example, the rate of burning of an alcohol fire in a small enclosure can be up to eight times greater than in the open (Thomas and Bullen, 1979; Takeda and Akita, 1981). There is evidence that the increase will be self-limiting, as the fuel vapours rising from the surface attenuate the radiation reaching the fuel (Thomas and Bullen, 1979; see also Section 5.1.1).

Given that flashover marks the beginning of the fully developed fire, this term must be defined more precisely in order that the factors which determine the duration of the growth period can be examined. Many definitions have appeared in the literature (Thomas, 1982), the most common of which are:

(i)   the transition from a localized fire to the general conflagration within the compartment when all fuel surfaces are burning;

(ii) the transition from a fuel controlled fire to a ventilation controlled fire; and

(iii) the sudden propagation of flame through the unburnt gases and vapours collected under the ceiling.

Martin and Wiersma (1979) point out that (ii) is the result of (i) and is not a fundamental definition. The third definition is based partly on an observation that flames often emerge from the window or other ventilation openings at around the time full room involvement commences, but as stated would seem to imply premixed burning. This is not correct, but it is likely that the burning of the smoke layer plays a significant role as one of several mechanisms that contribute to the transition to the fully developed fire. This is discussed in greater detail in the following sections.

The first definition (i) is the one that is being used in this text, although Thomas (1974b) has pointed out that it may not apply to very long or 'deep' compartments in which it may be physically impossible for all the fuel to become involved at the same time.

9.2   Growth to Flashover

9.2.1   Conditions necessary for flashover

Waterman (1968) studied the flashover phenomenon by carrying out a series of experiments in a compartment $3.64 \times 3.64 \times 2.43$ m high, in which items of furniture were burned. The onset of the fully developed fire was defined by the ignition of paper 'targets' located on the floor. He concluded that a heat flux of about 20 kW/m^2 at floor level was required for flashover to occur. While this may be sufficient to bring about spontaneous ignition of paper (Martin and Wiersma, 1979), it is well below that necessary for 'thick' pieces of wood and other combustible solids. However, it is more than enough to promote pilot ignition and rapid flame spread at the surface of most combustible materials (Sections 6.3 and 7.2.5). Waterman postulated that most of this heat flux came from the heated upper surfaces of the room, rather than directly from the flames above the burning fuel. He also observed that flashover was not achieved until the rate of burning exceeded 40 g/s.

Experimental studies of flashover carried out since this work have tended to define the 'onset of flashover' in terms of either Waterman's criterion (20 kW/m^2 at low level, e.g. Babrauskas, 1979) or a ceiling temperature of approximately 600°C, which has been observed in several studies (Hägglund *et al.*, 1974; Fang, 1975b) although Hägglund *et al.* (1974) actually defined flashover as the moment at which flames emerged from the opening. The ceiling height of their compartment was 2.7 m: if the mechanism of flashover involves heat radiated to the lower levels, it might be anticipated that the ceiling temperature at flashover would be a function of height. While there is limited evidence to support this conclusion (Heselden and Melinek (1975) report 450°C for 'flashover' in a small-scale experimental compartment 1.0 m high), it neglects other sources of thermal radiation.

In addition to the vertical flames above the fire, there are three sources of radiative flux, namely:

(i)   from the hot surfaces in the upper part of the enclosure;
(ii) from flames under the ceiling;
(iii) from hot combustion products trapped under the ceiling.

The relative importance of these will vary as a fire develops and which predominates at flashover will depend on the nature of the fuel and the nature and degree of ventilation. If a clean-burning fuel such as methanol is involved, the heat feedback will be from the upper surfaces; as the flames and combustion products are of low emissivity (Section 2.4.2). However, smoke is produced in almost all fires and ultimately thickness and temperature of the hot layer of combustion products under the ceiling will determine the behaviour (Section 2.4.3).

For convenience, two situations can be considered, as illustrated in Figure 9.3. In the first (Figure 9.3(a)), the fire plume is deflected to form a one-dimensional ceiling jet, thus modelling a corridor ceiling with no barrier to limit horizontal spread in the direction of flow. The depth of layer will then depend only on the width of the 'corridor' and the rate of flow of hot gas into the ceiling layer. If the fire is of sufficient size that the flames are deflected, then the nature of burning under the ceiling will depend on whether (i) the fire plume penetrating the ceiling layer contains sufficient or excess air (i.e. oxygen) to burn the fuel vapours, or (ii) insufficient clear air has been entrained into the fire plume, thus leading to low oxygen concentrations and high levels of unburnt fuel in the smoke layer. Hinkley *et al.* (1968) (Section 4.3.4) have studied the range of behaviour in the context of corridor flows, referring to conditions (i) and (ii) as 'fuel lean' and 'fuel rich', respectively. In the 'fuel lean' situation, more than enough air to burn the fuel gases has been entrained in the vertical flame and consequently the flame in the ceiling jet will be of limited extent and will burn close to the surface. With the 'fuel rich' situation, flaming occurs at the lower boundary of the layer where entrainment is taking place. The difference between these two extremes is shown clearly by the vertical temperature distributions below the ceiling (Figure 4.30). The net result of the horizontal deflection of the fire plume is to increase the radiative heat flux at floor level at points remote from the fire. This is shown schematically in Figure 9.4 (see Section 4.3.4). Hinkley *et al.* (1968) estimated from a detailed heat balance that long, fuel rich flames lost as much as 55% of the heat of combustion released in the layer by radiation. If a combustible lining is involved, the flame extension is increased, and the heat flux to the lower levels is correspondingly greater.

The second scenario, illustrated in Figure 9.3(b), shows the formation of a smoke layer under the ceiling of a compartment. Immediately after burning has become established, a fire plume will develop and grow in strength as the fire grows in size. It will rise to the ceiling to form a ceiling jet which will flow outwards until it reaches the

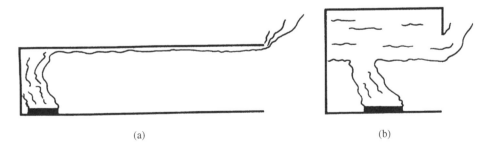

(a)   (b)

Figure 9.3   Interaction of a fire plume with a ceiling: (a) corridor ceiling, with unrestricted horizontal flow in one direction; (b) compartment ceiling, showing the formation of a smoke layer

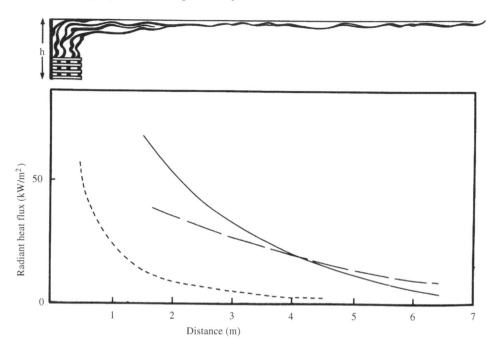

Figure 9.4   Radiative heat flux at floor level as a function of distance from a fire of constant heat output located at the closed end of a corridor. The dashed line (- - - - -) represents the distribution of heat flux from the same fire in the absence of the ceiling (vertical fire plume only). The other lines represent different values of h (the height of the ceiling above the crib base): ———, $h = 970$ mm; — — —, $h = 660$ mm. (From Thomas, 1975.) (Reproduced by permission of The Controller, HMSO. © Crown Copyright)

walls (Section 4.3.4)*. A smoke layer will quickly form under the ceiling, deepening as the fire continues to burn. As the layer descends below the top of any opening, such as a door or window, hot combustion products will flow from the compartment. The height of the smoke layer will stabilize at a level below the soffit of the opening, reducing the height through which cool air can be entrained into the vertical plume (Quintiere, 1976; Beyler, 1984b, 1986a; Zukoski, 1986, 1995): consequently, the temperature of the layer will progressively increase. The plume will continue to penetrate to the ceiling, maintaining a hot ceiling jet. Although the temperature of the layer is not strictly uniform (Babrauskas and Williamson, 1978), it is commonly assumed to be so, particularly in zone models (Quintiere, 1989, 1995b). As the walls heat up relatively slowly, the layer of gas close to the wall will cool and become negatively buoyant with respect to the surrounding gas (Cooper, 1984; Jaluria, 1995). A downward flow will be generated which will carry smoke into the clear, cool air below — when it will then become positively buoyant and rise towards the layer again. This will contribute to some mixing of smoke in the lower layer. The temperature will remain highest near the ceiling, and the ceiling and upper walls will be heated, initially by convection, although radiation

---

* If the ceiling is of unlimited extent, the outward flow will eventually cease as the gases cool and lose buoyancy when they will tend to 'fall' and mix with the air below.

will come to dominate as the temperature of the layer rises and the concentration of the smoke particles increases.

As the fire grows in size, the flame will eventually penetrate the smoke layer. When this happens, the upper part of the flame will entrain increasingly vitiated air, already contaminated with fire products. This will lead to an increase in the rate of accumulation of partially burnt fuel vapours and particulate smoke (Beyler, 1985; Zukoski, 1986). As the temperature of the layer continues to rise, it will radiate downwards with increasing intensity, enhancing not only the rate of burning of the existing fire (Equation (9.3)), but also the rate of flame spread over contiguous surfaces (Section 7.2.4). At the same time, the smoke layer will descend, albeit more slowly than in the early stages, encompassing an increasing proportion of the growing flame. This is the situation that is likely to exist at the onset of flashover, when the fire undergoes a rapid transition to its fully developed stage.

Before considering the subsequent events, it is important to consider the nature and composition of the upper layer at this stage. These have been examined by carrying out experiments in which a stable, buoyant diffusion flame has been allowed to burn under a hood, designed to mimic the situation described above in which the upper part of the flames penetrate the smoke layer (Figure 9.5). It is also an experimental idealization of the 'zone model' in which the upper and lower layers are considered separately. By varying the supply rate of gaseous fuel and the distance of the burner below the lower edge of the hood, Beyler (1984b) was able to examine the effect of varying the equivalence ratio (ϕ) on the composition and behaviour of the layer. This quantity (ϕ) is the ratio of fuel to air in the mixture entering the layer, normalized to the stoichiometric ratio: it is determined by the amount of 'clean' air that can be entrained into the vertical fire plume below the base of the hot layer. He found that the layer would begin to burn when the equivalence ratio exceeded approximately 1.8. The entrainment of hot smoke, containing a progressively decreasing concentration of oxygen, into the upper part of the flame inhibits the combustion process and causes the

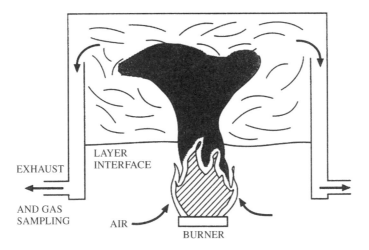

Figure 9.5 Diagram showing the two-layer system used by Beyler (1984b) to study the smoke layer that forms under the ceiling during a compartment fire (By permission of the Society of Fire Protection Engineers)

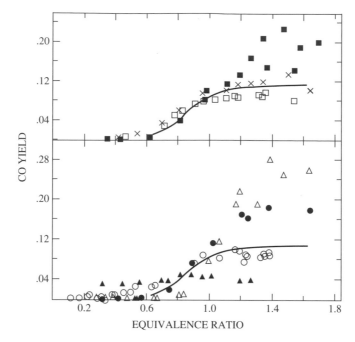

Figure 9.6 Normalized carbon monoxide yields as a function of equivalence ratio (ϕ) for propane (——); propene (○); hexanes (□); toluene (▲); methanol (△); ethanol (●); isopropanol (×); and acetone (■) (Beyler, 1986a). By permission of Elsevier

concentration of unburned and partially burnt fuel vapours in the layer to increase as ϕ increases. In particular, the level of carbon monoxide increases significantly as the equivalence ratio increases above 0.6 (Figure 9.6). Its yield increases asymptotically to a high level (>10% of its theoretical maximum value for $\phi > 1.2$). This corresponds to levels of CO of up to 6%, and is the source of the highly toxic smoke that can flow from the room of origin through an open door into the rest of the building during the period immediately preceding flashover (Beyler, 1986a; Zukoski *et al.*, 1991).

In a series of steady state experiments, Beyler (1984b) found that when the equivalence ratio exceeded 0.8 to 1.0, sufficient unburned and partially burned products had accumulated for independent flaming to occur at the lower face of the smoke layer, located close to the region where the main fire plume penetrated. If the rate of fuel supply was increased so that $\phi > 1.8$, then flame was observed to spread across the interface, independently of the original fire plume. Beyler was unable to pursue this further with his experimental facility, but he identified an equivalency ratio of this order as marking a significant stage in the development of a fire. The development of burning at the interface will cause the layer to become unstable and undergo mixing with air from below. It can be argued that this is part of the flashover process, and accounts for the observation that flames emerge from the openings 'when flashover occurs'.

There is now good evidence that a large proportion of the heat radiated to the lower part of a compartment during fire growth is from the layer of hot smoky combustion products (Croce, 1975; Quintiere, 1975a). This is discussed briefly by Tien *et al.* (1995), but earlier, Orloff *et al.* (1979) described a sophisticated model for calculating radiative output from the smoky gases under the ceiling of a poorly ventilated compartment, taking into account the stratification of temperature and combustion products within the smoke layer. The problem is complicated by the fact that although radiation is being emitted by H_2O, CO_2 and soot throughout the layer, the cooler, lower layers tend to absorb radiation emitted from above. Orloff *et al.* (1979) compared their theoretical model with the results of a full-scale test fire in which a slab of polyurethane foam was burned under conditions of restricted ventilation. They found excellent agreement between measured and predicted radiant heat flux levels at the floor during the development stage of the fire and from their calculations showed that only a small proportion (less than 15% of the radiant heat reaching the floor) came from the ceiling and upper walls of the enclosure; during the pre-flashover stage, radiation from the layer of smoky gases dominated. Of course, this represents a particular case. The contribution from the hot walls and ceiling would be greater if the layer was relatively thin and/or the yield of particulate smoke was significantly less, although the net effect would not be distinguishable.

Waterman's original criterion for the onset of flashover (20 kW/m^2 at floor level) can only be achieved when the fire has grown to a size sufficient to create conditions under the ceiling which can provide this heat flux. Normally the latter will rise monotonically as the fire grows, but if the fuel burns out or oxygen depletion occurs, then flashover may not be reached (see Figure 9.1). Oxygen depletion will be significant if the rate of ventilation to the compartment is insufficient to replenish oxygen as it is consumed by the developing fire. In an extreme case, a fast burning fire (e.g. one involving a hydrocarbon pool, or an upholstered chair bursting into flame after a prolonged period of smouldering) in a small, poorly ventilated compartment, may self-extinguish or (more likely) die down and burn slowly at a rate dictated by the ingress of air. Under these vitiated conditions, the combustion process will be incomplete and unburned and partially burned pyrolysis products will accumulate in the space. Bullen (1978) considered the conditions under which this would be important in domestic fires. On the assumption that the rate of burning will be controlled by the rate of ventilation (see Section 10.1), he showed that — theoretically — burning would continue, supported by leakage of air around reasonably well-fitting doors and windows, assuming availability and distribution of fuel. Under these circumstances, the fire would create additional ventilation (e.g. by burning through a door) which will permit a progressive increase in intensity. However, if the fire has been burning for a prolonged period under vitiated conditions, and additional ventilation is provided suddenly (e.g. by the *opening* of a door), the ingress of fresh air may be followed by rapid burning of the accumulation of unburned smoke and fuel vapours, producing a backdraught.

This process has been studied in detail by Fleischmann *et al.* (1994) (Figure 9.7) and has been reviewed by Chitty (1994). The most hazardous characteristic from the firefighter's point of view is the sudden eruption of flame which occurs when the fire is vented.* It should be noted that while backdraught may be followed by a fully

---

* Relatively small overpressures will be created, but occasionally forces are generated which are capable of causing structural damage. For this to occur it is likely that a significant amount of a flammable premixed smoke/air mixture has formed prior to its ignition, or that unusually high levels of turbulence were created before and during the burning process. The term 'smoke explosion' has been used by some to describe such events: these were reviewed by Croft (1980/81). An oft-quoted example is explosion and fire which occurred at Chatham Dockyard in 1975 (Croft, 1980/81; Woolley and Ames, 1975) (see Section 8.2).

(a)

(b)

Figure 9.7   Development of backdraught (Fleischmann *et al.*, 1994). A 70 kW methane flame was burned in a sealed chamber, the flame eventually self-extinguishing due to oxygen starvation. The vent was opened (on the right): a continuous ignition source was present towards the rear of the enclosure. (a) 5.6 s after opening the vent; (b) 7.1 s after opening the vent; (c) 8.0 s after the vent was opened. By permission.

developed fire, it is distinct from the above discussion of 'flashover' in which a sudden influx of air is *not* required (Drysdale, 1996): flashover can take place under well-ventilated conditions (e.g. see Figure 9.2).

Thomas *et al.* (1980) have shown how flashover can be regarded as a case of thermal instability within the compartment, analogous in many ways to the Semenov model

(c)

Figure 9.7   (*Continued*)

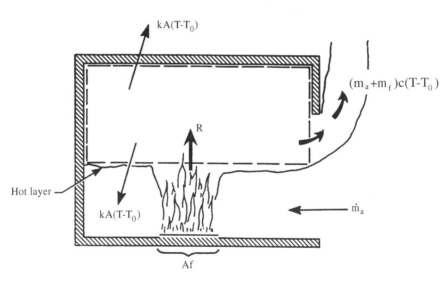

Figure 9.8   Diagram showing the energy balance for the hot layer in an enclosure fire (from Thomas *et al.*, 1980) Reproduced by permission of the Combustion Institute.

for spontaneous ignition (Section 6.1). The starting point is to assume that the rate of burning is a function of temperature but is limited by the rate of supply of air. (This limit is the ventilation controlled fire in which the rate of burning is determined by the maximum rate at which air can flow into the compartment during fully developed burning (Section 10.1).) These authors have considered a quasi-steady state model of the fire by considering the energy balance of the hot layer under the ceiling (Figure 9.8). The rates of heat generation and heat loss are compared, recalling that both are functions of temperature of the upper layer (T), thus:

$$\dot{Q}_{c}(T, t) = \dot{L}(T, t)$$
$$R = L \qquad (9.5)$$

where $\dot{Q}_{c}(T, t) = R$ is the rate of heat release — i.e. the rate of energy gain by the upper layer — and $\dot{L}(T, t) = L$ is the rate of heat loss as shown in Figure 9.9, where several simplifying assumptions are made, e.g.

$$\dot{L}(T, t) = h_{k}A_{T}(T - T_{0}) + (\dot{m}_{a} + \dot{m}_{f})c_{p}(T - T_{0}) \qquad (9.6)$$

where h_{k} is an effective heat transfer coefficient (referring to conduction through the compartment boundaries) and A_{T} is the appropriate internal surface area through which heat is lost (see McCaffrey *et al.*, 1981). For the simplest case (constant fuel bed area), the rate of heat release is determined by radiative heat transfer from the upper layer, and is thus a strongly non-linear function of temperature. However, it is limited by the rate of supply of air, as shown schematically in Figure 9.9 where the rate of heat release curve is compared with three representative heat loss curves, L_1, L_2 and L_3, which correspond (for example) to decreasing compartment sizes. Three types of intersection for the heat release (R) and heat loss (L) curves are possible, namely A, B and C. While B is unstable, A corresponds to a 'steady' ventilation controlled fire (Chapter 10), corresponding to a high rate of burning, and C represents the small localized fire which is not influenced to any significant extent by the heat feedback from the upper parts of the enclosure. With a growing fire, both R and L will change and can exhibit a critically in which any slight increase in rate of burning would cause

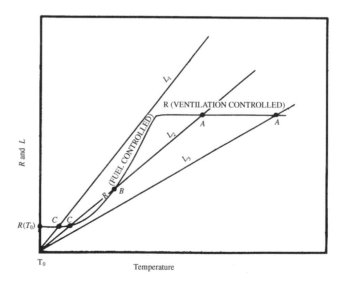

Figure 9.9 Flashover as a thermal instability. The line marked 'R' represents the rate of heat release within the compartment as a function of temperature, while the lines marked 'L' are representative of heat loss functions (reproduced by permission of the Combustion Institute from Thomas *et al.*, 1980)

a substantial jump in both temperature and rate of burning to point A. Thomas *et al.* (1980) drew attention to the fact that this behaviour is 'suggestive of the phenomenon of flashover' and take the model much further to show how various types of instability can be explained on the basis of such diagrams. Their prediction that 'oscillating' fires might occur has been confirmed experimentally (Takeda and Akita, 1981). More recently, Bishop *et al.* (1993) have shown how the techniques of non-linear dynamics can be used to analyse this model quantitatively and predict how flashover depends on parameters such as the ventilation conditions and the thermal characteristics of the compartment boundaries (Holborn *et al.*, 1993).

9.2.2   Fuel and ventilation conditions necessary for flashover

Although the fully developed fire will be discussed in detail in Chapter 10, one result is required here, namely that the rate of burning of a wood crib in a compartment under conditions of restricted ventilation obeys the relationship:

$$\dot{m} = K.A_w H^{1/2} \qquad \text{kg/s} \tag{9.7}$$

where A_w and H are the area and height of the ventilation opening, respectively, and K is a constant (approximately 0.09 kg/m$^{5/2}$.s). (This is the classic 'ventilation-controlled' fire originally studied by Kawagoe (1958).) Hägglund and his co-workers (1974) carried out a series of experiments in which they burned wood cribs in a compartment measuring 2.9 m $\times$ 3.75 m $\times$ 2.7 m high, monitoring the rate of burning continuously by weighing the fuel. They presented their data graphically as a plot of $\dot{m}$ versus $A_w H^{1/2}$ and found that those fires that 'flashed over' (i.e. flames out of the door and temperatures in excess of 600°C under the ceiling) occupied a narrowly defined region of the diagram. These are identified as solid symbols in Figure 9.10. Flashover was not observed for rates of burning less than about 80 g/s (twice that quoted by Waterman (1966)), although this limit increased as the ventilation was increased in accordance with the empirical expression:

$$\dot{m}_{\text{limit}} = 50.0 + 33.3\ A_w H^{1/2}\ \text{g/s} \tag{9.8}$$

While flashover was not observed for values of $A_w H^{1/2}$ less than 0.8 m$^{5/2}$, only a limited range was studied and the upper limit for Equation (9.8) was not determined. As will be shown in Section 10.1, the rate of burning becomes independent of ventilation at large values of $A_w H^{1/2}$ when the transition to the fully developed fire (i.e. 'flashover') may be ill-defined, or not occur (e.g. Takeda and Akita, 1981) (see Figure 10.12).

Although this work was confined to a single compartment of height 2.7 m, it does suggest a more general principle, that a limiting burning rate must be exceeded — and presumably maintained for a period of time — before flashover can occur. Indeed, single items of furniture can lead to flashover if their rates of burning are high enough (Klitgaard and Williamson, 1975; Babrauskas, 1979): thus, Babrauskas (1979) observed that flashover (defined as 20 kW/m^2 at floor level) was achieved in 280 s by a single modern chair consisting of block polyurethane foam covered by acrylic fake-fur burning in a compartment 2.8 m high. The burning rate 'peaked' at 150 g/s. A leather chair also exhibited a high rate of burning, peaking at 112 g/s, but too briefly for the flashover criterion to be met.

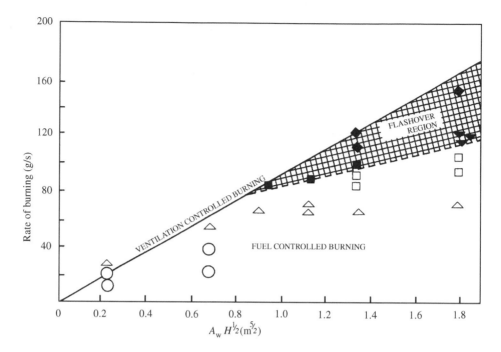

Figure 9.10   Compartment fire burning rate data of Hägglund *et al.* (1974) as a function of the
ventilation parameter $A_wH^{1/2}$. Fuel: wood cribs, mass 18–36 kg (○), 45–55 kg
(△), 60–74 kg (□), 80–91 kg (▽), 98–103 kg (◇). Solid symbols indicate flames
out of doorway and ceiling gas temperatures $\geq 600°C$ (from Hägglund *et al.*,
1974). (Quintiere, 1976, reproduced by permission of ASTM)

While the concept of a limiting burning rate for flashover to occur is based on
experimental observation, it would seem more logical to examine critical rates of
heat release within the compartment, particularly in view of Thomas' contention that
flashover represents a thermal instability (Thomas *et al.*, 1980). Babrauskas (1980b)
demonstrated how this could be done using a heat balance for the upper layer as given
in Equation (9.5) and a set of very simple assumptions, viz. at flashover, the upper
layer temperature is 600°C, and that heat losses are by radiation to the floor, and the
walls below the hot layer. He deduced the equation

$$\dot{Q}_{FO} = 600A_WH^{1/2} \text{ kW} \tag{9.9}$$

which he showed to correspond to 40% of the heat release associated with stoichiometric burning. However, this was not based on any experimental data.

McCaffrey *et al.* (1981) have taken this further by analysing a model developed
previously by Quintiere *et al.* (1978) in which a simple heat balance is applied to the
layer of hot gases below the ceiling (Figure 9.8). Combining Equations (9.5) and (9.6),
we have:

$$\dot{Q}_c = (\dot{m}_a + \dot{m}_f)c_p(T - T_0) + h_kA_T(T - T_0) \tag{9.10}$$

T and T_0 are the temperatures of the upper layer and the ambient atmosphere respectively. It is assumed that the layer is well mixed and its temperature is uniform.

Writing

$$\dot{m}_g = \dot{m}_a + \dot{m}_f \tag{9.11}$$

and rearranging Equation (9.10) gives:

$$\frac{\Delta T}{T_0} = \frac{\dot{Q}_c/(c_p T_0 \dot{m}_g)}{1 + h_k A_T/(c_p \dot{m}_g)} \tag{9.12}$$

The mass flowrate of gas leaving the compartment above the neutral plane (Figure 9.8: see also Section 10.1) can be approximated by the expression developed by Rockett (1976) for flow of air induced by a small fire:

$$\dot{m}_g = \frac{2}{3} C_d A_w H^{1/2} \rho_0 \left(2g \frac{T_0}{T} \left(1 - \frac{T_0}{T} \right) \right)^{1/2} \left(1 - \frac{h_0}{H} \right)^{3/2} \tag{9.13}$$

where h_0 is the height of the neutral plane and C_d is the discharge coefficient. For the present purposes (see below), Equation (9.13) may be reduced to the proportionality:

$$\dot{m}_g \propto g^{1/2} \rho_0 A_w H^{1/2} \tag{9.14}$$

where $g = 9.81$ m/s and ρ_0 is the density of ambient air. Using this relationship, Equation (9.12) can then be cast in general terms, giving $\Delta T/T_0$ as an unspecified function of two dimensionless groups, namely:

$$\frac{\Delta T}{T_0} = f \left(\frac{\dot{Q}_c}{g^{1/2}(c_p\rho_0)T_0 A_w H^{1/2}}, \frac{h_k A_T}{g^{1/2}(c_p\rho_0)A_w H^{1/2}} \right) \tag{9.15}$$

or

$$\frac{\Delta T}{T_0} = C \cdot X_1^N \cdot X_2^M \tag{9.16}$$

where X_1 and X_2 represent the two dimensionless groups in Equation (9.15), and the constant C and exponents N and M remain to be determined from experimental data.

McCaffrey *et al.* (1981) analysed data from more than 100 experimental fires (from eight series of tests involving several types of fuel) in which steady burning rates were achieved, but upper gas layer temperatures did not exceed 600°C. Above 600°C, flaming can occur intermittently within the layer and many of the assumptions inherent in the model break down. To enable these data to be cast in the form suggested by Equation (9.16), it is necessary to obtain appropriate values for h_k which depend on the duration of the fire and the thermal characteristics of the compartment boundary. For a fire which burns with a characteristic time (t_c) greater than the 'thermal penetration time' (t_p) of the boundary ($t_p = \delta^2/4\alpha$, where α is the thermal diffusivity ($k/\rho c$ and δ is the boundary thickness — see Equation (2.41) *et seq.*), h_k can be approximated by

$$h_k = k/\delta \tag{9.17}$$

where k is the thermal conductivity of the material from which the compartment boundaries have been constructed (i.e. heat is conducted in a quasi-steady state manner to the exterior). If, on the other hand, t_c is less than t_p, the boundary will be storing heat during the fire and little will be lost through the outer surface. Normally, this would require detailed solution of the transient heat conduction equations (Section 2.2.2) but a simplification can be achieved by replacing δ by $(\alpha t_c)^{1/2}$, the effective depth of the lining material which is heated significantly during the course of the fire (the 'thermal thickness'). In these circumstances,

$$h_k = \frac{k}{(\alpha t_c)^{1/2}} = \left(\frac{k\rho c}{t_c}\right)^{1/2} \tag{9.18}$$

For a compartment bounded by different lining materials, the overall value of h_k must be weighted according to the areas: thus, if the walls and ceiling (W,C) are of a different material to the floor (F), then, if $t_c > t_p$:

$$h_k = \frac{A_{W,C}}{A_T} \cdot \frac{k_{W,C}}{\delta_{W,C}} + \frac{A_F}{A_T} \cdot \frac{k_F}{\delta_F} \tag{9.19a}$$

but if $t_c < t_p$, then:

$$h_k = \frac{A_{W,C}}{A_T}\left(\frac{(k\rho c)_{W,C}}{t_c}\right)^{1/2} + \frac{A_F}{A_T}\left(\frac{(k\rho c)_F}{t_c}\right)^{1/2} \tag{9.19b}$$

where A_T is the total internal surface area.

The experimental data were found to fit satisfactorily to the form of Equation (9.16): multiple linear regression analysis on its logarithm allowed the constants N and M to be evaluated. Figure 9.11 shows the correlation between ΔT and the dimensionless variables X_1 and X_2 when the floor is included in the calculation of h_k and A_T. If the floor is excluded, the correlation is equally good, if slightly different. McCaffrey *et al.* (1981) suggest that the expression:

$$\Delta T = 480 \, X_1^{2/3} \cdot X_2^{-1/3} \tag{9.20}$$

(in which T_0 is taken to be 295 K) describes both situations adequately, and that providing the thermal characteristics of the floor and of the wall and ceiling linings are not too dissimilar, it does not make a significant difference if floor losses are ignored.

Equation (9.20) can be used to estimate the size of fire necessary for flashover to occur. If a temperature rise of 500 K is taken as a conservative criterion for the upper layer gas temperature at the onset of flashover (cf. Hägglund *et al.*, 1974) then substitution for X_1 and X_2 in Equation (9.20) (see Equation (9.15)) gives (after rearrangement):

$$\dot{Q}_c = \left[g^{1/2}(c_p\rho_0)T_0^2\left(\frac{\Delta T}{480}\right)^3\right]^{1/2} (h_k A_T A_w H^{1/2})^{1/2} \tag{9.21}$$

With $\Delta T = 500$ K and appropriate values for g, etc.,

$$\dot{Q}_{FO} = 610(h_k A_T A_w H^{1/2})^{1/2} \tag{9.22}$$

where h_K is in kW/m^2, A_T and A_W are in m^2 and H is in m where $\dot{Q}_{FO}$ (kW) is the rate of heat output necessary to produce a hot layer at approximately 500°C beneath the ceiling. The square root dependence indicates that if there is 100% increase in any of

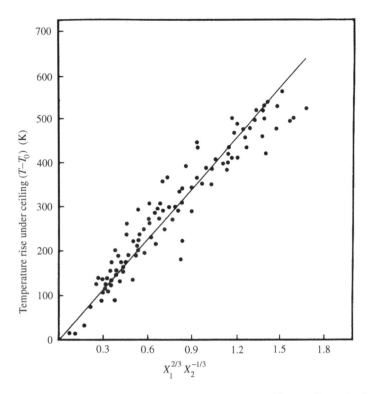

Figure 9.11 Correlation of the upper room gas temperatures with two dimensionless variables (no floor). The line represents the least squares fit of eight sets of data. (From McCaffrey *et al.*, 1981 by permission)

the parameters h_k, A_T or A_w, then the fire will have to increase in heat output by only 40% to achieve the flashover criterion as defined. The effect of changing the thermal properties of the lining material is shown in Figure 9.12. For highly insulating linings, such as fibre insulating board, or expanded polystyrene, the size of the fire sufficient to produce flashover is greatly reduced, even if the linings are assumed to be inert and any contribution to the rate of heat release is neglected.

These correlations are based on a limited set of data and would not be expected to apply to very rapidly growing fires in large compartments significantly different in size and shape from those involved in the correlation. Although two subsets of the data were obtained in small-scale experiments, the remainder refer to near-cubical compartments with heights in the range 2.4 ± 0.3 m. As compartment height is not specified *per se* in these equations, being incorporated into A_T, Equation (9.22) should not be applied to elongated geometries. Similarly, it is likely to break down for compartments with restricted ventilation in which the assumption of a uniform temperature within the layer is no longer tenable (Orloff *et al.*, 1979). Moreover, the data were obtained from fires set in the centre of the compartment. If the fire is next to a wall, or in a corner, the minimum burning rate for flashover is reduced for reasons developed in Section 4.3.4. This is illustrated by results obtained by Lee (1982) who studied experimental fires in full-scale and quarter-scale compartments. The relevant information is summarized

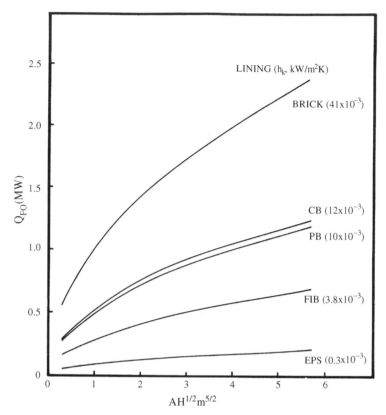

Figure 9.12   Rate of heat release necessary for flashover (according to Equation (9.22) in a compartment 4 m × 6 m × 2.4 m high as a function of $A_w H^{1/2}$ for different boundary materials. CB, chipboard: PB, plasterboard; FIB, fibre insulating board; EPS, expanded polystyrene. The lining materials were assumed to be inert and thermally thick, and the characteristic time of the fire was taken to be $t_c = 1000$ s. The values for h_k are suspect (see original references). (From McCaffrey *et al.*, 1981, by permission)

in Table 9.1: these refer to a full-scale compartment although most of the data were scaled up from the small-scale tests.

Mowrer and Williamson (1987) have shown how this effect can be quantified using plume theory (Morton *et al.*, 1956), although they predict much greater differences in the ratios $\dot{Q}_{FO}(\text{wall})/\dot{Q}_{FO}(\text{centre})$ and $\dot{Q}_{FO}(\text{corner})/\dot{Q}_{FO}(\text{centre})$ than were suggested by Lee (1982).

Table 9.1   Variation of $\dot{Q}_{FO}$ with location of the fire[a]

Location of fire	$\dot{Q}_{FO}/\text{kW}$
In the centre	475
Next to a wall	400
In a corner	340

[a] 3 m × 3 m × 2.3 m high compartment (Lee, 1982).

Equations (9.8) and (9.22) are compared in Figure 9.13 for Hägglund's experiments (Hägglund *et al.*, 1974), which form one of McCaffrey's databases (McCaffrey *et al.*, 1981). Line A represents $\dot{Q}_{FO} = \dot{m}_{limit} \cdot \Delta H_c$, where $\dot{m}_{limit}$ is calculated from Equation (9.8) and ΔH_c is taken to be the heat of combustion of the volatiles (~15 kJ/g, p. 185). Line B is calculated from Equation (9.22). The large difference between these arises because a very conservative figure is taken for the ceiling temperature at the onset of flashover ($\Delta T = 500$ K), whereas Hägglund *et al.* (1974) observed 600°C (i.e. $\Delta T \sim 600$ K). The agreement is greatly improved if ΔT is taken to be 600 K in the derivation of Equation (9.22) (line C in Figure 9.13). As would be expected Equation (9.22), as written, will underestimate the critical rate of heat production. This diagram also shows (line D) Babrauskas' Equation (9.9) (Babrauskas, 1980b): this was deduced theoretically, and falls below the correlation of McCaffrey *et al.* (1981) (Equation (9.22)), which is already conservative.

Equation (9.22) offers a method of identifying the maximum rate of heat release that may be permitted in a room with a view to avoiding flashover. However, its application to the assessment of the flashover potential of 'real' compartments requires information on the rates of heat release during the burning of the contents, both as separate items and in combination, as $\dot{Q}_{FO}$ may only be achieved if the fire has spread to involve more than one item. A simple logic diagram to illustrate this point is shown in Figure 9.14.

Before the development of the technique of Oxygen Consumption Calorimetry (Huggett, 1980), rates of heat release from individual items of furniture had to be

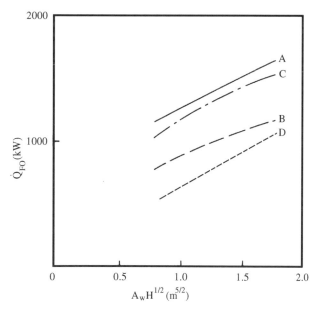

Figure 9.13 Comparison of Equations (9.8) and (9.22) for data of Hägglund *et al.* (1974). A, Equation (9.8), with the heat of combustion of wood equal to 15 kJ/g; B, Equation (9.22) with $h_k A_T = 2.03$ kW/K (including floor); C, as B, but with $\Delta T = 600$ K in Equation (9.21) (McCaffrey *et al.*, 1981); D, Equation (9.9), Babrauskas (1980b)

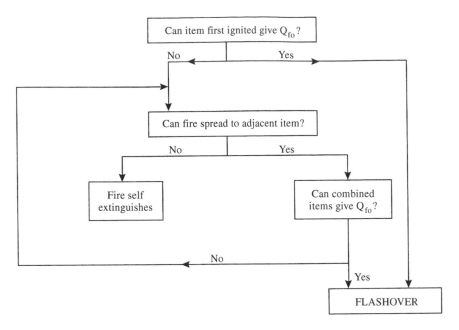

Figure 9.14 ·Logic diagram for flashover in a compartment

calculated from data on their mass loss rates (e.g. Quintiere, 1976), assuming an 'effective' heat of combustion (Section 1.2.3). It is now possible to measure the rate of heat release of full-scale items of furniture using large heat release calorimeters (Babrauskas *et al.*, 1982; Nordtest, 1991; Ames and Rogers, 1990; Babrauskas, 1992b). The method involves burning an item below a hood from which the fire products are removed through a duct (Figure 9.15(a)). The temperature and flowrate of the gases in the duct are measured continuously, as are the concentrations of oxygen, carbon dioxide and carbon monoxide. The rate of heat release is calculated on the basis of the oxygen deficiency, taking into account incomplete combustion of CO to CO_2 (Section 1.2.3). As the combustion products are removed continuously, there is no thermal feedback from the enclosing boundary or an accumulating smoke layer. However, Babrauskas (1980b) argues that the effect of thermal feedback will be small up to flashover, and Quintiere (1982) quotes evidence that the rate of burning shortly before flashover is 'nearly equal to the free burn value'. On the other hand, Alpert (1976) has reported an increase of 150–200% in the rate of burning of individual items just prior to flashover. These two statements are not necessarily incompatible and may simply reflect different definitions of the 'onset' of flashover. Inevitably, the rate of burning will increase as the flashover transition is entered. Recently, data from the CBUF Project — the European research programme into the combustion behaviour of upholstered furniture (Sundström, 1995) — have indicated that the heat release rate for a series of 27 items of upholstered furniture in the ISO Room Test (Figure 9.15(b)) (ISO, 1993b) is on average only 10% higher than in the Furniture Calorimeter (Nordtest, 1991). This is consistent with the observations of Babrauskas (1980b) and Quintiere (1982), but it should be noted that none of the CBUF items gave rise to flashover in the room test.

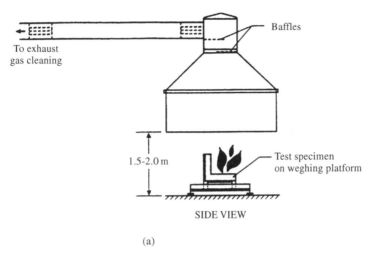

SIDE VIEW

(a)

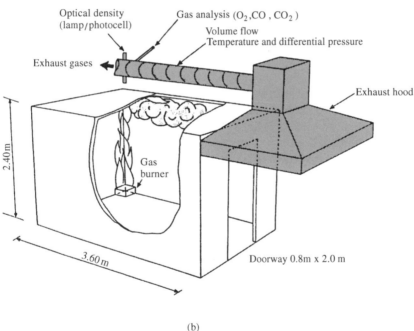

(b)

Figure 9.15   (a) The Nordtest furniture calorimeter (Nordtest, 1991). By permission of Elsevier
(b) The ISO Room-corner test. The gas burner was run according to the Nordtest
method for 10 min at 100 kW and then for another 10 min at 300 kW (ISO,
1993b). By permission of John Wiley & Sons

Examples from the CBUF Report of the heat release rate curves of two armchairs are
given in Figure 9.16: this illustrates the effect of changing from a high resilience foam
to a 'combustion modified' foam. The possibility of deducing full-scale fire behaviour
of items of upholstered furniture from data obtained from the Cone Calorimeter was
examined in some detail in the CBUF project (Sundström, 1995), but at the present

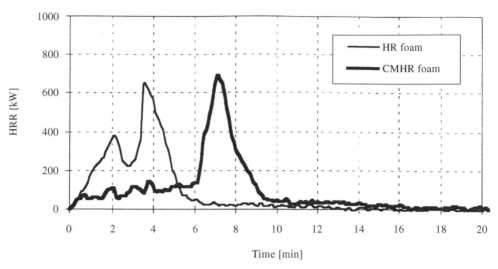

Figure 9.16   Heat release curves of two armchairs covered with cotton fabric, one filled with high resilience PU foam, the other with combustion modified HR foam (Sundström, 1995) © Interscience Communications

time it must be concluded that it is not possible to predict reliably the rate-of-heat-release curves from such bench-scale tests. Earlier, Babrauskas and Walton (1986) proposed a correlation method which was based on a rather limited, contemporary set of data, but in view of the large variety of upholstered items that are now in existence, there seems to be no reliable alternative to carrying out full-scale burns at the present time.

It will be noted that a significant period of steady state burning for a given item may not occur (e.g. see Figure 9.16), and the peak rate of heat release may not be sufficient to create the conditions for flashover: however, such information will be indicative of the potential of the item to contribute to the flashover process. The logic diagram of Figure 9.14 can be applied to assess the magnitude of the problem in a given space, recalling that flame spread to adjacent items may require close proximity (<1 m, according to Babrauskas (1982), see Section 9.2.4) or the involvement of a combustible floor covering. When two or more items are burning, there may be enhancement of the total rate of heat release due to cross-radiation (see Section 9.2.3). This cannot be generalized, but it has to be taken into account in the assessment of the flashover potential of a given space.

Data are now accumulating on the rate of heat release of a variety of items of furniture and building contents, including upholstered furniture, mattresses, pillows, wardrobes, television sets, curtains, waste containers, and Christmas trees (Babrauskas, 1995a). Some of these are summarized in Table 9.2, but the reader is referred to the source documents for full information.

The importance of wall linings in fire development has been recognized for several decades. The first 'reaction to fire' tests were designed to identify materials which would present a hazard with respect to life safety (e.g. BSI, 1987, 1989; ASTM, 1995a), but the results were apparatus-dependent (Emmons, 1974) and were quite unsuitable

Table 9.2 Maximum (peak) rates of heat release from some typical items (see also Babrauskas, 1995a)

Item	Maximum rate of heat release (kW)	Source
Chair, wooden frame, HR foam, cotton fabric	650	Sundström (1995)
Chair, wooden frame, CMHR foam, cotton fabric	700	Sundström (1995)
Latex foam pillow, 50/50 cotton/polyester fabric	117	Babrauskas (1984/85)
Wardrobe, 68 kg, 12.7 mm thick plywood	3500	Lawson *et al.* (1983)
Curtain (closed), 117 g/m^2 cotton/polyester	267	Moore (1978)
Curtain (open), 117 g/m^2 cotton/polyester	303	Moore (1978)
Christmas tree (dry), 7.0 kg	650	Ahonen *et al.* (1984)
Waste container (0.63 kg polyethylene) with empty milk cartons (0.41 kg)	13	Ahonen *et al.* (1984)

for engineering applications. They could be used only to 'rank' materials, and arbitrary limits of behaviour in the test had to be set empirically. The new generation of tests, spawned out of these concerns by the International Organization for Standardization (ISO, 199?, 1993a, 1997a), are now well established and, for the first time, provide data which can be used in engineering calculations. The principal tests are the Cone Calorimeter (Babrauskas, 1984, 1995b; ISO, 1993a) and the LIFT Apparatus developed by Quintiere (1981) (ISO, 1992; ASTM, 1990b). There has been considerable success in using data from the Cone Calorimeter (CC) to model upward flame spread on vertical surfaces (Section 7.3) which has prompted work to investigate if this test protocol could be used to assess the likely fire performance of wall lining materials. There have been two approaches: in one, the CC data are used as the basis for a correlation which can be used to rank materials, while in the second, the data are used in an appropriate model to predict the rate of fire development in a room-corner assembly. Both approaches have been studied with reference to standard room-corner tests such as Nordtest Method NT FIRE 025 (ISO, 1993b). Thus, Östman and Nussbaum (1988) found that the quantity

$$2.76 \times 10^6 \times \frac{t\sqrt{\rho}}{A} - 46 \quad \text{seconds}$$

(where t is the time to ignition (s) at 25 kW/m^2, A is the heat released during the peak period (J/m^2) at 50 kW/m^2 in the CC, and ρ is the density of the material (kg/m^3)) produced a ranking order for 13 materials which correlated very well with that based on the time to flashover when the same materials were tested according to NT FIRE

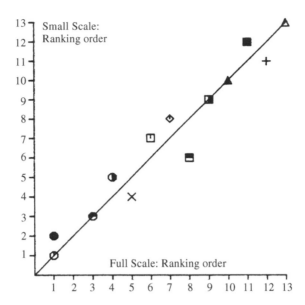

Figure 9.17 Comparison of the ranking orders obtained for wall-covering materials in the Cone Calorimeter and in the ISO Room Fire Test (Östman and Nussbaum, 1988) © Hemisphere

025 (Figure 9.17). With such a correlation, the Cone Calorimeter could be adopted as the means of ranking lining materials, avoiding the expense of full-scale testing. However, this method is less flexible than methods that use mathematical models to predict the rate of fire growth from bench-scale test data. A number of these have been developed to predict response in the room-corner tests (ISO, 1993b; ASTM, 1982) (e.g. Magnusson and Sundstrom, 1985; Cleary and Quintiere, 1991; Karlsson, 1993; Quintiere, 1993). Karlsson (1993) has shown how a graphical representation developed by Baroudi and Kokkala (1992) can be used to identify lining materials which will give flashover in the room-corner test, essentially by examining whether or not vertical flame spread will accelerate (Figure 9.18). For those materials which gave flashover in the room-corner test, it was found that

$$\lambda \tau \leq (1 - \sqrt{a})^2 \tag{9.23}$$

where $\lambda(\mathrm{s}^{-1})$ is the decay coefficient as shown in Figure 9.19, τ is the time to ignition (s) and $a = K\dot{Q}_{\max}(\dot{Q}_{\max}$ being the maximum rate of heat release), all at 50 kW/m² in the Cone Calorimeter. Continuing success with the development of such models will surely lead the way to a more rational approach to the selection of materials which can be used in fire safety engineering design and analysis.

9.2.3 Factors affecting time to flashover

As the duration of the pre-flashover fire has direct relevance to the safety of life (Equation (9.1) *et seq.*), it is important to know how the fuel and ventilation parameters affect the initial rate of growth. In a major research programme (the 'Home Fire

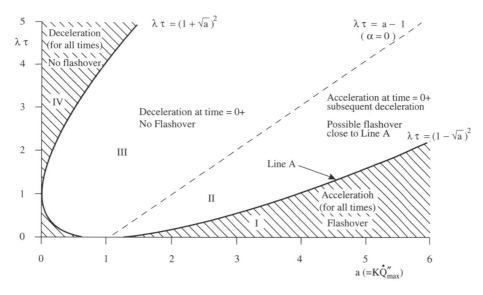

Figure 9.18 Regions of flame front acceleration and deceleration (see Equation (9.23)) (Karlsson, 1993). By permission of Elsevier

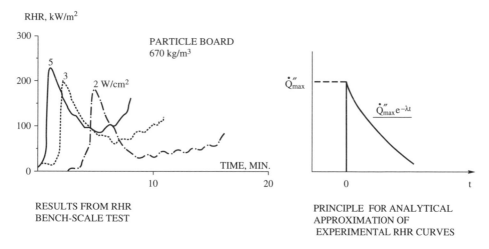

Figure 9.19 Representation of the rate of heat release from a material (Karlsson, 1993). By permission of Elsevier

Project') (e.g. Croce, 1975) carried out jointly by Harvard University and Factory Mutual Research Corporation, a wealth of data was obtained on the growth of fire in a full-scale, fully furnished bedroom. While this has encouraged further study of many of the factors recognized as contributing to fire growth (e.g. Quintiere *et al.*, 1981) and has enabled sophisticated mathematical models of fire behaviour to be developed (e.g. Mitler and Emmons, 1981; Mitler, 1985; Friedman, 1992; Quintiere, 1995b), there are so many potential variables it is difficult to determine which are the most significant parameters. A systematic study of all the variables requires that a very large number of experiments be carried out. The Fire Commission (W14) of the Conseil Internationale du Bâtiment (CIB) launched such a programme in the late 1960s in which nine laboratories around the world participated (Japan, The Netherlands, Australia, USA (FMRC and NBS), UK, Germany, Canada and Sweden). The secretariat, which was located at the Fire Research Station, UK, coordinated the experimental programme as well as the collection and analysis of the data (Heselden and Melinek, 1975). It was decided to carry out tests using small-scale compartments with wood cribs as the fuel bed, looking at the effects of changing each of the eight variables listed in Table 9.3. Each was studied at two levels, as shown. The number of individual tests carried out was $2^8 = 256$, which were allocated randomly around the nine laboratories: each one was duplicated and some additional tests were performed at FMRC and NBS. The times at which each of the following events occurred were noted in each experiment, namely:

$t_f =$ the time for flames to reach the ceiling;
$t_2 =$ the time to the final transition from slow to fast rate of spread over the
      upper surface of the crib(s); and
$t_3 =$ the time to flaming over the whole of the upper surfaces of the crib(s).

The last (t_3) was taken as the time to flashover (Heselden and Melinek, 1975). A multiple regression analysis of the results led to the following conclusions:

(i) t_3 was not significantly affected by the shape of the compartment;

(ii) t_3 was only slightly dependent on the size of the ventilation opening and the continuity of the fuel, although with the former there may have been some interaction with the laboratory conditions;

(iii) t_3 was much more sensitive to the position and the area of the ignition source, the height of the fuel bed, the bulk density of the fuel and the nature of the lining material. Of these, the last was *not* the most significant.

Table 9.3   CIB study of the pre-flashover fire[a]

Variable	Level 1	Level 2
Shape of compartment[b]	$1 \times 2 \times 1$	$2 \times 1 \times 1$
Position of ignition source[c]	Rear corner	Centre
Fuel height	160 mm	320 mm
Ventilation opening	Full width	Quarter width
Bulk density of fuel[d]	20 mm	60 mm
Fuel continuity	One large crib	21 small cribs
Lining (walls and ceiling)[e]	None	Hardboard
Ignition source area[c]	16 cm²	144 cm²

[a] The fuel was in the form of wood cribs (beech). The sticks were 20 mm square and the moisture content was ~10%.
[b] Figures refer to width × depth × height (all in metres).
[c] The ignition source was a tray of alcohol.
[d] Figures refer to stick spacing.
[e] The inner surface of the compartment (unlined) was asbestos millboard.

The effects of the latter group can be understood in terms of a simple fire growth model. The following refer to first-order effects (interactions are considered subsequently):

(a) *Ignition source* t_3 was shorter with a central ignition source as the area of the initial fire increases more rapidly. Similarly, the larger area of ignition gave a shorter t_3 as initially a larger area of fuel was involved.

(b) *Fuel height* With a high fuel bed, flames reach the ceiling much more rapidly, thereby promoting the spread of fire over the combustible surfaces at an earlier stage.

(c) *Bulk density* The cribs of low bulk density tend to spread fire more rapidly (Section 7.4), hence the diameter of the fire increases at a greater rate and flashover is achieved much earlier. In terms of a real fire, this might correspond to fire spread between adjacent items of low thermal capacity.

(d) *Lining material* A combustible lining material reduces the time to flashover, but this variable was not the most significant. Bruce (1953) noted that in a full-scale compartment fire with central ignition, combustible wall linings did not become involved until the fire was well advanced (after the flames touched the ceiling).

Several important interactive effects were noted, some of which are intuitively obvious. The most significant was that between the position of the ignition source and the nature of the wall lining material. If the lining is combustible and becomes involved as the result of direct ignition from a source in the corner, then the time to flashover is greatly reduced. A similar, but less dramatic, interaction exists between height and bulk density of the fuel bed. While these results refer to small-scale tests, relevance to 'real' fires is readily deduced.

However, not all factors that might be deemed to be important were examined. The most serious omission was the height of the ceiling, but in addition the effects of moisture content of the fuel, the relative humidity and air movement (draughts) within the enclosure were not considered. During the period 1973/74, three tests were carried out at Factory Mutual Research Corporation in association with Harvard University on the development of fire in a bedroom, each test furnished and ignited identically (Croce, 1975). These showed a variation in the times to flashover (Table 9.4), the greatest difference being between Test 1 and Tests 2 and 3. It was found that although the furnishings between Tests 1 and 2 were similar, they were not identical. The

Table 9.4 Times to flashover in the Factory Mutual 'bedroom fire tests' (Croce, 1975)

Test number	Date	Time to flashover (s)[a]
1	11 July 1973	1055
2	24 July 1974	429
3	31 July 1975	391

[a] Flashover in Test 1 was well defined. Those given for Tests 2 and 3 are averaged values of time to 'equivalent involvement' (e.g. ignition of a book lying on the floor) as the transition was not as 'dramatic'. (Croce, personal communication.)

furnishings for Test 3 were carefully selected to be the same as in Test 2, yet there was still a difference in the time to flashover. This was thought to be due largely to a difference in relative humidity—by a process of elimination. These results were the first to indicate that the early stages of a developing fire are very sensitive to random variations in a number of factors (such as relative humidity) to the extent that it may not be possible to predict the earliest stages of development of a fire with any certainty. Indeed, in the CBUF project (Sundström, 1995), the investigators found that for fires involving upholstered furniture, the stage from ignition to 50 kW was not reproducible, although great care was taken to try to achieve identical conditions in each replicate test. The stage from 50 kW to 400 kW was much more repeatable. The implication here is that once a fire has grown beyond a certain size (taken as 50 kW in this case), it is powerful enough to overcome minor stochastic influences.

Another factor which can affect the time to flashover is the thermal inertia ($k\rho c$) of the compartment boundary (cf. Figure 9.12). This was first reported in the 1950s, following some full-scale fires carried out at the Fire Research Station, UK, in a compartment 4.5 m square by 2.7 m high, containing wooden furniture at a density of 22 kg/m^2 (5 lb/ft^2). The time to flashover varied enormously with the density of the wall lining, as can be seen in Table 9.5 (Joint Fire Research Organisation, 1962). Similar results from small-scale compartment fires were obtained by Waterman (Pape and Waterman, 1979). Limited evidence of the effect was found in some extra experiments carried out at NBS as part of the CIB programme (Heselden and Melinek, 1975). Recent theoretical work by Thomas and Bullen (1979) suggests that the time to flashover is directly proportional to the square root of the thermal inertia, or a lesser dependence for a fast growing fire.

The importance of this effect on fire development in buildings can be overstated. The need to conserve energy has led to increased thermal insulation in all types of building, but it is only if the insulating material (of low $k\rho c$) is exposed as the internal surface of the enclosure that a marked effect on the time to flashover would be observed. Normally the insulation is protected by at least a layer of plasterboard so that, providing the thermal penetration time (t_p) of the protective layer is greater than the characteristic time of the fire (t_c) as defined by McCaffrey and co-workers (1981), then the insulation

Table 9.5 Variation of time to flashover with the density of the wall lining material (Joint Fire Research Organisation, 1962)

Nature of wall finish	Density[a] (kg/m^3)	Time to flashover (min)
Brick	1600	23.5
Lightweight concrete A	1360	23.0
Lightweight concrete B	800	17.0
Sprayed asbestos	320	8.0[b]
Fibre insulating board[c]	(~300)	6.75

[a] Details are not available on k and c to allow the thermal inertia to be calculated. However, density on its own is an adequate indicator.

[b] A repeat of this experiment with the same finish which had been completely dried during the first test gave a time to flashover of 4.5 min.

[c] This combustible lining is included for comparison. Its density is not known, but it is noted that the sprayed asbestos was chosen to have the same density and thermal conductivity as the fibre insulating board (Joint Fire Research Organisation, 1961).

will not affect the time to flashover significantly. On the other hand, insulation will tend to increase the 'severity' of a fully developed fire (Section 10.3).

9.2.4 Factors affecting fire growth

The CIB programme involved wood crib fires in small-scale compartments, and although fuel continuity was varied by comparing the behaviour of one single crib with that of a large number of small cribs, the results give little information about fire spread between isolated 'packets' of fuel in a real fire situation. Unless the item first ignited is capable of producing (and at least sustaining) the necessary heat output for flashover to occur, then development to flashover will require that other items of fuel become involved (Figure 9.14). Only in this way can the rate of burning increase.

Whether or not an adjacent item will ignite will depend on its separation from the item already burning. It may be sufficiently close and in a suitable configuration for direct flame impingement to occur (Section 4.3.4), but if this is not possible, fire can only spread by a mechanism involving radiant heat transfer. This was originally considered by Theobald (1968) who showed that the radiation from a fire involving a traditional upholstered chair could ignite cotton cloth 0.15 m away, while a burning wardrobe could ignite the same material 1.2 m away. Pape and Waterman (1979) have commented, on the basis of a series of full-scale tests, that fire will not spread from a single upholstered chair to a neighbouring one if the separation is more than 30 cm.

However, this statement is too general as the properties of both the burning item and the 'target' must determine the outcome. Fang (1975b) and Babrauskas (1982) have attempted to 'map' the distribution of radiant flux around items of burning furniture ranging from stacking chairs to storage cabinets. Not unexpectedly, it was found that the radiant heat flux at a distance was a function of the rate of burning. Items which burned rapidly were capable of giving substantial radiant heat fluxes at up to 1 m from their leading edge. In Figure 9.20, the distribution of radiant heat flux near a hospital couch (wicker frame, polyurethane foam, covered by polypropylene fabric) at peak burning, is shown as a function of distance from the front edge and height above the platform on which the couch was resting. In principle, this type of information could be obtained for all items of furniture, and compared with the 'ignitability' of target materials, perhaps expressed as the heat flux necessary for pilot ignition to occur within a certain period of time (say 40 s). While this will depend substantially on the thermal properties of the 'target' (Section 6.3), Babrauskas (1982) concluded that if the 'target item' was more than 1 m away from a burning upholstered chair, ignition would be unlikely. Even with the most rapidly burning item of furniture, a heat flux exceeding 20 kW/m^2 was not found beyond 0.88 m. Information of this type is necessary to allow an assessment to be made of the potential for fire to spread directly from one fuel 'package' to the next. While it may be possible to use the theoretical approach described in earlier chapters to calculate rates of burning and radiant heat flux from flames above burning items (Section 4.4.1), there is a continuing need at present to use empirical data. Additional information of this type is still required, but combined with ignitability data, it should be possible to examine flashover potential of a compartment from a more rational point of view.

However, fire growth from ignition may proceed by direct flame spread between contiguous items. The rate of growth will be strongly dependent on the configuration

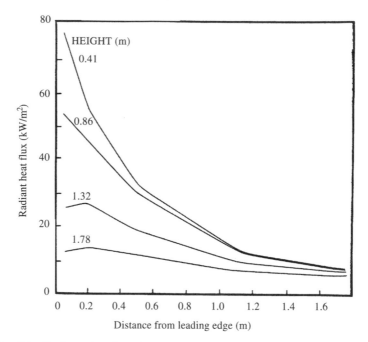

Figure 9.20   Distribution of radiant heat flux at various heights near a couch at maximum rate of burning (polyurethane foam with polypropylene fabric, wicker frame) (Babrauskas, 1982, by permission)

of the combustible surfaces. There are numerous scenarios which are associated with rapid flame spread, the most obvious of which is the combustible wall lining, but there are many others which are equally, if not more, hazardous. These include any configuration in which heat is conserved in the vicinity of the burning surface(s), e.g. by cross-radiation. Examples include ignition of combustible materials under a wooden floor where the intensity of burning will be enhanced by cross-radiation within the confined space (Popplewell, 1986); a fire in a corner of a room lined with combustible wall linings (see Section 7.2.4(b)); and ignition between two closely spaced items of furniture, such as a bed and a wardrobe (Morris and Hopkinson, 1976). Indeed, any confined geometry is likely to lead to rapid flame development, for two reasons: (a) conservation of heat in the vicinity of combustible surfaces; and (b) creation of flow dynamics which generate high rates of heat transfer, thus promoting rapid flame spread (e.g. Foley and Drysdale, 1995). Thus, fire development in a cavity with combustible linings and/or containing combustible materials (such as cables, etc.) will be rapid once it gets hold. Indeed, many multiple-fatality fires in buildings have been shown to have started (often unseen) in cavities; when the fire eventually breaks out, not only is it large, but it may have spread far beyond its origin. Examples include the fire at the Summerland Leisure Complex, Isle of Man, in 1973 (Silcock and Hinkley, 1974; Rasbash, 1991), the Beverley Hills Supper Club fire of 1977 (Best, 1978), the Bradford City Football Stadium fire of May 1985 (Popplewell, 1986), and the MGM Grand Hotel fire of 1981 (Best and Demers, 1982). Configurations that promote fast fire spread have also been implicated in multi-fatality fires. Fast fire growth is associated with high rack storage (e.g. see Sawyer, 1996) and can lead to major losses — for example, a fire in the Donnington Ordnance Depot in Shropshire, UK, was responsible for direct losses of £165M (Andrews, 1987). Historically, there have been few casualties simply because few people are present in such facilities, but in places of public assembly, there are many examples where people have become trapped and have succumbed because fire growth has been so rapid. Examples include the Stardust Nightclub fire in Dublin in 1981 (Rasbash, 1984), and the King's Cross Underground Station fire of November 1987 (Fennell, 1988; Drysdale, 1992). These fast-fire-spread scenarios are often associated with a combination of incorrect or careless use of combustible materials in configurations which are conducive to rapid fire development—the Stardust fire makes an interesting case study (Report of the Official Inquiry, 1982; Building Research Establishment, 1982)—but in the case of the King's Cross fire, the mechanism of spread on the wooden escalator had not been foreseen (the 'trench effect') and its rapidity was quite unexpected (see Section 7.2.1).

Other factors can play an important role in the development of flashover: thus thermoplastics can melt and flow, creating liquid pool fires which can spread to involve other combustible items. Under certain conditions, expanded polystyrene ceiling tiles, once ignited, will rain burning droplets of molten polymer on to the surfaces below. These features can be recognized but are difficult to quantify.

Despite these uncertainties, it is found that the rates of development of many fires approximate to a parabolic growth ('t^2 fire') after an initial incubation period (Heskestad, 1982), thus

$$\dot{Q} = \alpha_f (t - t_0)^2 \qquad (9.24)$$

where α_f is a fire-growth coefficient (kW/s^2) and t_0 is the length of the incubation period (s). This is shown schematically in Figure 9.21. The coefficient α_f appears to lie in the range 10^{-3} kW/s^2 for very slowly developing fires to 1 kW/s^2 for very fast fire growth Table 9.6. The incubation period (t_0) is difficult to quantify (see Section 9.2.3), and will depend *inter alia* on the nature of the ignition source, its location and the properties of the item first ignited. Data are now available on fire growth rates on a number of commodities and fuel arrays which may be expressed in these terms (Babrauskas, 1979, 1983a; National Fire Protection Association, 1993a). This is useful in engineering calculations relating, for example, to the response of ceiling-mounted heat detectors to a growing fire when it is necessary to take the growth rate into account (National Fire Protection Association, 1993a; Evans, 1995).

Table 9.6 Parameters used for 't-squared fires' (Evans, 1995)

Description	Typical scenario	α_f kW/s^2
Slow	Densely packed paper products[a]	0.00293
Medium	Traditional mattress/boxspring[a] Traditional armchair	0.01172
Fast	PU mattress (horizontal)[a] PE pallets, stacked 1 m high	0.0469
Ultrafast	High-rack storage PE rigid foam stacked 5 m high	0.1876

[a] National Fire Protection Association (1993a).

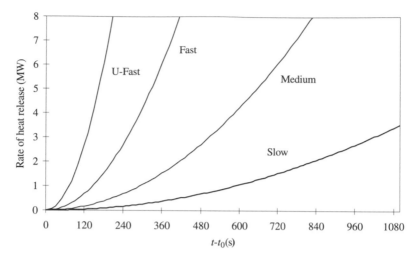

Figure 9.21   Parabolic fire growth from the end of the incubation period ($t = t_0$) according to Equation (9.24) for the fire growth coefficients given in Table 9.6

Problems

9.1   Estimate how long it would take for the concentration of oxygen in a poorly ventilated room, 5 m × 8 m × 3 m high, to fall to 10% by volume (the concentration below which flaming combustion is unlikely to continue under these circumstances) if a settee is burning on its own with an *average* rate of heat release of 300 kW. Assume that the average temperature of the room reaches 500 K and that there is no air entering the room during this period. (It also necessary to assume that $\Delta H_{c,ox}$ remains constant at 13 kJ/g of oxygen.)

9.2   How much oxygen will remain in this room at 10 minutes if the CMHR foam chair referred to in Figure 9.16 burns out in the closed room of Problem 9.1, without involving any other furniture or furnishings? Approximate the rate of heat release as 100 kW for six minutes and 300 kW for four minutes.

9.3   Calculate the size of fire necessary to give flashover in the following rooms, assuming that the basic structure comprises brick walls and concrete slab floor and ceiling:

(a)   4 m × 6 m × 2.5 m high with a single ventilation opening (a door) 0.9 m × 2 m high. The walls and ceiling are lined with plaster.

(b)   As (a) but with a ceiling height of 3.5 m.

(c)   As (a) but with the walls and ceiling lined with 20 mm thick fire-retardant fibre insulation board (ignore possible involvement of the lining).

(d)   As (a) but with one window open (area 2.0 m × 1.5 m high) (see Equation (10.37)).

Assume in all of these that the characteristic time of the fire is 10^3 s.

10
The Post-Flashover Compartment Fire

After flashover has occurred, the exposed surfaces of all combustible items in the room of origin will be burning and the rate of heat release will develop to a maximum, producing high temperatures (Figure 9.1). Typically, this may be as high as 1100 °C, but much higher temperatures can be obtained under certain conditions*. These will be maintained until the rate of generation of flammable volatiles begins to decrease as a result of fuel consumption. It is during the period of the fully developed fire that building elements may reach temperatures at which they may fail. Failure of a structural element may cause local or more general collapse of the building structure. The term is also applied to compartment boundaries, which may or may not be load bearing, yet 'fail' by permitting fire spread into adjacent spaces by flame penetration or excessive transmission of heat. This is the origin of the concept of 'fire compartmention' whose objective is to set a limit to the maximum loss that might be sustained in industrial or commercial premises, by dividing the building into fire compartments, separated by walls, or barriers of appropriate fire resistance (Section 10.4).

Traditionally, fire resistance is measured empirically as the time to failure of the element of structure in a standard furnace test (ISO, 1975; British Standards Institution, 1987b American Society for Testing and Materials, 1995b). This was originally designed to model the thermal environment to which an element of structure would be exposed in a 'real' fire. However, the validity of this method needs to be examined in the light of our knowledge of the behaviour of fully developed fires. Ventilation and the nature, distribution and quantity of fuel all have a significant effect on duration and severity: the concepts of 'fire severity' and 'fire resistance' in relation to real fires require careful definition. These will be introduced and discussed in subsequent sections.

10.1 Regimes of Burning

The first systematic study of the behaviour of the fully developed compartment fire was carried out in Japan in the late 1940s. Kawagoe and his co-workers (Kawagoe,

---

* Temperatures in excess of 1300–1400 °C, sufficient to cause the surface of bricks to fuse (melt), are occasionally encountered. For example, the Summit Rail Tunnel Fire (Department of Transport, 1986) produced sufficiently high temperatures to cause the faces of brick-lined ventilation shafts to fuse.

1958) measured the rate of burning of wood cribs contained within compartments with different sizes of ventilation opening. Full scale and reduced scale tests were carried out, the characteristic dimension of the smallest compartment being less than 1 m. The burning rate ($\dot{m}$) was found to depend strongly on the size and shape of the ventilation opening, the results correlating very well with the relationship

$$\dot{m} = 5.5A_wH^{1/2} \text{ kg/min}$$

$$= 0.9A_wH^{1/2} \text{ kg/s} \tag{10.1}$$

where A_w and H are the area (m^2) and height (m) of the ventilation opening respectively (Figure 10.1). However, the numerical constant is somewhat ill-defined (Thomas *et al.*, 1967; Thomas and Heselden, 1972) and it is found that the correlation only holds over a limited range of values of $A_wH^{1/2}$. The conventional interpretation is that within this range the rate of burning is controlled by the rate at which air can flow into the compartment: such a fire is said to be 'ventilation controlled'. However, if the ventilation opening is enlarged, a condition will be reached beyond which the rate of burning becomes independent of the size of the opening and is determined instead by the surface area and burning characteristics of the fuel. Indeed, it is found that the transition from ventilation to fuel control depends predominantly on the surface area

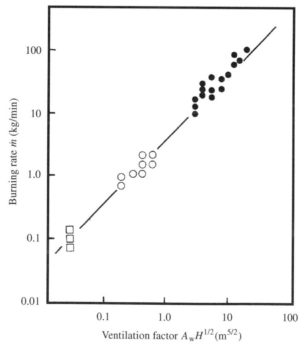

Figure 10.1   Mass burning rate of wood cribs in enclosures as a function of the ventilation parameter, $A_wH^{1/2}$ for ventilation-controlled fires (Equation (10.1)).   ● Full scale enclosures; ○ intermediate scale models; □ small scale models (Kawagoe and Sekine, 1963). (Reproduced by permission of Elsevier Applied Science Publishers Ltd)

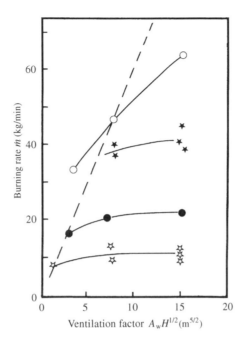

Figure 10.2 Variation of mass burning rate with $A_wH^{1/2}$ for large ventilation openings and different fire loads (wood cribs). ☆ 7.5 kg/m^2; ● 16 kg/m^2; ★ 30 kg/m^2; ○ 60 kg/m^2. Dashed line (— — —) represents Equation (10.1) for the ventilation-controlled fire (Thomas *et al.*, 1967). (Reproduced by permission of The Controller. HMSO. © Crown Copyright)

of combustible material: this was noted by Gross and Robertson (1965), emphasised by Thomas *et al*(1967) (Figure 10.2) and is consistent with the criticality of flashover (Thomas *et al.*, 1980; Bishop *et al.*, 1993).

The ventilation factor $A_wH^{1/2}$ was deduced semi-empirically by Kawagoe (1958): it can be derived by a theoretical analysis of the flow of gases in and out of a burning compartment. The following assumptions must be made:

(a) The gases in the compartment behave as if they are 'well stirred', i.e. their properties are uniform throughout the volume (support for this is to be found in Figure 10.3 in which it is shown that, except near the floor, temperature gradients virtually disappear at flashover (Croce, 1975; Nakaya *et al.*, 1986));

(b) There is no net flow created by buoyancy *within* the compartment;

(c) Hot gases leave the compartment above a neutral plane and cold air enters below it (Figure 10.4);

(d) The flow of gases *in* and *out* of the compartment is driven by buoyancy forces;

(e) There is no interaction between the inflowing and outflowing gases.

The compartment is thus modelled as a 'well-stirred combustion chamber' as shown in Figure 10.4. The induced flows are caused by the differences in pressure between the compartment and the outside, which are a direct consequence of the high internal

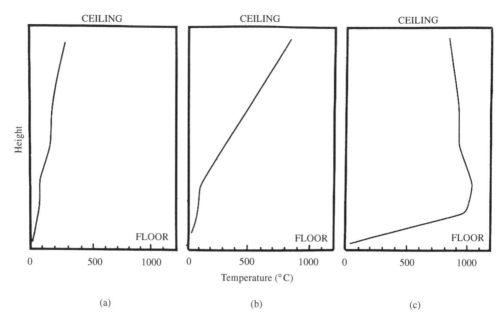

Figure 10.3 Vertical temperature distribution in a compartment fire (data from the Home Fire Project (Croce, 1975)): (a) pre-flashover (5.67 minutes after ignition); (b) incipient flashover (6.11 min); (c) 'time of observed flashover' (7.05 min) (Babrauskas and Williamson, 1978). Reproduced with permission

temperature: this has been studied experimentally by McCaffrey and Rockett (1977) whose results vindicate the theoretical treatment that follows.

The horizontal flow out of the compartment along any streamline above the neutral plane can be calculated from Bernoulli's equation if the pressures inside and outside are known (see Emmons, 1995). Following the treatment presented by Babrauskas and Williamson (1978), the pressure inside the compartment at a height y above the neutral plane (point 1 in Figure 10.4) will be

$$P_1 = P_0 - \rho_1 g y \qquad (10.2a)$$

where P_0 is the atmospheric pressure on the neutral plane (i.e. at $y = 0$). Just outside the ventilation opening, at point 2, the pressure of the issuing jet will be equal to the atmospheric pressure at that level, i.e.

$$P_2 = P_0 - \rho_0 g y \qquad (10.2b)$$

(Note that on the neutral plane $P_1 = P_2 = P_0$ and there is no net flow.) Bernoulli's equation can be used to relate the conditions at points 1 and 2 thus:

$$\frac{P_1}{\rho_1} + \frac{v_1^2}{2} = \frac{P_2}{\rho_2} + \frac{v_2^2}{2} \qquad (10.3)$$

where v_1 and v_2 are the net horizontal flowrates at the two points. Well away from the incoming jet, there will be no directional flow as the mixture inside the compartment is highly turbulent. Thus, v_1 will be zero, and Equation (10.3) can be written:

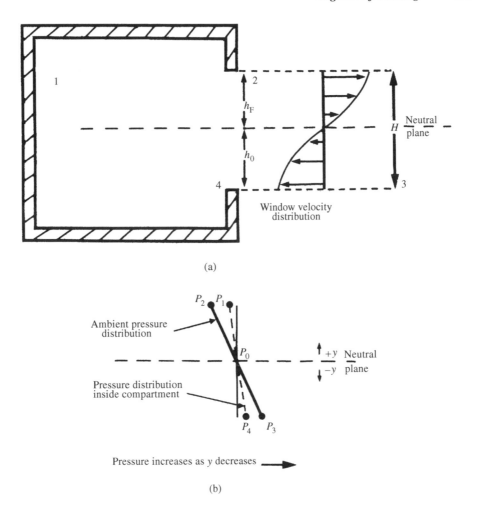

(a)

(b)

Figure 10.4 Buoyancy-driven flows through a ventilation opening during a fully developed fire. (a) Vertical section showing window flows; (b) pressure distributions outside (———) and inside (- - - - -) the compartment (after Babrauskas and Williamson, 1978). Reproduced with permission

$$\frac{P_0 - \rho_1 g y}{\rho_1} = \frac{P_0 - \rho_0 g y}{\rho_1} + \frac{v_2^2}{2} \tag{10.4}$$

in which it is assumed that the gas issuing from the opening at position 2 will be at the same temperature (and density, ρ_1) as the gas at position 1. This rearranges to give

$$v_2 = \left(\frac{2(\rho_0 - \rho_1) g y}{\rho_1} \right)^{1/2} \tag{10.5}$$

A similar analysis can be carried out for the inflowing air, where

$$P_3 = P_0 - \rho_0 g y \tag{10.6a}$$

and

$$P_4 = P_0 - \rho_4 g y, \tag{10.6b}$$

y being negative below the neutral plane. Using the subscripts 'F' (for the compartment gases) and '0' for the ambient air) we then have two equations:

$$v_F = \left(\frac{2(\rho_0 - \rho_F)g y}{\rho_F} \right)^{1/2} \tag{10.7}$$

and

$$v_0 = \left(\frac{2(\rho_F - \rho_0)g y}{\rho_0} \right)^{1/2} \tag{10.8}$$

which refer to the horizontal flows at any height y above or below the neutral plane respectively. These velocities are normally quite low, of the order of 5–10 m/s. Equations (10.7) and (10.8) can then be used to calculate the mass flowrates:

Inflow:
$$\dot{m}_{air} = C_d B \rho_0 \int_{-h_0}^{0} v_0 \, dy \tag{10.9}$$

Outflow:
$$\dot{m}_F = C_d B \rho_F \int_{0}^{h_F} v_F \, dy \tag{10.10}$$

where C_d is a discharge coefficient, B is the width of the window (m), $\dot{m}$ is the mass flow (kg/s) and $h_F + h_0 = H$, as shown in Figure 10.4. These lead to:

$$\dot{m}_{air} = \frac{2}{3} C_d B (h_0)^{3/2} \rho_0 \left(2g \frac{\rho_0 - \rho_F}{\rho_0} \right)^{1/2} \tag{10.11}$$

and

$$\dot{m}_F = \frac{2}{3} C_d B (h_F)^{3/2} \rho_F \left(2g \frac{\rho_0 - \rho_F}{\rho_0} \right)^{1/2} \tag{10.12}$$

If the overall chemical reaction taking place within the compartment can be expressed as

$$1 \text{ kg fuel} + r \text{ kg air} \rightarrow (1 + r) \text{ kg products}$$

or more generally, if the burning is non-stoichiometric

$$1 \text{ kg fuel} + \frac{r}{\phi} \text{ kg air} \rightarrow \left(1 + \frac{r}{\phi} \right) \text{ kg products}$$

where ϕ is a correction factor, then

$$\frac{\dot{m}_F}{\dot{m}_{air}} = \frac{1 + r/\phi}{r/\phi} = 1 + \frac{\phi}{r} \tag{10.13}$$

The height of the neutral plane (h_0) may be expressed as a fraction of the total height of the ventilation opening (H) by substituting $\dot{m}_{air}$ and $\dot{m}_F$ from Equations (10.11) and

(10.12) into Equation (10.13): writing $h_F = H - h_0$ and rearranging,

$$\frac{h_0}{H} = \frac{1}{1 + [(1 + (\phi/r))^2 \cdot \rho_0/\rho_F]^{1/3}}$$ (10.14)

Using typical values of ϕ, r and ρ_F, the ratio h_0/H works out to be 0.3–0.5, consistent with general observations of the fire plume emerging from the ventilation opening of a compartment (Figure 10.7(b) below). If it is assumed as an approximation that $\dot{m}_F = \dot{m}_{air}$ (i.e. $\phi/r = 0$), then substituting h_0 from Equation (10.14) into Equation (10.11) gives

$$\dot{m}_{air} \approx \frac{2}{3} A_w H^{1/2} C_d \rho_0 (2g)^{1/2} \left(\frac{(\rho_0 - \rho_F)/\rho_0}{[1 + (\rho_0/\rho_F)^{1/3}]^3} \right)^{1/2}$$ (10.15)

As the ratio ρ_0/ρ_F normally lies between 1.8 and 5 for postflashover fires (Babrauskas and Williamson, 1978) the square root of the density term can be approximated by 0.21. Then with $\rho_0 = 1.2$ kg/m^3, $C_d = 0.7$ (Prahl and Emmons, 1975) and $g = 9.81$ m/s^2, the rate of inflow of air can be approximated by

$$\dot{m}_{air} \approx 0.52 A_w H^{1/2} \text{ kg/s}$$ (10.16)

If stoichiometric burning occurs within the compartment (i.e. $\phi = 1$ in Equation (10.13)) then for wood the rate of burning must be

$$\dot{m}_b \approx \frac{0.52}{5.7} A_w H^{1/2} = 0.09 A_w H^{1/2} \text{ kg/s}$$

$$= 5.5 A_w H^{1/2} \text{ kg/min}$$ (10.17)

since the stoichiometric air requirement for the combustion of wood is approximately 5.7 kg air/kg wood (Section 4.4.3).*

The remarkable agreement with Kawagoe's original correlation (Equation (10.1)) must be regarded as fortuitous in view of the many simplifying assumptions that are made, but the emergence of the 'ventilation factor' $A_w H^{1/2}$ is significant.

However, by assuming that stoichiometric burning occurs within the compartment (Equation (10.17)), it is implied that the rate of burning is directly coupled to the rate of air inflow. This is difficult to understand as it is known that in a confined situation burning rate is increased by radiative heat feedback from the surroundings (Section 9.1 and Figure 9.2). It is not clear how this could be determined by the ventilation conditions. It seems likely that the relationship described in Equation (10.1) is fortuitous and applies only to wood cribs in which the burning surfaces are largely shielded from the influence of the compartment (see Section 5.2.2). Harmathy (1972, 1978) has developed this concept, proposing that the energy responsible for producing 'the volatiles' during the steady burning period comes principally from surface oxidation of the char within the structure of the crib. He analysed data from a large number of wood crib fires in compartments and showed that the results plotted as $\dot{m}/A_F$ versus $\rho g^{1/2} A_w H^{1/2}/A_f$ (where A_f is the surface area of the fuel) show a clear distinction

---

* As the char burns slowly in comparison, it would be more logical to take the stoichiometric air requirement for wood volatiles (4.6 kg/kg) which leads to $\dot{m}_b = 6.8 A_w H^{1/2}$ kg/min (see Section 4.4.3).

between the 'ventilation-controlled regime' and a 'fuel-controlled regime' in which $\dot{m}$ is independent of the ventilation factor (Figure 10.5). Harmathy has recommended the following be used to distinguish between ventilation- and fuel-controlled fires involving cellulosic (i.e. wood or wood-based) fuels:

Ventilation control:
$$\frac{\rho g^{1/2} A_w H^{1/2}}{A_f} < 0.235 \qquad (10.18a)$$

Fuel control:
$$\frac{\rho g^{1/2} A_w H^{1/2}}{A_f} > 0.290 \qquad (10.18b)$$

the transition being ill-defined (see Figure 10.5). However, as these apply to wood crib fires in compartments, their applicability to 'real' fires is not clear (see Figure 10.11 below). Harmathy has argued most cogently about the unusual properties of char-forming fuels such as wood, but his analysis is too restrictive, particularly as it takes no account of radiative feedback from the environment within the compartment (Thomas, 1975).

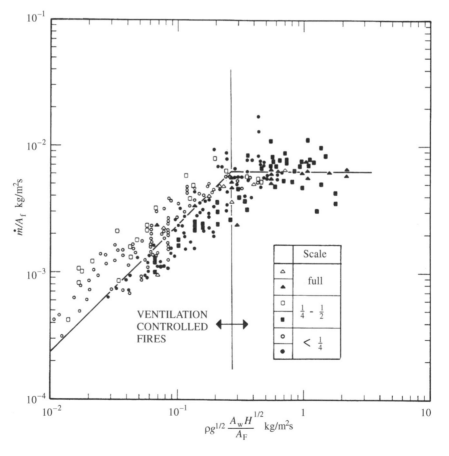

Figure 10.5 Identification of the transition point between ventilation-controlled and fuel-controlled burning for wood cribs, according to Harmathy (1972)

It is important to be able to distinguish between these two regimes as the fuel-controlled fire is generally less severe except in comparison with fires in which the ventilation is very poor. This is illustrated by data obtained in an international programme of small-scale tests co-ordinated by CIB and designed to investigate the factors which influence the behaviour of the fully-developed fire (Thomas and Heselden, 1972) (see Section 10.2). Figure 10.6 shows a plot of average gas temperature inside the compartment during a number of fully developed fires versus $A_T/A_w H^{1/2}$, where A_T is the area of the walls and ceiling of the compartment, excluding the ventilation area, A_w. Values of $A_T/A_w H^{1/2}$ less than 8–10 m$^{-1/2}$ correspond to fuel controlled fires.

In the fuel controlled regime, the excess air entering the compartment has the effect of moderating the temperature. The cross-over point between the two regimes will depend on the relationship between $\dot{m}$, the rate of volatilization of the fuel (kg/s), i.e. the 'burning rate', and the rate of inflow of air (kg/s). Ideally, if

$$\dot{m} > \dot{m}_{air}/r \text{ kg/s} \tag{10.19}$$

where r is the stoichiometric air/fuel ratio (Section 1.2.3), and $\dot{m}_{air}$ is given by Equation (10.16), then the burning is ventilation controlled. Similarly, the fire is fuel controlled if

$$\dot{m} < \dot{m}_{air}/r \text{ kg/s} \tag{10.20}$$

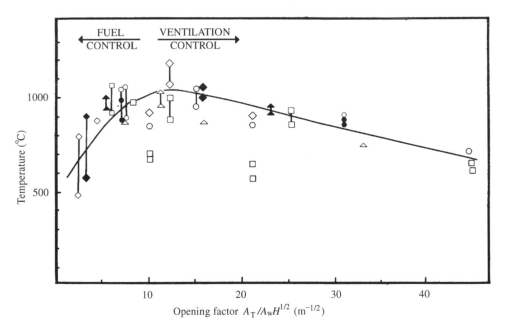

Figure 10.6 Average compartment temperatures during the steady burning period for wood crib fires in model enclosures as a function of the 'opening factor' $A_T/A_w H^{1/2}$. Symbols refer to different compartment shapes (see Table 9.3): ○ 1 × 2 × 1; △ 2 × 2 × 1; ◇ 2 × 1 × 1; □ 4 × 4 × 1. Solid points are means of 8–12 experiments (Thomas and Heselden, 1972). (Reproduced by permission of The Controller, HMSO. © Crown Copyright)

although this argument tacitly assumes that the rate of the reaction between the fuel volatiles and oxygen from the air is infinitely fast. That this is not the case is shown by the fact that flames may be seen issuing from the ventilation opening even under conditions when the fire is perceived to be under 'fuel control', i.e. the ratio of the rate of entry of air to the rate of release of volatiles is slightly greater than the stoichiometric ratio, r (Thomas *et al.*, 1967; Harmathy, 1978). The rate at which the volatiles are oxidized is finite and it takes a significant period of time for them to burn even when mixed with sufficient air. In effect, gases emerge from the opening while they are still burning. On the other hand, if $\dot{m} > \dot{m}_{air}/r$ (Equation (10.19)), excess vapours flow from the opening and burn as they mix with air outside the compartment (Section 10.6).

10.2   Fully-developed Fire Behaviour

Early work on compartment fires involved the use of wood cribs as the fuel bed, simply because this provided a means of achieving reproducible fires. However, as noted earlier, the burning surfaces are largely shielded from the environment within the compartment and consequently the rate of burning is relatively insensitive to the thermal environment. Thomas and Nilsson (1973) have pointed out that a third regime of burning must be considered in which the rate is controlled by the crib structure: this will occur only if the crib 'density' is high (Section 5.2.2(b)).

Thus the 'coupling' between burning rate and ingress of air, which is implied in Kawagoe's relationship (Equation (10.1)) is at least in part a consequence of the unusual burning properties of this type of fuel bed. When 'real' fire loads are used, particularly if these involve non-cellulosic materials, there is no reason why such coupling should exist. This point is emphasized in Figure 10.7 which shows marked

(a)   (b)

Figure 10.7   The effect of a large exposed fuel surface area on fire behaviour. (a) Fuel control regime, 15 kg/m². Fuel in the form of wood cribs, $A_f = 55$ m²: no external flaming. (b) Ventilation control regime, 7.5 kg/m². Fuel was fibre insulating board, lining the walls and ceiling, $A_f = 65$ m²; external flaming lasted for 5.5 minutes (Butcher *et al.*, 1968. Reproduced by permission of The Controller, HMSO. © Crown Copyright)

differences in behaviour between two fires in identical compartments, one involving twice as much fuel (wood) as the other. In the former, the wood was in the form of a crib, while in the other it was present as the wall lining material (Butcher *et al.*, 1968). The large exposed area of the fuel in the latter case produced flashover with flames emerging from the window, while the wood crib burned as a fuel-controlled fire. Harmathy's method (Equation (10.18)) does not distinguish between exposed surfaces and the internal, shielded surfaces of wood cribs.

Any realistic theoretical treatment of the post-flashover fire must consider rate of burning and rate of ventilation separately (Bullen, 1977a; Babrauskas and Williamson, 1978; 1979). Bullen (1977a) carried out an analysis of steady-state compartment fires involving liquid fuels in which detailed heat and mass balances were considered. While of limited application, this model is useful for demonstrating the significance of different parameters. Thus, the rate of burning is calculated from Equation (5.24), i.e.

$$\dot{m} = \frac{\dot{Q}_F'' + \dot{Q}_E'' - \dot{Q}_L''}{L_v} A_F \quad \text{kg/s} \qquad (10.21)$$

and the rate of air inflow from

$$\dot{m}_{air} = 0.5 \, A_w H^{1/2} \quad \text{kg/s} \qquad (10.16)$$

which are quite independent. If the burning was under fuel-rich conditions, Bullen assumed that all the air was 'burned' within the compartment, giving $\dot{m}_{air} \cdot \Delta H_{c,air}$ as the rate of heat release within the compartment, where $\Delta H_{c,air}$ is the heat of combustion per unit mass of air consumed (kJ/kg) (Section 1.2.3). The net heat flux to the fuel surface is estimated by assuming that the gas within the compartment is 'grey' and that the walls are also radiating. Two simultaneous equations in T_g (the gas temperature) and T_1 (the wall temperature) were solved, and $\dot{m}$ calculated from Equation (10.21). The effects of varying $\Delta H_{c,air}$ and the heat required to produce the volatiles (L_v) were calculated: the results are shown in Figures 10.8 and 10.9. In Figure 10.8 the variation of burning rate with $A_w H^{1/2}$ is shown for four liquid fuels (three of which are hypothetical). Curves (a) and (b) refer to two fuels with the same $\Delta H_{c,air}$ but with widely differing values of L_v. As would be expected, the fuel with the lower value of L_v (i.e. curve (a)) gives a much higher burning rate ($\dot{m}$). Similarly, reducing $\Delta H_{c,air}$ while keeping L_v constant (compare (a) and (d)) has a very dramatic effect on $\dot{m}$, although in fact the 'heat of combustion of air' does not vary much from 3000 kJ/kg (Table 1.13). However, this observation would have relevance to situations involving vitiated air (less than 21% oxygen).

The effect on steady state gas temperatures within the compartment is shown in Figure 10.9. Bullen's calculations show that the fuel with the lower value of L_v gives a lower temperature (cf. curves (a) and (b)). The reason for this is that the excess volatiles released from the fuel leave the compartment unburnt, thereby increasing the 'convective' heat loss and also the volume of outflow, which restricts the inflow of fresh air (see Equation (10.14)). Thus the rate of heat release within the compartment is reduced. (On the basis of the experimental work quoted earlier (Figure 10.6) T_g should fall with increasing $A_w H^{1/2}$ once the boundary between ventilation control and fuel control has been crossed ($A_w H^{1/2} \approx 6 \text{ m}^{5/2}$). However, this is not taken into account in the model.)

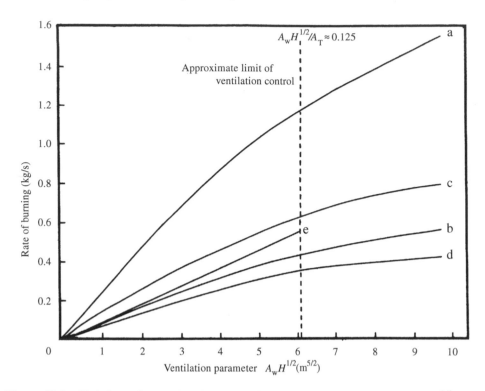

Figure 10.8   Variation of mass burning rate with the ventilation parameter $A_w H^{1/2}$ for a range of fuels: (a) $\Delta H_{c,air} = 3$ kJ/g, $L_v = 0.5$ kJ/g; (b) $\Delta H_{c,air} = 3$ kJ/g, $L_v = 2.0$ kJ/g; (c) $\Delta H_{c,air} = 2.5$ kJ/g, $L_v = 0.85$ kJ/g (values for industrial methylated spirits); (d) $\Delta H_{c,air} = 1.5$ kJ/g, $L_v = 0.5$ kJ/g; (e) mass burning rate according to Equation (10.1). The limit of ventilation control $A_w H^{1/2}/A_T \approx 0.125$ is shown as the vertical dashed line (Bullen. 1977a). (Reproduced by permission of the The Controller, HMSO. © Crown Copyright)

Bullen (1977a) also showed that of a number of other variables the thermal conductivity of the wall and the area of the fuel bed had the most significant effect on $\dot{m}$. The importance of radiation as the dominant mode of heat transfer in determining $\dot{m}$ was confirmed in a later publication by Bullen and Thomas (1979). They burned 'pool fires' of ethanol (IMS), polymethylmethacrylate, and polyethylene in a compartment 2 m wide × 1 m × 1 m, with three different ventilation openings, corresponding to $A_w H^{1/2}/A_T$ of 0.032, 0.067 and 0.14 m$^{1/2}$, and measured $\dot{m}$ (g/s), temperatures inside and outside the compartment, and the radiant heat flux at the ventilation opening. In Figure 10.10 the ratio $\dot{m}/A_w H^{1/2}$ (kg/m$^{5/2}$ s) — which Kawagoe found to be constant for crib fires — is plotted against fuel bed area (A_f m^2) for the ethanol fires, showing clearly that $\dot{m}$ is not determined by $A_w H^{1/2}$ alone (cf. Equation (10.1)) and that the effects of $A_w H^{1/2}$ and A_f are not independent. Although not immediately obvious, Figure 10.10 also shows that $\dot{m}$ increases as A_f is increased, as anticipated (compare the values of $A_w H^{1/2}/A_f$ assigned to each curve).

In Figure 10.11, data from a number of sources are plotted on the basis of the correlation proposed by Harmathy (1972) (Figure 10.5). This shows that the rates of

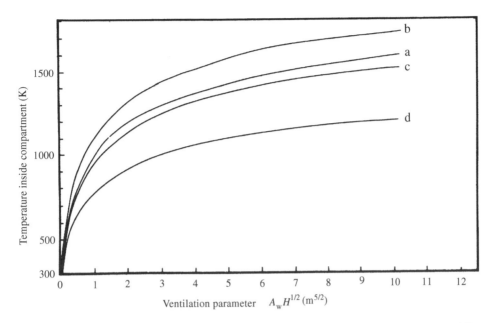

Figure 10.9 Compartment temperature as a function of the ventilation parameter $A_w H^{1/2}$ for the fuels indicated in Figure 10.8 (Bullen, 1977a). The limit of ventilation control is as shown in Figure 10.8. (Reproduced by permission of The Controller, HMSO. © Crown Copyright)

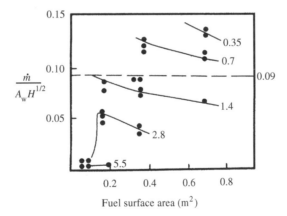

Figure 10.10 Variation of $\dot{m}/A_w H^{1/2}$ with fuel area (A_f) for ethanol pool fires in a small scale compartment (2 m wide × 1 m × 1 m). The numbers assigned to the curves are values of $A_w H^{1/2}/A_f$ (Bullen and Thomas, 1979). (Reproduced by permission of The Controller, HMSO. © Crown Copyright)

burning of noncellulosic materials can be substantially higher than that predicted by Equation (10.1) and suggests that flashover could be achieved with some non-cellulosic fuels with surface areas only one-tenth of that required for wood. The divergence of the rate of burning from 'Kawagoe behaviour' is illustrated schematically in Figure 10.12 (Bullen and Thomas, 1979). In the 'ventilation controlled regime', $\dot{m}$ can be greater

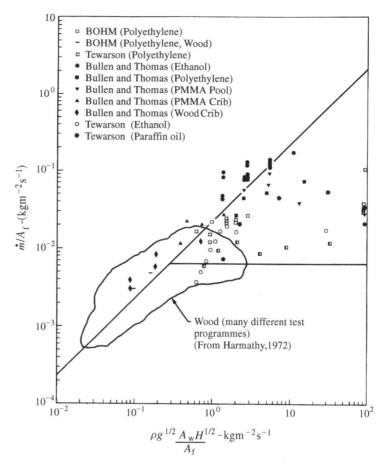

Figure 10.11 Comparison of the mass burning rates of different fuels in compartments, using Harmathy's (1972) correlation (Bullen and Thomas, 1979) (cf. Figure 10.6). Data from Bohm (1977), Tewarson (1972) and Bullen and Thomas (1979). (Reproduced by permission of The Controller, HMSO. © Crown Copyright)

than predicted by Equation (10.1) (the stoichiometric line), while for large values of $A_wH^{1/2}$, the rate of burning will be similar to that in the open, perhaps with some enhancement due to radiation feedback. The abrupt decrease in burning rate at high values of $AH^{1/2}$ is demonstrated clearly by experimental results of Takeda and Akita (1981).

Bullen and Thomas (1979) examined the influence of radiant heat flux within the compartment by plotting the burning rate ($\dot{m}$ g/s) of a number of fuels versus IA_f/L_v, where I (kW/m^2) is the intensity of radiant heat as measured at the ceiling of the compartment during steady burning (Figure 10.13). Although the data are scattered, there is a reasonable correlation about the line $\dot{m} = IA_f/L_v$, indicating that radiation dominates the heat transfer to the fuel surface. This relationship should be compared with Equation (10.21). Many of the data corresponding to high burning rates of ethanol fall below the line, but this is consistent with the observation that there is a substantial layer of vapour above the fuel surface which would attenuate the radiant heat reaching

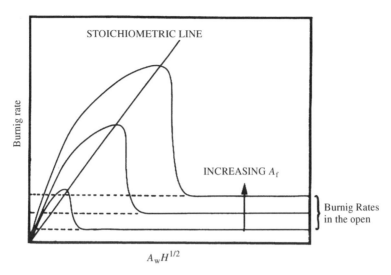

Figure 10.12   Schematic diagram showing the variation of mass burning rate with ventilation parameter $A_wH^{1/2}$ and fuel bed area A_f (Bullen and Thomas, 1979). (Reproduced by permission of The Controller, HMSO. © Crown Copyright)

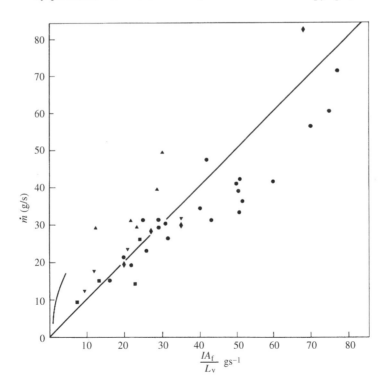

Figure 10.13   Correlation of mass burning rate ($\dot{m}$) with radiant intensity at ceiling level (I) (see text). ●, Ethanol ($L_v = 850$ J/g); ▼, PMMA pool (1600 J/g); ■, polyethylene (2200 J/g); ◆, wood (1340 J/g); ▲, PMMA crib; ———, ethanol in the open (Bullen and Thomas, 1979). (Reproduced by permission of The Controller, HMSO. © Crown Copyright)

the liquid (de Ris, 1979). The same comment applies to one single datum for poly-ethylene.

The behaviour in the ventilation controlled regime illustrated in Figure 10.12 is observed with many synthetic fuels and with cellulosic fuels which have extended surface areas, such as wall linings (Figure 10.7(b)). The rate at which air enters the compartment is insufficient to burn all the volatiles and the excess will be carried through the ventilation opening with the outflowing combustion products. This is normally accompanied by external flaming. Bullen and Thomas (1979) compared burning rates ($\dot{m}$) of their small-scale liquid and plastic fuel fires with rates of air inflow calculated using Equation (10.16). They defined an excess fuel factor (f_{ex}) which is zero for stoichiometric burning ($\dot{m}_{air} = r \cdot \dot{m}$) and positive if there are unburnt volatiles leaving the compartment (assuming that all the oxygen in the entrained air has been consumed within the compartment):

$$f_{ex} = 1 - \frac{\dot{m}_{air}}{r} \cdot \frac{1}{\dot{m}} \qquad (10.22)$$

Some of the results are quoted in Table 10.1: burning rates of the same fuel beds in the open are included for comparison. This illustrates clearly how the excess fuel factor depends on the area of the fuel surface, and the ventilation parameter $A_w H^{1/2}$, and also how the 'severity' of the fire (as judged by the maximum temperature under the ceiling) depends on both f_{ex} and $A_w H^{1/2}$.

Temperature profiles in the external flames from three of these small-scale compartment fires (Bullen and Thomas, 1979) are shown in Figure 10.14. The fuel was IMS and only the area of the fuel 'bed' was varied. The tip of the flame was assumed to be represented by the 550°C contour, although photographs taken during the experiments suggest that this might underestimate the height by ~10%. External flaming is an important mechanism of fire spread and can cause damage to the external load-bearing structural members (Section 10.7). There is a very rough correlation between the radiant heat flux to which the building facade is exposed and the excess fuel factor, f_{ex}. This is illustrated in Figure 10.15 in which the radiant heat flux is shown as a

Table 10.1   Pool fire burning rates in the open and in compartments (Bullen and Thomas, 1979)

Fuel[a] type	Fuel area	Open fire	Compartment fire $A_w H^{1/2}$					
			0.25			0.5		
	(m²)	$\dot{m}_{ave}$ (g/s)	$\dot{m}$ (g/s)	f_{ex}	$\theta °C^b$	$\dot{m}^c$ (g/s)	f_{ex}	$\theta °C^b$
IMS	0.186	3.5	19	0.41	880	26	0.09	1060
PE	0.186	3.7	10	0.25	980	NF	−0.9	—
PMMA	0.186	5.1	12	0.025	910	NF	−2.2	—
IMS	0.372	8.8	30	0.63	780	42	0.47	950
PE	0.372	8.4	14	0.48	890	26	0.45	1150
PMMA	0.372	7.4	21	0.45	820	31	0.30	1030

[a] IMS, industrial methylated spirits; PE polyethylene; PMMA, polymethylmethacrylate.
[b] $\theta °C$ is the maximum temperature rise over ambient under the ceiling.
[c] NF, no external flames.

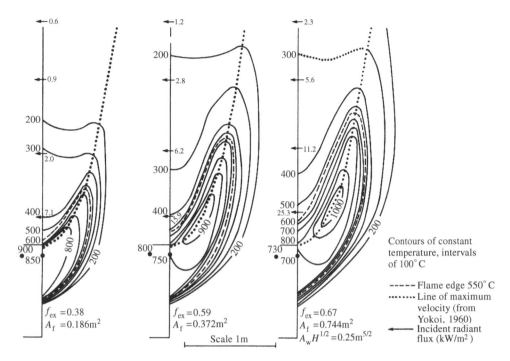

Figure 10.14   Effect of fuel area on external flame temperature profiles (Bullen and Thomas, 1979). (Reproduced by permission of The Controller, HMSO. © Crown Copyright)

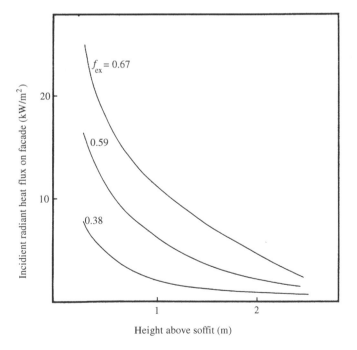

Figure 10.15   Variation of radiant flux in the plane of the opening of a small-scale compartment as a function of height above the soffit for the flames shown in Figure 10.14 (from which this diagram has been derived)

function of height above the soffit of the ventilation opening for the three small-scale compartment fires described in Figure 10.14. This should only be taken as indicative of large-scale behaviour, particularly as these results refer to IMS as the fuel which burns very cleanly. Fuels which generate a lot of smoke will give flames of lower temperature but high emissivity (see Table 5.4).

10.3   Temperatures Achieved in Fully-developed Fires

While there have been a number of efforts to develop a theoretical model for the fully-developed fire, only one significant attempt has been made to resolve the important parameters experimentally. This will be discussed briefly before the theoretical work is reviewed.

10.3.1   Experimental study of fully-developed fires in single compartments

A major experimental programme was undertaken in the 1960s to improve our understanding of the behaviour of the fully developed compartment fire (Thomas and Heselden, 1972; Thomas, 1972b). The object was to provide information which would allow the development of a rational approach to the problems of fire severity and fire resistance. Fire research laboratories of eight countries collaborated in this work, which was carried out under the auspices of the Conseil Internationale du Bâtiment (CIB). It involved over 400 experiments with small scale compartments, using wood cribs as the fuel, and varying the compartment size and shape, the ventilation parameter $A_w H^{1/2}$ and the fire load density. The effect of wind was also considered.

The conclusions of this study have had wide-reaching implications *vis-à-vis* compartment fire testing but need not be enumerated in detail here. However, three are of relevance to the present discussion:

(i)   The ratio $(\dot{m}/A_w H^{1/2})$ is not a constant as indicated in Equation (10.1) but depends on compartment shape (depth (D) and width (W)) and scale (particularly A_T, the internal surface area *excluding* the ventilation opening and the floor). A correlation between $(\dot{m}/A_w H^{1/2})$ $(D/W)^{1/2}$ and $A_T/A_w H^{1/2}$ was found, indicating the importance of the internal dimensions. The dependence of burning rate on $A_w H^{1/2}/A_T$ (the 'opening factor') was derived independently by Odeen (1963), and has been incorporated into theoretical models of compartment fires (e.g. Pettersson *et al.*, 1976).

(ii) The intensity of radiation emitted through the ventilation opening (as viewed from directly in front of the compartment) is sensitive to shape (particularly the ratio of compartment width to height) for all ventilation sizes. It correlates strongly with the rate of burning, except for small ventilation openings.

(iii) All other things being equal, the maximum temperature for a given compartment fire scenario is observed just inside the ventilation-controlled regime (Figure 10.6), and corresponds to $A_T/A_w H^{1/2} \approx 8$–15 (see (i) above).

This CIB programme has provided a substantial quantity of data which has been used to develop models relating to fire resistance requirements (Law, 1971) and behaviour of flames outside ventilation openings (Thomas and Law, 1974).

10.3.2 Mathematical models for compartment fire temperatures

Several research groups have turned their attention to developing ways of predicting the likely temperature–time history of a potential compartment fire (Kawagoe and Sekine, 1963; Pettersson *et al.*, 1976; Babrauskas and Williamson, 1978). The ultimate objective of such an exercise is to be able to specify for design purposes the thermal stress to which elements of structure would be exposed in the event of fire in a particular space, thus providing an alternative to the current building codes and regulations. In Sweden, Pettersson and co-workers (Pettersson *et al.*, 1976) developed a method of calculating the 'fire resistance requirements' which has been adopted by the regulatory authorities. In this section, that part of the model which relates to the temperature–time history of the fire is discussed: the second stage of the procedure will be considered in Section 10.5.

As we are concerned only with that period of the fire during which structural damage can occur, the pre-flashover stage may be neglected as average temperatures are relatively low. Thus, in any subsequent calculations, $t = 0$ refers to the start of the fully developed fire. The compartment is regarded as a calorimeter and its temperature obtained by solving the following heat balance (Figure 10.16):

$$\dot{q}_C = \dot{q}_L + \dot{q}_W + \dot{q}_R + \dot{q}_B \tag{10.23}$$

where:

$\dot{q}_C$ = rate of heat release due to combustion;
$\dot{q}_L$ = rate of heat loss due to replacement of hot gases by cold;

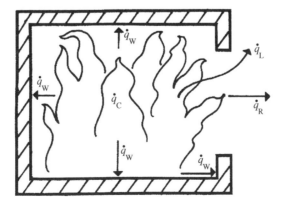

Figure 10.16   Heat losses during a fully developed compartment fire (Equation (10.23)) (after Pettersson *et al.*, 1976. Reproduced by permission of The Swedish Institute of Steel Construction)

$\dot{q}_W$ = rate of heat loss through the walls, ceiling and floor;
$\dot{q}_R$ = rate of heat loss by radiation through the openings;
$\dot{q}_B$ = rate of heat storage in the gas volume (neglect).

The following assumptions are made to simplify the model:

(i)   combustion is complete and takes place entirely within the confines of the compartment;

(ii) the temperature is uniform within the compartment at all times (cf. Figure 10.3);

(iii) a single surface heat transfer coefficient may be used for the entire inner surface of the compartment; and

(iv) the heat flow to and through the compartment boundaries is unidimensional, i.e. corners and edges are ignored and the boundaries are assumed to be 'infinite slabs'.

The terms in Equation (10.23) are as follows:

$\dot{q}_C$ — rate of heat release

Pettersson *et al.*, (1976) assume that the fire will be ventilation-controlled and that the Kawagoe relationship (Equation (10.1)) can be applied directly. Should the fire happen to be in the fuel-controlled regime, then this assumption will lead to an overestimate of the rate of burning (cf. Figure 10.2). The following expression is taken for the rate of heat release:

$$\dot{q}_C = 0.09 A_w H^{1/2} \cdot \Delta H_c \tag{10.24}$$

where ΔH_c is the heat of combustion of the fuel. ΔH_c is taken to be the heat of combustion of wood (18.8 MJ/kg), and in the subsequent calculations other materials are expressed in terms of 'wood equivalents'. Moreover, $\dot{q}_C$ is assumed to remain constant from $t = 0$ (i.e. immediately after flashover) until all the fuel has been consumed. This ignores the fact that any char produced will burn much more slowly than implied by Equation (10.24). (See footnote, p. 331.)

$\dot{q}_R$ — heat loss by radiation through the openings

From the Stefan–Boltzmann law (Section 2.4):

$$\dot{q}_R = A_w \varepsilon_F \sigma (T_g^4 - T_0^4) \text{ kW} \tag{10.25}$$

where A_w is the total area of the openings (m^2), T_g is the gas temperature within the compartment (K), and T_0 is the outside (ambient) temperature (K).

As $T_g \gg T_0$, this can be written:

$$\dot{q}_R = A_w \varepsilon_F \sigma T_g^4 \tag{10.26}$$

where ε_F is the effective emissivity of the gases within the compartment. This can be calculated from

$$\varepsilon_F = 1 - \exp(-Kx_F) \tag{10.27}$$

where x_F is the flame thickness (m) (normally taken as the depth of the room) and K is the emission coefficient (m^{-1}) (see Table 2.10). Pettersson *et al.* (1976) use $K = 1.1 \; m^{-1}$, quoting work by Hägglund and Persson (1976a) on wood crib fires.

$\dot{q}_L$ — heat loss due to convective flow

$$\dot{q}_L = \dot{m}_F c_p (T_g - T_0) \tag{10.28}$$

where $\dot{m}_F$ is the rate of outflow of fire gases (see Equation (10.10)). Assuming that $\dot{m}_F \approx \dot{m}_{air}$ (i.e. ignoring fuel volatilization), then letting $\chi = \dot{m}_{air}/A_w H^{1/2}$ (which is assumed to be approximately constant at $\sim 0.5 \; kg/m^{5/2}$. s, see Equations (10.15) and (10.16)).

$$\dot{q}_L = \chi c_p (T_g - T_0) A_w H^{1/2} \tag{10.29}$$

$\dot{q}_w$ — heat loss through the compartment boundaries

The rate of heat loss through the boundaries will depend on both the gas temperature within the compartment (T_g) and the internal surface temperature (T_i). Conductive heat transfer into the boundary must be solved numerically, as described briefly in Section 2.2.3(b). The enclosing boundary of the compartment is divided into n layers, each of thickness Δx (see Figure 10.17). A series of equations can then be written, one expression for each layer.
Exposed surface layer:

$$\Delta x c_1 \rho \frac{\Delta T_1}{\Delta t} = \frac{T_g - T_1}{\dfrac{1}{\gamma_i} + \dfrac{\Delta x}{2k_1}} - \frac{T_1 - T_2}{\dfrac{\Delta x}{2k_1} + \dfrac{\Delta x}{2k_2}} \tag{10.30}$$

Inner layer 'j' (describing $n-2$ layers):

$$\Delta x c_j \rho \frac{\Delta T_j}{\Delta t} = \frac{T_{j-1} - T_j}{\dfrac{\Delta x}{2k_{j-1}} + \dfrac{\Delta x}{2k_j}} - \frac{T_j - T_{j+1}}{\dfrac{\Delta x}{2k_j} + \dfrac{\Delta x}{2k_{j+1}}} \tag{10.31}$$

Outer layer:

$$\Delta x c_n \rho \frac{\Delta T_n}{\Delta t} = \frac{T_{n-1} - T_n}{\dfrac{\Delta x}{2k_{n-1}} + \dfrac{\Delta x}{2k_n}} - \frac{T_n - T_0}{\dfrac{\Delta x}{2k_n} + \dfrac{1}{\gamma_u}} \tag{10.32}$$

where T_i is replaced by T_1, the temperature of the innermost layer (to which Equation (10.30) refers). Both c and k are functions of temperature and have to be assigned values according to the local temperature. T_g and T_0 are the compartment gas

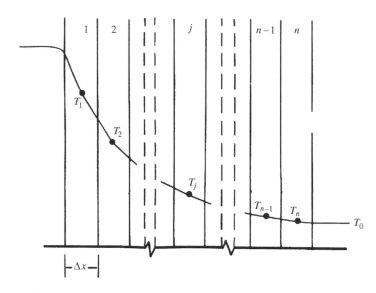

Figure 10.17 Transient heat conduction through the compartment boundaries. Boundary assumed to be an infinite slab, divided into elements 1–*n* (Pettersson *et al.*, 1976. Reproduced by permission of The Swedish Institute of Steel Construction)

temperature (K) and the ambient temperature (K) respectively. The two heat transfer coefficients γ_i and γ_u are given by Pettersson *et al.* (1976):

$$\gamma_i = \frac{\varepsilon_r \sigma}{T_g - T_i} \left(T_g^4 - T_i^4\right) + 0.023 \text{ kW/m}^2.\text{K} \tag{10.33}$$

where ε_r is the resultant emissivity

$$\varepsilon_r = \left(\frac{1}{\varepsilon_F} + \frac{1}{\varepsilon_i} - 1\right)^{-1} \tag{10.34}$$

where the subscript i refers to the inner surface (equivalent to $j = 1$ in Figure 10.17) and

$$\gamma_u = 3.3 \times 10^{-5} T_u - 3.09 \times 10^{-4} \text{ kW/m}^2.\text{K} \tag{10.35}$$

where T_u is the temperature of the outside surface (K) (equivalent to $j = n$). (Pettersson uses the symbols α_i and α_u instead of γ_i and γ_u in Equations (10.33) and (10.35). The change has been made to avoid confusion with α, which is used here for thermal diffusivity.)

The first-order difference Equations (10.30)–(10.32) are solved numerically for each time step, and corresponding values of $\dot{q}_w$ calculated from

$$\dot{q}_w = (A_t - A_w) \left(\frac{1}{\gamma_i} + \frac{\Delta x}{2k_1}\right)^{-1} (T_g - T_i) \tag{10.36}$$

where A_t is the total area of the boundary surfaces (walls, ceiling and floor), including the area of the ventilation openings, A_w m². Enclosing boundaries of different materials can be dealt with by a simple adaptation of these equations.

Calculation of the ventilation parameter $(A_w H^{1/2})$

When there is more than one opening in the walls of a compartment, the ventilation parameter is given by

$$A_\Sigma H_m^{1/2} = \sum_i A_i H_i^{1/2} \qquad (10.37)$$

where A_Σ is the sum of the areas of the openings, and H_m is a mean height, as defined by Equation (10.37). This is unlikely to apply to unusual ventilation conditions, such as several small openings at different heights (Babrauskas and Williamson, 1978) or if an opening exists in the roof. The latter situation is discussed by Pettersson *et al.* (1976) who provide a monogram which allows a corrected ventilation parameter to be calculated.

Calculation of $T_g(t)$

Now that the terms in the heat balance equation have been identified, substitution of Equations (10.24), (10.25), (10.29) and (10.36) into Equation (10.23) gives:

$$T_g = \frac{\dot{q}_c + 0.09c_p A_w H^{1/2} T_0 + (A_t - A_w)\left[\dfrac{1}{\gamma_i} + \dfrac{\Delta x}{2k}\right]^{-1}(T_g - T_1) - \dot{q}_R}{0.09c_p A_w H^{1/2} + (A_t - A_w)\left[\dfrac{1}{\gamma_i} + \dfrac{\Delta x}{2k}\right]^{-1}} \qquad (10.38)$$

T_g is calculated by numerical integration, e.g. using the Runge-Kutta procedure (Margenau and Murphy, 1956). T_1 depends on T_g and is obtained by solving the set of equations described above, iterating on T_g five times for each time step. For each solution, the value of T_g used is that calculated at the end of the previous time interval using Equation (10.38). Note that $\dot{q}_R$, γ_i and c_p are also functions of T_g. The time interval is chosen to be short enough for the surface heat transfer coefficients to be regarded as constant during each increment. The duration of burning is taken to be $M_f/0.09A_w H^{1/2}$ s, where M_f is the fire load in kg 'wood equivalent'. After this time, $\dot{q}_c$ is set equal to zero.

The validity of the assumptions relating to the heat losses in this model was tested by using data on the burning rates in experimental compartment fires to predict temperature–time histories using Equation (10.38). These were then compared with the measured temperature–time curves, and were found to be in satisfactory agreement (Figure 10.18).

Typical temperature–time curves derived from the above equations are shown in Figure 10.19 for a 'standard compartment' constructed from materials of 'average thermal properties'. Each set of curves shown corresponds to a different 'opening factor' $A_w H^{1/2}/A_t$ m$^{1/2}$ (Section 10.3.1) while individual curves correspond to different 'fire loads' expressed in terms of their net head of combustion, $q_f = M.\Delta H$. If several fuels are involved, then

$$q_t = \sum_i M_i \Delta H_i \qquad (10.39)$$

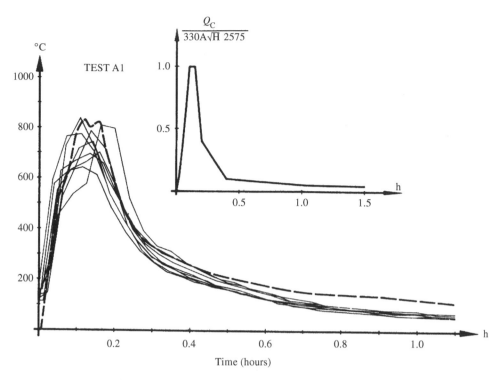

Figure 10.18 Gas temperature–time curves in full-scale fires. Solid lines represent experimental results from a number of full-scale tests using furniture as the fire load (96 MJ/m^2). Opening factor $A_w H^{1/2}/A_t = 0.068$ m$^{1/2}$. Dashed line (- - - - -) is the temperature–time curve calculated using the measured rate of burning ($\dot{m}$) to give $\dot{q}_c$ as a function of time (see inset) which in turn was incorporated into Equation (10.38) (Pettersson *et al.*, 1976. Reproduced by permission of The Swedish Institute of Steel Construction)

The data are also presented in Table 10.2. The effect of changing the thermal properties of compartment boundaries is illustrated in Figure 10.20: insulating materials like lightweight concrete tend to give hotter fires. (This is confirmed experimentally by Latham *et al.* (1987).) In Pettersson's method, this can be incorporated by applying an empirical correction factor to q_f and $A_w H^{1/2}/A_t$ before selecting the appropriate temperature–time curve. Application of these data to the calculation of fire resistance requirements for elements of structure is discussed below (Section 10.5).

As indicated above, Pettersson's model assumes ventilation controlled burning throughout the fully developed stage, until the fuel is consumed (Equation (10.24) *et seq.*), and consequently will overestimate the burning rate of a fire for which fuel control conditions actually exist. It is likely that this will be reflected by an overestimate of the 'severity' of the fire with respect to its potential for damaging the structure of the building. Babrauskas and Williamson (1978, 1979) have developed a more sophisticated model which requires as input details of the fuel nature and distribution, but incorporates the means of choosing whether the fire is ventilation-controlled or fuel-controlled at every stage. The rate of pyrolysis of the fuel must be known at all

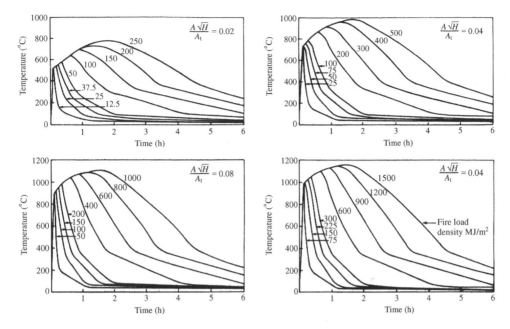

Figure 10.19 Theoretical temperature–time curves for compartment fires with different fire load densities (MJ/m²) and opening factors, $A_w H^{1/2}/A_t$ (m$^{1/2}$) (Pettersson *et al.*, 1976. Reproduced by permission of the The Swedish Institute of Steel Construction). (Pettersson's 'Standard compartment', wall thickness 0.2 m, $k = 0.8$ W/m.k, $\rho c = 1700$ kJ/m³.K)

times (i.e. $\dot{m}$) so that the rate of heat release ($\dot{q}_c$) can be calculated as either

$$\dot{q}_c = \dot{m} \cdot \Delta H_c \quad \text{(fuel controlled)} \qquad (10.40)$$

or

$$\dot{q}_c = \dot{m}_{air} \cdot \frac{\Delta H_c}{r} \quad \text{(ventilation controlled)} \qquad (10.41)$$

It is assumed that $\dot{q}_c$ is released entirely within the compartment and that $\dot{m}_{air}$ can still be calculated from Equations (10.15) and (10.16), even under fuel-controlled conditions when it may no longer be applicable. However, the model has great flexibility and has been shown to correlate well with some full-scale compartment fire tests.

The problem with the methods described above (Kawagoe and Sekine, 1963; Pettersson *et al.*, 1976; Babrauskas and Williamson, 1978) is that they all require lengthy computation. It can be argued that in view of the many assumptions and uncertainties associated with the compartment fire model, such sophistication is premature. Thus, Lie (1974, 1995) has suggested that if the objective is to develop a method of calculating fire resistance requirements, then it is necessary only to find a fire temperature–time curve 'whose effect, with reasonable probability, will not be exceeded during the use of the building'. With this in mind, he developed an expression based on a series of temperature–time curves computed by Kawagoe and Sekine (1963) which could be used as an approximation for the most severe fire that is likely to

Table 10.2 Compartment temperature–time curves according to Pettersson *et al.* (1976) (by permission) (Pettersson's 'Standard compartment', wall thickness 0.2 m, $k = 0.8$ W/m.K, $\varrho c = 1700$ kJ/m^3.K)

$A_w\sqrt{H}/A_t = 0.04$ m$^{1/2}$

t (h)	Fire load (MJ/m^2)							
	25	50	75	100	200	300	400	500
0.05	528	528	528	528	528	528	528	528
0.10	742	742	742	742	742	742	742	742
0.15	423	733	746	746	746	746	746	746
0.20	359	697	750	750	750	750	750	750
0.25	268	594	761	761	761	761	761	761
0.30	163	478	732	777	777	777	777	777
0.35	161	439	668	792	792	792	792	792
0.40	154	390	595	758	806	806	806	806
0.45	148	338	515	706	820	820	820	820
0.50	141	282	481	647	832	832	832	832
0.55	134	269	442	584	843	843	843	843
0.60	127	254	400	534	854	854	854	854
0.65	120	238	356	509	864	864	864	864
0.70	113	222	309	481	860	874	874	874
0.75	106	205	298	453	839	883	883	883
0.80	99	190	285	423	816	891	891	891
0.85	92	173	273	392	789	899	899	899
0.90	85	156	261	360	761	907	907	907
0.95	77	138	248	326	730	914	914	914
1.00	69	119	236	315	698	907	920	920
1.10	53	88	210	293	627	886	933	933
1.20	51	83	185	270	585	861	944	944
1.30	48	79	160	248	551	831	951	954
1.40	46	75	132	226	515	801	943	964
1.50	44	71	104	203	478	765	934	972
1.60	43	68	94	181	440	727	923	980
1.70	41	65	90	158	400	685	903	974
1.80	40	62	86	134	360	644	881	967
1.90	39	60	82	109	345	612	858	958
2.00	38	58	79	101	331	580	833	948

$A_w\sqrt{H}/A_t = 0.06$ m$^{1/2}$

t (h)	Fire load (MJ/m^2)							
	37.5	75	112.5	150	300	450	600	750
0.05	602	602	602	602	602	602	602	602
0.10	854	854	854	854	854	854	854	854
0.15	481	845	858	858	858	858	858	858
0.20	403	802	862	862	862	862	862	862
0.25	296	673	873	873	873	873	873	873
0.30	211	537	836	888	888	888	888	888
0.35	173	489	759	903	903	903	903	903
0.40	165	432	671	860	916	916	916	916
0.45	158	370	574	798	928	928	928	928
0.50	150	304	532	727	940	940	940	940
0.55	142	289	485	650	950	950	950	950
0.60	135	271	436	589	960	960	960	960
0.65	127	253	383	558	969	969	969	969
0.70	118	235	328	525	961	977	977	977
0.75	111	216	314	491	934	985	985	985
0.80	103	197	301	456	904	992	992	992
0.85	95	179	287	419	873	999	999	999
0.90	86	160	273	381	839	1005	1005	1005
0.95	78	139	259	342	803	1011	1011	1011
1.00	68	118	245	329	763	1001	1017	1017
1.10	50	82	216	304	678	973	1027	1027
1.20	47	77	188	279	628	941	1036	1036
1.30	45	73	159	254	587	905	1040	1045
1.40	43	69	128	229	545	866	1028	1052
1.50	41	65	96	204	502	824	1014	1059
1.60	40	62	86	180	458	779	999	1065
1.70	38	59	82	154	412	729	973	1055
1.80	37	57	78	127	366	681	946	1043
1.90	36	54	74	99	350	644	918	1031
2.00	35	52	71	91	335	606	887	1017

$A_w\sqrt{H}/A_t = 0.08$ m$^{1/2}$

t (h)	Fire load (MJ/m^2)							
	50	100	150	200	400	600	800	1000
0.05	649	649	649	649	649	649	649	649
0.10	928	928	928	928	928	928	928	928
0.15	520	917	931	931	931	931	931	931
0.20	432	868	934	934	934	934	934	934
0.25	313	732	944	944	944	944	944	944
0.30	180	574	901	959	959	959	959	959
0.35	179	520	815	972	972	972	972	972
0.40	171	456	718	922	984	984	984	984
0.45	163	387	609	852	995	995	995	995
0.50	155	315	562	775	1005	1005	1005	1005
0.55	146	298	509	688	1015	1015	1015	1015
0.60	138	279	454	620	1023	1023	1023	1023
0.65	129	260	396	585	1031	1031	1031	1031
0.70	121	240	336	548	1021	1039	1039	1039
0.75	112	220	321	510	989	1046	1046	1046
0.80	104	200	306	471	955	1052	1052	1052
0.85	95	180	292	431	919	1058	1058	1058
0.90	86	159	277	389	881	1063	1063	1063
0.95	77	138	262	347	841	1068	1068	1068
1.00	67	114	247	334	799	1055	1073	1073
1.10	47	76	216	307	704	1022	1081	1081
1.20	44	71	187	281	649	984	1089	1089
1.30	42	67	156	255	603	944	1091	1095
1.40	40	63	123	228	558	900	1076	1101
1.50	38	60	89	201	511	854	1058	1107
1.60	37	56	79	176	463	805	1040	1111
1.70	36	54	74	149	414	750	1011	1099
1.80	34	51	71	120	365	697	980	1085
1.90	33	49	68	91	349	657	948	1070
2.00	32	47	64	82	334	616	915	1055

$A_w\sqrt{H}/A_t = 0.12$ m$^{1/2}$

t (h)	Fire load (MJ/m^2)							
	75	150	225	300	600	900	1200	1500
0.05	708	708	708	708	708	708	708	708
0.10	1020	1020	1020	1020	1020	1020	1020	1020
0.15	571	1008	1021	1021	1021	1021	1021	1021
0.20	465	950	1023	1023	1023	1023	1023	1023
0.25	332	797	1031	1031	1031	1031	1031	1031
0.30	186	619	978	1044	1044	1044	1044	1044
0.35	183	554	880	1055	1055	1055	1055	1055
0.40	175	482	772	994	1065	1065	1065	1065
0.45	167	405	648	914	1074	1074	1074	1074
0.50	158	324	593	827	1082	1082	1082	1082
0.55	149	305	534	730	1090	1090	1090	1090
0.60	140	285	472	653	1097	1097	1097	1097
0.65	131	265	407	612	1103	1103	1103	1103
0.70	122	243	341	571	1089	1108	1108	1108
0.75	113	222	326	529	1051	1114	1114	1114
0.80	104	200	309	485	1012	1118	1118	1118
0.85	95	179	294	441	970	1123	1123	1123
0.90	85	157	278	395	927	1127	1127	1127
0.95	75	133	263	349	881	1130	1130	1130
1.00	63	108	247	335	834	1115	1134	1134
1.10	41	66	214	306	728	1074	1140	1140
1.20	39	61	183	279	667	1030	1145	1145
1.30	37	57	150	252	618	984	1145	1150
1.40	35	54	115	224	568	935	1126	1154
1.50	34	51	78	197	517	883	1105	1158
1.60	33	48	67	170	465	829	1083	1161
1.70	32	46	63	141	412	770	1050	1146
1.80	31	44	59	110	359	712	1015	1129
1.90	30	42	56	79	344	668	979	1112
2.00	29	41	54	69	330	624	942	1094

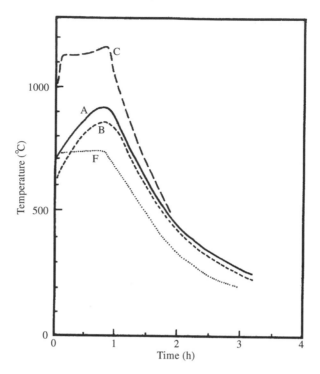

Figure 10.20   Theoretical temperature–time curves for fully-developed fires in compartments with different boundaries: A, materials with thermal properties corresponding to the average values for concrete, brick and lightweight concrete; B, concrete ($\rho \approx 500$ kg/m^3); F, 80% uninsulated steel sheeting, 20% concrete. (In all cases, the fire load was 250 MJ/m^2 and $A_\mathrm{w}H^{1/2}/A_\mathrm{t} = 0.04$ m$^{1/2}$.) (Pettersson *et al.*, 1976. Reproduced by permission of The Swedish Institute of Steel Construction)

occur in a particular compartment. Defining the opening factor $F = A_\mathrm{w}H^{1/2}/A_\mathrm{t}$ m$^{1/2}$, the temperature–time curve is given by:

$$T_\mathrm{g} = 250(10F)^{0.1/F^{0.3}} \cdot \exp(-F^2 t) \cdot [3(1 - \exp(-0.6t)) - (1 - \exp(-3t))$$
$$+ 4(1 - \exp(-12t))] + C(600/F)^{0.5} \qquad (10.42)$$

provided that there is an unlimited supply of fuel and that the conditions $t \le (0.08/F) + 1$ and $0.01 \le F < 0.15$ are met. If $t > 0.08/F + 1$, and/or $F > 0.15$, then the values of T_g corresponding to $t = 0.08/F + 1$ and/or $F = 0.15$ should be taken for the period up until the fuel is consumed. T_g is the gas temperature within the compartment ($^\circ$C) and t is the time after 'flashover' in *hours; C* is a constant, taking values of 0 and 1 for heavy ($\rho \ge 1600$ kg/m^3) and light ($\rho < 1600$ kg/m^3) materials, respectively. The duration of the fire is given by the expression:

$$\tau = M'_\mathrm{f}A_\mathrm{t}/330A_\mathrm{w}H^{1/2} = M'_\mathrm{f}/330F \text{ (h)} \qquad (10.43)$$

where M'_f is the fire load per unit internal surface area (kg/m^2). This is similar to the criterion used by Pettersson *et al.* (1976). When $t > \tau$, Equation (10.42) is

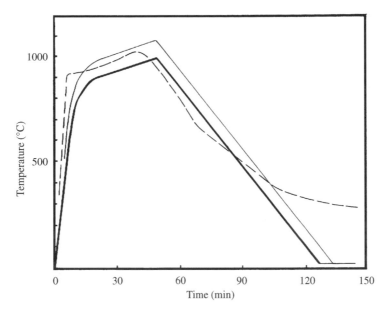

Figure 10.21 Comparison between theoretical temperature–time curves developed by Pettersson *et al.* (1976) (- - - -) and Lie (1974) (——$C = 1$); ——$C = 0$ (Equation 10.42)). Fire load 250 MJ/m². opening factor 0.08 m$^{1/2}$

replaced by

$$T_g = -600 \left(\frac{t}{\tau} - 1 \right) + T_\tau \tag{10.44}$$

where T_τ is the value of T_g at $t = \tau$. When $T \leq 20°$C. (the initial temperature), Equation (10.44) is replaced by $T_g = 20°$C.

Although this expression incorporates none of the refinements of Pettersson's and Babrauskas' models, it can be used to obtain a rough sketch of the course of a fire. Typical curves are compared in Figure 10.21 with temperature–time curves taken from Table 10.2.

10.4 Fire Resistance and Fire Severity

As stated at the start of this chapter, the term 'fire resistance' is associated with the ability of an element of building construction to continue to perform its function as a barrier or structural component during the course of a fire. Conventionally, it is determined by testing a full-scale sample (under load, if appropriate) to failure as it is subjected to a 'standard fire', defined by the temperature–time variation of the fire gases within a large furnace. The concept was first introduced in 1916, based on observations of the temperatures of wood fires used in early *ad hoc* testing (Babrauskas and Williamson, 1980a,b). The standard curve has changed only slightly over the years. In the USA it is specified by a set of data points (American Society for Testing and Materials, 1995b) (Table 10.3), although it is more common to define it mathematically,

Table 10.3 Definition of the standard temperature–time curve according to ASTM E119 (American Society for Testing and Materials, 1995b)

Time (min)	Temperature (ASTM E119) (°C)	Temperature (BS 476 Part 8)[a] (°C)
5	538	583
10	704	683
30	843	846
60	927	950
120	1010	1054
240	1093	1157
≥480	1260	1261[b]

[a] Calculated from Equation (10.45) (British Standards Institution, 1987b).
[b] At 480 minutes.

thus:

$$T = T_0 + 345 \log(0.133t + 1) \qquad (10.45)$$

where T_0 and T are the temperatures (°C or K) at time $t = 0$ and $t = t$ (s) respectively (British Standards Institution, 1987b): this is plotted in Figure 10.22(a). A very similar curve has been adopted by the International Organization for Standardization (ISO, 1975).

In the standard test, the required temperature–time curve is achieved by programming the temperature of the test furnace through control of the rate of fuel supply (gas or oil). Unfortunately, as the measured temperature refers to the gases within the furnace, the actual 'fire exposure' created is sensitive to the physical properties (and emissivity) of the furnace walls. It is believed that the dominant mode of heat transfer to the specimen is by radiation from the walls (Paulsen and Hadvig, 1977). If these have a low thermal inertia, then the surface temperature will increase rapidly (Figure 2.10), resulting in a more severe fire exposure than in another, otherwise identical furnace whose walls are constructed from a denser material (Malhotra, 1982). Indeed, it is likely that no two fire resistance furnaces are capable of giving the same fire exposure. Such anomalies have been noted in the past, but it is relatively recently that specifications have been proposed for the thermal properties of the furnace boundaries (Bohm, 1982).

It is normal to equate the 'fire resistance' of an element of structure with the time to a specified failure in the standard test, although the exposure in a real fire can be very different from that in the standard test. A quantitative measure of this can be seen in Figure 10.22(b) and (c). In the former, the standard temperature–time curve (Equation (10.45)) is compared with the temperatures measured in a series of full-scale compartment fires with different fire loads and ventilation openings; the fuel was in the form of wood cribs. In Figure 10.22(c), the effect of introducing a thermoplastic (polypropylene) is shown.

When the disparity between the standard test and 'real fires' was recognized in the 1920s, there was considerable discussion about the validity of continuing to use the standard curve. However, Ingberg (1928) carried out a series of tests which identified 'fire load' (combustible content per unit floor area) as an important factor in determining

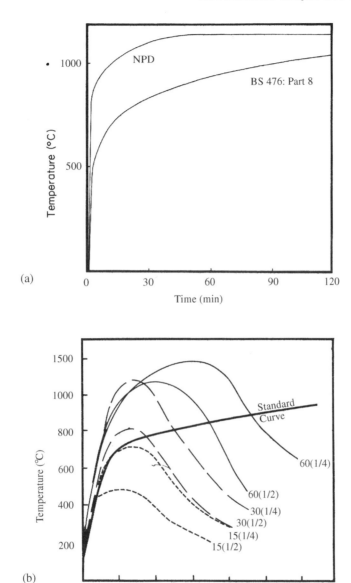

Figure 10.22 (a) Comparison between the 'standard' temperature–time curve BS 476 Part 21, and the hydrocarbon fire curve adopted by the Norwegian Petroleum Directorate (Shipp, 1983). (b) Comparison of the standard temperature–time curve with the temperatures measured during compartment fires. The fire load density is given in kg/m^2 and the ventilation as a fraction of one wall. 60(1/2) implies 60 kg/m^2 and 50% of one wall open (Butcher *et al.*, 1966. Reproduced by permission of The Controller, HMSO. © Crown copyright). (c) Average combustion gas temperature inside a compartment for a fire load density of 15 kg/m^2 and 1/4 ventilation, for wood (■) and wood + polypropylene (◆). The standard temperature–time curve is shown for comparison (Latham *et al.*, 1987)

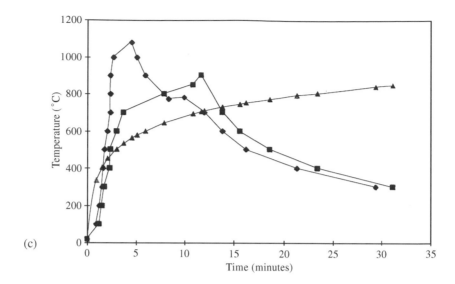

(c)

Figure 10.22   (*Continued*)

the potential fire severity. He proposed that the 'severity' of a fire could be related to the fire resistance requirement, using what came to be known as the 'equal area hypothesis', in which it is assumed that if the areas under the temperature–time curves (above a baseline of 300°C) of two fires are equal, then the severities are equal. If one of these 'fires' is the standard temperature–time curve, then 'severity' and 'fire resistance' can be equated. This is illustrated in Figure 10.23.

On this basis, Ingberg developed a table relating fire load and 'fire severity' (Table 10.4) from which the fire resistance requireme¡.t of a particular compartment can be obtained directly from the measured, or anticipated, fire load. Little information on fire load densities was available to Ingberg, but surveys have been carried out subsequently and are available in the literature (e.g. Pettersson *et al.*, 1976; Conseil Internationale du Bâtiment, 1986; National Fire Protection Association, 1997). Following this hypothesis, fire resistance requirements can be specified for an occupancy type if the potential (or actual) fire load is known. However, in addition to the practical question of whether or not the fire load will change significantly during the lifetime of the building, there is no theoretical justification for the hypothesis. The dependence of radiative heat flux on T^4 makes simple scaling impossible when heat transfer is dominated by radiation: for example, 10 minutes at 900°C will *not* have the same effect as 20 minutes at 600°C (cf. Figure 10.23). Furthermore, in the last few decades changes in building design and construction has resulted in buildings of much lower thermal mass, and consequently much more responsive to heat transfer from a fire.

These difficulties were tacitly ignored, and the apparent success of fire load as a surrogate for 'fire severity' was accepted by default: apparently, no building failures have occurred which have shown the method to be unsafe. Although it cannot be defended scientifically or technically, the philosophy is embedded into the common approach to structural fire safety, through the standard fire resistance test. A number of attempts have been made to relate 'real fires' to the standard tests, using the concept

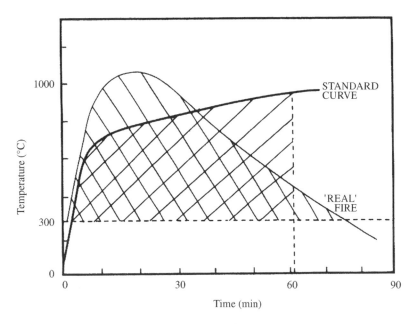

Figure 10.23   Ingberg's equal area hypothesis. If the shaded areas are equal then the fires are said to have equal 'severity'. Thus the fire resistance requirement for the 'real' fire shown corresponds to ~60 min exposure to the standard test

Table 10.4   Ingberg's fuel-load—fire-severity relationship

Combustible content[a] (wood equivalent)		Equivalent[a,b] (kJ/m^2 × 10^{-6})	Standard fire duration (h)
lb/ft^2	kg/m^2		
10	49	0.90	1
15	73	1.34	1.5
20	98	1.80	2
30	146	2.69	3
40	195	3.59	4.5
50	244	4.49	6
60	293	5.39	7.5

[a] Calculated on the basis of *floor* area.
[b] Heat of combustion of wood taken as 18.4 kJ/g.a

of 'equivalent fire severity'. This has to be based on a comparison of the response of an element of structure in the real fire and in the test furnace. Law (1971) sought such a relationship by analysing the thermal responses of insulated steel columns exposed to the standard temperature–time curve and to real fires. A steel temperature of 550°C was taken as marking a critical point in the fire exposure, above which significant weakening of the element had to be assumed. The real fire temperature was modelled as a constant temperature, equal to the peak fire temperature, sustained for a period $\tau = M_f/\dot{m}$, where M_f is the total fire load (kg wood equivalent) and $\dot{m}$ is the mean burning rate (kg/min). Drawing on data gathered in the CIB study of fully developed compartment fires (Thomas and Heselden, 1972), Law found the following correlation to hold:

$$t_f = K' \frac{M_f}{(A_w A_T)^{1/2}} \qquad (10.46)$$

where t_f is the required fire resistance (minutes) (Figure 10.24). With the areas in m², K' is a constant whose value is close to unity. It should be noted that this expression is based on an analysis of insulated columns and will not hold for exposed steelwork.

Law's analysis — unlike that of Ingberg — showed clearly that ventilation to the fire was an important factor in determining 'severity', and could not be ignored. However, as comparative studies of the behaviour of elements of structure in different fire resistance furnaces has shown, the thermal properties of the compartment boundaries can

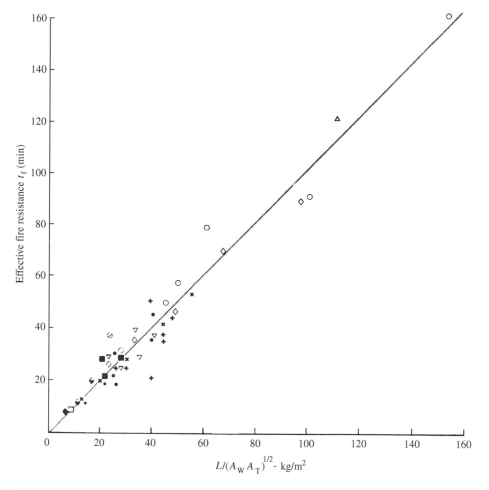

Figure 10.24 Law's (1971) correlation between fire resistance requirement (t_f) and $L/(A_w A_T)^{1/2}$, where L is the fire load (kg), A_w is the area of the ventilation and A_T is the total internal surface area of the compartment (m²) (Equation (10.46)) (Reproduced by permission of The Controller, HMSO. © Crown Copyright)

have a significant effect. This is demonstrated clearly in the results of Butcher *et al.* (1966) and Latham *et al.* (1987). Pettersson developed the following relationship:

$$t_f = 0.31C \frac{M_f}{(A_t A_w \sqrt{H})^{1/2}} \tag{10.47}$$

where C is a factor which depends on the thermal inertia of the compartment boundaries (Table 10.5). This is based on the analysis developed by Pettersson and co-workers (1976). A similar formula is given in DIN 18230:

$$t_f = 0.087(M_f/A_F)m.w.c \tag{10.48}$$

where m is a dimensionless burning factor (taken as unity for wood), w is dependent on the ratio A_w/A_F and c depends on the thermal inertia of the compartment boundaries (see Table 10.6).

These formulae are compared by Harmathy (1987), who has introduced the concept of 'normalized heat load' (H) (Harmathy and Mehaffey, 1982) which is defined as:

$$H = \frac{1}{\sqrt{k\rho c_p}} \int_0^\tau \dot{q}'' \cdot dt \tag{10.49}$$

where $\dot{q}''$ is the heat flux through the building element (kW/m^2), t is time (s), and τ is the fire duration (s). The 'normalization' is with respect to the thermal inertia of the wall lining materials.

These five formulae (Table 10.7) give widely differing results. While Ingberg's method may be rejected out of hand for the reasons given above, it is not clear how the others can best be interpreted. Law's formula does not bring the nature of the compartment boundaries into account, but on the other hand the inclusion of relevant correction terms into the DIN method and that of Pettersson is based on very limited data. The methods of Law and Harmathy give more conservative figures than the others

Table 10.5 Values of C relevant to Equation (10.47)

$\sqrt{(k\rho c_p)}$ (J/m^{2}s$^{1/2}$K)	C (h.m$^{3/4}$/kg)
≤ 720	0.09
720–2520	0.07
≥ 2520	0.05

Table 10.6 Values for Equation (10.48)

A_w/A_F	w	$\sqrt{(k\rho c_p)}$ J/m^{2}s$^{1/2}$K	c
0–0.05	3.2	0–720	0.25
0.05–0.10	2.0	720–2520	0.20
0.10–0.15	1.5	>2520	0.15
0.15–0.20	1.2		

Table 10.7 Comparison of the formulae for equivalent fire exposure (Harmathy, 1987)

Method	Prediction (hours)[a]
Ingberg (1928)	0.62
Law (1971)	1.22
Pettersson (CIB, 1986)	1.02
DIN (1978)	1.04
Harmathy (1987)	1.29

[a] The values are as follows: $A_F = 20$ m^2, $A_t = 85$ m^2, $H = 1.2$ m, $A_w = 1.8$ m^2, thermal inertia of the boundaries $(k\rho c) = 10^6$ J/m^{2}s$^{1/2}$K and fire load density $= 30$ kg/m^2.

and may provide useful methods for estimating fire resistance requirements (based on the standard test), but may be too conservative. The alternative is to abandon the furnace test completely and rely entirely on calculating the amount of fire protection necessary on the basis of a predicted temperature–time curve (Pettersson *et al.*, 1976). This is discussed in the following section. However, it must be emphasized that fire resistance as defined above refers only to compartment fires involving combustible solids. Flammable liquid fires, particularly those encountered in the open in the refining and petrochemical industries, exhibit totally different characteristics. A hydrocarbon spill fire can engulf an item of process plant in flame within seconds, thus exposing the structure to dangerous levels of heating very quickly. The standard fire resistance test is not relevant in this situation, and consequently there has been a move to develop a more appropriate test in which the temperature–time curve rises very much more steeply than indicated by Equation (10.45). The 'hydrocarbon temperature–time curve' adopted by the Norwegian Petroleum Directorate is shown in Figure 10.22(a) (Shipp, 1983).

10.5   Methods of Calculating Fire Resistance

It has been shown in previous sections how 'fire severity' has come to be identified with fire resistance requirements, and thus with the results of the standard tests. While this seems to have proved satisfactory in the past, there are numerous problems associated with fire resistance testing, including the obvious ones of cost, lack of interlaboratory reproducibility, and the questionable relevance of the standard temperature–time curve to real fire exposure, which point to the need for a more rational approach to structural fire protection (Milke, 1995). This requires failure criteria to be identified and incorporated into a suitable mathematical model. Failure of most structural components in a fire can be related to the loss of strength at high temperature. Thus, steel begins to lose strength above $\sim 550°$C ($>1000°$F), but how quickly structural steel will fail will depend on many factors: indeed, a steel column or beam can continue to perform its function at significantly higher temperatures. This will depend, *inter alia*, on the loading, the support conditions and the degree of restraint, the dependence of the material properties on temperature, and the temperature gradient within the section (British Standards Institution, 1990).

In general, a bare steel member will tend to fail early if exposed to a high heat flux, although the time to failure will depend on its thermal mass (see below). Structural members may be protected from fire by encasing them with a suitable insulating material (Elliot, 1974; National Fire Protection Association, 1997), the thickness of which will determine the 'fire resistance' of the assembly, as defined in the standard test (British Standards Institution, 1987b; American Society for Testing and Materials, 1995b). In reinforced concrete, the reinforcing bars are protected by the concrete although this will itself lose strength as dehydration and loss of water of crystallization occur. Spalling may also occur under fire conditions and expose the reinforcing bars. The effect of temperature on building materials in general is discussed at some length by Lie (1972), Abrams (1979) and Harmathy (1995).

A discussion of the behavioural response of structures to fire conditions is outwith the scope of this text, but to illustrate a number of important points, it will be assumed that there is a critical temperature above which a specific steel member in a structure can be regarded as significantly weakened. The critical temperature is not a constant, but for the sake of this discussion, 550°C will be taken as the criterion for the onset of failure. The 'fire resistance' of a structural element may then be obtained from basic heat transfer calculations, if the temperature–time curve for the fire exposure is known. The standard test may be taken as a simple example (Equation (10.45)). Consider a hollow, square cross-section column, of outside dimensions 0.2 m × 0.2 m, constructed from 0.015 m thick plate, as shown in Figure 10.25, exposed in the column furnace. To enable the fire resistance of this column to be calculated, the following must be known:

(a) the thermal capacity of the column;
(b) the mode of heat transfer within the furnace; and
(c) the emissivity of the furnace and the column.

This is a problem of transient heat transfer, requiring numerical solution. Manual, iterative calculation is possible, but the problem may best be approached using a spreadsheet, or writing a short computer program. It may be reduced to a single dimension by assuming that the column is of infinite length and heated uniformly. If the steel behaves as a 'lumped thermal capacity' (Section 2.2.2) then the heat transferred to the

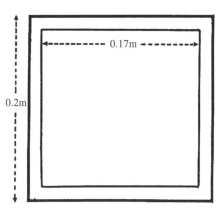

Figure 10.25 Cross-section of the hollow column to which Equations (10.50a), (10.50b) and (10.51b) and Figure 10.26 apply

column (for each unit of length) during a short time interval Δt is:

$$Q_{in} = \varepsilon_r \sigma A_s [(T_F^{j+1})^4 - (T_s^j)^4] \Delta t + h A_s (T_F^{j+1} - T_s^j) \Delta t \qquad (10.50a)$$

This quantity of heat causes the temperature of the steel to rise by an amount determined by the thermal capacity of the column (per unit length):

$$Q_{stored} = V_s \rho_s c_s (T_s^{j+1} - T_s^j) \qquad (10.50b)$$

where A_s and V_s are the surface area (m^2) and volume (m^3) of the column per unit length, T_s^{j+1} and T_s^j are the temperatures (K) of the steel at the end of the $(j + 1)$th and jth time intervals respectively, and T_F^{j+1} is the temperature of the furnace (or the fire) at the end of the $(j + 1)$th time interval. Equations (10.50a) and (10.50b) can be combined to give a value for the steel temperature at the end of the $(j + 1)$th timestep, as follows:

$$T_s^{j+1} = T_s^j + \frac{A_s}{V_s} \left\{ \frac{\varepsilon_r \sigma}{\rho_s c_s} \left[\left(T_F^{j+1}\right)^4 - \left(T_s^j\right)^4 \right] + \frac{h}{\rho_s c_s} \left[T_F^{j+1} - T_s^j \right] \right\} \Delta t \qquad (10.51a)$$

The importance of the ratio A_s/V_s should be noted: it is equal to the ratio of the exposed perimeter to the cross sectional area of the element of structure ('H_p/A') which is used in current design guidance in the UK to compare the heating rates of structural elements of different section (e.g. Latham *et al.*, 1987; British Standards Institution, 1990). In the USA, the ratio W/D is used in a similar manner (Law, 1972; Stanzak and Lie, 1973; Milke, 1996), where W is the mass of steel per unit length (i.e. $\rho_s V_s$), and D is the heated perimeter (i.e. H_p) which is numerically equal to A_s. As the density of steel is constant, this ratio is equivalent to $(A_s/V_s)^{-1}$. The greater the ratio W/D (the smaller A_s/V_s), the greater will be the fire resistance: this can be deduced from Equation (10.51a).

Numerical values may be substituted into Equation (10.51a). Thus, the resultant emissivity ε_r is calculated from the emissivities of the furnace and the steel (ε_F and ε_s respectively) using Equation (10.34): typically, $\varepsilon_F = 0.7$ and $\varepsilon_s = 0.9$, so that $\varepsilon_r = 0.65$, and the convective heat transfer coefficient, h, can be taken as 0.023 kW/m^2.K (Pettersson *et al.*, 1976). The density (ρ_s) and thermal capacity (c_s) of steel are 7850 kg/m^3 and 0.54 kJ/kg.K respectively and are assumed to be independent of temperature. The following equation is obtained by substituting these numerical values into Equation (10.51a):

$$T_s^{j+1} = T_s^j + \{6.266 \times 10^{-13}[(T_F^{j+1})^4 - (T_s^j)^4] + 3.910 \times 10^{-4}(T_F^{j+1} - T_s^j)\} \Delta t \qquad (10.51b)$$

By choosing a suitably small time interval, Δt, and iterating the calculation from $t = 0$ (i.e. $j = 0$), values of T_s^{j+1} can be worked out from the furnace temperature T_F^{j+1} (Equation (10.45)) and the steel temperature at the end of the previous time interval, T_s^j. The iteration must be continued until $T_s^{j+1} > 550°C$ (823 K); this is shown in detail in Table 10.8 with $\Delta t = 120$ s, and the results are plotted in Figure 10.26. This gives the time to achieve 550°C as 16.0 min. While these iterations may be carried out manually when only a small number are required, it is most convenient to use a spreadsheet.

Table 10.8 Calculation of fire resistance of a hollow square column
(Figure 10.25) using Equation (10.51b)

$$\Delta t = 120 \text{ s}$$

$$A\Delta t = 6.266 \times 10^{-13}[(T_F^{j+1})^4 - (T_s^j)^4]\Delta t$$

$$B\Delta t = 3.910 \times 10^{-4}[T_F^{j+1} - T_s^j]\Delta t$$

$$T_F = T_o + 345 \log(0.133t + 1)$$

$$T_o = 293 \text{ K}$$

Time step j	Time t (s)	T_F (K)	$A\Delta t$ (K)	$B\Delta t$ (K)	T_s (K)	T_s (°C)
0	0	293	0	0	293	20
1	120	717	19.32	19.90	332	59
2	240	816	32.42	22.70	387	114
3	360	876	42.59	22.90	453	180
4	480	918	50.24	21.83	525	252
5	600	951	55.79	19.99	601	328
6	720	978	59.00	17.70	677	405
7	840	1002	59.96	15.23	753	480
8	960	1021	57.59	12.59	823	550
9	1080	1038	52.83	10.10	886	613

The steel reaches 550°C (823 K) after eight iterations, at 16 minutes.

The calculation given in Table 10.8 underestimates the time to failure because the steel temperature at the end of one timestep (j) is held constant during the duration of the next timestep ($j + 1$). The effect is to overestimate the heat transfer to the steel because the value of T_F is taken as the value at the *end* of the timestep ($j + 1$). Thus the steel temperature is overestimated, as may be deduced by inspecting Equation (10.50a). A more accurate result can be achieved by reducing the timestep substantially, a task that is easily undertaken if a spreadsheet (or computer program) is used. The effect of reducing the timestep to 30 s (which quadruples the number if iterations required) is shown on Figure 10.26. Using a spreadsheet, it can be shown that the time to achieve 550°C is increased to 17.3 minutes. A reasonably accurate result can be obtained manually with only eight or nine iterations if the furnace temperature during each time interval is taken as the average $(T_F^j + T_F^{j+1})/2$: this is also shown in Figure 10.26, and is applied later in Table 10.10 for a similar problem. (Note that if the column had been solid, rather than hollow, the same method may be used to show that it would take 38.5 minutes to reach 550°C.)

A simple modification can be made to the above model to allow for the temperature-dependence of c_s, etc., but it is much more difficult to improve the heat transfer model. The column will receive heat radiated from the gases *and* the internal boundaries within the furnace. As the latter will inevitably lag behind the gas temperature (Section 10.4), and thus influence the general level of radiant heat within the furnace, it is necessary to incorporate the thermal response of the furnace walls into the model. This renders the model much more complex, and amenable only to more sophisticated numerical techniques.

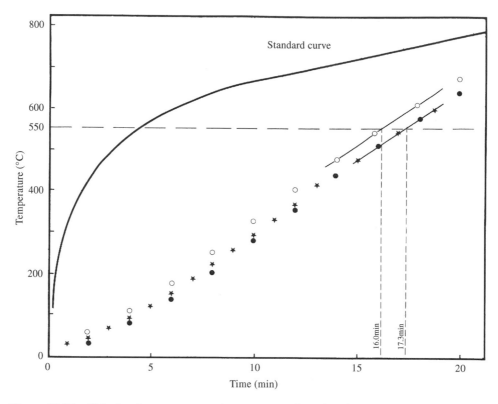

Figure 10.26 Calculated temperature–time curves for the hollow column shown in Figure 10.25, exposed to the standard curve. Equation (10.51b) and Table 10.8 apply. $\circ$, Data from Table 10.8 ($\Delta t = 120$ s); $\star$, $\Delta t = 30$ s; $\bullet$ $\Delta t = 120$ s, but at each iteration, $T_{\mathrm{F}} = (T_{\mathrm{F}}^{j} + T_{\mathrm{F}}^{j-1})/2$

The time to failure of an insulated column can also be deduced by an iterative procedure, but simplifying assumptions make the calculation more straightforward. If the Biot number (Section 2.2.2) is large it may be assumed that the temperature of the exposed surface of the insulation is equal to that of the furnace at all times during the test. Moreover, it is also assumed that in the iteration, quasi steady state is achieved over each time increment Δt, so that the rate of heat transfer through the insulation is given by

$$\dot{q}'' = \frac{k}{L}(T_{\mathrm{F}}^{j+1} - T_{\mathrm{s}}^{j}) \tag{10.52}$$

where k is the thermal conductivity of the insulation material (kW/m.K) and L is its thickness. The energy entering the steel per unit length of column in time interval Δt is then

$$Q_{\mathrm{in}} = \frac{k}{L}(T_{\mathrm{F}}^{j+1} - T_{\mathrm{s}}^{j}) \cdot A_{\mathrm{i}} \cdot \Delta t \tag{10.53}$$

where A_{i} is the internal surface area of the insulation per unit length of column through which heat is being transferred, assuming perfect thermal contact. The temperature rise of the steel during the time interval Δt is then:

$$T_s^{j+1} - T_s^j = \frac{Q_{in}}{V_s \rho_s c_s} \qquad (10.54)$$

The calculation must be iterated until $T_s^{j+1} \geq 550°C$ (or until T_s^{j+1} begins to fall), using the procedure laid down in Table 10.8.

Using the same failure criterion, the 'fire resistance' of structural steel members exposed to real fires can be calculated if the temperature–time curve is known. The models developed by Pettersson *et al.* (1976) and Babrauskas and Williamson (1978, 1979), as well as the formula of Lie (1974) (Equation (10.42)), can all be used to provide input to iterative calculations as described above. In fact, Pettersson has developed the method beyond this point to analyse the stability of the structural element (under load) at its maximum temperature, thus requiring a knowledge of the temperature-dependence of those factors which determine the strength of the steel (modulus of elasticity, yield stress, etc.). This avoids defining a specific 'failure temperature' such as 550°C which, as we have already noted, is only indicative of the temperature at which the element will begin to lose strength significantly. This aspect cannot be pursued further here.

The temperature of a steel element as a function of time during a fire can be calculated using Equation (10.51) in which T_F can be taken either from Pettersson's model (interpolation of the data presented in Table 10.2) or from Lie's formulae (Equations (10.42) and (10.43)). Pettersson *et al.* (1976) recommend certain values for ε_r which they refer to as a 'resultant emissivity' although they apply to different configurations of structural element: they are in fact composite terms incorporating both an emissivity and a geometric factor. These are quoted in Table 10.9.

Consider the following example which is given by Pettersson *et al.* (1976). A steel girder carrying floor units on its bottom flange, as shown in Figure 10.27, is exposed to a fully developed compartment fire in the room below. The temperature rise in the steel in a time interval Δt is given by:

$$\Delta T_s = \frac{\gamma_i}{\rho_s c_s} \cdot \frac{A_s}{V_s} (T_F^{j+1} - T_s^j) \Delta T \qquad (10.55)$$

where γ_i is given by Equation (10.33), with $T_g = T_F^{j+1}$ and $T_i = T_s^j$. Normally, A_s/V_s is the ratio of the perimeter exposed to fire to the cross-sectional area of the member

Table 10.9 Resultant emissivities ε_r (Pettersson *et al.*, 1976)

Type of construction	ε_r
1. Column exposed to fire on all sides	0.7
2. Column outside façade	0.3
3. Floor girder with floor slab of concrete, only the underside of the bottom flange being directly exposed to fire	0.5
4. Floor girder with floor slab on the top flange	
I section girder, width/depth ratio < 0.5	0.7
I section girder, width/depth ratio > 0.5	0.5
Box girder and lattice girder	0.7

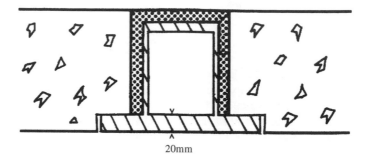

20mm

Figure 10.27   Steel floor girder carrying precast concrete floor units on the bottom flange. Table 10.10 applies. (Pettersson *et al.*, 1976. Reproduced by permission of The Swedish Institute of Steel Construction)

but, as only one face is exposed, A_s is taken as the exposed area and V_s the volume of the lower flange per unit length of girder. In this case, $A_s/V_s = 1/0.02 = 50$ m^{-1}, and with ρ_s and c_s as given above, Equation (10.55) can be written:

$$T_s^{j+1} = T_s^j + 1.18 \times 10^{-2}\gamma_i(T_F^{j+1} - T_s^j)\Delta t \qquad (10.56)$$

Table 10.10   Calculation of the steel temperature–time curve for the floor girder in Figure 10.27 exposed to a 'real fire' temperature–time curve (Pettersson *et al.*, 1976) (see Figure 10.28)

Line	Time (s)	T_F (°C)	T_F(ave) (°C)	γ_i (kW/m^{2}K)	$T_F - T_s$ (C)	ΔT (C)	T_s (°C)
1	0	20					20
2			334.5	0.038	315	25	
3	180	649					45
4			788.5	0.085	743	135	
5	360	928					180
6			922.5	0.122	742	193	
7	540	917					373
8			892.5	0.142	519	156	
9	720	868					529
10			800.0	0.147	271	85	
11	900	732					614
12			653.0	0.133	39	11	
13	1080	574					625
14			547.0	0.117	−78	−19	
15	1260	520					606
16			488.0	0.105	−118	−26	
17	1440	456					579
18			421.5	0.092	−158	−31	
19	1620	387					549
20			351.0	0.080	−198	−33	
21	1800	315					515

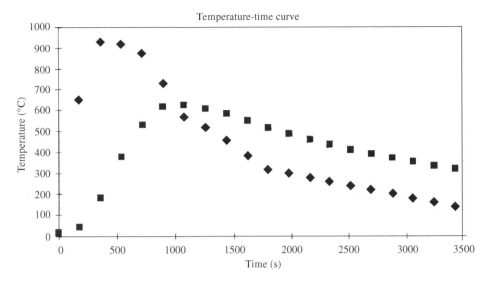

Figure 10.28   Temperature–time curve for the lower flange shown in Figure 10.27, (◆) exposed to a fire in the room below (100 MJ/m², $A_w H^{1/2}/A_t = 0.08$ m$^{1/2}$)(■)

The fire temperature–time curve is required before the calculation can be carried out. For a compartment with a fire load corresponding to 100 MJ/m² and an opening factor $A_w H^{1/2}/A_t = 0.08$ m$^{1/2}$, the temperature–time curve may be found in the tables given by Pettersson *et al.* (1976) (see Table 10.2). The temperature of the steel flange may then be calculated from Equation (10.56) using the procedure outlined above, with $\Delta t = 180$ s and taking T_F to be the average value for each timestep. The calculation is shown in Table 10.10 and T_F and T_s are plotted as functions of time in Figure 10.28. This shows clearly that the steel flange reaches 550°C after 1080 s (18 minutes), during the decay period of the fire. Whether or not failure will occur will depend on conditions of loading and the details of the design of the beam.

10.6   Projection of Flames from Burning Compartments

External flaming which accompanies most fully developed compartment fires not only provides a mechanism for fire spread to upper floors, perhaps involving combustible cladding (Oleszkiewicz, 1989; Lougheed and Yung, 1993) but also poses a threat to the stability of external structural members. To enable an assessment of this problem to be made, the general behaviour of such flames needs to be understood. Bullen and Thomas (1979) related flame size (and radiative heat flux to the building façade) to an excess fuel factor f_{ex}, given by Equation (10.22) (Figures 10.14 and 10.15), but the correlation has not yet been explored sufficiently to be applied generally to estimating flame sizes. Indeed, there is evidence that even with an 'excess fuel factor' of zero (stoichiometric) or less, some external flaming can occur due to the time taken for the flame gases to burn out (cf. the 'fuel lean' deflected flames described by Hinkley *et al.* (1968) (Section 9.2.1)).

Consequently, it is necessary to rely on an empirical correlation derived by Thomas and Law (1972) from data on flame projection from burning buildings by Yokoi (1960), Seigel (1969) and others (Thomas and Heselden, 1972). The data were correlated using dimensional analysis of the buoyant plume, and led to the following expressions which were found to apply approximately to crib fires in enclosures in the absence of wind. The scatter of data is large.

$$z + H = 12.8(\dot{m}/B)^{2/3} \tag{10.57}$$

$$x/H = 0.454/n^{0.53} \tag{10.58}$$

The symbols are identified in Figure 10.29 and are as follows:

H,B: height and width of the ventilation opening respectively (m);
z: height of flame tip above the window soffit (m);
$\dot{m}$:   rate of burning (kg/s);
x: horizontal reach from face of building (m); and
n: $2B/H$ (the 'shape factor').

The width of the flame may be assumed to be equal to B, although some spreading is likely to occur for wide openings when the exit velocity is relatively low. The horizontal projection (x in Figure 10.29) is strongly influenced by the velocity of this emerging jet, which in turn is dependent on the 'shape factor' n (Equation (10.58)). This behaviour has also been reported by Lougheed and Yung (1993) in a study of the radiation exposure experienced by adjacent structures.

The tip of the flame is identified as the point on the flame axis, where $T_f = 540°C$ (Figure 10.14). Law and O'Brien (1981) have incorporated these into a design guide for protection of exterior steel structures, and recommend that the burning rate ($\dot{m}$) is taken from either

$$\dot{m} = \frac{M}{1200} \text{ kg/s} \tag{10.59}$$

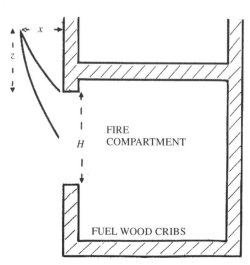

Figure 10.29   Projection of flame from a compartment during fully developed burning. Symbols refer to Equations (10.57) and (10.58)

where M is the fire load (on the assumption that most fires burn out in 20 minutes), or (Thomas and Heselden, 1972):

$$\frac{\dot{m}}{A_{w}H^{1/2}} \cdot \left(\frac{D}{W}\right)^{1/2} = 0.18 \left[1 - \exp\left(-\frac{0.036A_{T}}{A_{w}H^{1/2}}\right)\right] \qquad (10.60)$$

where D and W are the depth and width of the compartment, whichever gives the lower value.

However, the correlations (Equations (10.57) and (10.58)) break down under the following circumstances (Thomas and Law, 1972):

(a) if there are substantial heat losses from the projecting flame to the façade of the building;

(b) if there is a wind (the flame will be deflected and reduced in length);*

(c) if there is a fire on a lower floor (flames will lengthen due to oxygen depletion by the rising combustion products. Merging of flames from different floors can occur — cf. the São Paulo fires of 1972 and 1974);

(d) if the fuel bed has a very large surface area ($\dot{m}$ will be higher than anticipated by Equations (10.59) and (10.60) and the flames will be longer (Figure 10.7(b))); and

(e) if the fuel is non-cellulosic and has a low value of L_{v} (the heat required to produce the volatiles) — as (d).

Situations (d) and (e) correspond to large excess fuel factors (Equation (10.22)) and indicate the need to incorporate this parameter into the flame height correlations (Figure 10.14).

The question of the effect of the presence of non-cellulosic fuels on the size of external flames has not been addressed. As hydrocarbon polymers have a much higher air requirement then cellulosic fuels, it is almost inevitable that external flaming will be more significant, all other things being equal. It should also be remembered that thermoplastics will tend to burn as pools in the fully developed fire, with the potential to produce large areas of burning surface.

10.7 Spread of Fire from a Compartment

The spread of fire from one compartment into neighbouring compartments, ultimately to involve an entire building, has received little direct experimental study. Available information has been reviewed by Quintiere (1979). Observation of the spread of test fires in buildings earmarked for demolition has led to the concept of a 'volumetric fire growth rate', i.e.

$$\frac{\mathrm{d}V_{F}}{\mathrm{d}t} = k'V_{F} \qquad (10.61)$$

where V_{F}, the volume of the fire, is a measure of its extent at any moment of time. This indicates that the fire grows at an exponential rate ($V_{F} \propto \exp(k't)$), provided that

---

* The effect of wind on the behaviour of the external plume is addressed in a recent paper by Sugawa *et al.* (1997).

the 'fuel bed' is uniform: a similar assumption based on fire area is made frequently regarding the growth of fire within a compartment (Section 9.2.3).

One configuration that has been investigated is the spread of fire from a room into a corridor. This requires that flames and hot gases escape from the fire compartment — as they will do after flashover — thereby igniting wall and ceiling linings (if combustible) and heating the combustible surfaces in the lower part of the corridor to the firepoint condition. Quintiere and his co-workers (Quintiere, 1974) have studied this scenario in some detail, particularly in relation to the behaviour of carpets in corridors and their contribution to fire spread. Flashover in the corridor leads to a further restriction of air flowing to the room of origin, resulting in fuel-rich burning which will result in an increase in the quantity of smoke produced. There will be very substantial flame extension below the ceiling under these circumstances (Section 4.3.4; Hinkley and Wraight, 1969) and it is possible for the unburnt volatiles to burn vigorously when they mix with air, perhaps at a point remote from the room of origin, e.g. where the corridor enters a stairwell.

Other routes by which fire can spread from the room of origin include loss of integrity of compartment boundaries (including closed doors) as a result of prolonged fire exposure, sufficient to overcome the 'resistance' of the barrier, or by way of penetrations through the boundaries which are required to carry services (water, electricity, etc.) (Quintiere, 1979). Such fire spread is enhanced by the positive pressure differential which exists above the neutral plane in the fire compartment (Figure 10.4), effectively forcing flame and hot gases through any cracks or openings in the compartment boundary (Section 11.2.1(b)).

External fire spread by flames projecting from windows has been referred to in the previous section. Convective and radiative heating by the fire plume may break windows and ignite combustible materials on the floor above. The use of horizontal projections between floors to deflect the flames away from the façade of the building has been investigated. Early small-scale work in Japan (Yokoi, 1960) showed that a relatively small projection was sufficient if the shape factor ($2B/H$) was small, but quite substantial projections were necessary where wide windows are involved and the thrust of the flame is less. In general, this principle has been confirmed in large-scale tests (Ashton and Malhotra, 1960; Moulen, 1971; Harmathy, 1979).

Problems

10.1 Calculate the fire resistance of a load-bearing unprotected hollow steel column of square cross-section (250 mm × 250 mm) and thickness 25 mm assuming the standard temperature–time curve to be given by Equation (10.45). Use the same data as for Equation (10.51b).

10.2 If the column described in Problem 10.1 is protected by 25 mm thick insulation board, what temperature will the steel achieve in 30 min? Assume that the thermal conductivity of the board is $k = 0.1$ W/m K. (Calculation of fire resistance requires at least a programmable calculator.)

10.3 The column described in Problem 10.1 is located within a compartment measuring 8 m × 6 m × 3 m high with a ventilation opening 2 m high and 4 m wide on one long wall. The fire load corresponds to 150 MJ/m^2. Using Table 10.2, ascertain whether or not the column would continue to perform its load-bearing function for the duration of the fire. What is the maximum temperature achieved by the steel?

10.4 What is the temperature that would be achieved by the steel after 30 minutes if the column of Problem 10.2 is exposed to the fire of Problem 10.3?

10.5 Use Law's formula (Equation (10.46)) to estimate the fire resistance required for an insulated column exposed to the fire described in Problem 10.3. Assume that the heat of combustion of wood is 19.5 MJ/kg.

10.6 Calculate how long the column described in Problem 10.1 would survive if it became engulfed in flames arising from a large spillage of hydrocarbon liquid (e.g. crude oil). For simplicity, assume a constant flame temperature of 900°C and take the emissivity to be 1.0. Compare your answer with that of Problem 10.1.

11

The Production and Movement of Smoke

Gross *et al.* (1967) define smoke as 'the gaseous products of burning organic materials in which small solid and liquid particles are also dispersed'. This is wider than most common definitions, one of which refers to smoke as 'the visible volatile products from burning materials' (*Shorter Oxford English Dictionary*), but it has its limitations. In particular, it does not address the fact that what the observer sees as 'smoke' will contain a substantial quantity of air which has been entrained into the fire plume. The definition given in NFPA 92B (National Fire Protection Association, 1993c) and quoted by Milke (1995b) is much more comprehensive in that it includes entrained air*. However, it is the combination of obscuration and toxicity that presents the greatest threat to the occupants of a building involved in fire. Indeed, statistics collected in the UK and the USA suggest that more than 50% of all fatalities can be attributed to the inhalation of '(particulate) smoke and toxic gas' (Home Office, 1995; National Fire Protection Association, 1997).

With very few exceptions (Section 11.1), particulate smoke is produced in all fires. The effect of reduced visibility is to delay escape and increase the duration of exposure of the occupants of a building to the products of combustion. These will contain a highly complex mixture of partially burned species, some of which are capable of causing individuals to be overcome. This may arise due to short exposure to high concentrations of narcotic gases such as carbon monoxide, or to long duration exposure to lower concentrations. What is important is the *dose* of the gas inhaled — in simple terms, the concentration-time product (Ct). If Ct exceeds a critical value (the effective dose which causes incapacitation or death) (Purser, 1995), then the person is unlikely to escape unaided. The length of exposure will be increased if the visibility is poor, or if the combustion products contain eye and/or respiratory irritants (e.g. hydrogen chloride). The possible effects of low oxygen concentrations and high temperatures must also be considered.

Although it is clearly unreasonable to do so, particulate 'smoke' is traditionally dealt with separately from the gaseous combustion products. Existing standard methods

---

* Smoke: 'The airborne solid and liquid particulates and gases evolved when a material undergoes pyrolysis or combustion. Together with the quantity of air that is entrained or otherwise mixed into the mass' (NFPA, 1993c).

of test for 'smoke' measure only the potential for materials to produce particulate matter under the specific test conditions (American Society for Testing and Materials, 1993b; 1994a). Completely different tests, all involving animal exposure to fire gases, have been considered for toxicity (Kimmerle, 1974; Potts and Lederer, 1977; Deutsches Institut für Normung, 1979), although it is arguable that the problem cannot be resolved in this manner (e.g. Punderson, 1981). The number of gaseous products formed, and their distribution, are extremely sensitive to relatively minor changes in the 'fire scenario' and it is unrealistic to suppose that combustible materials may be classified by a 'toxic potential' as measured in a single fire test. The distribution of combustion products, even in the proposed standard tests (e.g. Potts and Lederer, 1977; Deutsches Institut für Normung, 1979) shows wide variation (Woolley and Fardell, 1982). Of course, this is not to say that certain materials might not be restricted from sensitive areas if their combustion products are known to contain harmful species. The most obvious example is polyvinyl chloride which, on heating, releases hydrogen chloride gas (Woolley, 1971). As this is a severe respiratory irritant, it would be wise to avoid the use of this material if at all possible in hospital wards caring for patients with respiratory problems.

However, there is some indication that meaningful measurements of the yield of particulate smoke from different materials can be made using small-scale test procedures. The yield can be quantified by measuring the optical density of the smoke under specified conditions. Optical density correlates directly with visibility (Rasbash, 1967) and is a gross property which, unlike toxicity, is relatively insensitive to the precise chemical nature of the individual constituents. Yields are still sensitive to the mode and condition of burning, but it is quite conceivable that they may be predicted from small-scale tests once the mechanism of smoke production is better understood. Rasbash and Phillips (1978) reviewed the existing 'standard' smoke tests and concluded that 'reasonable' agreement could be expected provided that the procedure involved measurement of optical density of smoke which had been allowed to accumulate in a volume or chamber (e.g. ASTM, 1994a). They observed that tests in which the optical density of the smoke was measured continuously as it issued from a combustion chamber or an experimental fire facility (e.g. at the end of a corridor (Woolley *et al.*, 1979/80)) tended to give highly variable results and to underestimate the smoke yield (Drysdale and Abdul-Rahim, 1985). This is the experience of others (Christian and Waterman, 1971; Rasbash and Pratt, 1980; Paul, 1983; Quintiere, 1983). However, the source of this problem lies in the assumption that is implicit in analysing these dynamic measurements, viz. that plug flow exists, with uniformity of concentration, temperature and velocity. If the smoke is removed from the fire apparatus through a duct, ensuring good mixing to give plug flow, then optical density measurements taken across the diameter of the duct give smoke yields in good agreement with 'static' measurements, in which the smoke has been allowed to accumulate within a known volume (Atkinson and Drysdale, 1989). This method for measuring smoke is used routinely in the Cone Calorimeter (American Society for Testing and Materials, 1990b; Babrauskas, 1995b) and can also be used in the larger-scale facilities such as Nordtest NT Fire 025 (Nordtest, 1986).

It is advantageous to express the yield of smoke in terms of the mass of material burnt — or more specifically, in terms of unit mass of volatiles released (Seader and Chien, 1974; Rasbash and Phillips, 1978). In principle, this would allow the total quantity of smoke that would be released during a fire to be predicted as a 'smoke load' (see Section 11.1.3) and its rate of production to be calculated if the rate of burning is known. This approach requires that the method by which the smoke is generated in the test procedure is representative of a real fire: however, in view of the wide range of conditions that exist during the course of any fire, this is unlikely to be achieved. It is known that smoke yield is sensitive to the ventilation conditions which exist during the course of a fire (Drysdale and Abdul-Rahim, 1985). Agreement between the smoke yields in small-scale and large-scale tests, where it exists, may be fortuitous as the standard tests involve precisely defined and unusually constant conditions (Quintiere, 1983).

Experiments based on the ASTM E662 apparatus have shown clearly how sensitive smoke yield is to a range of experimental variables, including radiant heat flux, oxygen concentration, ventilation conditions and sample orientation and geometry. All of these vary as a fire progresses. However, there are two distinct parts to the smoke problem, relating to (i) the early stages of a fire, when it is small and there is only a single item involved, and (ii) the post-flashover fire when all items within a compartment are burning, perhaps under ventilation-controlled conditions. The former is relevant to fire detection while the latter represents the limiting case in which large quantities of smoke are being produced, capable of rendering escapeways impassable at points remote from the fire. There is evidence that small-scale test results can be used to predict the amount of smoke generated in the early stages of the fire, but additional information is necessary before the problem of the post-flashover fire can be resolved satisfactorily.

The first part of this chapter (Section 11.1) is devoted to a summary of smoke measurement, while later sections provide a brief review of the subject of smoke movement. The latter is of growing importance in relation to smoke control and general fire safety in large buildings, and it is included here for completeness.

11.1 Production and Measurement of Smoke

Particulate smoke is a product of incomplete combustion. It is generated in both smouldering and flaming combustion, although the nature of the particles and their modes of formation are very different. Smoke from smouldering is similar to that obtained when any carbon-based material is heated to temperatures at which there is chemical degradation and evolution of volatiles. The high molecular weight fractions condense as they mix with cool air to give an aerosol consisting of minute droplets of tar and high-boiling liquids. These will tend to coalesce under still air conditions to give a distribution of particle size with a mass median diameter of the order of 1 µm (Bankston *et al.*, 1978) and will deposit on surfaces to give an oily residue.

Smoke from flaming combustion is different in nature and consists almost entirely of solid particles. While a small proportion of these may be produced by ablation of a solid under conditions of high heat flux, most are formed in the gas phase as a result of incomplete combustion and high temperature pyrolysis reactions at low oxygen concentrations. (This is reviewed by Rasbash and Drysdale (1982)). Particulate matter can be generated even if the original fuel is a gas or a liquid.

Both types of smoke are combustible, due at least in part to the unburned and partially burned fuel vapours which are inevitably present along with the particles:

under certain circumstances, explosions or explosion-like events can occur (Croft, 1980/81). One of the best documented cases is the 'Chatham mattress fire' which involved prolonged smouldering of latex rubber mattresses in a large warehouse (Woolley and Ames, 1975) (Section 8.2). A related phenomenon is backdraught (Section 9.2.1) which involves rapid burning when ventilation is provided to a compartment in which fuel-rich smoke has accumulated during conditions of poorly ventilated flaming combustion (Fleischmann *et al.*, 1994).

11.1.1 Production of smoke particles

Under conditions of complete combustion, fuel will be converted into stable gaseous products (Section 1.2.3), but this is rarely, if ever, achieved in diffusion flames. In a typical fire, mixing involves buoyancy-induced turbulent flows in which substantial temperature and concentration gradients exist (Section 4.3). In regions of low oxygen concentration a proportion of the volatiles may undergo a series of pyrolysis reactions which lead to the formation of unsaturated molecular species, including acetylene (ethyne) which under these conditions undergoes quasi-polymerization reactions (Section 1.1.1) to produce aromatic species, the simplest of which is benzene (C_6H_6). Benzene is the precursor of polycyclic hydrocarbons which in turn grow to produce minute particles of soot within the flame (Glassman, 1989; Griffiths and Barnard, 1995).

Although a discussion of the chemical mechanism of smoke formation is outwith the scope of this text, some of the possible steps in this complex process are shown in Figure 11.1 (from Glassman, 1989). All these steps are reversible, and any of the intermediate species may undergo oxidation although, once formed, the aromatic 'nucleus' (C_6H_6) is remarkably stable at high temperatures and resistant to oxidation. It is the presence of 'soot particles' within the flame that gives the diffusion flame its characteristic yellow luminosity (Section 2.4.3) and provides the principal mechanism for radiative heat loss. The concentration of soot particles determines the effective emission coefficient (K in Equation (2.83), Section 2.4.3). It is difficult to measure, but can be related to the soot volume fraction (f_v) which is the fraction of the flame volume occupied by soot particles (Pagni and Bard, 1979; Tien *et al.*, 1995). These minute particles (10–100 nm diameter) may oxidize within the flame (Wagner, 1979) but if the temperature and oxygen concentration are not high enough they will tend to grow in size and agglomerate to give substantially larger particles which will escape from the high temperature environment of the flame as 'smoke'. Smoke particles are complex agglomerations of minute soot particles which have come together to form chains and clusters, growing to an overall size in excess of 1 μm (Figure 11.2). When their size becomes of the same order as the wavelength of light (0.3–0.7 μm), they cause obscuration and reduced visibility by a combination of absorption and light scatter.

The tendency of a fuel to produce smoke may be assessed by measuring its 'smoke point'. It is defined as the minimum laminar flame height (or fuel mass flowrate) at which smoke first escapes from the flame tip. It has been shown to correlate very well with incompleteness of combustion and smoke yield (Tewarson, 1995), and also with the radiant fraction emitted from turbulent buoyant diffusion flames (Markstein, 1984).

The chemical composition and structure of the parent fuel are of considerable importance (Rasbash and Drysdale, 1982; Glassman, 1989). A small number of pure fuels (namely carbon monoxide, formaldehyde, metaldehyde, formic acid and methyl

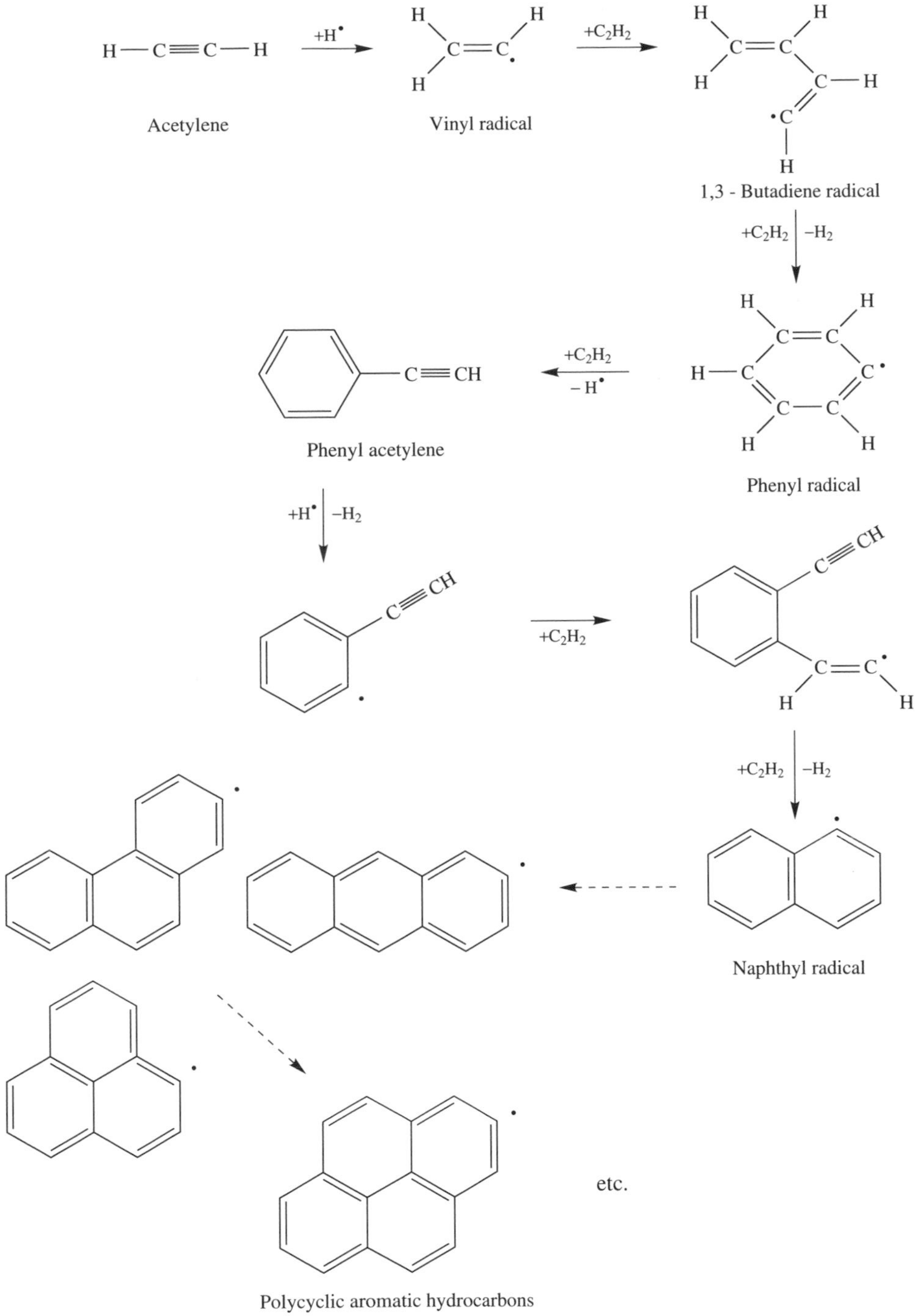

Figure 11.1 The chemical pathway to 'soot' (after Glassman, 1989)

AGGLOMERATE VIEWED AT TWO ANGLES

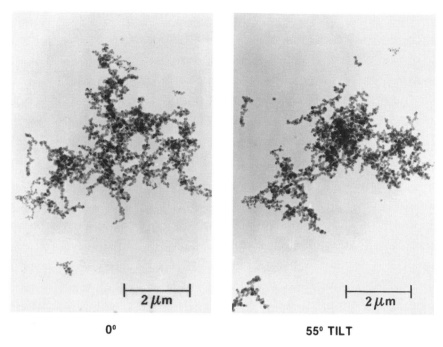

0°   55° TILT

Figure 11.2   Transmission electron micrograph of a smoke particle. The overall size of the agglomerate is about 6 μm, and the diameter of the individual spherules is about 0.03 μm (from Mulholland, 1995). Photograph by Eric B. Steel, Chemical Science and Technology Laboratory, National Institute of Standards and Technology

alcohol) burn with non-luminous flames and do not produce smoke. Other fuels, burning under identical conditions, give substantial yields, depending on their chemical nature. Oxygenated fuels such as ethyl alcohol and acetone give much less smoke than the hydrocarbons from which they are derived (Table 11.1), while for the hydrocarbons, there is a progression in the tendency to produce smoke with the introduction of branching (cf. *n*-alkanes and iso-alkanes), unsaturation (cf. alkanes, alkenes and alkynes) and aromatic character (cf. alkanes, aromatics and polynuclear aromatics). These observations on pure fuels are consistent with observations of the smoke production potential of solids (see Table 11.4 below). Under free burning conditions, oxygenated fuels such as wood and polymethylmethacrylate give substantially less smoke than hydrocarbon polymers such as polyethylene and polystyrene. Of the latter pair, polystyrene produces by far the greater yield as the volatiles from this polymer consist largely of styrene and its oligomers, which contain the aromatic nucleus (Table 11.1).

In the next section, we shall consider briefly how smoke yields can be measured and enlarge upon the dichotomy that exists between standard methods of test and smoke production in 'real fires' to which reference has already been made.

Table 11.1 Smoke-forming tendency of simple fuels

Fuel Type		Typical example	Smoke point[a] (m)
Alcohols	Ethanol	C_2H_5OH	0.227
Ketones	Acetone	$CH_3.CO.CH_3$	0.205
n-Alkanes	n-Hexane	$CH_3(CH_2)_4.CH_3$	0.118
iso-Alkanes	2,3-Dimethyl butane	$(CH_3)_2.CH.CH.(CH_3)_2$	0.089
Alkenes	Propene	$CH_3.CH=CH_2$	0.029
Alkynes	Propyne	$CH_3.C{\equiv}CH$	—
Aromatics	Styrene	(structure: benzene ring with $CH=CH_2$)	0.006
Polynuclear aromatics	Naphthalene	(structure: naphthalene)	0.005

[a] A. Tewarson, personal communication (1986).

11.1.2 Measurement of particulate smoke

The yield of particulate smoke from a burning material may be assessed by one of the following methods:

(a) Filtering the smoke and determining the weight of particulate matter (suitable for small-scale tests only) (ASTM, 1989);

(b) collecting the smoke in a known volume and determining its optical density (small and medium scale only) (Rasbash and Phillips, 1978; ASTM, 1994a; ISO, 1990); and

(c) allowing the smoke to flow along a duct, measuring its optical density where plug flow has been established, and integrating over time to obtain a measure of the total (particulate) smoke yield (Atkinson and Drysdale, 1989; ASTM, 1993d; Babrauskas, 1995b).

The quantity of particulate smoke produced in a test must be expressed as a yield, i.e. the amount generated per unit mass of fuel burned. For the gravimetric method (a), the mass of smoke particles from a known quantity of material is measured, and the yield may be quoted as mg particles/g fuel. However, methods (b) and (c) rely on measuring the optical density of the 'smoke' under the conditions of the test, involving some dilution by air. The smoke yield has to be related to the *concentration* of particles and the volume in which they have been dispersed. Fortunately, as will be shown below, optical density is directly proportional to the particle concentration (to within satisfactory limits of accuracy) and may be used as a surrogate measurement.

The definition of smoke given earlier specifically includes air that has been entrained Thus, in the calculation of the flowrate of smoke 15 m above a 6.4 MW fire (see Section 4.4.4), the 'smoke' was almost entirely entrained air. The 'mass' (hence volume) of smoke increases with height according to Equation (4.48), in inverse proportion to the concentration of the smoke particles. Clearly, if optical density is to be used to quantify 'smoke yield', the volume in which the smoke particles are dispersed must be known.

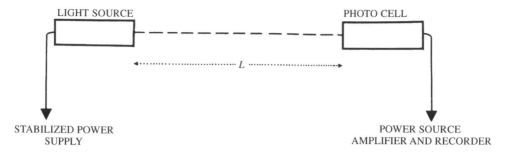

Figure 11.3   Representation of the apparatus required to measure optical density

Optical density may be determined by measuring the attenuation of a beam of light passing through the smoke (see Figure 11.3). If in the absence of smoke the intensity of light falling on the photocell is I_0, then in the presence of smoke, the reduced intensity (I) will be given by the Lambert-Beer Law (Equation (2.80)):

$$I = I_0 \exp(-\kappa C L) \tag{11.1}$$

where κ is the extinction coefficient, C is the mass concentration of smoke particles and L is the path length of the optical beam passing through the smoke. Optical density is then defined in terms of either natural logarithms (Babrauskas, 1995b):

$$D_e = -\log_e \left(\frac{I}{I_0} \right) = \kappa C L \tag{11.2}$$

or common logarithms (Rasbash and Phillips, 1978; Rasbash, 1995):

$$D_{10} = -10 \log_{10} \left(\frac{I}{I_0} \right) = \frac{10}{2.303} \cdot \kappa C L \tag{11.3}$$

where the factor of 10 is introduced to be consistent with the measurement of the attenuation of sound and electrical signals (decibels, or db).

Comparing these two expressions with Equation (11.1) shows that optical density is directly proportional to the concentration of smoke and is a much more satisfactory measurement than 'percentage obscuration', given by:

$$\frac{I_0 - I}{I_0} \times 100$$

(see Table 11.2).

Strictly speaking, Equations (11.2) and (11.3) apply to monochromatic light, but they have been used satisfactorily with white light (Rasbash, 1967). The absorption coefficient is dependent on wavelength, and on the particle size distribution which is likely to change as the smoke 'ages' with time (Seader and Chien, 1975). (The ageing process involves coalescence of particles, so that the number density will decrease. Ionization smoke detectors are particularly sensitive to the number of particles per unit volume and respond best to 'young smoke'. They can respond to particles as small as 10 nm in diameter, while scatter and obscuration devices do not respond until the particle size is of the same order as the wavelength of light (O(100 nm)).

Table 11.2   Relationship between percentage obscuration and optical density

Percentage Obscuration	Optical density (Equation (11.2))	Optical density (db) (Equation (11.3))
10	0.11	0.46
50	0.69	3.01
90	2.30	10.00
95	3.00	13.01
99	4.61	20.00

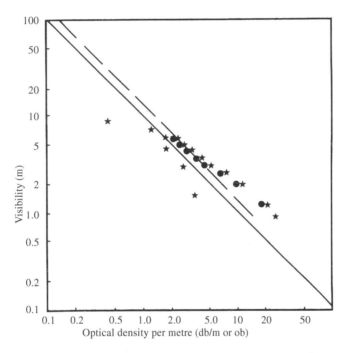

Figure 11.4   Relationship between visibility and optical density (per metre pathlength) for objects illuminated by scattered light. ●, Rasbash (1951); ⋆ Malhotra (1967); − − −, Jin (1970, 1971); ——, simple correlation recommended by Butcher and Parnell (1979). By permission of E&F Spon Ltd from Butcher and Parnell, *Smoke Control in Fire Safety Design.*

  Rasbash (1967) showed that the optical density, expressed in terms of unit path length (i.e. D_{10}/L db/m), correlates reasonably well with visibility, 1 db/m corresponding approximately to a visibility of 10 m (Figure 11.4). This is for general visibility through smoke. For visibility of 'exit' signs in buildings, there is a substantial difference between those signs that are 'front illuminated' and those that are 'back illuminated'. On the basis of results of Jin (1978), Butcher and Parnell (1979) suggest that a sign that is self-illuminated can be seen at 2.5 times the distance of one that relies on surface illumination.

11.1.3   Methods of test for smoke production potential

In view of the serious nature of the problems presented by smoke generation in fires, it is hardly suprising that there has been considerable pressure from those sectors of local and national government concerned with public safety for a method of test whereby materials can be rated according to their potential to produce smoke. However, in searching for such a test, it has been tacitly assumed that smoke yield is a property of the material and that the 'fire environment' has only a second-order effect. This assumption is incorrect: relatively little attention has been paid to the fact that the smoke yield is strongly dependent on the 'fire scenario' — in particular the ventilation conditions that exist at the time, and whether the fire is pre- or post-flashover.

There are many smoke tests in existence (see Table 11.3). Of these, the 'NBS Smoke Density Chamber' (ASTM, 1994a) has probably been the most widely used. It involves exposing a 75 mm square sample of material in a vertical configuration to a nominally constant radiant heat flux of 25 kW/m². Flaming combustion is initiated and sustained by an array of six small pilot flames impinging on the face of the sample. The smoke is confined within the volume of the apparatus (0.51 m³) and the optical density measured by means of a vertical light beam, $L = 0.914$ m. The standard specifies that the results are expressed as the maximum 'specific optical density', defined as

$$D_m = \frac{1}{10} \frac{D_{10}V}{LA_s} \qquad (11.4)$$

where D_{10} is defined as in Equation (11.3), V is the volume of the test chamber (m³), and A_s is the exposed area of the sample (m²).

Typical examples for flaming and non-flaming degradation are given in Table 11.4. However, as only the area of the sample is specified, in principle D_m needs to be determined as a function of thickness. The reproducibility and repeatability of this test are better than many others although a variability of $\pm 25\%$ is still to be anticipated

Table 11.3   Summary of existing smoke tests

Name	Type	Reference
Rohm-Haas XP-2	F,O,S	ASTM,[a] 1993b
NBS test	R,O,S	ASTM, 1994a
Arapahoe test	F,G	ASTM, 1989
Steiner tunnel	F,O,D	ASTM, 1995a
Radiant panel test	R,O,D	ASTM, 1995b
OSU calorimeter	R,O,D	ASTM, 1993c
ISO dual chamber	R,O,S	ISO[b], 1990
Cone calorimeter	R,O,D	ASTM, 1993d

Key: F = sample exposed to flame only; R = sample exposed to radiant heat flux (flame may or may not be present); O = smoke determined by obscuration of a light beam; G = smoke determined gravimetrically; S = smoke allowed to accumulate in a known volume; D = smoke measured as it flows from the test apparatus.
[a] American Society for Testing and Materials.
[b] International Organization for Standardization.

Table 11.4 Comparison of smoke yields from flaming and nonflaming exposure in the NBS Smoke Chamber Test (ASTM, 1994a)

	Specific optical density[a]	
	Flaming	Non-flaming
Polyethylene	62	414
Polypropylene	96	555
Polystyrene	717	418
Polymethylmethacrylate	98	122
Polyurethane	684	426
Polyvinylchloride	445	306
Polycarbonate	370	41

[a] The maximum specific optical density is defined in Equation (11.4). Although apparently dimensionless, it has units (bels/m).m^3/m^2, where 1 bel/m corresponds to a tenfold reduction in light intensity over a pathlength of 1 metre.

(ASTM, 1994a). Rasbash and Phillips (1978) carried out a series of measurements of the smoke produced from a range of materials, generating the smoke using a furnace modelled on that used in the original NBS test (ASTM, 1994a), but collecting it in a much larger volume (13 m^3). The results were expressed in terms of a 'smoke potential' D_0:

$$D_0 = \frac{D_{10}}{L} \cdot \frac{V}{W_l} \text{ (db/m).m}^3\text{/g} \qquad (11.5)$$

where W_l is the mass of material (g) vaporized (burned) during the test. There are a number of advantages in expressing the results in this way. D_0 is the volume of smoke of unit optical density per metre (i.e. visibility 10 m) that would be produced from one gram of material (in a standardized test). Rasbash suggested that that the term 'obscura' (ob) be used for db/m (Rasbash and Phillips, 1978) not only to emphasize the significance of db/m, but also to aid comprehension of the units of D_0: values of D_0 for some typical materials under conditions of flaming and non-flaming degradation are given in Table 11.5. Most common materials give less smoke with flaming combustion, although there are some notable exceptions (Tables 11.4 and 11.5).

This method of expressing 'smoke production potential' has not been widely adopted, despite its relative ease of application. For example, consider what will happen if a polyurethane cushion weighing 0.5 kg burns in a room of volume 50 m^3. If 15% by weight of residual char remains after combustion (i.e. $W_l = 0.85 \times 0.5 = 0.425$ kg) the optical density of the smoke which accummulates in the room will be

$$\frac{D_{10}}{L} = \frac{D_0 \cdot W_l}{V} = \frac{0.96 \times 425}{50} = 8.16 \text{ ob}$$

if mixing is complete. (According to Figure 11.4, this corresponds to a visibility of slightly less than 2 m.) In this calculation, the product $D_0 \cdot W_l$ (408 ob.m^3) is a measure of the total amount of smoke released from the polyurethane during the fire, expressed as the volume which has an optical density D/L of unity (1 ob). This may

Table 11.5   Smoke potentials of different materials (Rasbash and Phillips, 1978)

	Smoke potential (ob.m^3/g)[a]	
	Flaming	Non-flaming
Fibre insulation board	0.6	1.8
Chipboard	0.37	1.9
Hardboard	0.35	1.7
Birch plywood	0.17	1.7
External plywood	0.18	1.5
α-Cellulose	0.22	2.4
Rigid PVC	1.7	1.8
Extruded ABS	3.3	4.2
Rigid polyurethane foam	4.2	1.7
Flexible polyurethane foam	0.96	5.1
Plasterboard	0.042	0.39

[a] Smoke potential (D_0) is defined in Equation (11.5). The unit 1 ob (i.e. 1 db/m) corresponds to a tenfold reduction in light intensity over a pathlength of 10 m.

be generalized to define the total 'smoke load' of a room or compartment as:

$$V = \sum D_0 \cdot W_l$$

Provided that D_0 and W_l are known for each combustible item present, this could be used to compare the smoke production potential of different compartments within a building, although it gives no information on rates of production (see below). A similar method based on specific optical density (D_m) was proposed by Robertson (1975).

It is now generally accepted that the yield of smoke must be determined on a mass loss basis, rather than total mass or exposed surface area of the specimen tested. In the ISO test (ISO, 1990), the smoke yield is expressed as the 'mass optical density':

$$\text{MOD} = \frac{V}{LW_l} \left(\log_{10} \frac{100}{T} \right) \tag{11.6}$$

where T is the percent light transmission ($100 \times I/I_0$). If this is compared with Equations (11.3) and (11.5), it is seen to differ from the 'smoke potential' by a factor of 10: in Equation (11.6), optical density has the unit 'bel', compared with 'decibel' in Equation (11.3): otherwise, the two approaches are identical.

However, since the development of the Cone Calorimeter and large-scale experimental rigs which use oxygen consumption calorimetry to measure rate of heat release, there has been a move to standardize smoke measurement using natural logarithms in which optical density is defined as D_e (Equation (11.2)). This has been incorporated into the software produced for the dynamic measurement of smoke associated with these devices. The optical density of the smoke is measured continuously as it passes through the extract duct at a linear flowrate that is monitored continuously. Using the cross-sectional area of the duct, the volumetric flow rate ($\dot{V}$), corrected to ambient temperature, is calculated. By combining this with the rate of mass loss ($\dot{m}$) from the specimen under test, the specific extinction area (SEA) is obtained:

$$\text{SEA} = \frac{D_e}{L} \cdot \frac{\dot{V}}{\dot{W_l}} \tag{11.7}$$

As the mass loss is expressed in kg, SEA has units m^2/kg, but only because the unit of optical density per metre is lumped together with the units of volume. It can be converted into MOD (Equation (11.6)) or 'smoke potential' D_0 (Equation (11.5)) by multiplying by 4.34×10^{-4} (i.e. $10^{-3}/\log_e 10$) or 4.34×10^{-3}, respectively. Measurement of smoke is discussed in detail by Östman (1992) and Whiteley (1994).

The relationship between the smoke yields determined in standard tests and those obtained under real fire conditions has received comparatively little attention (see, however, Heskestad and Hovde, 1994). Most of the tests are carried out to allow the sample to burn under well-ventilated conditions (e.g. the Cone Calorimeter), but with tests such as the NBS Chamber (American Society for Testing and Materials, 1994a), it is necessary to limit the amount of material subjected to test because of the limited air available in the chamber. In fact, smoke yield is very sensitive to the conditions of burning. This has been demonstrated in the small-scale tests by varying the imposed heat flux and the orientation of the fuel, in addition to comparing the smoke from flaming combustion and non-flaming thermal decomposition (pyrolysis). In 'real fires', the full range of orientations are possible, and the imposed heat flux will exceed 30 kW/m^2 in the post-flashover fire. However, of greater significance is the fact that during the fully-developed fire the fuel volatiles are reacting in a hot atmosphere in which the oxygen concentration may be low. There have been a few studies which have shown that under these conditions the smoke yields may be greater than that predicted in the well-ventilated small-scale tests by as much as a factor of 4–6 (Drysdale and Abdul-Rahim, 1985) and perhaps by as much as a factor of 10 (Fleischmann *et al.*, 1990). On the other hand, vigorous free burning under well-ventilated conditions (i.e. pre-flashover) can produce less smoke than is predicted by the small-scale test (Rasbash and Pratt, 1980) (see Figure 11.5). Factors such as these need to be borne in mind when smoke test data are being considered as a means of assessing performance in real life. Heskestad and Hovde (1994) compared results on smoke yield from the Cone Calorimeter with data gathered from the ISO and CSTB rooms, and found no overall, simple correlation. They noted a reduction in the smoke yield from the room tests when the smoke layer began to burn, providing 'secondary combustion', but had some success in correlating the ratio (total smoke produced)/(total heat released) for cone data with that for data from the CSTB room.

While these effects have still to be quantified, it is often overlooked that the *rate* of smoke production is a major factor in determining how rapidly smoke will spread from the fire area to affect other spaces and escape routes. The rate of burning of a material must be considered with its smoke production potential. Consider the flexible polyurethane foam examined by Rasbash and Phillips (1978) (Table 11.5). Its smoke potential (0.96 ob m^3/g) is only 2–3 times greater than that for the wood products fibre insulating board, chipboard and hardboard, yet subjective impressions of fires involving these materials would suggest that polyurethane is in a different league as regards smoke production. This is due to the rapid development of fire on the foamed

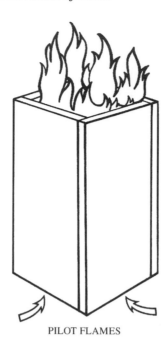

PILOT FLAMES

Figure 11.5   Chimney configuration used to generate smoke from freely burning combustible solids by Rasbash and Pratt (1980) (Less smoke is produced than predicted by small-scale tests.)

plastic (Section 6.3.2) and the subsequent high burning rate (see Table 5.8) which will produce unacceptable levels of smoke much more rapidly than the cellulosic materials. Indeed, any material or configuration of materials which can develop fire quickly and burn vigorously presents a threat to life safety.

It has been suggested that a proper assessment of the hazard due to smoke should include a weighting factor to penalize those materials which release smoke quickly (e.g. Routley and Skipper, 1980). Babrauskas (1984) has suggested the adoption of a smoke parameter based on data obtained in the Cone Calorimeter. This is the product of the average SEA for the first 5 min of flaming combustion and the maximum rate of release over the same period (measured at several imposed heat fluxes between 20 and 50 kW/m^2).

11.2   Smoke Movement

People die in fires in buildings because they are unable to reach a place of safety before being overcome by conditions created by the fire. This has been discussed in the context of escape route design by Marchant (1976) who distinguished the components of the escape time, namely t_p (the time from ignition to perception that a fire exists), t_a (time from perception to the start of the escape action) and t_{rs} (time taken to move to a place of relative safety) (Equation (9.1)). If

$$t_p + t_a + t_{rs} < t_u \tag{9.1}$$

where t_u is the time from ignition to the production of untenable conditions, then escape will be successful. While t_p and t_a are largely dependent on state of awareness, t_{rs} is influenced by many factors, including the agility of the individual, the building geometry and the degree to which the person is affected by the products of the fire. It was argued earlier (Chapter 9) that the chance of successful escape can be enhanced by providing means of early detection (reducing t_p) and by avoiding materials and configurations which would lead to rapid growth (increasing t_u). Alternatively, measures can be taken to reduce t_a and t_{rs}. These would include provision of well-marked and well-designed escape routes which could be kept free of smoke for as long as it takes to evacuate the building.

The substantial majority of fire fatalities can be attributed to the inhalation of smoke and toxic gases (Home Office, 1995; National Fire Protection Association, 1997). While many of these individuals may have been overcome by combustion products while asleep or otherwise incapacitated in the room of origin, a large number are found at points remote from the fire, perhaps attempting to effect their escape. Clearly, any life safety system designed into a building must incorporate means of protecting the occupants from smoke as they make their way to safety. For this to be achieved, the behaviour and movement of smoke within a building must be understood. In the following discussion, 'smoke' refers to the suspension of particles in air which is also contaminated by gaseous combustion products, some of which are toxic. As the smoke flows away from the seat of the fire, entraining more air, the concentration of these products (particles and vapours) fall proportionately, as does the temperature.

During the early stages of a fire in a single compartment, a smoke layer will form below the ceiling and descend at a rate determined by the rate of burning (rate of heat release) and (initially) by the height of the ceiling. As the smoke layer descends, the mass flow into the layer will decrease as the height of the free plume decreases. This logic was applied in Section 4.4.4 in the example relating to smoke extraction, and formed the basis of the computer programme developed by Cooper and Stroup (1985) to calculate the 'available safe egress time' from a single compartment (see also Cooper, 1995). This is the simplest of the 'zone models' (see Section 4.4.5). Since the 1980s, a plethora of this type of model has appeared (Friedman, 1992), some dealing with multiple interconnecting spaces in which the behaviour of smoke may be determined by some factors which are unconnected with the fire itself. These are forces which are responsible for the movement of smoke in a building, and will be discussed before a brief review of smoke control systems is presented.

11.2.1 *Forces responsible for smoke movement*

Smoke, and indeed any fluid, will move under the influence of forces which will be manifest as pressure gradients within the bulk of the fluid. For the movement of smoke within a building, such forces are created by:

(a) buoyancy created directly by the fire;
(b) buoyancy arising from differences between internal and external ambient temperature;
(c) effects of external wind and air movement; and
(d) the air handling system within the building.

Each of these will be considered in turn as all must be taken into account if the direction and rate of movement of the smoke are to be predicted.

(a) *Pressure generated directly by the fire*   Burning in a compartment generates high temperatures which produce the buoyancy forces responsible for hot fire gases being expelled through the upper portion of any ventilation opening (Figures 10.4 and 10.7) or through any other suitable leakage paths. If the temperature inside the fire compartment is known, then from Section 10.2 (Equations (10.2a) and (10.2b)), the value of Δp at any height (y) above the neutral plane can be calculated as $(P_1 - P_2)$. Thus, if the fire temperature is 850°C (1123 K) and the ambient (external) temperature is 20°C (293 K) then, taking densities from Table 11.6 below, the pressure difference $y = 1.5$ m above the neutral plane will be $c.13$ N/m^2. This will be proportionately greater for taller compartments, and is the source of the positive pressure that will force smoke, and ultimately flames, through unstopped service penetrations at high level.

These calculations are in general agreement with measurements made by Fang (1980) in a study carried out to ascertain the magnitude of fire pressures that would have to be overcome to protect escape routes from smoke by pressurization (Section 11.3.3).

(b) *Pressure differences due to natural buoyancy forces*   In addition to the natural buoyancy forces generated by the fire itself, the stack effect must be taken into account in tall buildings. The concept of buoyancy was introduced and developed in Section 4.3.1. As long as the smoke is at a higher temperature than the surrounding air, it will rise, the buoyancy force per unit volume being given by the product $g(\rho_0 - \rho)$ where ρ_0 is the density of the ambient air, ρ is the density of the smoke or smoke-laden air, and g is the gravitational acceleration constant. The fire provides the energy necessary to drive these buoyant flows which will dominate movement in the vicinity of the fire (Section 4.3). However, as one moves away from the locus of the fire, temperatures fall and the associated buoyancy force will diminish to a point where other forces can begin to dominate behaviour.

In tall buildings containing vertical spaces (stairwells, elevator shafts, etc.) temperature differences between the inside and the outside will give rise to buoyancy-induced pressure differences known as the stack effect. If the temperature inside the building is uniform and greater than the external (ambient) temperature, then there will be a natural tendency for air to be drawn in at the lowest levels and expelled at the highest (see Figure 11.6(a)) (this should be compared with the arguments presented with Figure 10.4). Consider first a vertical column of air contained within a shaft of height H which is open only at the bottom (Figure 11.7) and take the internal and external temperatures to be T_i and T_0 respectively. If the external pressure at ground level is p_0 then the pressures at height H inside and outside the building will be:

$$p_i(H) = p_0 - \rho_i g H \tag{11.8}$$

and

$$p_0(H) = p_0 - \rho_0 g H \tag{11.9}$$

where ρ_i and ρ_0 are the densities of air at T_i and T_0 respectively. Thus the pressure difference between the inside and outside at the top of the building will be:

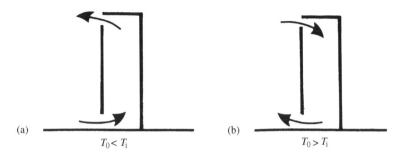

(a) $T_0 < T_i$

(b) $T_0 > T_i$

Figure 11.6   The stack effect in tall buildings. (a) external temperature (T_0) < internal temperature (T_i): (b) $T_0 > T_i$, showing accompanying flows

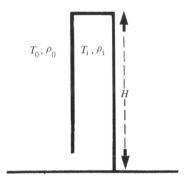

Figure 11.7   The origin of the stack effect (no flow)

$$\Delta p = (\rho_0 - \rho_i)gH \qquad (11.10)$$

i.e. the pressure will be greater within the building than outside it.

If the shaft has openings at the top and the bottom then, provided that $T_i > T_0$, there will be a net upward flow of air and a neutral plane will establish itself where $p_0 = p_i$. Carrying out the same analysis as before, then at any height h *above* the neutral plane, the pressure difference between the inside and outside will be:

$$\Delta p = (\rho_0 - \rho_i)gh \qquad (11.11)$$

In a real building, of course, the openings will be distributed as a large number of small leakage areas, around windows and doors, even if these are closed. Above the neutral plane, air (smoke) will tend to flow outwards from the shaft while below there will be a net inward flow, creating an upward movement within the shaft (Figure 11.6(a)). However, if the external temperature is greater than that inside the building, as would be the case in air-conditioned buildings in hot climates, then the situation will be reversed (Figure 11.6(b)). Air will tend to flow towards the bottom of such shafts, and should fire occur, the initial movement of smoke could be contrary to expectations.

The pressure difference that will exist between inside and outside at height H in Figure 11.7 can be calculated by assuming ideal behaviour, i.e. the relationship

$pV = nRT$ is obeyed: if the molecular weight of 'smoke' approximates to that of air (0.0289 kg/mol), then from Equation (11.10) (see Equation (1.10)).

$$\Delta p = 3.46 \times 10^3 \left(\frac{1}{T_0} - \frac{1}{T_i} \right) H \tag{11.12}$$

For a 30 m high building for which $T_0 = 273$ K and $T_i = 293$ K, $\Delta p = 26$ N/m^2. However, the distribution of leakage paths to the outside will determine where the neutral plane will lie. This is required for calculations relating to smoke movement, and can be calculated if the sizes of the openings are known. If h_1 and h_2 are the distances from the neutral plane to the lower and upper openings respectively, Figure 11.6(a),

$$\frac{h_1}{h_2} = \frac{A_2^2}{A_1^2} \cdot \frac{T_1}{T_0} \tag{11.13}$$

where A_1 and A_2 are the cross-sectional areas of the lower and upper openings, respectively. This expression was originally derived for leakage via the top and bottom of the door to a fire compartment. Note that the greater value of A_2 the higher the neutral plane will be above the lower opening.

The significance of the stack effect is that it can move relatively cool smoke around a high-rise building very efficiently, moving it rapidly to some areas, while protecting others (e.g. Figure 11.8). A fire in the lower part of a high-rise building can cause very rapid smoke logging of the upper levels of the building. There are many cases in which this has happened (Butcher and Parnell, 1979; Best and Demers, 1982)).

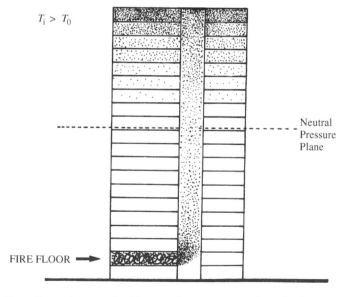

Figure 11.8   The effect of the stack effect on the movement of smoke in a high-rise building ($T_i > T_0$). Below the neutral plane there is a tendency for air to be drawn into the central core, while above it, there is a net outward flow (cf. Figure 11.6(a))

(c) *Pressure differences generated by wind*   Natural wind can generate pressure distributions over the envelope of a building which are capable of influencing smoke movement within. The external pressure distribution depends on numerous factors including the speed and direction of the wind and the height and geometry of the building. The magnitude of the effects may be sufficient to override the other forces (natural and artificial) which influence smoke movement. Generally speaking, wind blowing against a building will produce higher pressures on the windward side and will tend to create air movement within the building towards the leeward side where the pressure is lower. The magnitude of the pressure difference is proportional to the square of the wind velocity (Benjamin *et al.* 1977). The pressure at the surface of a building is given by:

$$p_w = \tfrac{1}{2} C_w \rho_o u^2 \ \text{N/m}^2 \qquad (11.14)$$

where C_w is a dimensionless pressure coefficient (ranging from $+0.8$ to -0.8, for windward and leeward walls respectively), ρ_o is the outside air density (kg/m^3) and u is the wind velocity (m/s). High windspeeds can generate pressure differentials which are sufficient to overcome other forces, both natural and those imposed artifically to control smoke movement (Section 11.3) (Klote, 1995).

The pressure distribution over the surface of a building is influenced strongly by the juxtaposition of the neighbouring buildings and by the geometry of the building itself. A common situation is a single-storey structure, such as a shopping mall, associated with a multi-storey building, e.g. an office block (Figure 11.9). The wind patterns around this particular geometry of building can be extremely complex and the pressure distribution over the surface of the roof of the shopping mall will vary considerably with changes in wind speed and direction (Marchant, 1984). Thus, while it is possible to protect a single-storey shopping mall from smoke logging during a fire by relying entirely on natural ventilation through smoke vents in the roof, the location and types of vent must be selected on the basis of the pressure distributions that are to be expected on the mall roof. If there is likely to be an area of relatively high pressure in the vicinity of the smoke vent under certain wind conditions, then natural venting will not be a reliable method of removing smoke from the shopping mall (see Section 11.2.3). However, it should be noted that for most other situations, the wind induces low pressures on roofs which will enhance the effectiveness of vents.

(d) *Pressure differences caused by air handling systems*   Many modern buildings contain air handling systems for the purpose of heating, ventilation and air-conditioning (HVAC). While the fans are idle, the ductwork can act as a network of channels through which smoke will move under the influence of the forces discussed above, including particularly the stack effect in multi-storey buildings. This will promote spread throughout the building, an effect which can be even greater if the system is running when fire breaks out. This situation can be avoided by automatic shutdown, activated by smoke detectors, should fire occur anywhere in that part of the building served by the HVAC system. Alternatively, on a different level of sophistication, the HVAC system can be designed to 'handle' the smoke, removing it from the building, while protecting other spaces by remote operation of isolation valves. This requires the facility to reverse flows of air within the system and would demand expert supervision and management. This type of dual-purpose system has been installed in some large

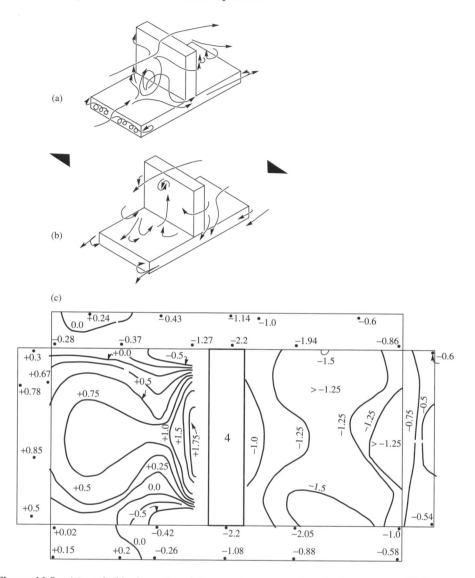

Figure 11.9   (a) and (b) show the airflow patterns around a single-storey mall/office tower complex for one wind direction. (c) The wind-induced pressure patterns over the surface of the building for the airflow shown in (a) and (b). Overpressures in N/m^2 (Marchant, 1984, by permission)

department stores in the UK. The role of HVAC systems in smoke movement is discussed briefly by Benjamin *et al.* (1977) and Klote (1995).

11.2.2   *Rate of smoke production in fires*

As will be discussed in Section 11.3, there are various ways in which smoke can be 'managed' in large buildings to allow the occupants to escape, and the firefighters to find and extinguish the fire quickly and safely. The smoke may be extracted at a rate sufficient to keep the bottom of the smoke layer above head height (Sections 11.3.1 and 11.3.2), but to be able to design the smoke extraction system, it is necessary to be able to calculate the volume of smoke that has to be handled. In this context, the optical density of the smoke — while it may be useful to know — is not required for the calculations.

During the early stages of a compartment fire, when burning is localized, the combustion products will be progressively diluted as they rise vertically in the buoyant plume until deflected by the ceiling (Section 4.3.3). The hot smoke will then flow horizontally as a ceiling jet until it finds a discontinuity and can continue its upward movement, or as is more likely, until it encounters a vertical barrier, such as a wall, which will prevent further travel and cause the smoke layer to back up and deepen, confined by the ceiling and the compartment walls (cf. Section 9.2.1). The rate at which this layer thickens will depend partly on the rate of burning, but predominantly on the amount of air that is entrained into the fire plume before it enters the layer.

In Section 4.4.4, it was shown how Equation (4.48) could be used to calculate the mass flowrate of 'smoke' at any height above a fire whose rate of heat release is $\dot{Q}_c$ (kW):

$$\dot{M} = 0.071 \dot{Q}_c^{1/3} z^{5/3} \left[1 + 0.026 \dot{Q}_c^{2/3} z^{-5/3} \right] \tag{11.15}$$

where z is the height (in m) above the virtual source. If z is the height of the smoke layer, then this equation gives the rate of addition of smoke to the layer. If it is required to extract smoke at roof level to maintain the level of the layer at z m above the fire, then Equation (11.15) gives the 'rate of production of smoke' which must be balanced by the rate of extraction (see Figure 11.10). As discussed earlier, it is necessary to specify an appropriate 'design fire' which is based on the nature and distribution of combustible items present in the space. The temperature of the layer is to be calculated from the equation:

$$\Delta T = \frac{\dot{Q}_c}{\dot{M} C_p} \tag{11.16}$$

where c_p is the heat capacity of air ($c.1.0$ kJ/kg.K) and ΔT is the temperature excess (i.e. above ambient). The latter information is required to specify the working temperature of the extract fan.

The concentration of smoke in the layer is not required for smoke venting calculations, but it can be estimated if the rate of burning and the 'smoke yield' are known. To take the example from Section 4.4.4, if it is assumed that the principal fuel is polyurethane foam, then taking $\Delta H_c = 25$ kJ/g (an estimate) and $\dot{Q}_c = 6.4 \times 10^3$ kW, the rate of burning becomes $\dot{Q}_c / \Delta H_c = 256$ g/s. With $D_0 = c.1$ ob.m^3/g from Table 11.5, the rate of production of smoke is equivalent to 256 ob.m^3/s. It was shown in Section 4.4.4 that the rate of addition of smoke to the layer was equivalent to 131 m^3/s, indicating that the optical density per metre in the layer is $256/131 = 1.95$ ob, corresponding to a visibility of $c.5$ m (Figure 11.4).

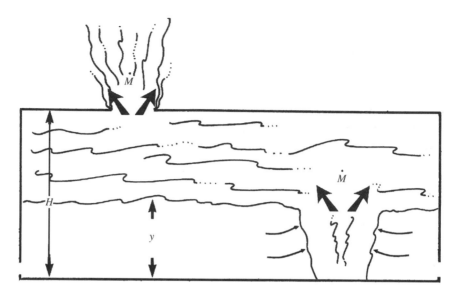

Figure 11.10   Venting of the hot gases from a fire. When the mass flowrate ($\dot{M}$) through the vent equals the mass flowrate from the fire plume into the smoke layer, the height of the smoke layer will stabilize at y

Thomas *et al.* (1963) developed the following expression which has been widely used in the UK for venting calculations, in which it was assumed that the total amount of air entering the rising plume was proportional to its 'surface area':

$$\dot{M} = 0.096 P_{\mathrm{f}} \rho_0 y^{3/2} (gT_0/T_{\mathrm{f}})^{1/2} \qquad (11.17)$$

where P_{f} is the perimeter of the fire (m), y is the distance between the floor and the underside of the smoke layer below the ceiling (i.e. the height of clear air, in metres) (Figure 11.10). ρ_0 is the density of ambient air (kg/m^3) and T_0 and T_{f} are the temperatures of the ambient air and the fire plume respectively (K). Taking $\rho_0 = 1.22$ kg/m^3, $T_0 = 290$ K and $T_{\mathrm{f}} = 1100$ K, this becomes

$$\dot{M} = 0.188 P_{\mathrm{f}} y^{3/2} \text{ kg/s} \qquad (11.18)$$

Equation (11.18) is valid only for fires in which the ratio of flame height to fire diameter is of the order of one or less, and in principle applies only up to the flame tip, i.e. it can only be applied when the flames reach into the smoke layer. Nevertheless, it is claimed to give adequate predictions above the flame tip. A typical application where this expression has been used is in estimating the rate of smoke extraction that would be required to avoid smoke logging of large public areas with relatively low ceilings, such as shopping malls. As before, the objective would be to prevent the smoke layer descending below a critical height (say $y = 2$ m) (Figure 11.10): from Equation (11.18) the rate of extraction would have to be:

$$\dot{M} = 0.188 P_{\mathrm{f}} \cdot (2)^{3/2} = 0.53 P_{\mathrm{f}} \text{ kg/s} \qquad (11.19)$$

Table 11.6   Density of air as a function
of temperature

Temperature (K)	Density (kg/m³)
280	1.26
290	1.22
300	1.18
500	0.70
700	0.50
1100	0.32

or

$$V_s = \frac{0.53 P_f}{\rho_s} \tag{11.20}$$

where V_s is the volumetric flowrate and ρ_s is the density of the smoke at the point of extraction (the smoke vent). As density varies inversely with temperature, substantial volumes may have to be extracted. (Table 11.6 gives the density of air at temperatures up to 1100 K: this is a good approximation to 'smoke'.)

Clearly, a number of parameters need to be known to enable V_s to be calculated (P_f, y and T_s, the temperature of the smoke at the vent). The volumetric flow through the vent (which must not be less than V_s) will depend on the vent area, T_s, and the buoyancy head existing under the steady-state conditions assumed for design purposes. This will be a function of T_s and the depth of the layer of hot smoke (($H - y$) in Figure 11.10). Thomas, Hinkley and co-workers (1963, 1964) (Butcher and Parnell, 1979) carried out a detailed analysis of the problem using scaled experimental models (Section 4.4.5) and developed a series of equations which may be used to calculate the required vent areas. These have been presented in the form of nomograms for use by the designer (Thomas *et al.*, 1963; Butcher and Parnell, 1979). It is required that there are sufficient air inlets near floor level to replenish the hot gases as they are vented.

Dividing Equation (11.17) by ρ_s, the density of the smoke layer, to give the volumetric rate of production (as in Equation (11.20)) and expressing $V_s = A(H - y)$, where A is the floor area of the enclosure, the resulting equation may be integrated to give:

$$t = 20.8 \frac{A}{P_f} \frac{T_0}{T_s} \left(\frac{T_f}{g T_0} \right)^{1/2} \left(\frac{1}{y^{1/2}} - \frac{1}{H^{1/2}} \right) \tag{11.21a}$$

which Butcher and Parnell (1979) quote as

$$t = \frac{20A}{P_f g^{1/2}} \left(\frac{1}{y^{1/2}} - \frac{1}{H^{1/2}} \right) \tag{11.21b}$$

assuming $T_s \approx 300°C$. This gives the time it would take for the smoke layer to descend to a height y m above the ground, assuming a fire of perimeter P_f burning in an enclosure of floor area A and height H. Clearly, the vents must open within this period of time if the smoke layer is to be arrested at a height y metres. Figure 11.11 shows the variation of y with t for a fire with a 6 m perimeter in a 5 m high enclosure of

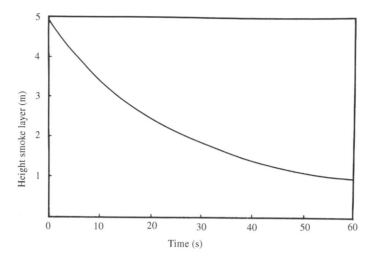

Figure 11.11   Decrease in the height of the lower boundary of the smoke layer with time for a fire of perimeter 6 m in a compartment 5 m high and of 100 m² floor area. It is assumed that the fire is burning steadily from $t = 0$. (Calculated from Equation (11.21b), from Hinkley, 1971)

area 100 m². This could be taken to represent a small hospital ward with a bed fully alight (ignoring the initial growth period).

If by means of suitable venting the smoke layer descends and is maintained at a height y above the floor, then the optical density within the layer will be given approximately by:

$$D = \frac{10^3 D_0 \cdot \dot{m} \cdot \rho_s}{0.188 P_f \cdot y^{3/2}} \qquad (11.22)$$

if $\dot{m}$ is in kg/s and D_0 is an effective standard optical density relating to the materials and the conditions of burning. Apart from the effect of density (ρ_s) any effect of temperature on the obscuration potential of the smoke is tacitly ignored. This equation could be used to estimate the opacity of the smoke reaching a point remote from a fire, provided that the extent of further entrainment and dilution could be quantified.

Regarding the prediction of smoke movement in complex spaces, several modelling techniques are now available. The most successful are based on Computational Fluid Dynamics, a tool which is becoming increasingly popular for this type of work. However, as was pointed out in Section 4.4.5, there is still a need to verify that the predictions are correct. More experimental data are required before any given model—zone or field—can be used outside the relatively narrow range of conditions for which it was developed. Nevertheless, it may be assumed that this will be the way forward once the problem of reliable verification has been resolved.

11.3   Smoke Control Systems

If it is considered necessary to prevent lethal concentrations of smoke accumulating in certain areas of a building to protect its occupants, there are two basic approaches that may be adopted at the design stage: either the smoke must be contained, or it must be extracted. Smoke containment may be achieved by the use of physical barriers such as walls, windows and doors, as well as smoke curtains or other purpose-designed smoke barriers. Service penetrations and other shafts between compartments must be smoke stopped. Containment may also be achieved by the use of pressurization to provide adequate pressure to resist the flow of smoke, or by opposed flow which overcomes the flow of smoke by creating a flow of air in the opposite direction. Both of these principles are used in the design of pressurized stairs (Section 11.3.3).

Smoke extraction is also referred to as smoke dilution, smoke clearance or smoke removal. The extraction system may be specified in terms of the number of air changes per hour, sufficient perhaps to ensure that tenable conditions in an area will be regained within a reasonable period. Alternatively, extracting smoke may be designed to retain the smoke above head height, or reduce the concentration of lethal products on escape routes to acceptable levels. In this instance, the extraction rate would be based on a specified design fire, as previously discussed: this may vary with time. Calculations would be carried out on smoke production rates and achievement of untenable conditions in occupied areas. Smoke might be removed by natural venting or by mechanical means, but in either event a sufficient supply of make-up air has to be arranged in order for the extraction system to work as designed.

Some particular issues in smoke control design in buildings are discussed below.

11.3.1 Smoke control in large spaces

Large, undivided single-storey buildings are normally designed around the operations which they are to house, e.g. assembly lines, etc., where subdivision would hinder efficient operation. Such buildings may be fully sprinklered so that any fire would be of limited extent, but smoke can still spread throughout the building. On arrival at the fireground, the first action of the fire brigade may be to vent the smoke by creating an opening in the roof, thereby improving visibility above floor level. If automatic vents of appropriate size open either by means of a fusible link or through a control system activated by smoke detectors, smoke logging of the building will not occur and fire brigade access to the seat of the fire will be much easier. In addition, there is no accumulation of hot smoke below the ceiling which could enhance the growth to flashover in the vicinity of the original fire. However, to be set against these points are the following:

(a) If the smoke vents are to operate effectively, there must be adequate ventilation at ground level to permit the hot gases to escape freely. It can be argued that this will enhance the fire growth and allow more rapid spread, which has been taken as an argument to reject fire venting as a general policy.

(b) If combined with sprinklers, early operation of the vents may divert the flow of hot gases away from the sprinkler head(s), thereby preventing their activation.

As with all these matters, there cannot be a definitive answer regarding the value of venting versus sprinklers: each case is different and must be judged on its merits. If there are good reasons for wanting to limit the spread of smoke within a large undivided space then a combination of automatic vents and smoke 'curtains' to create smoke reservoirs underneath the ceiling might be an appropriate solution (Figure 11.12(a)). The number, size and location of vents necessary for efficient venting can be calculated using the procedures described briefly in the previous section (e.g. Butcher and Parnell, 1979). The theoretical background is to be found in the original Fire Research Technical Papers (Thomas *et al.*, 1963; Thomas and Hinkley, 1964). Some of the factors that must be considered are:

(i)   the size of the fire;
(ii) the height of the building;
(iii) the type of roof; and
(iv) the pressure distribution over the roof.

Of these, (i) and (ii) have already been discussed. The type of roof (flat, pitched, north-light, etc.) will determine the need for smoke 'curtains' or screens which will not only limit the spread but also allow the smoke to build up a buoyancy head below the ceiling which will enhance the flow through the vent (Figure 11.12(a)). If there is a positive pressure on the roof, created by the wind, then the effectiveness of the vent will at best be reduced: if this pressure is too great, the vent may operate in reverse (Figure 11.12(b)).

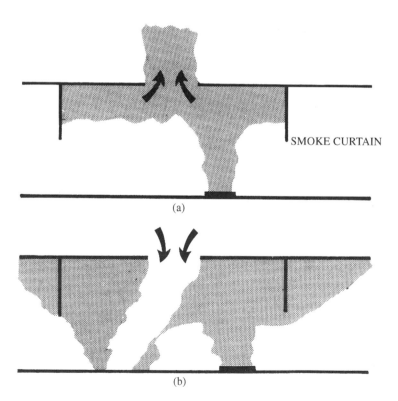

SMOKE CURTAIN

(a)

(b)

Figure 11.12   The operation of vents in association with smoke curtains located to form a 'reservoir': (a) neutral or negative pressure at roof level; (b) positive pressure at roof level (see Figure 11.9). In practice, it is better to have several smaller vents rather than one large one. (See Figure 11.13 (Butcher and Parnell, 1979))

11.3.2 Smoke control in shopping centres

While the general rule is to keep smoke out of escape routes, there are certain situations in which this is not practicable. The most common example is the modern single-storeyed shopping mall which is designed to allow free movement of people within a large enclosed space from which they can enter shops. The mall itself, if suitably designed and its contents controlled, will not present a fire hazard, but should a fire develop in any of the shops, the mall could become smoke-logged very quickly, making rapid egress impossible. There are two methods of controlling this situation (Hinkley, 1971):

(i)   vent the smoke directly from the shop to the outside;

(ii) provide smoke 'reservoirs' fitted with automatic ventilators in the ceiling of the shopping mall.

The first of these is desirable if the shop is large, as by the time the smoke enters the mall it will have cooled considerably and will have lost much of the buoyancy necessary to allow natural venting to occur. Smoke reservoirs similar to those shown in Figure 11.12 should be formed within the shop to encourage direct venting (Butcher and Parnell, 1979). However, these must be large enough to prevent smoke entering the mall. Provision of an emergency smoke door or shutter dividing the shop from the mall is not practicable as the escape routes must remain clear. Consequently, it is physically possible to apply this method only to the largest shops. If there are sprinklers installed, their interaction with the smoke layer may cause some cooling and entrainment of smoke in the downward spray pattern, bringing cool smoke to ground level. However, this is not as great a problem as it might seem at first sight. If the smoke layer is greater than 1 m thick and is hot enough to activate the sprinkler, then the smoke will retain sufficient buoyancy to flow through the vents, at least until the sprinklers begin to reduce the intensity of the fire below (Bullen, 1977b). Also — initially at least — some of the smoke entrained in the sprinkler spray will be drawn back into the fire near floor level. The entrainment (and cooling) can be reduced by using a lower water pressure (Bullen, 1997b).

With all but the largest shops, it is inevitable that smoke will enter the mall, which must therefore contain its own smoke control system. If the smoke flows unhindered under the ceiling of the mall, it will cool and begin to descend towards the floor, particularly near the exits where it can mix with inflowing fresh air. Under these conditions, smoke-logging could occur very quickly and vents would not operate efficiently as the layer of smoke would be too thin (Figure 11.13). Consequently, the shopping mall must also be provided with smoke reservoirs which are deep enough (>1 m) to provide the buoyancy head required for natural venting. However, if the mall is associated with a high-rise building — or indeed with any multi-storey structure — then care must be taken that adverse pressure distributions over the roof structure will not negate the venting action (Figure 11.9). If this is a problem then power-operated fans may have to be used to assist the venting process.

An important general principle regarding all smoke control problems is that the total volume of smoke should be kept to a minimum. The provision of deep smoke reservoirs helps to achieve this. High temperatures (and buoyancy) are maintained

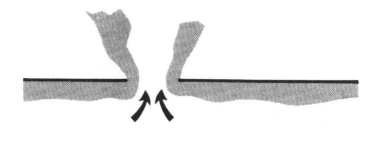

Figure 11.13   Operation of a vent in the absence of a substantial buoyant head (thin smoke layer). The same effect will occur with a deep layer if the area of the vent is too large

and less 'smoke' has to be 'handled' and removed from the building. The problem is exacerbated in multi-storey atrium-like shopping malls where a fire in a shop at ground level will produce a plume which has a long way to rise before entering the smoke layer under the roof. The extra air entrainment that will occur during the vertical movement will greatly increase the effective smoke volume (Morgan *et al.*, 1976). The magnitude of this effect can be limited by restricting the initial width of the vertical plume, but the volume of the smoke reservoir and the area of the smoke vents have to be considerably greater than for a single-storey shopping mall.

11.3.3   Smoke control on protected escape routes

By definition, smoke must not enter protected escape routes. Smoke doors between the fire and the protected escape route will help to keep smoke back but this relies on (a) the door being closed at the time of the fire, and (b) persons using the door to reach the escape route not keeping the doors open for a prolonged period. However, smoke will migrate with the natural movement of air within the building, and pressure differentials may encourage movement into the escapeways. One way to overcome this is to pressurize the escape route sufficiently so that even under the most unfavourable conditions smoke will not enter as there is a net flow of air from the escape route into the adjacent spaces. This technique has received a great deal of attention in the UK, Canada, USA and Australia, and has been adopted in several modern high-rise buildings around the world. The principles are illustrated in Figure 11.14.

Early studies involved measuring the pressure differentials that can exist across doorways, etc., within a building as a result of the stack effect and the wind. It was shown that differentials of 25–50 N/m² would be sufficient to override the worst conditions that might arise naturally, although with a very tall building the stack effect might overcome even this. If the leakage of air around all doors, windows, etc., in the escape route is calculated, then the capacity of the fan necessary to maintain the desired over-pressure can be found. In the British Standard Code of Practice (Butcher and Parnell, 1979; British Standards Institution, 1978) a pressure differential of 50 N/m² is called

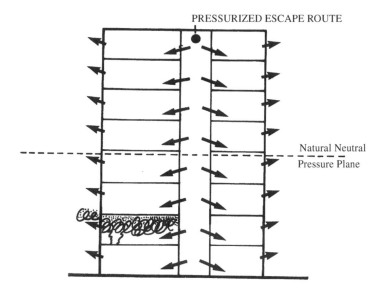

PRESSURIZED ESCAPE ROUTE

Natural Neutral
Pressure Plane

Figure 11.14 Pressurization as a means of keeping escape routes clear of smoke. Compare this with Figure 11.8

for under emergency conditions. This may be provided by a continuously operating fan which is integrated into the ventilation system, although normally providing a much lower overpressure. Leakage around doors into the remainder of the building will produce slight overpressures in the corridors leading to the protected escape route. This will help to reduce (but not prevent) the entry of smoke into these spaces.

The pressurization system must be designed to cope with one door being open continuously, providing an airflow through the open door of 0.75 m/s, although lower values are permitted if there is a lobby between the escape route (e.g. stairwell) and the corridor. This technique is described in detail by Butcher and Parnell (1979).

The subject of smoke movement and control has been studied extensively in the UK, USA, Canada and Australia. The various codes of practice that have been developed are based largely on the results of fundamental, small-scale studies which have been scaled up using established modelling procedures. Reviews of current practice may be found in the *SFPE Handbook* (Klote, 1995; Milke, 1995b) (USA) and in *CIBSE Guide E* (Chartered Institute of Building Services Engineers, 1997) (UK).

Answers to Selected Problems

1.1 CO_2: 1.798 kg/m^3; C_3H_8: 1.798 kg/m^3; C_4H_{10}: 2.370 kg/m^3.

1.2 3.32 m^3.

1.3 (a) 3.26 mm Hg; (b) 28.41 mm Hg; (c) 61.38 mm Hg.

1.4 n-Hexane: 3.93 mm Hg; n-decane: 1.71 mm Hg.

1.5 iso-Octane: 2.46 mm Hg; n-dodecane: 0.19 mm Hg.

1.6 −103.8 kJ/mol.

1.7 −159.5 kJ/mol.

1.8 −1844 kJ/mol.

1.9 −2976 kJ/mol.

1.10 (a) −41.3 kJ/g pentane; (b) −2.89 kJ/g air.

1.11 12.37 g air/g propane; 12.1 g air/g pentane; 12.27 g air/g decane.

1.12 (a) 6308°C (dissociation would be dominant, preventing such high temperatures being achieved); (b) 2194°C (dissociation would be significant and keep the temperature below 2000°C); (c) 1373°C (dissociation relatively unimportant).

2.1 320 W/m^2 (to 3 d.p., 319.861 W/m^2).

2.2 Taking 319.861 W/m^2, the temperatures are 60.020°C and 59.983°C, respectively (effectively both at 60°C). Bi = 0.001 (definitely thermally thin). For the 50 mm thick barrier, the rate of heat transfer is 318.615 W/m^2, and the temperatures are 60.17°C and 59.83°C. The associated Bi = 0.009, also thermally thin. Conclusion? Because the thermal conductivity of steel is so high, substantial sections may be treated as 'thermally thin'. (See Chapter 10.)

2.3 1.71 kW (1.27 kW through the brick, 0.44 kW through the window).

2.4 (a) 0.34 kW through window; (b) 0.79 kW through brickwork; (c) 0.20 kW through brickwork.

2.5 239°C (it is likely that the brickwork would fail before a steady state is reached).

2.6 963°C (the fibre insulation board would ignite and burn long before a steady state is reached).

2.7 The numerical solution gives 47.36°C, while the analytical solution is 46.81°C.

2.8 6 seconds.

2.9   202°C.

2.10 (a) 60°C; (b) 50°C; (c) 50°C (approximate values only, obtained by interpolating data from Figure 2.6(a) and (b)).

2.11 (a) 50°C; (b) 34°C; (c) 20°C (approximate values only). Compare these results with those from Problem 2.10.

2.12 The polyurethane foam and the fibre insulation board can be treated as semi-infinite solids.

2.13 (a) 54°C; (b) 115°C.

2.14 (a) 0.85 kW/m^2; (b) 1.1 kW/m^2; (c) 1.8 kW/m^2.

2.15 0.07.

2.16 13.1 kW/m^2.

2.17 10.7 kW/m^2.

2.18 The convective and radiative heat losses respectively are: at 200°C, 3.0 kW and 4.8 kW; at 400°C, 8.2 kW and 19.8 kW; at 600°C, 14.5 kW and 56.0 kW; and at 800°C, 21.7 kW and 127.8 kW.

2.19 649.5 K (taking $h = 11$ W/m^2K).

3.1   4.35%.

3.2   1614 K.

3.3   (a) 1.7%; (b) 3.4%.

3.4   (a) 43.5%; (b) 31%.

3.5   (a) 31.6% CO_2; (b) 19.9% CF_3Br. Much less CF_3Br is found to be sufficient because this species acts as a chemical suppressant.

3.6   At 200°C, $L = 1.8\%$ and $U = 10.7\%$: at 400°C, $L = 1.5\%$ and $U = 12.1\%$.

3.7   Between -26°C and 6.3°C (assuming limits unchanged from their values at 25°C).

3.8   Between 32 and 235 mm Hg (however, refer to Figure 3.8(a)).

4.1   19.9 m and 9.7 m.

4.2   (a) 654 kW (b) 1672 kW

4.3   (a) 0.612 (b) 1.566

4.4   Using Equation 40, (a) 2.06 m: (b) 3.46 m. Using Equation 41, (a) 2.02 m; (b) 3.41 m. The flame volumes are (a) 0.54 m^3, and (b) 1.39 m^3.

4.5   1.5 Hz predicted for both.

4.6   From Table 4.2, (a) 72°C; (b) 60°C. From Equation 4.35, (a) 71°C; (b) 60°C. If the virtual source is taken into account, then these become 70°C and 57°C respectively.

4.7   (a) 239.70°C; (b) 90°C and 64°C.

4.8   At 4 m, $u = 1.18$ m/s: at 8 m, $u = 0.797$ m/s. The response times are 359 s and 1069 s respectively. If the RTI is reduced to 25 m1/2.s1/2, the response times are 87.5 s and 267 s respectively.

4.9   0.9 MW if fire directly below detector. Worst situation gives 2.2 MW. If fire close to one wall or in a corner expect minimum size to be 1.1 MW and 0.55 MW respectively.

4.10 (a) 372 kW; (b) 186 kW; and (c) 93 kW

4.11 With 5 m centres, (a) 64 kW; (b) 32 kW; and (c) 16 kW. With 2.5 m centres, (a) 32 kW; (b) 16 kW; and (c) 8 kW.

4.12 (a) 73 kg/s; (b) 110 kg/s.
4.13 The values of ΔT are (a) 87 K and (b) 36 K

5.1 Using ΔH_c (n-hexane) $= 45$ kJ/g (see Table 1.13), rate of heat release $= 9.6$ MW.
5.2 Rate of heat release $= 1.4$ MW. Flame height ~ 2.1 m. For radiation calculation, need H_2O and CO_2 concentrations in the flame, to enable Hottell's method to be used.
5.3 Assuming that the oil is consumed in 120 minutes, average rate of burning $= 20.85$ kg/s (0.118 kg/m^2.s); average rate of heat release $= 939$ MW.
5.4 12.5 kW/m^2
5.5 The limiting regression rates (Table 5.1) give 0.0224 kg/m^2.s and 0.083 kg/m^2.s for methanol and heptane, respectively. (Table 5.2 gives 0.017 for methanol). The 'ideal' values of Tewarson are 0.032 kg/m^2.s and 0.093 kg/m^2.s respectively. The latter assume no heat losses, and are therefore slightly higher.
5.6 (a) Ethanol, 0.00253 kg/m^2.s; (b) Hexane, 0.0427 kg/m^2.s; Benzene, 0.063 kg/m^2.s.
5.7 Assuming that the liquids burn at their limiting values, the diameters of the trays are predicted to be 1.54 m and 0.459 m for alcohol and heptane respectively. According to Equation (5.11), the corresponding burning rates are too low. Arbitrarily doubling the diameters gives burning rates closer to the values required.

For ethanol, this gives 825 kW, and will require 18.5 kg of fuel (c. 23 litres). (A closer estimate could be obtained if required.)

For heptane, this gives 765 kW, and requires 10.2 kg of fuel (c. 14.6 litres). (Note: combustion will be less than 100% efficient, particularly for the heptane.)
5.8 From Equation (5.27), (a) 4 mm thick, $\dot{m}'' = 9$ g/m^{2}s; (b) 2 mm thick, $\dot{m}'' < 0$. i.e. it cannot maintain this surface temperature. See Section 6.3.2.

6.1 13°C. Equation (3.3e) can be used to check the effect of temperature on the lower limit. The effect is not significant.
6.2 Closed cup flashpoint is less than 32°C.
6.3 56°C.
6.4 2.7% by volume.
6.5 Ignoring radiant heat losses, $t_i = 4.9$ s. Including radiant heat losses (calculated with $T_{\text{fabric}} = 200°C$ as an average) $t_i = 11.2$ s.
6.6 As in Problem 6.5 $t_i = 2.2$ s and $t_i = 4.2$ s.
6.7 These answers are approximate: (a) 270 s; (b) 100 s. They can be derived by extracting the relevant data from Table 2.2 (inverse error function) and using a spreadsheet.
6.8 The radiative loss (320°C $= 593$ K) is 7 kW/m^2. The depth of the heated layer can be approximated by $\sqrt{(\alpha t)}$, and the temperature gradient immediately below the surface estimated as $(320 - 20)/\sqrt{(\alpha t)}$. This gives the conductive losses as (a) 8.9 kW/m^2 and (b) 14.6 kW/m^2.

7.1 $V = 0.57$ mm/s (from Section 7.2.2)
7.2 Steady-state surface temperature is 116°C (ignoring re-radiation) so that $V = 0.13$ mm/s. (Equation 7.7) Re-radiation is significant and should be included in the calculation.

7.3 Steady-state surface temperature is 65.8°C (ignoring re-radiation). Thus, $V = 0.07$ mm/s.

7.4 From Figure 9.4, 2.75 m (no ceiling); 5.9 m (ceiling 950 mm above crib base); and >7 m (ceiling 660 mm above crib base).

8.1 The data plotted according to Equation (8.1) lie on a slight curve as Bi < 10 for the smaller cubes. A slight extrapolation gives $r_0 \approx 0.87$ m at 40°C for a 'slab'. The data are inadequate.

8.2 Long extrapolation required to give $r_0 \approx 3.4$ m at 40°C for a 'slab'. Spontaneous ignition of this material would be most unlikely.

9.1 Take air density at 500 K = 0.70 kg/m³ (Table 11.6), and $\Delta H_{ox} = 13$ kJ/g, to give $t = 646$ s. (The effect of subsequent cooling is ignored here.)

9.2 This involves the reverse procedure: 9.9% oxygen (by mass) remains.

9.3 (a) 1.8 MW (average kpc for plaster + brick taken for walls and ceiling); (b) 1.9 MW; (c) 1.2 MW (fibreboard thermally thick); (d) 2.8 MW.

10.1 23.2 minutes (120 s timestep), 23.5 (30 s timestep).

10.2 130°C at 30 minutes. 'Fire resistance' works out at 146 minutes.

10.3 550°C at $14\frac{1}{2}$ minutes. Maximum temperature 820°C.

10.4 76°C at 30 minutes.

10.5 45 minutes.

10.6 10.25 minutes (with a timestep of 30 s).

References

Abrams, M.S. (1979). 'Behaviour of inorganic materials in fire', in *Design of Buildings for Fire Safety*, ASTM STP 685 (eds E.E. Smith and T.Z. Harmathy), pp. 14–75. American Society for Testing and Materials, Philadelphia.

Ahonen, A., Kokkala, M., and Weckman, H. (1984). 'Burning characteristics of potential ignition sources of room fires', *VTT Research Report 285*, Valtion Teknillinen Turkimuskeskus, Espoo, Finland.

Akita, K. (1972). 'Some problems of flame spread along a liquid surface'. *14th Symposium (International) on Combustion*, pp. 1075–1083. The Combustion Institute, Pittsburgh.

Akita, K., and Yumoto, Y. (1965). 'Heat transfer in small pools and rates of burning of liquid methanol'. *10th Symposium (International) on Combustion*, pp. 943–948. The Combustion Institute, Pittsburgh.

Alderson, S., and Breden, L. (1976). 'Evaluation of the fire performance of carpet underlayments'. National Bureau of Standards, NBSIR 76–1018.

Alpert, R.L. (1972). 'Calculation of response time of ceiling-mounted fire detectors'. *Fire Technology*, **8**, 181–195.

Alpert, R.L. (1975). 'Turbulent ceiling jet induced by large scale fires'. *Combustion Science and Technology*, **11**, 197–213.

Alpert, R.L. (1976). In 'The Third Full-Scale Bedroom Fire Test of the Home Fire Project (July 30, 1975). Vol. II — Analysis of Test Results (ed. A.T. Modak). *FMRC Technical Report Serial No. 21011.7, RC–B–48.*

Alpert, R.L., and Ward, E.J. (1984). 'Evaluation of unsprinklered fire hazards'. *Fire Safety Journal*, **7**, 127–143.

Alvares, N.J. (1975). 'Some experiments to delineate the conditions for flashover in enclosure fires'. *International Symposium on Fire Safety of Combustible Materials*, pp. 375–383. University of Edinburgh.

Alvarez, N.J. (1985) Personal communication to Zukoski (1995).

American Society for Testing and Materials (1982). 'Proposed standard method for room fire test of wall and ceiling materials, and assemblies'. 1982 Annual Book of ASTM Standards, Part 18, ASTM, Philadelphia.

American Society for Testing and Materials (1989). 'Method for gravimetric determination of smoke particulates from combustion of plastic materials'. ASTM D4100–89.

American Society for Testing and Materials (1990a). 'Standard test method for flashpoint and firepoint by the Cleveland Open Cup'. ASTM D92–90.

American Society for Testing and Materials (1990b). 'Standard test method for determining material ignition and flame spread properties'. ASTM E1321-90, ASTM, Philadelphia.

American Society for Testing and Materials (1993a). 'Standard test method for flammability of finished textile floor covering materials'. ASTM D2859–93.

American Society for Testing and Materials (1993b). 'Standard method of test for density of smoke from the burning or decomposition of plastics'. ANSI/ASTM D2843–93.

American Society for Testing and Materials (1993c). 'Proposed test method for heat and visible smoke release rates for materials and products'. ASTM E906–93.

American Society for Testing and Materials (1993d). 'Standard method of test for heat and visible smoke release rates for materials and products using an oxygen consumption calorimeter'. ASTM E1354, ASTM, Philadelphia.

American Society for Testing and Materials (1994a). 'Standard test method for specific optical density of smoke generated by solid materials'. ANSI/ASTM E662–94.

American Society for Testing and Materials (1994b). 'Standard test method for flashpoint by the Pensky–Martens Closed Tester'. ASTM D93–94.

American Society for Testing and Materials (1994c). 'Standard test method for surface flammability of materials using a radiant heat energy source'. ASTM E162–94.

American Society for Testing and Materials (1995a). 'Standard test method for surface burning characteristics of building materials'. ASTM E84–95.

American Society for Testing and Materials (1995b). 'Standard methods of fire tests of building construction and materials'. ASTM E119–95.

American Society for Testing and Materials (19??). 'Standard guide for room fire experiments'. E603-77(?).

Ames, S. and Rogers, S. (1990) 'Large and small-scale fire calorimetry assessment of upholstered furniture', Proceedings of Interflam'90, Interscience Communications, London, Pages 221–232.

Andrews, C.L. (1987). 'Rebuilding a warehouse complex'. *Fire Prevention*, No. 207, 28–32.

Andrews, G.E., and Bradley, D. (1972). 'Determination of burning velocities — a critical review'. *Combustion and Flame*, **18**, 133–153.

Anon. (1975). 'Fatal mattress store fire at Chatham Dockyard'. *Fire*, **67**, 388.

Anthony, E.J., and Greaney, D. (1979). 'The safety of hot, self-heating materials'. *Combustion Science and Technology*, **21**, 79–85.

Ashton, L.A., and Malhotra, H.L. (1960). 'External walls of buildings. I — The protection of openings against spread of fire from storey to storey'. *Fire Research Note No. 836*.

Atkinson, G.T., and Drysdale, D.D. (1989). 'A note on the measurement of smoke yields'. *Fire Safety Journal*, **15**, 331–335.

Atkinson, G.T., Drysdale, D.D. and Wu, Y. (1995). 'Fire driven flow in an inclined trench'. *Fire Safety Journal*, **25**, 141–158.

Atreya, A. (1995). 'Convection heat transfer', in *SFPE Handbook of Fire Protection Engineering*, 2nd Edition (eds P.J. Di Nenno *et al.*), pp. 1.39–1.64. Society of Fire Protection Engineers, Boston.

Atreya, A., and Abu-Zaid, M. (1991). 'Effect of environmental variables on piloted ignition'. *Proceedings of the Third International Symposium on Fire Safety Science*, pp. 177–186. Elsevier Applied Science, Barking.

Atreya, A., Carpentier, C., and Harkleroad, M. (1986). 'Effect of sample orientation on piloted ignition and flame spread'. *Proceedings of the First International Symposium on Fire Safety Science*, pp. 97–109. Hemisphere Publishing Corporation, Washington.

Babrauskas, V. (1979). 'Full scale burning behaviour of upholstered chairs'. National Bureau of Standards, *NBS Technical Note No. 1103*.

Babrauskas, V. (1980a). 'Flame lengths under ceilings'. *Fire and Materials*, **4**, 119–126.

Babrauskas, V. (1980b). 'Estimating room flashover potential'. *Fire Technology*, **16**, 94–104.

Babrauskas, V. (1982). 'Will the second item ignite?' *Fire Safety Journal*, **4**, 281–292.

Babrauskas, V. (1983a). 'Upholstered furniture heat release rates: measurements and estimation'. *J. Fire Sciences*, **1**, 9–32.

Babrauskas, V. (1983b). 'Estimating large pool fire burning rates'. *Fire Technology*, **19**, 251–261.

Babrauskas, V. (1984). 'Development of the cone calorimeter — a bench scale heat release rate apparatus based on oxygen consumption'. *Fire and Materials*, **8**, 81–95.

Babrauskas, V. (1984/85). 'Pillow burning rates'. *Fire Safety Journal*, **8**, 199–200.

Babrauskas, V. (1992a). 'From Bunsen burner to heat release rate calorimeter', in *Heat Release in Fires* (eds V. Babrauskas and S.J. Grayson), pp. 7–29. Elsevier Applied Science, Barking.

Babrauskas, V. (1992b). 'Full-scale heat release rate measurements', in *Heat Release in Fires* (eds V. Babrauskas and S.J. Grayson), pp. 93–111. Elsevier Applied Science, Barking.

Babrauskas, V. (1995a). 'Burning rates', in *SFPE Handbook of Fire Protection Engineering*, 2nd Edition (eds P.J. Di Nenno *et al.*), pp. 3.1–3.15. Society of Fire Protection Engineers, Boston.

Babrauskas, V. (1995b). 'The cone calorimeter', in *SFPE Handbook of Fire Protection Engineering*, 2nd Edition (eds P.J. Di Nenno *et al.*), pp. 3.37–3.52. Society of Fire Protection Engineers, Boston.

Babrauskas, V., Lawson, J.R., Walton, W.D., and Twilley, W.H. (1982). 'Upholstered furniture heat release rates measured with a furniture calorimeter', NBSIR 82–2604, National Bureau of Standards, Gaithersburg, MD.

Babrauskas, V., and Parker, W. (1987). 'Ignitability measurements with the cone calorimeter'. *Fire and Materials*, **11**, 31–43.

Babrauskas, V., and Peacock, R. (1992). 'Heat release rate: the single most important variable in fire hazard'. *Fire Safety Journal*, **18**, 255–272.

Babrauskas, V., and Walton, W.D. (1986). 'A simplified characterisation for upholstered furniture heat release rates'. *Fire Safety Journal*, **11**, 181–192.

Babrauskas, V., and Williamson, R.B. (1978). 'Post flashover compartment fires: basis of a theoretical model'. *Fire and Materials*, **2**, 39–53.

Babrauskas, V., and Williamson, R.B. (1979). 'Post flashover compartment fires: application of a theoretical model'. *Fire and Materials*, **3**, 1–7.

Babrauskas, V., and Williamson, R.B. (1980a). 'The historical basis of fire resistance testing — Part I'. *Fire Technology*, **14**, 184–194.

Babrauskas, V., and Williamson, R.B. (1980b). 'The historical basis of fire resistance testing — Part II'. *Fire Technology*, **14**, 304–316.

Back, E.L. (1981/82). 'Auto-ignition in hygroscopic organic materials'. *Fire Safety Journal*, **4**, 185–196.

Back, G., Beyler, C.L., Di Nenno, P., and Tatum, P. (1994). 'Wall incident heat flux distributions resulting from an adjacent fire'. *Proceedings of the Fourth International Symposium on Fire Safety Science*, pp. 241–252. International Association for Fire Safety Science, Boston, MA.

Baker, R.R. (1977). 'Combustion and thermal decomposition regions inside a burning cigarette'. *Combustion and Flame*, **30**, 21–32.

Bamford, C.H., Crank, J., and Malan, D.H. (1946). 'The burning of wood'. *Proceedings of the Cambridge Philosophical Society*, **42**, 166–182.

Bankston, C.P., Cassanova, R.A., Powell, E.A., and Zinn, B.T. (1978). 'Review of smoke particulate properties data for burning natural and synthetic materials'. National Bureau of Standards, *NBS–GCR–78–147*.

Baroudi, D., and Kokkala, M. (1992). 'Analysis of upward flame spread', *VTT Publication No. 89*, Technical Research Centre of Finland, Espoo.

Bartknecht, W. (1981). *Explosions: Course, Prevention, Protection*. Springer-Verlag, Berlin.

Baulch, D.L., and Drysdale, D.D. (1974). 'An evaluation of the rate data for the reaction $CO + OH \rightarrow CO_2 + H$'. *Combustion and Flame*, **23**, 215–225.

Beever, P.F. (1990). 'Estimating the response of thermal detectors'. *J. Fire Protection Engineering*, **2**, 11–24.

Beever, P.F. (1995) 'Self heating and spontaneous combustion', in *SFPE Handbook of Fire Protection Engineering*, 2nd Edition (eds P.J. Di Nenno *et al.*), pp. 2.180–2.189. Society of Fire Protection Engineers, Boston.

Benjamin, I.A., Fung, F., and Roth, L. (1977). 'Control of smoke movement in buildings: a review'. National Bureau of Standards, *NBSIR 77–1209*.

Best, R.L. (1978). 'Tragedy in Kentucky'. *Fire Journal*, **72**, 18–35.

Best, R.L., and Demers, D.P. (1982). 'Fire at the MGM Grand'. *Fire Journal*, **76**, 19–37.

Beyler, C.L. (1984a). 'A design method for flaming fire detection'. *Fire Technology*, **20**(4), 5–16.

Beyler, C.L. (1984b). 'Ignition and burning of a layer of incomplete combustion products'. *Combustion Science and Technology*, **39**, 287–303.

Beyler, C.L. (1986a). 'Major species production by diffusion flames in a two-layer compartment fire environment'. *Fire Safety Journal*, **10**, 47–56.

Beyler, C.L. (1986b). 'Fire plumes and ceiling jets'. *Fire Safety Journal*, **11**, 53–75.

Beyler, C.L., (1992). 'A unified model of fire suppression'. *J. Fire Protection Engineering*, **4**, 5–15.

Beyler, C.L., and Hirschler, M.M. (1995). 'Thermal decomposition of polymers', in *SFPE Handbook of Fire Protection Engineering*, 2nd Edition (eds P.J. Di Nenno *et al.*), pp. 1.99–1.119. Society of Fire Protection Engineers, Boston.

Billmeyer, F.W. (1971). *Textbook of Polymer Science*. 2nd Edition. Wiley, New York.

Bishop, S.R., Holborn, P.G., Beard, A.N., and Drysdale, D.D. (1993). 'Nonlinear dynamics of flashover in compartment fires'. *Fire Safety Journal*, **21**, 11–45.

Blinov, V.I., and Khudiakov, G.N. (1957). 'The burning of liquid pools'. *Doklady Akademi Nauk SSSR*, **113**, 1094.

Blinov, V.I., and Khudiakov, G.N. (1961). *Diffusive Burning of Liquids* (English translation by US Army Engineering Research and Development Laboratories, T–1490a–c. ASTIA, AD 296 762).

Block, J.A. (1970). 'A theoretical and experimental study of non-propagating free-burning fires'. PhD Thesis, Harvard University.

Block, J.A. (1971). 'A theoretical and experimental study of non-propagating free-burning fires'. *13th Symposium (International) on Combustion*, pp. 971–978. The Combustion Institute, Pittsburgh.

Bluhme, D.A. (1987). 'ISO ignitability test and proposed ignition criteria'. *Fire and Materials*, **11**, 195–199.

Board of Trade (1966). 'Carriage of dangerous goods in ships'. Report of the Standing Advisory Committee, Board of Trade, HMSO, London.

Boddington, T., Gray, P., and Harvey, D.I. (1971). 'Thermal theory of spontaneous ignition: criticality in bodies of arbitrary shape'. *Philosophical Transactions of the Royal Society*, **A270**, 467–506.

Bohm, B. (1977). PhD Thesis, Technical University of Denmark.

Bohm, B. (1982). 'Calculated thermal exposure of steel structures in fire test furnaces — a non-ε_{res} approach'. Laboratory of Heating and Air Conditioning, Technical University of Denmark, CIB W14/82/73 (DK).

Botha, J.P., and Spalding, D.B. (1954). 'The laminar flame speed of propane/air mixtures with heat extraction from the flame'. *Proceedings of the Royal Society*, **A225**, 71–96.

Bouhafid, A., Vantelon, J.P., Joulain, P., and Fernandez-Pello, A.C. (1988). 'On the structure at the base of a pool fire'. *22nd Symposium (International) on Combustion*, Combustion Institute, Pittsburgh, PA, pp. 1291–1298.

Bowes, P.C. (1971). 'Application of the theory of thermal explosion to the self-heating and ignition of organic materials'. *Fire Research Note No. 867*.

Bowes, P.C. (1974). 'Fires in oil soaked lagging'. Building Research Establishment, *BRE Current Paper CP 35/74*.

Bowes, P.C. (1984). *Self-heating: Evaluating and Controlling the Hazards*, HMSO, London.

Bowes, P.C., and Cameron, A.J. (1971). 'Self-heating and ignition of chemically activated carbon'. *J. Applied Chemistry and Biotechnology*, **21**, 244–250.

Bowes, P.C., and Townsend, S.E. (1962). 'Ignition of combustible dusts on hot surfaces'. *British Journal of Applied Physics*, **13**, 105–114.

Bowman, C.T. (1975). 'Non-equilibrium radical concentrations in shock-initiated methane oxidation'. *15th Symposium (International) on Combustion*, pp. 869–882. The Combustion Institute, Pittsburgh.

Brandrup, J., and Immergut, E.H. (eds) (1975). *Polymer Handbook*, 2nd Edition. Wiley, New York.

Brannigan, F.L. (ed.) (1980). 'Fire Investigation Handbook'. *National Bureau of Standards Handbook No. 134*, US Department of Commerce.

Brenden, J.J. (1967). 'Effect of fire retardant and other salts on pyrolysis products of Ponderosa pine at 280°C and 350°C'. *US Forest Service Research Paper FPL 80*. US Department of Agriculture.

Brenton, J.R., Thomas, G.O., and Al-Hassam, T. (1994). 'Small scale studies of water spray dynamics during explosion mitigation tests'. *Proceedings of the Institution of Chemical Engineers Symposium Series, Hazards XII: European Advances in Process Safety Conference No. 134*, I. Chem E., Rugby, pp. 393–403.

British Standards Institution (1959). 'Method of test for flameproof materials'. BS 3119:1959.

British Standards Institution (1976). 'Method of test for flammability of vertically oriented textile fabrics and fabric assemblies subjected to a small igniting flame'. BS 5438: 1976.

British Standards Institution (1977). 'Electrical apparatus for potentially explosive atmospheres. Part 7. Intrinsic safety 'i''. BS 5501 Part 7: 1977.

British Standards Institution (1978). 'Code of Practice for Fire precautions in the Design of Buildings. Part 4: Smoke control in protected escape routes using pressurisation'. BS 5588: 1978.

British Standards Institution (1979). 'Method of test for the ignitability by smokers' material of upholstered composites for seating'. BS 5852 Part 1: 1979.

British Standards Institution (1980). 'Method of test for determination of the punking behaviour of phenol-formaldehyde foam'. BS5946: 1980.

British Standards Institution (1982). 'Flashpoint by the Abel apparatus (Statutory Method: Petroleum (Consolidation) Act 1928)'. BS 2000 Part 33: 1982.

British Standards Institution (1987a). 'Fire tests on building materials and structures. Part 7. Method for classification of the surface spread of flame of products'. BS 476 Part 7: 1987.

British Standards Institution (1987b). 'Fire tests on building materials and structures. Part 20. Method for determination of the fire resistance of elements of construction (general principles); Part 21. Method for determination of the fire resistance of loadbearing elements of construction; Part 22. Method for determination of the fire resistance of non-loadbearing elements of construction'. BS 476 Parts 20–22: 1987.

British Standards Institution (1989). 'Fire tests on building materials and structures. Part 6. Method of test for fire propagation for products'. BS 476 Part 6: 1989.

British Standards Institution (1990). 'Structural use of steelwork in building. Code of practice for fire resistant design'. BS 5950: Part 8:1990.

British Standards Institution (1993). 'Fire tests on building materials and structures. Part 15. Methods of measuring the rate-of-heat release of products'. BS 476: Part 15: 1993.

British Standards Institution (1997). 'Draft for Development: Fire Safety Engineering in Buildings. Part 1. Guide to the application of fire safety engineering principles'. BSI DD 240: Part 1: 1997.

Brosmer, M.A., and Tien, C.L. (1987). 'Radiative energy blockage in large pool fires'. *Combustion Science and Technology*, **51**, 21–37.

Browne, F.L., and Brenden, J.J. (1964). 'Heats of combustion of the volatile pyrolysis products of Ponderosa pine'. *US Forest Service Research Paper FPL 19*, US Department of Agriculture.

Bruce, H.D. (1953). 'Experimental dwelling room fires'. Report No. D 1941, Forest Products Laboratory, US Department of Agriculture, Madison, Wisconsin.

Bryan, J.L. (1974). *Fire Suppression and Detection Systems*. Glencoe Press, Beverley Hills.

Budavari, S. (ed.) (1996). *The Merck Index: an Encyclopedia of Chemicals, Drugs and Biologicals*, 12th Edition. Merck Co. Inc., Whitehouse Station, NJ.

Building Research Establishment, Garston, Watford (1982). Video: 'An anatomy of a fire'.

Building Research Establishment, Garston, Watford (1987). Video: 'Fire at Valley Parade, Bradford'.

Building Research Establishment, Garston, Watford (1989). Video: 'The Front Room Fire'.

Bullen, M.L. (1977a). 'A combined overall and surface energy balance for fully developed ventilation controlled liquid fuel fires in compartments'. *Fire Research*, **1**, 171–185.

Bullen, M.L. (1977b). 'The effect of a sprinkler on the stability of a smoke layer beneath a ceiling'. *Fire Technology*, **13**, 21–34.

Bullen, M.L. (1978). 'The ventilation required to permit growth of a room fire'. Building Research Establishment, *BRE CP 41/78*.

Bullen, M.L., and Thomas, P.H. (1979). 'Compartment fires with non-cellulosic fuels'. *17th Symposium (International) on Combustion*, pp. 1139–1148. The Combustion Institute, Pittsburgh.

Burgess, D.S., and Zabetakis, M.G. (1973). 'Detonation of a flammable cloud following a propane pipeline break'. US Bureau of Mines, RI 7752.

Burgess, D.S., Strasser, A., and Grumer, J. (1961). 'Diffusive burning of liquids in open trays'. *Fire Research Abstracts and Reviews*, **3**, 177–192.

Burgoyne, J.H., and Cohen, L. (1954). 'The effect of drop size on flame propagation in liquid aerosols'. *Proceedings of the Royal Society (London)*, **A225**, 375–392.

Burgoyne, J.H., and Katan, L.L. (1947). 'Fires in open tanks of petroleum products: some fundamental aspects'. *J. Institute of Petroleum*, **33**, 158–185.

Burgoyne, J.H., and Roberts, A.F. (1968). 'The spread of flame across a liquid surface, Part 2'. *Proceedings of the Royal Society (London)*, **A308**, 55–68.

Burgoyne, J.H., and Williams-Leir, G. (1949). 'Inflammability of liquids'. *Fuel*, **28**, 145–149.

Burgoyne, J.H., Newitt, D.M., and Thomas, A. (1954). 'Explosion characteristics of lubricating oil mist in crank cases'. *The Engineer, London*, **198**, 165.

Burgoyne, J.H., Roberts, A.F., and Alexander, J.L. (1967). 'The significance of open flash points'. *J. Institute of Petroleum*, **53**, 338–341.

Burgoyne, J.F., Roberts, A.F., and Quinton, P.G. (1968). 'The spread of flame across a liquid surface, Part 1'. *Proceedings of the Royal Society (London)*, **A308**, 39–54.

Burke, S.P., and Schumann, T.E.W. (1928). 'Diffusion flames'. *Industrial and Engineering Chemistry*, **20**, 998–1004.

Butcher, E.G., and Parnell, A.C. (1979). *Smoke Control in Fire Safety Design*. E. & F. N. Spon Ltd, London.

Butcher, E.G., Bedford, G.K., and Fardell, P.J. (1968). 'Further experiments on temperatures reached by steel in building fires'. *Joint Fire Research Organization Symposium No. 2*, HMSO, London, pp. 2–17.

Butcher, E.G., Chitty, T.B., and Ashton, L.A. (1966). 'The temperature attained by steel in building fires'. *Fire Research Station Technical Paper No. 15*, HMSO, London.

Butler, C.P. (1971). 'Notes on charring rates in wood'. *Fire Research Note No. 896*.

Calcraft, A.M., Green, R.J.S., and McRoberts, T.S. (1975). 'Burning plastics: smoke formation'. *International Symposium on Fire Safety of Combustible Materials, University of Edinburgh*, pp. 253–257.

Carslaw, H.S., and Jaeger, J.C. (1959). *Conduction of Heat in Solids*. 2nd Edition, Oxford Science Publications.

Cetegen, B.M., Zukoski, E.E., and Kubota, T. (1984). 'Entrainment into the near and far field of fire plumes'. *Combustion Science and Technology*, **39**, 305–331.

Chartered Institute of Building Services Engineers (1997). *CIBSE Guide E: Fire Engineering*. CIBSE, London.

Chitty, R. (1994). 'A survey of backdraught', *Fire Research and Development Group. Publication No. 5/94*, Home Office, London.

Chitty, R., and Cox, G. (1979). 'A method of measuring combustion intermittency in fires'. *Fire and Materials*, **3**, 238–242.

Christian, W.J., and Waterman, T.E. (1971). 'The ability of small scale tests to predict full-scale smoke production'. *Fire Technology*, **7**, 332–344.

Clancey, V.J. (1974). 'The evaporation and dispersal of flammable liquid spillages'. *5th Symposium on Chemical Process Hazards*. Institution of Chemical Engineers, Rugby, 80–98.

Cleary, T.G., and Quintiere, J.G. (1991). 'A framework for utilizing fire property tests'. *Proceedings of the 3rd International Symposium on Fire Safety Science*, Elsevier Science Publishers, Barking, pp. 647–656.

Conseil Internationale du Bâtiment W14 (1983). 'A conceptual approach towards a probability-based design guide on structural fire safety'. *Fire Safety Journal*, **6**, 24–79.

Conseil Internationale du Bâtiment W14 (1986). 'Design Guide: Structural Fire Safety'. *Fire Safety Journal*, **10**, 79–137.

Cook, S.J., Cullis, C.F., and Good, A.J. (1977). 'The measurement of the flammability of mists'. *Combustion and Flame*, **30**, 309–317.

Cooke, R.A., and Ide, R.H. (1985). *Principles of Fire Investigation*. Institution of Fire Engineers, Leicester.

Cooper, L.Y. (1984) *Combustion Science and Technology*, **40**, 19–34.

Cooper, L.Y. (1995). 'Compartment fire-generated environment and smoke filling', in *SFPE Handbook of Fire Protection Engineering*, 2nd Edition (eds P.J. Di Nenno *et al.*), pp. 3.174–3.196. Society of Fire Protection Engineers, Boston.

Cooper, L. Y. and Stroup, D. W. (1985) 'ASET — a computer program for calculating Available Safe Egress Time'. *Fire Safety Journal*, **9**, 29–45.

Corlett, R.C. (1968). 'Gas fires with pool-like boundary conditions'. *Combustion and Flame*, **12**, 19–32.

Corlett, R.C. (1970). 'Gas fires with pool-like boundary conditions: further results and interpretation'. *Combustion and Flame*, **14**, 351–360.

Corlett, R.C. (1974). 'Velocity distributions in fires', in *Heat Transfer in Fires* (ed. P.L. Blackshear), pp. 239–255. John Wiley and Sons, New York.

Coward, H.F., and Jones, G.W. (1952). 'Limits of flammability of gases and vapors'. *US Bureau of Mines Bulletin 503*.

Cox, G. (1983). 'A field model of fire and its application to nuclear containment problems'. *Proceedings of the CSNI Specialist Meeting on Interaction of Fire and Explosion with Ventilation Systems in Nuclear Facilities*, Los Alamos Report LA-9911C.

Cox, G. (1995). 'Basic considerations', in *Combustion Fundamentals of Fire* (ed. G. Cox), pp. 1–30. Academic Press, London.

Cox, G. and Chitty, R. (1980). 'A study of the deterministic properties of unbounded fire plumes'. *Combustion and Flame*, **39**, 191–209.

Cox, G., and Chitty, R. (1985). 'Some source-dependent effects of unbounded fires'. *Combustion and Flame*, **60**, 219–232.

Cox, G., Chitty, R. and Kumar, S. (1989). 'Fire modelling and the King's Cross Fire Investigation'. *Fire Safety Journal*, **15**, 103–106.

Cox, G., Kumar, S., Cumber, P., Thomson, V., and Porter, A. (1990). 'Fire simulation in the design evaluation process: an exemplification of the use of a computer field model'. *Interflam'90: Proceedings of the 5th International Fire Conference*, Interscience Communications, London, pp. 55–66.

Crauford, N. L., Liew, S. L., and Moss, J. B. (1985). 'Experimental and numerical simulation of a buoyant fire'. *Combustion and Flame*, **61**, 63–77.

Croce, P.A. (ed.) (1975). 'A study of room fire development: the second full scale bedroom fire test of the Home Fire Project'. Factory Mutual Research Corporation Serial 21011.4, RC 75–T–31.

Croft, W.M. (1980/81). 'Fire involving explosions—a literature review'. *Fire Safety Journal*, **3**, 3–24.

Cullen, The Hon. Lord. (1989) *The Public Inquiry into the Piper Apha Disaster* (HMSO, London)

Cullis, C.F., and Hirschler, M.M. (1981). *The Combustion of Organic Polymers*. Clarendon Press, Oxford.

Damkohler, G. (1940). 'Influence of turbulence on the velocity of flames in gas mixtures'. *Z. Electrochem.*, **46**, 601–626.

Deepak, D., and Drysdale, D.D. (1983). 'Flammability of solids: an apparatus to measure the critical mass flux at the firepoint'. *Fire Safety Journal*, **5**, 167–169.

De Haan, J.D. (1997). *Kirk's Fire Investigation*, 4th Edition. Brady/Prentice Hall, Englewood Cliffs, New Jersey.

Delichatsios, M.A., and Orloff, L. (1988). 'Effects of turbulence on flame radiation from diffusion flames'. *22nd Symposium (International) on Combustion*, pp. 1271–1279. The Combustion Institute, Pittsburgh.

Delichatsios, M.A., Panagiotou, Th., and Kiley, F. (1991). 'The use of time to ignition data for characterising the thermal inertia and the minimum (critical) heat flux for ignition or pyrolysis'. *Combustion and Flame*, **84**, 323–332.

Department of Transport (1986) "*Report on the Derailment and Fire that occurred on 20th December 1984 at Summit Tunnel in the London Midland Region of British Railways*" (HMSO, London).

de Ris, J. and Cheng, X-F. (1994) 'The rôle of smoke point in material flammability testing'. Proceedings of the Fourth International Symposium on Fire Safety Science (IAFSS, Boston) pp. 301–312.

de Ris, J.N. (1969). 'Spread of a laminar diffusion flame'. *12th Symposium (International) on Combustion*, pp. 241–252. The Combustion Institute, Pittsburgh.

de Ris, J.N. (1973). 'Modeling techniques for prediction of fires'. *Applied Polymer Symposium No. 22*, pp. 185–193. John Wiley and Sons, New York.

de Ris, J.N. (1979). 'Fire radiation—a review'. *17th Symposium (International) on Combustion*, pp. 1003–1016. The Combustion Institute, Pittsburgh.

di Blasi, C., Crescitelli, S., Russo, G., and Fernandez-Pello, A.C. (1988). 'Model of the flow assisted spread of flames over a thin charring combustible'. *22nd Symposium (International) on Combustion*, pp. 1205–1212. The Combustion Institute, Pittsburgh.

Di Nenno, P.J., *et al.* (eds) (1995). *SEPE Handbook of Fire Protection Engineering*, 2nd Edition, Section 1. Society of Fire Protection Engineers, Boston.

Di Nenno, P.J. (1997). 'Direct halon replacement agents and systems', in *NFPA Handbook*, 18th Edition (ed. A.E. Cote), Section 6, Chapter 19, pp. 6-297–6.330. National Fire Protection Association, Quiney, MA.

DIN 18230 (1978). 'Structural fire protection in industrial building construction: Part 1: Required fire resistance period; Part 2: Determination of the burning factor *m*'. Deutsches Institut für Normung.

DIN 53436 (1979). 'Erzeugung thermischer Zersetsungsprodukte von Werkstoffen unter Luftzufuhr und ihre toxikologische Prufung. Verfahren zur thermischen Zersetzung'. DIN 53436. Deutsches Institut für Normung.

Dixon-Lewis, G. (1967). 'Flame structure and flame reaction kinetics. I. Solution of conservation equations and application to rich hydrogen-oxygen flames'. *Proceedings of the Royal Society (London)*, **A289**, 495–513.

Dixon-Lewis, G., and Williams, A. (1967). 'Some observations on the combustion of methane in premixed flames'. *11th Symposium (International) on Combustion*. pp. 951–958. The Combustion Institute, Pittsburgh.

Dixon-Lewis, G.L., and Williams, D.J. (1977). 'Oxidation of hydrogen and carbon monoxide', in *Comprehensive Chemical Kinetics*, Volume 17 (eds C.H. Bamford and C.F.H. Tipper). Elsevier, Amsterdam.

Dorofeev, S.B., *et al.* (1995). 'Fireballs from deflagration and detonation of heterogeneous fuel-rich clouds'. *Fire Safety Journal*, **25**, 323–336.

Dosanjh, S.S., Pagni, P.J., and Fernandez-Pello, A.C. (1987). 'Forced co-current smouldering combustion'. *Combustion and Flame*, **68**, 131–142.

Drysdale, D.D. (1980). 'Aspects of smouldering combustion'. *Fire Prevention Science and Technology, No. 23*, pp. 18–28. Fire Protection Association, London.

Drysdale, D.D. (1981). Unpublished results.

Drysdale, D.D. (1983). 'Ignition: the material, the source and subsequent fire growth'. Society of Fire Protection Engineers, Technology Report 83–5. (Presented at the 1983 SFPE Seminar on Fire Protection Engineering, Kansas City, Missouri, May 1983.)

Drysdale, D.D. (1985a). 'Fire behaviour of cellular polymers'. *Cellular Polymers*, **4**, 405–419.

Drysdale, D.D., (1985b). *Introduction to Fire Dynamics*, 1st Edition. John Wiley and Sons.

Drysdale, D.D. (ed.) (1992). 'Special Issue: The King's Cross Underground Fire'. *Fire Safety Journal*, **18**, 1–121.

Drysdale, D.D. (1996). 'The flashover phenomenon'. *Fire Engineers Journal*, **56**, 18–23.

Drysdale, D.D., and Abdul-Rahim, F.F. (1985). 'Smoke production in fires. I. Small scale experiments'. In *Fire Safety: Science and Engineering*, ASTM STP 882, ed. T.Z. Harmathy, American Society for Testing and Materials, Philadelphia, PA, pp. 285–300.

Drysdale, D.D., and Kemp, N. (1982). 'Prevention and suppression of flammable and explosive atmospheres'. in *Factories: Planning, Design and Modernisation* (ed. J. Drury). Architectural Press, London, pp. 266–286.

Drysdale, D.D., and Macmillan, A.J.R. (1992). 'Flame spread on inclined surfaces'. *Fire Safety Journal*, **18**, 245–254.

Drysdale, D.D., and Thomson, H.E. (1989). 'Flammability of plastics II. Critical mass flux at the firepoint'. *Fire Safety Journal*, **14**, 179–188.

Drysdale, D.D., and Thomson, H.E. (1990). 'Ignition of polyurethane foams: a comparison between modified and unmodified foams', in *Flame Retardants'90* (ed. The British Plastics Federation), pp. 191–205. Elsevier Applied Science, Barking.

D'Souza, M.V., and McGuire, J.H. (1977). 'ASTM E-84 and the flammability of foamed thermosetting plastics'. *Fire Technology*, **13**, 85–94.

Dugger, G.L., Weast, R.C., and Heimel, S. (1955). 'Flame velocity and preflame reaction in heated propane-air mixtures'. *Industrial and Engineering Chemistry*, **47**, 114–116.

Dusinberre, G.M. (1961). *Heat Transfer Calculation by Finite Difference*. International Textbook Company, Scranton.

Eckhoff, R.K. (1997). *Dust Explosions in the Process Industries*, 2nd Edition. Butterworth-Heinemann, Oxford.

Edwards, D.K. (1985) in *Handbook of heat transfer fundamentals* (Mc Graw Hill, New York) pp. 14–15.

Egerton, A.C., Gugan, K., and Weinberg, F.J. (1963). 'The mechanism of smouldering in cigarettes'. *Combustion and Flame*, **7**, 63–78.

Elliot, D.A. (1974). 'Fire and steel construction: protection of structural steelwork'. *CONSTRADO Publication 4/74*, Constructional Steel Research and Development Organisation, London.

Emmons, H.W. (1965). 'Fundamental problems of the free-burning fire'. *10th Symposium (International) on Combustion*, pp. 951–964. The Combustion Institute, Pittsburgh.

Emmons, H.W. (1974). *Scientific American*, **231**, 21–27.

Emmons, H.W. (1995). 'Vent flows', in *SFPE Handbook of Fire Protection Engineering*, 2nd Edition (eds P.J. Di Nenno *et al.*), pp. 2.40–2.49. Society of Fire Protection Engineers, Boston.

Evans, D.D. (1995). 'Ceiling jet flows', in *SFPE Handbook of Fire Protection Engineering*, 2nd Edition (eds P.J. Di Nenno *et al.*), pp. 2.32–2.39, Society of Fire Protection Engineers, Boston.

Evans, D.D., and Stroup, D.W. (1986). 'Methods to calculate the response time of heat and smoke detectors installed below large unobstructed ceilings'. *Fire Technology*, **22**(1), 54–66.

Evers, E., and Waterhouse, A. (1978). 'A complete model for analysing smoke movement in buildings'. Building Research Establishment, BRE CP 69/78.

Fan, W.C., Hua, J.S., and Liao, G.X. (1995). 'Experimental study of the premonitory phenomena of boilover in liquid pool fires supported on water'. *J. Loss Prevention in Industry*, **8**, 221–227.

Fang, J.B. (1975a). 'Fire build-up in a room and the role of interior finish materials'. National Bureau of Standards, *NBS Technical Note No. 879*.

Fang, J.B. (1975b). 'Measurement of the behaviour of incidental fires in a compartment'. National Bureau of Standards, NBSIR 75–679.

Fang, J.B. (1980). 'Static pressures produced by room fires', *National Bureau of Standards NBSIR 80–1984*.

Fennell, D. (1988). *Investigation into the King's Cross Underground Fire*. HMSO, London.

Fernandez-Pello, A.C. (1977a). 'Downward flame spread under the influence of externally applied thermal radiation'. *Combustion Science and Technology*, **17**, 1–9.

Fernandez-Pello, A.C. (1977b). 'Upward laminar flame spread under the influence of externally applied thermal radiation'. *Combustion Science and Technology*, **17**, 87–98.

Fernandez-Pello, A.C. (1984). 'Flame spread modelling'. *Combustion Science and Technology*, **39**, 119–134.

Fernandez-Pello, A.C. (1995). 'The solid phase', in *Combustion Fundamentals of Fire* (ed. G. Cox), pp. 31–100. Academic Press, London.

Fernandez-Pello, A.C., and Hirano, T. (1983). 'Controlling mechanisms of flame spread'. *Combustion Science and Technology*. **32**, 1–31.

Fernandez-Pello, A.C., and Santoro, R.J. (1980). 'On the dominant mode of heat transfer in downward flame spread'. *17th Symposium (International) on Combustion*, pp. 1201–1209. The Combustion Institute, Pittsburgh.

Fernandez-Pello, A.C., and Williams, F.A. (1974). 'Laminar flame spread over PMMA surfaces'. *15th Symposium (International) on Combustion*, pp. 217–231. The Combustion Institute, Pittsburgh.

Field, P. (1982). *Dust Explosions*. Volume 4 of *Handbook of Powder Technology* (eds J.C. Williams and T. Allen). Elsevier Science Publishers, Amsterdam.

Fine, D.H., Gray, P., and Mackinven, R. (1969). 'Experimental measurements of self-heating in the explosive decomposition of diethylperoxide'. *12th Symposium (International) on Combustion*, pp. 545–555. The Combustion Institute, Pittsburgh.

Fire Protection Association (1972). 'Fire and related properties of industrial chemicals'. Fire Protection Association, London.

Fleischmann, C.M., Dod, R.L., Brown, N.J., Novakov, T., Mowrer, F.W., and Williamson, R.B. (1990). 'The use of medium scale experiments to determine smoke characteristics', in *Characterisation and Toxicity of Smoke*, ASTM STP 1082 (ed. H.K. Hasegawa), pp. 147–164. American Society for Testing and Materials, Philadelphia.

Fleischmann, C.M., Pagni, P.J., and Williamson, R.B. (1994). 'Quantitative backdraft experiments'. *Proceedings of the 4th International Symposium on Fire Safety Science*, pp. 337–348. International Association for Fire Safety Science, Boston, MA.

Foley, M., and Drysdale, D.D. (1995). 'Heat transfer from flames between vertical parallel walls'. *Fire Safety Journal*, **24**, 53–73.

Fons, W.L. (1946). 'Analysis of fire spread in light forest fuels'. *J. Agricultural Research*, **72**, 93–121.

Frank-Kamenetskii, D.A. (1939), 'Temperature distribution in reaction vessel and stationary theory of thermal explosion'. *Journal of Physical Chemistry (USSR)*, **13**, 738–755.

Friedman, R. (1968). 'A survey of knowledge about idealized fire spread over surfaces'. *Fire Research Abstracts and Reviews*, **10**, 1–8.

Friedman, R. (1971). 'Aerothermodynamics and modeling techniques for prediction of plastic burning rates'. *J. Fire and Flammability*, **2**, 240–256.

Friedman, R. (1975). 'Behaviour of fires in compartments'. *International Symposium on Fire Safety of Combustible Materials*, pp. 100–113, Edinburgh University.

Friedman, R. (1977). 'Ignition and burning of solids', in *Fire Standards and Safety. ASTM STP 614* (ed. A.F. Robertson), pp. 91–111. American Society for Testing and Materials, Philadelphia.

Friedman, R. (1989). *Principles of Fire Protection Chemistry*, 2nd Edition. National Fire Protection Association, Quincy, MA.

Friedman, R. (1992). 'An international survey of computer models for fire and smoke'. *J. Fire Protection Engineering*, **4**, 83–92.

Friedman, R. (1995). 'Chemical equilibrium', in *SFPE Handbook of Fire Protection Engineering*, 2nd Edition (eds P.J. Di Nenno *et al.*), pp. 1.88–1.98. Society of Fire Protection Engineers, Boston.

Fristrom, R.M., and Westenberg, A.A. (1965). *Flame Structure*. McGraw-Hill, New York.

Fung, F. (1973). 'Evaluation of a pressurised stairwell smoke control system for a twelve storey apartment building'. National Bureau of Standards, NBSIR 73–277.

Gagnon, R.M. (1997). 'Ultra-high-speed suppression systems for explosive hazards'. in *NFPA Handbook*, 18th Edition, pp. 6.198–6.215. National Fire Protection Association, Quincy, MA).

Gann, R.G., Earl, W.L., Manka, M.J., and Miles, L.B. (1981). 'Mechanism of cellulose smouldering retardance by sulphur'. *18th Symposium (International) on Combustion*, pp. 571–578. The Combustion Institute, Pittsburgh.

Gaydon, A.G., and Wolfhard, H.G. (1979). *Flame: Their Structure, Radiation and Temperature*, 4th Edition. Chapman and Hall, London.

Glassman, I. (1989). 'Soot formation in combustion process'. *22nd Symposium (International) on Combustion*, pp. 295–311. The Combustion Institute, Pittsburgh.

Glassman, I., and Dryer, F. (1980/81). 'Flame spreading across liquid fuels'. *Fire Safety Journal*, **3**, 123–138.

Gottuk, D.T., Roby, R.J., Peatross, M.J., and Beyler, C.L. (1992). *J. Fire Protection Engineering*, **4**, 133–150.

Grant, G.B., and Drysdale, D.D. (1995). 'Numerical modelling of early fire spread in warehouse fires'. *Fire Safety Journal*, **24**, 247–278.

Grant, G.B., and Drysdale, D.D. (1997). 'The suppression and extinction of class "A" fires using water sprays'. Fire Research and Development Group, Publication No. 1/97. (To be published in *Progress in Energy and Combustion Science*).

Gray, P., and Lee, P.R. (1967). 'Thermal explosion theory,' in *Oxidation and Combustion Reviews*, **2** (ed. C.F.H. Tipper), pp. 1–183. Elsevier, Amsterdam.

Gray, W.A., and Muller, R. (1974). *Engineering Calculations in Radiative Heat Transfer*. Pergamon Press, Oxford.

Greenwood, C.T., and Banks, W. (1968). *Synthetic High Polymers*. Oliver and Boyd, Edinburgh.

Greenwood, C.T., and Milne, E.A. (1968). *Natural High Polymers*. Oliver and Boyd, Edinburgh.

Griffiths, J.F. and Barnard, J.A. (1995) *Flame and Combustion, 3rd Edition*, Blackie Academic and Professional.

Gross, D. (1962). 'Experiments on the burning of cross piles of wood'. *Journal of Research, National Bureau of Standards*, **66C**, 99–105.

Gross, D. (1989). 'Measurement of flame lengths under ceilings'. *Fire Safety Journal*, **15**, 31–44.

Gross, D., and Robertson, A.F. (1965). 'Experimental fires in enclosures'. *10th Symposium (International) on Combustion*, pp. 931–942. The Combustion Institute, Pittsburgh.

Gross, D., Loftus, J.J., and Robertson, A.F. (1967). 'Method for measuring smoke from burning materials'. *Symposium on 'Fire Test Methods — Restraint and Smoke', 1966*, ASTM STP 422 (ed. A.F. Robertson), pp. 166–204. American Society for Testing and Materials, Philadelphia.

Grosshandler, W.L., and Modak, A.T. (1981). 'Radiation from non-homogeneous combustion products'. *18th Symposium (International) on Combustion*, pp. 601–609. The Combustion Institute, Pittsburgh.

Gugan, K. (1976). 'Technical lessons of Flixborough'. *The Chemical Engineer*, May 1976.

Gugan, K. (1979). *Unconfined Vapour Cloud Explosions*. Institution of Chemical Engineers, Rugby.

Hadvig, S., and Paulsen, O.R. (1976). 'One dimensional charring rates in wood'. *J. Fire and Flammability*, **7**, 433–449.

Hägglund, B., and Persson, L.E. (1976a). 'An experimental study of the radiation from wood flames'. *FoU-Brand*, **1**, 2–6.

Hägglund, B., and Persson, L.E. (1976b). 'The heat radiation from petroleum fires'. *FOA Report. C20126-D6 (A3)*. Forsvarets Forskningsanstalt, Stockholm.

Hägglund, B., Jansson, R., and Onnermark, B. (1974). 'Fire development in residential rooms after ignition from nuclear explosions'. *FOA Report C 20016–D6 (A3)*. Forsvarets Forskningsanstalt, Stockholm.

Hall, A.R. (1973). 'Pool burning: a review', in *Oxidation and Combustion Reviews*, **6** (ed. C.F.H. Tipper), pp. 169–225. Elsevier, Amsterdam.

Hall, C. (1981). *Polymeric Materials: an Introduction for Technologists and Scientists*. Macmillan Press Ltd, London.

Hall, H. (1925). 'Oil tank fire boilover'. *Mechanical Engineering*, **47**, 540.

Hamilton, D.C., and Morgan, W.R. (1952). *NACA Technical Note TN-2836*.

Hamins, A., Yang, T.C., and Kashiwagi, T. (1992). 'An experimental investigation of the pulsation frequency of flames'. *24th Symposium (International) on Combustion*, pp. 1695–1702. The Combustion Institute, Pittsburgh.

Harmathy, T.Z. (1972). 'A new look at compartment fires'. Parts I and II. *Fire Technology*, **8**, 196–219; 326–351.

Harmathy, T.Z. (1978). 'Mechanism of burning of fully-developed compartment fires'. *Combustion and Flame*, **31**, 265–273.

Harmathy, T.Z. (1979). 'Design to cope with fully developed compartment fires', in *Design of Buildings for Fire Safety* (eds E.E. Smith and T.Z. Harmathy). American Society for Testing and Materials, STP 685, pp. 198–276.

Harmathy, T.Z. (1987). 'On the equivalent fire exposure'. *Fire and Materials*, **11**, 95–104.

Harmathy, T.Z. (1995). 'Properties of building materials', in *SFPE Handbook of Fire Protection Engineering*, 2nd Edition (eds P.J. Di Nenno *et al.*), pp. 1.141–1.155. Society of Fire Protection Engineers, Boston.

Harmathy, T.Z., and Mehaffey, J.F. (1982). 'Normalised heat load: a key parameter in fire safety design'. *Fire and Materials*, **6**, 27–31.

Harris, R.J. (1983). *The Investigation and Control of Gas Explosions in Buildings and Heating Plant*. E. & F. N. Spon Ltd, London.

Harrison, A.J., and Eyre, J.A. (1987). 'The effect of obstacle arrays on the combustion of large premixed gas/air clouds'. *Combustion Science and Technology*, **52**, 121–137.

Hasegawa, K. (1989). 'Experimental study on the mechanism of hot zone formation in open tank fires'. *Proceedings of the 2nd International Symposium on Fire Safety Science*, pp. 221–230. Hemisphere Publishing Co. New York.

Hasemi, Y. (1984). 'Experimental wall heat transfer correlations for the analysis of upward wall flame spread'. *Fire Science and Technology*, **4**, 75.

Hasemi, Y. (1988). 'Deterministic properties of turbulent flames and implications for fire growth'. *Interflam'88: Proceedings of the 4th International Fire Conference*, pp. 45–52. Interscience Communications, London.

Hasemi, Y., and Nishihata, M. (1989). 'Fuel shape effect on the deterministic properties of turbulent diffusion flames'. *Proceedings of the 2nd International Symposium on Fire Safety Science*, pp. 275–286. Hemisphere Publishing Co., New York.

Hasemi, Y., and Takunaga, T. (1984). 'Some experimental aspects of turbulent diffusion flames and buoyant plumes from fire sources against a wall and in a corner of walls'. *Combustion Science and Technology*, **40**, 1–17.

Hasemi, Y., Yoshida, M., Nohara, A., and Nakabayashi, T. (1991). 'Unsteady state upward flame spreading velocity along vertical combustible solid and influence of external radiation on the flame spread'. *Proceedings of the 3rd International Symposium on Fire Safety Science*, pp. 197–206. Elsevier Applied Science, Barking.

Hawthorne, W.R., Weddell, D.S., and Hottel, H.C. (1949). *3rd Symposium (International) on Combustion*, pp. 266–288. Williams and Wilkins, Baltimore.

Health and Safety Executive (1980). 'Flame arresters and explosion reliefs'. *Health and Safety Series Booklet HS(G)11*, HMSO, London.

Hertzberg, M. (1982). 'The flammability limits of gases, vapours and dusts: theory and experiment', in *Symposium on Fuel-Air Explosions*, pp. 3–47, University of Waterloo Press, Waterloo, Ontatio.

Hertzberg, M., Johnson, A.L., Kuchta, J.M., and Furno, A.L. (1981). 'The spectral radiance, growth, flame temperatures and flammability behaviour of large scale, spherical combustion waves'. *16th Symposium (International) on Combustion*, pp. 767–776. The Combustion Institute, Pittsburgh.

Heselden, A.J.M., and Baldwin, R. (1976). 'The movement and control of smoke on escape routes in buildings', *Building Research Establishment, Current Paper CP13/76*.

Heselden, A.J.M., and Melinek, S.J. (1975). 'The early stages of fire growth in a compartment. A co-operative research programme of the CIB (Commission W14). First Phase'. *Fire Research Note No. 1029.*

Heskestad, A.W., and Hovde, P.J. (1994). 'Assessment of smoke production from building products'. *Proceedings of the 4th International Symposium on Fire Safety Science*, pp. 527–538. International Association for Fire Safety Science, Boston, MA.

Heskestad, G. (1972). 'Similarity relations for the initial convective flow generated by fire'. American Society of Mechanical Engineers, Winter Annual Meeting, New York, November 26–30.

Heskestad, G. (1975). 'Physical modelling of fire'. *J. Fire and Flammability*, **6**, 253–273.

Heskestad, G. (1976). Unpublished results.

Heskestad, G. (1981). 'Peak gas velocities and flame heights of buoyancy-controlled turbulent diffusion flames'. *18th Symposium (International) on Combustion*, pp. 951–960. The Combustion Institute, Pittsburgh.

Heskestad, G. (1982). 'Engineering relations for fire plumes'. *Society of Fire Protection Engineers, Technology Report 82–8.*

Heskestad, G. (1983). 'Luminous heights of turbulent diffusion flames'. *Fire Safety Journal*, **5**, 103–108.

Heskestad, G. (1986). 'Fire plume entrainment according to two competing assumptions'. *21st Symposium (International) on Combustion*, pp. 111–120. The Combustion Institute, Pittsburgh.

Heskestad, G. (1989). 'Note on maximum rise of fire plumes in temperature-stratified ambients'. *Fire Safety Journal*, **15**, 271–276.

Heskestad, G. (1995). 'Fire plumes', in *SFPE Handbook of Fire Protection Engineering*, 2nd Edition (eds P.J. Di Nenno *et al.*), pp. 2.9–2.19. Society of Fire Protection Engineers, Boston.

Heskestad, G., and Bill, R.G. (1988). 'Quantification of thermal responsiveness of automatic sprinklers, including conduction effects'. *Fire Safety Journal*, **14**, 113–125.

Heskestad, G., and Delichatsios, M.A. (1978). 'The initial convective flow in fire'. *17th Symposium (International) on Combustion*, pp. 1113–1123. The Combustion Institute, Pittsburgh.

Heskestad, G., and Smith, H. (1976). 'Investigation of a new sprinkler sensitivity approval test: the plunge test'. *FMRC Serial No. 22485*. Factory Mutual Research Corporation, Norwood, MA.

Hinkley, P.L. (1971). 'Some notes on the control of smoke in enclosed shopping centres'. *Fire Research Note No. 875.*

Hinkley, P.L., and Wraight, H.G.H. (1969). 'The contribution of flames under ceilings to fire spread in compartments. Part II. Combustible ceiling linings'. *Fire Research Note No. 743*.

Hinkley, P.L., Wraight, H.G.H., and Theobald, C.R. (1968). 'The contribution of flames under ceilings to fire spread in compartments. Part 1: Incombustible ceilings'. *Fire Research Note 712*.

Hirano, T., and Tazawa, K. (1978). 'A further study of the effects of external thermal radiation on flame spread over paper'. *Combustion and Flame*, **32**, 95–105.

Hirano, T., Noreikis, S.E., and Waterman, T.E. (1974). 'Postulations of flame spread mechanisms'. *Combustion and Flame*, **22**, 353–363.

Hirst, R., Savage, N., and Booth, K. (1981/82). 'Measurement of inerting concentrations'. *Fire Safety Journal*, **4**, 147–158.

Hjertager, B.H. (1993). 'Computer modelling of turbulent gas explosions in complex 2D and 3D geometries'. *J. Hazardous Materials*, **34**, 173–197.

Hoffmann, N., Galea, E.R., and Markatos, N. C. (1989). *Applied Mathematical Modelling*, **13**, 298.

Holborn, P.G., Bishop, S.R., Drysdale, D.D., and Beard, A.N. (1993). 'Experimental and theoretical models of flashover'. *Fire Safety Journal*, **21**, 257–266.

Hollyhead, R. (1996). 'Ignition of flammable gases and liquids by cigarettes'. *Science and Justice*, **36**, 257–266.

Holman, J.R. (1976). *Heat Transfer*, 4th Edition. McGraw-Hill, New York.

Holmes, F.H. (1975). 'Flammability testing of apparel fabrics'. *International Symposium on Fire Safety of Combustible Materials*, pp. 317–324. Edinburgh University.

Holmstedt, G., and Persson, H. (1985). 'Spray fire tests with hydraulic fluids'. *Proceedings of the 1st International Symposium on Fire Safety Science*, pp. 869–879. Hemisphere Publishing Co., New York.

Holve, D.J., and Sawyer, R.F. (1974). 'Measurement of burning polymer flame structure and mass transfer numbers'. Technical Report No. ME–74–2, Department of Mechanical Engineering, University of California, Berkeley.

Home Office (1995). *UK Fire and Loss Statistics 1993*.

Hottel, H.C. (1930). 'Radiant heat transmission'. *Mechanical Engineering*, 52.

Hottel, H.C. (1959). 'Review: Certain laws governing the diffusive burning of liquids', by Blinov and Khudiakov (1957). (*Dokl. Akad. Nauk SSSR*, **113**, 1096.) *Fire Research Abstracts and Reviews*, **1**, 41–43.

Hottel, H.C. (1961). 'Fire modelling', in *International Symposium on 'The use of models in fire research'* (ed. W.G. Berl), National Academy of Sciences, Publication 786 (Washington), p. 32.

Hottel, H.C. and Egbert, R.B. (1942). *American Institution of Chemical Engineers Transaction*, **38**.

Hottel, H.C., and Hawthorne, W.R. (1949). *3rd Symposium (International) on Combustion*, pp. 255–266. Williams and Wilkins, Baltimore.

Hottel, H.C., and Sarofim, A.F. (1967). *Radiative Transfer*. McGraw-Hill, New York.

Huggett, C. (1973). 'Habitable atmospheres which do not support combustion'. *Combustion and Flame*, **20**, 140–142.

Huggett, C. (1980). 'Estimation of rate of heat release by means of oxygen consumption measurements'. *Fire and Materials*, **4**, 61–65.

Huggett, C., von Elbe, G., and Haggerty, W. (1966). 'The combustibility of materials in O_2/He and O_2/N_2 atmospheres'. *Report SAM–TR–66–85*.

Inamura, T., Saito, K., and Tagivi, K.A. (1992). 'A study of boilover in liquid pool fires supported on water. Part 2. Effects of in-depth radiation absorption'. *Combustion Science and Technology*, **86**, 105.

Ingberg, S.H. (1928). 'Fire loads'. *Quarterly Journal of The National Fire Protection Association*, **22**, 43–61.

Iqbal, N., and Quintiere, J.G. (1994). 'Flame heat fluxes in PMMA pool fires'. *J. Fire Protection Engineering*, **6**, 153–162.

ISO (1975). Fire Resistance Tests. Elements of Building Construction. ISO 834, International Organization for Standardization, Geneva.

ISO (1980). Draft proposal ISO/DP 5924. International Organization for Standardization, Geneva.

ISO (1981). Fire Tests — Building Materials: Corner wall/room type test. ISO/TC92 N581, International Organization for Standardization, Geneva.

ISO (1990). 'Fire Tests. Smoke generated by building products — dual chamber test'. ISO 5924, International Organization for Standardization, Geneva.

ISO (1992). 'Fire Tests. Reaction to Fire. Lateral ignition and flame spread of building products'. ISO 5658, International Organization for Standardization, Geneva.

ISO (1993a). 'Fire Tests. Reaction to Fire. Rate of heat release from building products'. ISO 5660, International Organization for Standardization, Geneva.

ISO (1993b). 'Fire tests — full-scale room test for surface products'. ISO 9705, International Organization for Standardization, Geneva.

ISO (1997a). 'Fire Tests. Reaction to Fire. Ignitability of building products'. ISO 5657, International Organization for Standardisation, Geneva.

ISO (1997b). 'Guidance on ignitability'. ISO DTR 11925-1, prepared by ISO TC92/SC1/WG2), International Organization for Standardization, Geneva.

Ito, A., Masuda, D., and Saito, K. (1991). 'A study of flame spread over alcohols using holographic interferometry'. *Combustion and Flame*, **83**, 375–389.

Jackman, L.A., Nolan, P.F., Gardiner, A.J., and Morgan, H.P. (1992a) 'Mathematical model of the interaction of sprinkler spray drops with fire gases'. *Proceedings of the First International Conference on Fire Suppression Research*, 5–8 May 1992, Stockholm, pp. 209–227.

Jackman, L.A., Nolan, P.F., and Morgan, H.P. (1992b). 'Characterisation of water drops from a sprinkler spray'. *Proceedings of the First International Conference on Fire Suppression Research*, 5–8 May 1992, Stockholm, pp. 159–184.

Jaluria, Y. (1995) 'Natural convection wall flows'. SFPE Handbook of Fire Protection Engineering, 2nd Edition (Ed P.J. Di Nenno *et al.*), pp. 2.50–2.70, Society of Fire Protection Engineers, Boston.

Janssens, M.L. (1991a). 'Piloted ignition of wood: a review'. *Fire and Materials*, **15**, 151–167.

Janssens, M.L. (1991b). 'Measuring rate of heat release by oxygen consumption'. *Fire Technology*, **27**, 234–249.

Janssens, M.L. (1993). 'Cone calorimeter measurements of the heat of gasification of wood'. *Interflam'93: Proceedings of the Sixth International Fire Conference*, pp. 549–558. Interscience Communications, London.

Janssens, M.L. (1995). 'Calorimetry', in *SFPE Handbook of Fire Protection Engineering*, 2nd Edition (eds P.J. Di Nenno *et al.*), pp. 3.16–3.36. Society of Fire Protection Engineers, Boston.

Janssens, M.L., and Parker, W.J. (1992). 'Oxygen consumption calorimetry', in *Heat Release in Fires* (eds V. Babrauskas and S.J. Grayson), pp. 31–59. Elsevier Applied Science, Barking.

Jin, T. (1970). 'Visibility through fire smoke'. *Building Research Institute, Tokyo, Report No. 30*.

Jin, T. (1971). 'Visibility through fire smoke'. *Building Research Institute, Tokyo, Report No. 33*.

Jin, T. (1978). 'Visibility through fire smoke'. *J. Fire and Flammability*, **9**, 135–155.

Joint Fire Prevention Committee (1978). 'Report of the Technical Sub-Committee on the Fire Risks of New Materials, Central Fire Brigades' Advisory Council for England, Wales and Scotland', Home Office Fire Department, London.

Joint Fire Research Organisation (1961). *Fire Research 1960: Report of the Director*. HMSO, London.

Joint Fire Research Organisation (1962). *Fire Research 1961: Report of the Director*. HMSO, London.

Jost, W. (1939). *Explosions — und Verbrennungsvorgägne in Gasen*. Springer-Verlag, Berlin.

Kandola, B.S. (1995). 'Introduction to the mechanics of fluids', in *SFPE Handbook of Fire Protection Engineering*, 2nd Edition (eds P.J. Di Nenno *et al.*), pp. 1.1–1.24. Society of Fire Protection Engineers, Boston.

Kanury, A.M. (1972). 'Ignition of cellulosic materials: a review'. *Fire Research Abstracts and Reviews*, **14**, 24–52.

Kanury, A.M. (1975). *Introduction to Combustion Phenomena*. Gordon and Breach, London.

Kanury, A. M. (1995a). 'Ignition of liquid fuels', in *SFPE Handbook of Fire Protection Engineering*, 2nd Edition (eds P.J. Di Nenno *et al.*), pp. 2.160–2.170. Society of Fire Protection Engineers, Boston.

Kanury, A. M. (1995b). 'Flaming ignition of solid fuels', *SFPE Handbook of Fire Protection Engineering*, 2nd Edition (eds P.J. Di Nenno *et al.*), pp. 2.190–2.204. Society of Fire Protection Engineers, Boston.

Karlekar, B.V., and Desmond, R.M. (1979). *Heat Transfer*, 2nd Edition, West Publishing Co., St Paul, Minnesota.

Karlsson, B. (1993). 'A mathematical model for calculating heat release rate in the room corner test'. *Fire Safety Journal*, **20**, 93–113.

Kashiwagi, T. (1976). 'A study of flame spread over a porous material under external radiation fluxes'. *15th Symposium (International) on Combustion*, pp. 255–265. The Combustion Institute, Pittsburgh.

Kashiwagi, T. (1979). 'Effects of attenuation of radiation on surface temperature for radiative ignition'. *Combustion Science and Technology*, **20**, 225–234.

Kashiwagi, T. (1994). 'Polymer combustion and flammability — role of the condensed phase'. *25th Symposium (International) on Combustion*, pp. 1423–1437. The Combustion Institute, Pittsburgh.

Kashiwagi, T., and Newman, D.L. (1976). 'Flame spread over an inclined thin fuel surface'. *Combustion and Flame*, **26**, 163–177.

Kawagoe, K. (1958). 'Fire behaviour in rooms'. *Report No. 27*, Building Research Institute, Tokyo.

Kawagoe, K., and Sekine, T. (1963). 'Estimation of temperature-time curves in rooms'. *Occasional Report No. 11*, Building Research Institute, Tokyo.

Kaye, G.W.C., and Laby, T.H. (1986). *Tables of Physical and Chemical Constants and Some Mathematical Functions*. 15th Edition, Longman, London and New York.

Kent, J.H., Prado, G., and Wagner, H.Gg. (1981). 'Soot formation in a laminar diffusion flame'. *18th Symposium (International) on Combustion*, pp. 1117–1126. The Combustion Institute, Pittsburgh.

Kimmerle, G. (1974). 'Aspects and methodology for the evaluation of toxicological parameters during fire exposure'. *J. Fire and Flammability/Combustion Toxicology*, **1**, 4–51.

Kinbara, T., Endo, H., and Saga, S. (1968). 'Downward propagation of smouldering combustion through solid materials'. *11th Symposium (International) on Combustion*, pp. 525–531. The Combustion Institute, Pittsburgh.

Klitgaard, P.F., and Williamson, R.B. (1975). 'Impact of contents on building fires'. *J. Fire and Flammability/Consumer Product Flammability Supplement*, **2**, 84–113.

Klote, J.H. (1995). 'Smoke control', SFPE Handbook of Fire Protection Engineering, 2nd Edition, (eds P.J. Di Nenno *et al.*), pp. 4.230–4.245. Society of Fire Protection Engineers. Boston.

Kokkala, M.A. (1993). 'Characteristics of a flame in an open corner of walls'. *Interflam'93: Proceedings of the 6th International Fire Conference*, pp. 13–29. Interscience Communications, London.

Kokkala, M., and Rinkenen, W.J. (1987). 'Some observations on the shape of impinging diffusion flames', *VTT Research Report 461*, Valtion Teknillinen Turkimuskeskus, Espoo, Finland.

Koseki, H. (1993/94). 'Boilover and crude oil fire'. *J. Applied Fire Science*, **3**, 243–272.

Krause Jr., R.F., and Gann, R.G. (1980). 'Rate of heat release measurements using oxygen consumption'. *J. Fire and Flammability*, **12**, 117–130.

Kreith, F. (1976). *Principles of Heat Transfer*, 3rd Edition. Intext Educational Publishers, New York.

Kumar, S., Heywood, G.M., and Liew, S.K. (1997). 'Superdrop modelling of a sprinkler spray in a two-phase CFD-particle tracking model'. *Proceedings of the 5th International Symposium on Fire Safety Science*, pp. 889–900. International Association for Fire Safety Science, Boston, MA.

Kung, H.C., and Stravrianides, P. (1982). "Buoyant plumes of large scale pool fires". *19th Symposium (International) on Combustion* pps 905–912. (The Combustion Institute, Pittsburgh, PA)

Landau, L.D., and Lifshitz, E.M. (1987). *Fluid Mechanics*. Pergamon Press, Oxford.

Lastrina, F.A. (1970). 'Flame spread over solid fuel beds: solid and gas phase energy considerations'. PhD Thesis, Stevens Institute of Technology.

Latham, D.J., Kirby, B.R., and Thomson, G. (1987). 'The temperatures attained by unprotected structural steelwork in experimental natural fires'. *Fire Safety Journal*, **12**, 139–152.

Laurendeau, N.M. (1982). 'Thermal ignition of methane–air mixtures by hot surfaces: a critical examination'. *Combustion and Flame*, **46**, 29–49.

Law, M. (1963). 'Heat radiation from fires, and building separation'. *Fire Research Technical Paper No. 5*, HMSO, London.

Law, M. (1971). 'A relationship between fire grading and building design and contents'. *Fire Research Note No. 877*.

Law, M. (1972). 'Nomograms for the fire protection of structural steelwork'. *Fire Prevention Science and Technology No. 3*, Fire Protection Association, London.

Law, M., and O'Brien, T. (1981). *Fire Safety of Bare External Structural Steel*. Constructional Steel Research and Development Organisation, London.

Lawson, D.I. (1954). 'Fire and the atomic bomb'. *Fire Research Bulletin No. 1*, HMSO, London.

Lawson, D.I., and Simms, D.L. (1952). 'The ignition of wood by radiation'. *British J. Applied Physics*, **3**, 288–292.

Lawson, J.R., Walton, W.D., and Twilley, W.H. (1983). 'Fire performance of furnishings in the NBS furniture calorimeter Part 1', NBSIR83–2727, National Bureau of Standards, Gaithersburg, MD.

Le Chatelier, H., and Boudouard, O. (1898). 'Limits of flammability of gaseous mixtures'. *Bulletin de la Société Chimique (Paris)*, **19**, 485.

Lee, B.T. (1982). 'Quarter scale modeling of room fire tests of interior finish'. National Bureau of Standards, NBSIR 81–2453.

Lees, F.P. (1996). *Loss Prevention in the process Industries: Hazard Identification Assessment and Control*, 2nd Edition, in three volumes. Butterworth–Heinemann, Oxford.

Lewis, B. (1954). *Selected Combustion Problems*. AGARD (Butterworths), p. 177.

Lewis, B., and von Elbe, G. (1987). *Combustion, Flames and Explosions of Gases*, 3rd Edition. Academic Press, Orlando, FL.

Lide, D.R. (ed.) (1993/94). *Handbook of Chemistry and Physics*, 74th Edition. Chemical Rubber Company, Ohio.

Lie, T.T. (1972). *Fires and Buildings*. Applied Science Publishers, London.

Lie, T.T. (1974). 'Characteristic temperature curves for various fire severities'. *Fire Technology*, **10**, 315–326.

Lie, T.T. (1995). 'Fire temperature—time relationships', in *SFPE Handbook of Fire Protection Engineering*, 2nd Edition. (eds P.J. Di Nenno *et al.*), pp. 4.167–4.173. Society of Fire Protection Engineers, Boston.

Linnett, J.W., and Simpson, C.J.S.M. (1957). 'Limits of inflammability', *Sixth Symposium (International) on Combustion*, pp. 20–27. Reinhold, New York.

Lougheed, G.D., and Yung, D. (1993). 'Exposure to adjacent structures from flames issuing from a compartment opening'. *Interflam'93: Proceedings of the Sixth International Fire Conference*, pp. 297–306. Interscience Communications, London.

Lovachev, L.A., Babkin, V.S., Bunev, V.A., V'yun, A.V., Krivulin, V.N., and Baratov, A.N. (1973). 'Flammability limits: an invited review'. *Combustion and Flame*, **20**, 259–289.

Luo, M., and Beck, V.R. (1994). 'The fire environment in a multi-room building — comparison of predicted and experimental results'. *Fire Safety Journal*, **23**, 413–438.

Lyons, J.W. (1970). *The Chemistry and Uses of Fire Retardants*. John Wiley, New York.

McAlevy, R.F., and Magee, R.S. (1969). 'The mechanism of flame spreading over the surface of igniting condensed phase materials'. *12th Symposium (International) on Combustion*, pp. 215–227.

McCaffrey, B.J. (1979). 'Purely buoyant diffusion flames: some experimental results'. National Bureau of Standards, NBSIR 79–1910.

McCaffrey, B.J. (1995). 'Flame height'. in *SFPE Handbook of Fire Protection Engineering*, 2nd Edition (eds P.J. Di Nenno *et al.*), pp. 2.1–2.8. Society of Fire Protection Engineers, Boston.

McCaffrey, B.J., and Rockett, J.A (1977). 'Static pressure measurements of enclosure fires', *J. Research of the National Bureau of Standards*, **82**, 107–117.

McCaffrey, B.J., Quintiere, J.G., and Harkleroad, M.F. (1981). 'Estimating room temperatures and the likelihood of flashover using fire test data correlations'. *Fire Technology*, **17**, 98–119; **18**, 122.

McCarter, R.J. (1976). 'Smouldering of flexible polyurethane foam'. *J. Consumer Product Flammability*, **3**, 128–140.

McCarter, R.J. (1977). 'Smouldering combustion of cotton and rayon'. *J. Consumer Product Flammability*, **4**, 346–357.

McCarter, R.J. (1978). 'Smouldering combustion of wood fibres: cause and prevention'. *J. Fire and Flammability*, **9**, 119–126.

McGuire, J.H. (1953). 'Heat transfer by radiation'. *Fire Research Special Report No. 2*. HMSO, London.

Mckinven, R., Hensel, J.G., and Glassman, I. (1970). 'Influence of laboratory parameters on flame spread over liquid surfaces'. *Combustion Science and Technology*, **1**, 293–306.

Madorsky, S.L. (1964). *Thermal Degradation of Organic Polymers*. Interscience, John Wiley, New York.

Magee, R.S., and McAlevy, R.F. (1971). 'The mechanism of flame spread'. *J. Fire and Flammability*, **2**, 271–297.

Magnusson, S.-E., and Sundstrom, B. (1985). 'Combustible linings and room fire growth — a first analysis', *Fire Safety Science and Engineering*, ASTM STP 882, (ed. T.Z. Harmathy), pp. 45–69. American Society for Testing and Materials, Philadelphia.

Malhotra, H.L. (1967). 'Movement of smoke on escape routes. Instrumentation and effect of smoke on visibility'. *Fire Research Note Nos. 651, 652, and 653*.

Malhotra, H.L. (1982). *Design of Fire-resisting Structures*. Surrey University Press.

Marchant, E.W. (1976). 'Some aspects of human behaviour and escape route design'. *5th International Fire Protection Seminar*, Karlsruhe, September.

Marchant, E.W. (1984). 'Effect of wind on smoke movement and smoke control systems'. *Fire Safety Journal*, **7**, 55–63.

Margenau, H., and Murphy, G.M. (1956). *The Mathematics of Chemistry and Physics*. Van Nostrand, Princeton, NJ.

Markstein, G.H. (1975). 'Radiative energy transfer from gaseous diffusion flames'. *15th Symposium (International) on Combustion*, pp. 1285–1294. The Combustion Institute, Pittsburgh.

Markstein, G.H. (1976). 'Radiative energy transfer from turbulent diffusion flames'. *Combustion and Flame*, **27**, 51–63.

Markstein, G.H. (1979). 'Radiative properties of plastics fires'. *17th Symposium (International) on Combustion*, pp. 1053–1062. The Combustion Institute, Pittsburgh.

Markstein, G.H. (1984). "Effects of turbulence on flame radiation from diffusion flames". *20th Symposium (International) on Combustion* pps 1055. (The Combustion Institute, Pittsburgh, PA)

Markstein, G.H., and de Ris, J. (1972). 'Upward fire spread over textiles'. *14th Symposium (International) on Combustion*, pp. 1085–1097. The Combustion Institute, Pittsburgh.

Markstein, G.H., and de Ris, J.N. (1975). 'Flame spread along fuel edges'. *J. Fire and Flammability*, **6**, 140–154.

Martin, S. (1965). 'Diffusion-controlled ignition of cellulosic materials by intense radiant energy'. *10th Symposium (International) on Combustion*, pp. 877–896. The Combustion Institute, Pittsburgh.

Martin, S.B., and Wiersma, S.J. (1979). 'An experimental study of flashover criteria for compartment fires'. Final Report, Products Research Committee PRC No. P–77–3–1, Stanford Research Institute, International Project No. PYC 6496.

Mayer, E. (1957). 'A theory of flame propagation limits due to heat loss'. *Combustion and Flame*, **1**, 438–452.

Meidl, J. (1970). *Explosive and Toxic Hazardous Materials*. Glencoe Press, Fire Science Series.

Middleton, J.F. (1983). 'Developments in flame detectors'. *Fire Safety Journal*, **6**, 175–182.

Mikkola, E. (1992). 'Ignitability of solid materials', in *Heat Release in Fires* (eds V. Babrauskas and S.J. Grayson), pp.225–232. Elsevier Applied Science, Barking.

Mikkola, E., and Wichman, I.S. (1989). 'On the thermal ignition of combustible materials'. *Fire and Materials*, **14**, 87–96.

Milke, J.A. (1995b). 'Smoke management in covered malls and atria', in *SFPE Handbook of Fire Protection Engineering*, 2nd Edition (eds P.J. Di Nenno *et al.*), pp. 4.246–4.258. Society of Fire Protection Engineers, Boston.

Milke, J.A. (1996). 'Analytical methods for determining fire resistance of steel members', in *SFPE Handbook of Fire Protection Engineering*, 2nd Edition (eds P.J. Di Nenno *et al.*), pp. 4.174–4.201. Society of Fire Protection Engineers, Boston.

Miller, F.J., and Ross, H.D. (1992). 'Further observations of flame spread over laboratory-scale alcohol pools'. *24th Symposium (International) on Combustion*, pp. 1703–1711. The Combustion Institute, Pittsburgh.

Ministry of Aviation (1962). *Report of the Working Party on Aviation Kerosene and Wide-Cut Gasoline* (HMSO).

Miron, Y., and Lazzara, C.P. (1988). 'Hot surface ignition temperatures of dust layers'. *Fire and Materials*, **12**, 115–126.

Mitchell, N.D. (1951). *Quarterly Journal of the National Fire Protection Association*, **45**, 165.

Mitler, H.E. (1985). 'The Harvard fire model'. *Fire Safety Journal*, **9**, 7–16.

Mitler, H.E., and Emmons, H.W. (1981). Documentation for CFC V, the Fifth Harvard Computer Fire Code, Home Fire Project Technical Report No. 45, Harvard University.

Modak, A.T., and Croce, P.A. (1977). 'Plastic pool fires'. *Combustion and Flame*, **30**, 251–265.

Montreal Protocol on Substances that Deplete the Ozone Layer, UNEP, Nairobi, 1987.

Moore, L.D. (1978). 'Full scale burning behaviour of curtains and drapes', NBSIR78–1448, National Bureau of Standards, Gaithersburg, MD.

Moore, W.J. (1972). *Physical Chemistry*, 5th Edition. Longman, London.

Morgan, H.P., Marshall, N.R., and Goldstone, B.M. (1976). 'Smoke hazards in covered multi-level shopping malls. Some studies using a model 2-storey mall'. Building Research Establishment, BRE CP 45/76.

Morris, W.A., and Hopkinson, J.S. (1976). 'Fire behaviour of foamed plastics ceilings used in dwellings'. Building Research Establishment, BRE CP 73/76.

Morton, B.R., Taylor, G., and Turner, J.S. (1956). 'Turbulent gravitational convection from maintained and instantaneous sources'. *Proceedings of the Royal Society (London)*, **A234**, 1–23.

Moulen, A.W. (1971). 'Horizontal projections in the prevention of spread of fire from storey to storey'. *Report TR52/75/397*, Commonwealth Experimental Building Station, Australia.

Moulen, A.W., and Grubits, S.J. (1979). 'Flammability testing of carpets'. Technical Record 449, Experimental Building Station, Department of Housing and Construction, Australia.

Moussa, N.A., Toong, T.Y., and Backer, S. (1973). 'An experimental investigation of flame-spreading mechanisms over textile materials'. *Combustion Science and Technology*, **8**, 165–175.

Moussa, N.A., Toong, T.Y., and Garris, C.A. (1977). 'Mechanisms of smouldering of cellulosic materials'. *16th Symposium (International) on Combustion)*, pp. 1447–1457. The Combustion Institute, Pittsburgh.

Mowrer, F.W. (1990). 'Lag times associated with fire detection and suppression'. *Fire Technology*, **26**, 244–265.

Mowrer, F.W., and Williamson, R.B. (1987). 'Estimating room temperatures from fires along walls and in corners'. *Fire Technology*, **23**, 133–145.

Mudan, K.S. (1984). 'Thermal radiation hazards from hydrocarbon pool fires'. *Progress in Energy and Combustion Science*, **10**, 59–80.

Mudan, K.S., and Croce, P.A. (1995). 'Fire hazard calculations for large open hydrocarbon fires', in *SFPE Handbook of Fire Protection Engineering*, 2nd Edition (eds P.J. Di Nenno *et al.*), pp. 3.197–3.240. Society of Fire Protection Engineers, Boston.

Mulhollund, G.W. (1995). 'Smoke production and properties', *SFPE Handbook of Fire Protection Engineering*, 2nd Edition (eds P.J. Di Nenno *et al.*), pp. 2.217–2.227, Society of Fire Protection Engineers, Boston.

Mullins, B.P., and Penner, S.S. (1959). *Explosions, Detonations, Flammability and Ignition*. Pergamon Press, London.

Murrell, J., and Rawlins, P. (1996). 'Fire hazard of multi-layer paint surfaces'. *Proceedings of the 4th International Conference on Fire and Materials*. Interscience Communications, London.

Nakaya, I., Tanaka, T., Yoshida, K., and Steckler, K. (1986). 'Doorway flow induced by a propane fire'. *Fire Safety Journal*, **10**, 185–195.

National Bureau of Standards (1972). 'The mechanisms of pyrolysis oxidation and burning of organic materials'. *Proceedings of the 4th Materials Research Symposium*, Gaithersburg, Maryland.

National Fire Protection Association (1978). 'Intrinsically safe process control equipment for use in hazardous locations'. NFPA No. UL913.

National Fire Protection Association (1993a). *NFPA 72, National Fire Alarm Code*. NFPA, Quincy, MA.

National Fire Protection Association (1993b). *NFPA 70 National Electrical Code: Hazardous Area Classification*. NFPA, Quincy, MA.

National Fire Protection Association (1993c). *NFPA 92B Guide for Smoke Management Systems in Malls, Atria and Large Areas*. NFPA, Quincy, MA.

National Fire Protection Association (1994). 'Fire hazards in oxygen enriched atmospheres'. NFPA No. 53.

National Fire Protection Association (1995). 'Standard method of test for critical radiant flux of floor covering systems using a radiant heat energy source'. NFPA No. 253.

National Fire Protection Association (1997). *NFPA Handbook*, 18th Edition. NFPA, Quincy, MA.

Neoh, K.G., Howard, J.B., and Sarofim, A.F. (1984). 'Effect of oxidation on the physical structure of soot'. *20th Symposium (International) on Combustion*, pp. 951–957. The Combustion Institute, Pittsburgh.

Newman, J.S. (1988). 'Principles for fire detection'. *Fire Technology*, **24**, 116–127.

Nordtest (1986). 'Surface products: Room fire tests in full scale', Nordtest method NT Fire 025.

Nordtest (1991). 'Upholstered furniture: burning behaviour — full scale test. Edition 2', Nordtest method NT Fire 032.

Odeen, K. (1963). 'Theoretical study of fire characteristics in enclosed spaces'. *Bulletin No. 10*, Division of Building Construction, Royal Institute of Technology, Stockholm.

Ohlemiller, T.J. (1990). 'Smouldering combustion propagation through a permeable horizontal fuel layer'; and 'Forced smoulder propagation and the transition to flaming in cellulosic insulation'. *Combustion and Flame*, **81**, 341–353; 354–365.

Ohlemiller, T.J. (1991). 'Smouldering combustion propagation on solid wood'. *Proceedings of the 3rd International Symposium on Fire Safety Science*, pp. 565–574. Elsevier Science Publishers, Barking.

Ohlemiller, T.J., and Lucca, D.A. (1983). 'An experimental comparison of forward and reverse smoulder propagation in permeable fuel beds'. *Combustion and Flame*, **54**, 131–148.

Oleskiewicz, I. (1989). Heat transfer from a window fire plume to a building facade. *Proceedings of the Winter Annual Meeting of the American Society of Mechanical Engineers*, San Francisco, HTD-volume 123, pages 163–170.

Ohlemiller, T.J., and Rogers, F.E. (1978). 'A survey of several factors influencing smouldering combustion in flexible and rigid polymer foams'. *J. Fire and Flammability*, **9**, 489–509.

Ohtani, H., Hirano, T., and Akita, K. (1981). 'Experimental study of bottom surface combustion of polymethylmethacrylate'. *18th Symposium (International) on Combustion*, pp. 591–599. The Combustion Institute, Pittsburgh.

Open University (1973). 'Giant Molecules'. Science Foundation Course S100 Unit 13.

Orloff, L., and de Ris, J. (1972). 'Cellular and turbulent ceiling fires'. *Combustion and Flame*, **18**, 389–401.

Orloff, L., and de Ris, J.N. (1982). 'Froude modelling of pool fires'. *19th Symposium (International) on Combustion*, pp. 885–895. The Combustion Institute, Pittsburgh.

Orloff, L., de Ris, J., and Markstein, G.H. (1974). 'Upward turbulent fire spread and burning of fuel surface'. *15th Symposium (International) on Combustion*, pp. 183–192. The Combustion Institute, Pittsburgh.

Orloff, L., Modak, A.T., and Alpert, R.L. (1976). 'Burning of large scale vertical surfaces'. *16th Symposium (International) on Combustion*, pp. 1345–1354. The Combustion Institute, Pittsburgh.

Orloff, L., Modak, A.T., and Markstein, G.H. (1979). 'Radiation from smoke layers'. *17th Symposium (International) on Combustion*. pp. 1029–1038. The Combustion Institute, Pittsburgh.

Östman, B.A.L. (1992). 'Smoke and soot', in *Heat Release in Fires* (eds V. Babrauskas and S.J. Grayson), pp. 233–250. Elsevier Applied Science, London.

Östman, B.A.L., and Nussbaum, R.M. (1988). 'Correlation between small-scale rate of heat release and full-scale room flashover for surface linings'. *Proceedings of the 2nd International Symposium on Fire Safety Science*, pp. 823–832. Hemisphere Publishing Co., New York.

Pagni, P.J. (1990). 'Some unanswered questions in fluid mechanics'. *Applied Mechanics Review*, **43**, 153–170.

Pagni, P.J., and Bard, S. (1979). 'Particulate volume fractions in diffusion flames'. *17th Symposium (International) on Combustion*, pp. 1017–1025. The Combustion Institute, Pittsburgh.

Palmer, K.N. (1957). 'Smouldering combustion of dusts and fibrous materials'. *Combustion and Flame*, **1**, 129–154.

Palmer, K.N. (1973). *Dust Explosions and Fires*. Chapman and Hall, London.

Pape, R., and Waterman, T.E. (1979). 'Understanding and modeling preflashover compartment fires', in *Design of Building for Fire Safety* (eds E.E. Smith and T.Z. Harmathy), American Society for Testing and Materials, STP 685.

Parker, W.J. (1972). 'Flame spread model for cellulosic materials'. *J. Fire and Flammability*, **3**, 254–269.

Paul, K.T. (1979). Private communication.

Paul, K.T. (1983). 'Measurement of smoke in large scale fires'. *Fire Safety Journal*, **5**, 89–102.

Paulsen, O.R., and Hadvig, S. (1977). 'Heat transfer in fire test furnaces'. *J. Fire and Flammability*, **8**, 423–442.

Perdue, G.R. (1956). 'Spontaneous combustion: how it is caused and how to prevent it occurring'. *Power Laundry*, **95**, 599.

Perry, R.H., Green, D.W., and Maloney, J.O. (eds) (1984). *Perry's Chemical Engineers' Handbook*. McGraw-Hill, New York.

Petrella, R.V. (1979). 'The mass burning rate and mass transfer number of selected polymers, wood and organic liquids'. *Polym. Plast. Technol. Eng.*, **13**, 83–103.

Pettersson, O., Magnuson, S.E., and Thor, J. (1976). 'Fire engineering design of structures'. Swedish Institute of Steel Construction, *Publication 50*.

Petty, S.E. (1983). 'Combustion of crude oil on water'. *Fire Safety Journal*, **5**, 123–134.

Phillips, H. (1965). 'Flame in a buoyant methane layer'. *10th Symposium (International) on Combustion*, pp. 1277–1283. The Combustion Institute, Pittsburgh.

Pipkin, O.A., and Sliepcevich, C.M. (1964). 'The effect of wind on buoyant diffusion flames'. *Industrial and Engineering Chemistry: Fundamentals*, **3**, 147–154.

Pitts, D.R., and Sissom, L.E. (1977). *Schaum's Outline Series: Theory and Problems of Heat Transfer*. McGraw-Hill, New York.

Pitts, W.M. (1991). 'Wind effects on fire'. *Progress in Energy and Combustion Science*, **17**, 83–134.

Pitts, W.M. (1994). 'Application of thermodynamic and detailed chemical kinetic modelling to understanding combustion product generation in enclosure fires'. *Fire Safety Journal*, **23**, 272–303.

Potts, W.J., and Lederer, T.S. (1977). 'A method for comparing testing of smoke toxicity'. *J. Combustion Toxicology*, **4**, 114–162.

Powell, F. (1969). 'Ignition of gases and vapours: a review of ignition of flammable gases and vapours by friction and impact'. *Industrial and Engineering Chemistry*, **61**, 29–37.

Prahl, J., and Emmons, H.W. (1975). 'Fire induced flow through an opening'. *Combustion and Flame*, **25**, 369–385.

Prime, D.M. (1981). Private communication.

Prime, D.M. (1982). 'Ignitibility and leather upholstery'. *Cabinet Maker and Retail Furnisher*, 26 March, p. 39.

Products Research Committee (1980). 'Fire research on cellular plastics: the final report of the Products Research Committee'.

Punderson, J.O. (1981). 'A closer look at cause and effect in fire fatalities—the role of toxic fumes'. *Fire and Materials*, **5**, 41–46.

Puri, R., and Santoro, R.J. (1991). 'The role of soot particle formation on the production of carbon monoxide in fires'. *Proceedings of the 3rd International Symposium on Fire Safety Science*, pp. 595–604. Elsevier Science Publishers, Barking.

Purser, D.A. (1995). 'Toxicity assessment of combustion products', in *SFPE Handbook of Fire Protection Engineering*, 2nd Edition (eds P.J. Di Nenno *et al.*), pp. 2.85–2.146. Society of Fire Protection Engineers, Boston.

Quintiere, J.G. (1975a). 'Some observations on building corridor fires'. *15th Symposium (International) on Combustion*, pp. 163–174. The Combustion Institute, Pittsburgh.

Quintiere, J.G. (1975b). 'The application and interpretation of a test method to determine the hazard of floor covering fire spread in building corridors'. *International Symposium on 'Fire Safety of Combustible Materials'*, Edinburgh University, pp. 355–366.

Quintiere, J.G. (1976). 'Growth of fire in building compartments', in *Fire Standards and Safety* (ed. A.F. Robertson). American Society for Testing and Materials, STP 614, pp. 131–167.

Quintiere, J.G. (1979). 'The spread of fire from a compartment: a review', in *Design of Buildings for Fire Safety* (eds E.E. Smith and T.Z. Harmathy). American Society for Testing and Materials, STP 685, pp. 139–168.

Quintiere, J.G. (1981). 'A simplified theory for generalising results from a radiant panel rate of flame spread apparatus'. *Fire and Materials*, **5**, 52–60.

Quintiere, J. (1982). 'An assessment of correlations between laboratory and full-scale experiments for the FAA Aircraft Fire Safety Programme, part 2. Rate of energy release in fires', NBSIR82-2536, National Bureau of Standards, Gaithersburg, MD.

Quintiere, J.G. (1983). 'An assessment of correlations between laboratory and full-scale experiments for the FAA aircraft fire safety program, Part 1: Smoke'. NBSIR 82–2508.

Quintiere, J.G. (1989). 'Fundamentals of enclosure fire zone models'. *J. Fire Protection Engineering*, **1**, 99–119.

Quintiere, J.G. (1995a). 'Surface flame spread', in *SFPE Handbook of Fire Protection Engineering*, 2nd Edition (eds P.J. Di Nenno *et al.*), pp. 2.205–2.216. Society of Fire Protection Engineers, Boston.

Quintiere, J.G. (1995b). 'Compartment fire modelling', in *SFPE Handbook of Fire Protection Engineering*, 2nd Edition (eds P.J. Di Nenno *et al.*), pp. 3.122–3.133. Society of Fire Protection Engineers, Boston.

Quintiere, J., Harkleroad, M., and Hasemi, Y. (1985). 'Wall flames and implications for upward flame spread', AIAA Paper 85–0456, American Institute of Astronautics and Aeronautics.

Quintiere, J.G., McCaffrey, B.J., and Kashiwagi, T. (1978). 'A scaling study of a corridor subjected to a room fire'. *Combustion Science and Technology*, **18**, 1–19.

Quintiere, J.G., Rinkinen, W.J., and Jones, W.W. (1981). 'The effect of room openings on fire plume entrainment'. *Combustion Science and Technology*, **26**, 193–201.

Rae, D., Singh, B., and Damson, R. (1964). *Safety in Mines Research Report No. 224*. HMSO, London.

Raj, P.P.K., Moussa, A.N., and Aravamudan, K. (1979). 'Experiments involving pool and vapour fires from spills of liquefied natural gas on water'. *US Coast Guard Report No. CG–D–55–79*.

Rasbash, D.J. (1951). 'The efficiency of hand lamps in smoke'. *Institute of Fire Engineers Quarterly*, **11**, 46–52.

Rasbash, D.J. (1962). 'The extinction of fires by water spray'. *Fire Research Abstracts and Reviews*, **4**, 28–53.

Rasbash, D.J. (1967). 'Smoke and toxic products produced at fires'. *Transactions and Journal of the Plastics Institute, Conference Supplement No. 2*, pp. 55–62 (January 1967).

Rasbash, D.J. (1969). 'The relief of gas and vapour explosions in domestic structures'. *The Structural Engineer*, October.

Rasbash, D.J. (1975). 'Relevance of firepoint theory to the assessment of fire behaviour of combustible materials'. *International Symposium on Fire Safety of Combustible Materials*, Edinburgh University, pp. 169–178.

Rasbash, D.J. (1976). 'Theory in the evaluation of fire properties of combustible materials'. *Proceedings of 5th International Fire Protection Seminar (Karlsruhe, September)*, pp. 113–130.

Rasbash, D.J. (ed.) (1984). 'Special Issue: the Stardust Club Fire, Dublin, 1991'. *Fire Safety Journal*, **7**, 205–295.

Rasbash, D.J. (1986). 'Quantification of explosion parameters for combustible fuel–air mixtures'. *Fire Safety Journal*, **11**, 113–125.

Rasbash, D.J. (1991). 'Major fire disasters involving flashover'. *Fire Safety Journal*, **17**, 85–93.

Rasbash, D.J., and Drysdale, D.D. (1982). 'Fundamentals of smoke production'. *Fire Safety Journal*, **5**, 77–86.

Rasbash, D.J., and Drysdale, D.D. (1983). 'Theory of fire and fire processes'. *Fire and Materials*, **7**, 79–88.

Rasbash, D.J., and Phillips, R.P. (1978). 'Quantification of smoke produced at fires. Test methods for smoke and methods of expressing smoke evolution'. *Fire and Materials*, **2**, 102–109.

Rasbash, D.J., and Pratt, B.T. (1980). 'Estimation of the smoke produced in fires'. *Fire Safety Journal*, **2**, 23–37.

Rasbash, D.J., and Rogowski, Z.W. (1960). 'Relief of explosions in duct systems'. *1st Symposium on Chemical Process Hazards*, pp. 58–65. Institution of Chemical Engineers, London.

Rasbash, D.J., and Rogowski, Z.W. (1962). 'Venting gaseous explosions in duct systems. IV. The effect of obstructions'. *Fire Research Note No. 490.*

Rasbash, D.J., Drysdale, D.D., and Deepak, D. (1986). 'Critical heat and mass transfer at pilot ignitions and extinction of a material'. *Fire Safety Journal*, **10**, 1–10.

Rasbash, D.J., Palmer, K.N., Rogowski, Z.W., and Ames, S.A. (1970). 'Gas explosions in multiple compartments', *Fire Research Note No. 847.*

Rasbash, D.J., Rogowski, Z.W., and Stark, G.W.V. (1956). 'Properties of fires of liquids'. *Fuel*, **31**, 94–107.

Ricou, F.P., and Spalding, B.P. (1961). 'Measurements of entrainment by axi-symmetric jets'. *J. Fluid Mechanics*, **11**, 21–32.

Roberts, A.F. (1964a). 'Calorific values of partially decomposed wood samples'. *Combustion and Flame*, **8**, 245–246.

Roberts, A.F. (1964b). 'Ultimate analysis of partially decomposed wood samples'. *Combustion and Flame*, **8**, 345–346.

Roberts, A.F. (1970). 'A review of kinetic data for the pyrolysis of wood and related substances'. *Combustion and Flame*, **14**, 261–272.

Roberts, A.F. (1971a). 'Problems associated with the theoretical analysis of the burning of wood'. *13th Symposium (International) on Combustion*, pp. 893–903. The Combustion Institute, Pittsburgh.

Roberts, A.F. (1971b). 'The heat of reaction during the pyrolysis of wood'. *Combustion and Flame*, **17**, 79–86.

Roberts, A.F. (1981/82). 'Thermal radiation hazards from releases of LPG from pressurised storage'. *Fire Safety Journal*, **4**, 197–212.

Roberts, A.F., and Quince, B.W. (1973). 'A limiting condition for the burning of flammable liquids'. *Combustion and Flame*, **20**, 245–251.

Roberts, P., Smith, D.B., and Wood, D.R. (1980). 'Flammability of paraffin hydrocarbons in confined and unconfined conditions'. *Chemical Process Hazards VII, I. Chem. E. Symposium Series No. 58*, pp. 157–169. Institution of Chemical Engineers, Rugby.

Robertson, A.F. (1975). 'Estimating smoke production during building fires'. *Fire Technology*, **11**, 80–94.

Robertson, R.B. (1976). 'Spacing in chemical plant design against loss by fire'. *Symposium on Process Industry Hazards—Accidental Release: Assessment, Containment and Control.* Institution of Chemical Engineers Symposium Series No. 47, pp. 157–173.

Robinson, C., and Smith, D.B. (1984). 'The autoignition temperature of methane'. *J. Hazardous Materials*, **8**, 199–203.

Rockett, J.A. (1976). 'Fire induced gas flow in an enclosure'. *Combustion Science and Technology*, **12**, 165–175.

Rockett, J.A., and Milke, J.A. (1995). 'Conduction of heat in solids', in *SFPE Handbook of Fire Protection Engineering*, 2nd Edition (eds P.J. Di Nenno *et al.*), pp. 1.25–1.38. Society of Fire Protection Engineers, Boston.

Rogers, F.E., and Ohlemiller, T.J. (1978). 'Minimising smoulder tendency in flexible polyurethanes'. *J. Consumer Product Flammability*, **5**, 59–67.

Rohsenow, W.M., and Choi, H.Y. (1961). *Heat, Mass and Momentum Transfer*. Prentice-Hall, London.

Rothbaum, H.P. (1963). 'Spontaneous combustion of hay'. *J. Applied Chemistry (London)*, **13**, 291–302.

Routley, A.F., and Skipper, R.S. (1980). 'A new approach to testing materials in the NBS Smoke Chamber'. *Fire and Materials*, **4**, 98–103.

Roy, D. (1997). 'The cost of fires: a review of the information available', Home Office Publication Unit, London.

Royal, J.H. (1970). 'The influence of fuel bed thickness on flame spreading rate'. Honours Report, Stevens Institute of Technology.

Saito, K., Quintiere, J.G., and Williams, F.A. (1985). 'Upward turbulent flame spread'. *Proceedings of the 1st International Symposium on Fire Safety Science*, pp. 75–86. Hemisphere Publishing Co., Washington.

Saito, K., Williams, F.A., Wichman, I.S., and Quintiere, J.G. (1989). 'Upward turbulent flame spread on wood under external radiation'. *Transactions of the ASME*, **111**, 438–445.

Salig, R.J. (1981). 'The smouldering behaviour of polyurethane cushionings and its relevanace to home furnishing fires', M.Sc. Dissertation, Massachusetts Institute of Technology.

Sapko, M., Furno, A., and Kuchta, J. (1976). US Bureau of Mines, RI 8176.

Sato, K., and Sega, S. (1988). 'Smoulder spread in a horizontal layer of cellulosic powder'. *Proceedings of the 2nd International Symposium on Fire Safety Science*, pp. 87–96. Hemisphere Publishing Co., New York.

Sato, T., and Kunimoto, T. (1969). *Mem. Faculty of Engineering, Kyoto University*, **31**, 47.

Sawyer, R.F., and Fristrom, R.M. (1971). 'Flame inhibition and chemistry'. AGARD Conference No. 84, AGARD CP 84–71.

Sawyer, S.F. (1996). 'Fire Investigation Report: Bulk Merchandising Store, Quincy, MA, May 23, 1995', National Fire Protection Association, Quincy, MA.

Schifiliti, R.P., Meacham, B.J., and Custer, R.L.P. (1995). 'Design of detection systems', in *SFPE Handbook of Fire Protection Engineering*, 2nd Edition (eds P.J. Di Nenno *et al.*), pp. 4.1–4.29. Society of Fire Protection Engineers, Boston.

Schnell, L.G. (1996). 'Flashover training in Sweden'. *Fire Engineers Journal*, **56**, 25–28.

Schult, D.A., Matkowski, B.J., Volpert, V.A., and Fernandez-Pello, A.C. (1995). 'Propagation and extinction of forced opposed flow smoulder waves'. *Combustion and Flame*, **101**, 471–490.

Seader, J.D., and Chien, W.P. (1974). 'Mass optical density as a correlating parameter for the NBS Smoke Chamber'. *J. Fire and Flammability*, **5**, 151–163.

Seader, J.D., and Chien, W.P. (1975). 'Physical aspects of smoke development in an NBS smoke density chamber'. *J. Fire and Flammability*, **6**, 294–310.

Seigel, L.G. (1969). 'The projection of flames from burning buildings'. *Fire Technology*, **5**, 43–51.

Semenov, N.N. (1928). 'Theories of combustion processes', *Z. Phys. Chem.*, **48**, 571–582.

Setchkin, N.P. (1954). 'Self-ignition temperatures of combustible liquids'. *J. Research, National Bureau of Standards*, **53**, 49–66.

Shields, T.J., Silcock, G.W., and Murray, J. (1993). 'The effects of geometry and ignition mode on ignition times obtained using a cone calorimeter and ISO ignitability apparatus'. *Fire and Materials*, **17**, 25–32.

Shipp, M. (1983). 'A hydrocarbon fire standard: an assessment of existing information'. Department of Energy, Offshore Energy Technology Board, TO/R/8294.

Shokri, M., and Beyler, C.L. (1989). 'Radiation from large pool fires'. *J. Fire Protection Engineering*, **1**, 141–150.

Silcock, A., and Hinkley, P.L. (1974). 'Report on the spread of fire at Summerland in Douglas, Isle of Man, 2 August 1973', Building Research Establishment BRE CP 74/74.

Simcox, S., Wilkes, N.S., and Jones, I.P. (1992). 'Computer simulation of the flows of hot gases from the fire at King's Cross Underground Station'. *Fire Safety Journal*, **18**, 49–73.

Simmons, R.F. (1995a). 'Premixed burning', in *SFPE Handbook of Fire Protection Engineering*, 2nd Edition (eds P.J. Di Nenno *et al.*), pp. 1.131–1.140. Society of Fire Protection Engineers, Boston.

Simmons, R.F. (1995b). 'Fire chemistry', in *Combustion Fundamentals of Fire* (ed. G. Cox), pp. 405–473. Academic Press, London.

Simmons, R.F., and Wolfhard, H.G. (1955). 'The influence of methyl bromide on flames — 1. Premixed flames'. *Transactions of the Faraday Society*, **51**, 1211–1217.

Simms, D.L. (1963). 'On the pilot ignition of wood by radiation'. *Combustion and Flame*, **7**, 253–261.

Simms, D.L., and Hird, D. (1958). 'On the pilot ignition of materials by radiation'. *Fire Research Note No. 365*.

Sirignano, W.A., and Glassman, I. (1970). 'Flame spreading above liquid fuels: surface tension-driven flows'. *Combustion Science and Technology*, **1**, 307–312.

Smith, D.A. (1992). 'Measurements of flame length and flame angle in an inclined trench'. *Fire Safety Journal*, **18**, 231–244.

Smith, D., and Cox, G. (1992). 'Major chemical species in buoyant turbulent diffusion flames'. *Combustion and Flame*, **91**, 226–238.

Smith, E.E. (1972). 'Heat release rate of building materials', in *Ignition, Heat Release and Noncombustibility of Materials* (ed. A.F. Robertson). American Society for Testing and Materials, STP 502, pp. 119–134.

Sorenson, S.C., Savage, L.D., and Strehlow, R.A. (1975). 'Flammability limits — a new technique'. *Combustion and Flame*, **24**, 347–355.

Spalding, D.B. (1955). *Some Fundamentals of Combustion*. Butterworths, London.

Spalding, D.B. (1957). 'A theory of inflammability limits and flame quenching'. *Proceedings of the Royal Society*, **A240**, 83–100.

Stanzak, W.W., and Lie, T.T. (1973). 'Fire tests on protected steel columns with different cross sections', National Research Council of Canada, Ottawa.

Sterner, E.S., and Wickstrom, U. (1990). 'TASEF — Temperature Analysis of Structures Exposed to Fire (Users Manual)', Report 1990:05. Swedish National Testing Institute, Borås, Sweden.

Steward, F.R. (1970). 'Prediction of the height of turbulent diffusion buoyant flames'. *Combustion Science and Technology, 2, 203–212.*

Steward, F.R. (1974a). 'Radiative heat transfer associated with fire problems', in *Heat Transfer in Fires* (ed. P.L. Blackshear), pp. 273–314. John Wiley and Sons, New York.

Steward, F.R. (1974b). 'Ignition characteristics of cellulosic materials', in *Heat Transfer in Fires* (ed. P.L. Blackshear), pp. 379–407. John Wiley and Sons, New York.

Strehlow, R.A. (1973). 'Unconfined vapour cloud explosions — an overview'. *14th Symposium (International) on Combustion*, pp. 1189–1200. The Combustion Institute, Pittsburgh.

Stull, D.R. (1971). 'Chemical thermodynamics and fire problems'. *Fire Research Abstracts and Reviews*, **13**, 161–186.

Stull, D.R. (1977). 'Fundamentals of fire and explosion'. *AIChE Monograph Series No. 10, Volume 73*. American Institute of Chemical Engineers, New York.

Sugawa, O, Momita, D and Takahashi, W (1997). 'Flow behaviour of ejected fire flame/plume effected by external side wind'. *Proceedings of the Fifth International Symposium on Fire Safety Science* pp. 249–260 (International Association for Fire Safety Science, Boston, Ma).

Sundström, B. (ed.) (1995). 'Fire Safety of Upholstered Furniture — the Final Report on the CBUF Research Programme', European Commission Measurement and Testing Report EUR 16477 EN.

Sundström, B. (1984). Room fire test in full-scale for surface products (Report SP-RAPP 1984: 16), Swedish National Testing Institute, Borås, Sweden.

Sutton, O.G. (1953). *Micrometerology*. McGraw-Hill, New York.

Suzuki, M., Dobashi, R., and Hirano, T. (1994). 'Behaviour of fire spreading downward over thick paper'. *25th Symposium (International) on Combustion*, pp. 1439–1446. The Combustion Institute, Pittsburgh.

Takeda, H., and Akita, K. (1981). 'Critical phenomena in compartment fires with liquid fuels'. *18th Symposium (International) on Combustion*, pp. 519–527. The Combustion Institute, Pittsburgh.

Tan, S.H. (1967). 'Flare system design simplified'. *Hydrocarbon Processing*, **46**, 172–176.

Terai, T., and Nitta, K. (1975). *Proceedings Symposia of the Architectural Institute, Japan.* (See Zukoski *et al.*, 1981a.)

Tewarson, A. (1972). 'Some observations on experimental fires in enclosures. Part II. Ethyl alcohol and paraffin oil'. *Combustion and Flame*, **19**, 363–371.

Tewarson, A. (1980). 'Heat release in fires'. *Fire and Materials*, **4**, 185–191.

Tewarson, A. (1982). 'Experimental evaluation of flammability parameters of polymeric materials', in *Flame Retardant Polymeric Materials. Volume 3* (eds M. Lewin, S.M. Atlas, and E.M. Pearce), pp. 97–153. Plenum Press, New York and London.

Tewarson, A. (1995). 'Generation of heat and chemical compounds in fires', in *SFPE Handbook of Fire Protection Engineering*, 2nd Edition (eds P.J. Di Nenno *et al.*), pp. 3.53–3.124. Society of Fire Protection Engineers, Boston.

Tewarson, A., and Ogden, S.D. (1992). 'Fire behaviour of PMMA'. *Combustion and Flame*, **89**, 237–259.

Tewarson, A., and Pion, R.F. (1976). 'Flammability of plastics. I. Burning intensity'. *Combustion and Flame*, **26**, 85–103.

Tewarson, A., and Pion, R.F. (1978). Factory Mutual Research Serial UI 1AGR1. RC.

Tewarson, A., Lee, J.L., and Pion, R.F. (1981). 'The influence of oxygen concentration on fuel parameters for fire modelling'. *18th Symposium (International) on Combustion*, pp. 563–570. The Combustion Institute, Pittsburgh.

Theobald, C.R. (1968). 'The critical distance for ignition from some items of furniture'. *Fire Research Note No. 736.*

Thomas, P.H. (1958). 'On the thermal conduction equation for self-heating materials with surface cooling'. *Transactions of the Faraday Society*, **54**, 60–65.

Thomas, P.H. (1960). 'Some approximations in the theory of self-heating and thermal explosion'. *Transactions of the Faraday Society*, **56**, 833–839.

Thomas, P.H. (1965). 'Fire spread in wooden cribs. Part 3. The effect of wind'. *Fire Research Note No. 600.*

Thomas, P.H. (1972a). 'Self heating and thermal ignition: a guide to its theory and applications', in *Ignition, Heat Release and Non-Combustibility* (ed. A.F. Robertson), ASTM STP 502, pp. 56–82. American Society for Testing and Materials, Philadelphia.

Thomas, P.H. (1972b). 'Behaviour of fires in enclosures — some recent progress'. *14th Symposium (International) on Combustion*, pp. 1007–1020. The Combustion Institute, Pittsburgh.

Thomas, P.H. (1973). 'An approximate theory of "hot spot" criticality'. *Combustion and Flame*, **21**, 99–109.

Thomas, P.H. (1974a). 'Effects of fuel geometry in fires'. Building Research Establishment CP 29/74.

Thomas, P.H. (1974b). 'Fires in enclosures', in *Heat Transfer in Fires*, (ed. P.L. Blackshear), pp. 73–94. John Wiley and Sons, New York.

Thomas, P.H. (1975a). 'Old and new looks at compartment fires'. *Fire Technology*, **11**, 42–47.

Thomas, P.H. (1975b). 'Factors affecting ignition of combustible materials and their behaviour in fire', *International Symposium on Fire Safety of Combustible Materials*, Edinburgh University, 84–99.

Thomas, P.H. (1981). 'Testing products and materials for their contribution to flashover in rooms'. *Fire and Materials*, **5**, 103–111.

Thomas, P.H. (1982). 'Modelling of compartment fires'. *6th International Fire Protection Seminar, Karlsruhe*, pp. 29–46. Vereinigung zur Forderung des Deutschen Brandschutzes, eV.

Thomas, P.H., and Bowes, P.C. (1961a). 'Some aspects of the self-heating and ignition of solid cellulosic materials'. *British J. of Applied Physics*, **12**, 222–229.

Thomas, P.H., and Bowes, P.C. (1961b). 'Thermal ignition in a slab with one face at a constant high temperature'. *Transactions of the Faraday Society*, **57**, 2007–2016.

Thomas, P.H., and Bullen, M.L. (1979). 'On the role of *kpc* of room lining materials in the growth of room fires'. *Fire and Materials*, **3**, 68–73.

Thomas, P.H., and Heselden, A.J.M. (1972). 'Fully developed fires in single compartments. A cooperative research programme of the Conseil Internationale du Bâtiment'. Conseil Internationale du Bâtiment Report No. 20, *Fire Research Note No. 923*.

Thomas, P.H., and Hinkley, P.L. (1964). 'Design of roof venting systems for single storey buildings'. *Fire Research Technical Paper No. 10*, HMSO, London.

Thomas, P.H., and Law, M. (1972). 'The projection of flames from burning buildings'. *Fire Research Note, No. 921*.

Thomas, P.H., and Law, M. (1974). 'The projection of flames from buildings on fire'. *Fire Prevention Science and Technology No. 10*. Fire Protection Association, London.

Thomas, P.H., and Nilsson, L. (1973). 'Fully developed compartment fires: new correlations of burning rates'. *Fire Research Note No. 979*.

Thomas, P.H., and Webster, C.T. (1960). 'Some experiments on the burning of fabrics and the height of buoyant diffusion flames'. *Fire Research Note No. 420*.

Thomas, P.H., Baldwin, R., and Heselden, A.J.M. (1965a). 'Buoyant diffusion flames: some measurements of air entrainment, heat transfer and flame merging'. *10th Symposium (International) on Combustion*, pp. 983–996. The Combustion Institute, Pittsburgh.

Thomas, P.H., Bullen, M.L., Quintiere, J.G., and McCaffrey, B.J. (1980). 'Flashover and instabilities in fire behaviour'. *Combustion and Flame*, **38**, 159–171.

Thomas, P.H., Heselden, A.J.M., and Law, M. (1967a). 'Fully developed compartment fires: two kinds of behaviour'. *Fire Research Technical Paper No. 18*, HMSO, London.

Thomas, P.H., Hinkley, P.L., Theobald, C.R., and Simms, D.L. (1963). 'Investigations into the flow of hot gases in roof venting'. *Fire Research Technical Paper No. 7*, HMSO, London.

Thomas, P.H., Simms, D.L., and Law, M. (1967b). 'The rate of burning of wood'. *Fire Research Note No. 657*.

Thomas, P.H., Simms, D.L., and Wraight, H. (1964). 'Fire spread in wooden cribs. Part 1'. *Fire Research Note No. 537*.

Thomas, P.H., Simms, D.L., and Wraight, H. (1965b). 'Fire spread in wooden cribs. Part 2. Heat transfer experiments in still air'. *Fire Research Note No. 799*.

Thomas, P.H., Webster, C.T., and Raftery, M.M. (1961). 'Some experiments on buoyant diffusion flames'. *Combustion and Flame*, **5**, 359–367.

Thomson, H.E., and Drysdale, D.D. (1987). 'Flammability of polymers I. Ignition temperatures'. *Fire and Materials*, **11**, 163–172.

Thomson, H.E., and Drysdale, D.D. (1988). 'Critical mass flowrate at the firepoint of plastics'. *Proceedings of the 2nd International Symposium on Fire Safety Science*, pp. 67–76. Hemisphere Publishing Corporation, Washington.

Thomson, H.E., Drysdale, D.D., and Beyler, C.L. (1988). 'An experimental evaluation of critical surface temperature as a criterion for piloted ignition of solid finds'. *Fire Safety Journal*, **13**, 185–196.

Thorne, P.F. (1976). 'Flashpoints of mixtures of flammable and non-flammable liquids'. *Fire and Materials*, **1**, 134–140.

Tien, C.L., Lee, K.Y., and Stretton, A.J. (1995). 'Radiation heat transfer', in *SFPE Handbook of Fire Protection Engineering*, 2nd Edition (eds P.J. Di Nenno *et al.*), pp. 1-65–1-79. Society of Fire Protection Engineers, Boston.

Toal, B.R., Silcock, G.W.H., and Shields, T.J. (1989). 'An examination of piloted ignition characteristics of cellulosic materials using the ISO ignitability test', *Fire and Materials*, **14**, 197–106.

Torero, J.L., and Fernandez-Pello, A.C. (1995). 'Natural convection smoulder of polyurethane foam: upward convection'. *Fire Safety Journal*, **24**, 35–52.

Torero, J.L., and Fernandez-Pello, A.C. (1996). 'Forward smoulder of polyurethane foam in a forced air flow'. *Combustion and Flame*, **106**, 89–109.

Torero, J.L., Fernandez-Pello, A.C., and Kitano, M. (1994). 'Downward smoulder of polyurethane foam'. *Proceedings of the 4th International Symposium on Fire Safety Science*, pp. 409–420. International Association for Fire Safety Science, Boston, MA.

Tritton, D.J. (1988). *Physical Fluid Dynamics*. Oxford Science Publishers, Oxford.

Tu, K.-M., and Quintiere, J.G. (1991). 'Wall flame heights with external radiation'. *Fire Technology*, **27**, 195–203.

US Department of Commerce (1971a). 'Standard for the flammability of children's sleepwear'. DOC FF 3–71. *Federal Register*, **36**, 14062–14073.

US Department of Commerce (1971b). *JANAF Thermochemical Tables*, 2nd Edition, NSRDS-NBS 37.

Vandevelde, P., van Hees, P., Twilt, L., van Mierlo, R.J.M., and van de Leur, P. H. E. (1996). 'Reaction to Fire of Construction Products. Area A: Test Methods', CI/SfB (K4), European Commission Directorate General XII, Brussels.

Vervalin, C.H. (ed.) (1973). *Fire Protection Manual for Hydrocarbon Processing Plants*, 2nd Edition. Gulf Publishers, Houston, Texas.

Viennean, H. (1964). 'Mixing controlled flame heights from circular jets'. B.Sc. Thesis, Department of Chemical Engineering, University of New Brunswick, Fredericton, NB.

Vytenis, R.J., and Welker, J.R. (1975). 'End-grain ignition of wood'. *J. Fire and Flammability*, **6**, 355–361.

Wade, S.H. (1942). 'Evaporation of liquids in currents of air'. *Transactions of the Institution of Chemical Engineers*, **20**, 1–14.

Wagner, H. Gg. (1979). 'Soot formation in combustion'. *17th Symposium (International) on Combustion*, pp. 3–19. The Combustion Institute, Pittsburgh.

Walker, I.K. (1967). 'The role of water in the spontaneous heating of solids'. *Fire Research Abstracts and Reviews*, **9**, 5–22.

Wall, L.A. (1972). 'The pyrolysis of polymers', in *The Mechanisms of Pyrolysis, Oxidation and Burning of Organic Materials* (ed. L. A. Wall) NBS Special Publication 357, pp. 47–60. National Bureau of Standards, Washington, DC.

Walton, W.D., and Thomas, P.H. (1995). 'Estimating temperatures in compartment fire', in *SFPE Handbook of Fire Protection Engineering*, 2nd Edition (eds P.J. Di Nenno *et al.*), pp. 3.134–3.147. Society of Fire Protection Engineers, Boston.

Waterman, T.E. (1966). 'Determination of fire conditions supporting room flashover'. *Final Report IITRI Project M6131, DASA 1886*, Defense Atomic Support Agency, Washington, DC.

Waterman, T.E. (1968). 'Room flashover — criteria and synthesis'. *Fire Technology*, **4**, 25–31.

Weast, R.C. (ed.) (1974/75). *Handbook of Chemistry and Physics*, 54th Edition. Chemical Rubber Company, Ohio.

Weatherford, W.D., and Sheppard, D.M. (1965). 'Basic studies of the mechanism of ignition of cellulosic materials'. *10th Symposium (International) on Combustion*, pp. 897–910. The Combustion Institute, Pittsburgh.

Weckman, E.J., and Sobiesiak, X.X. (1988). 'The oscillatory behaviour of medium scale pool fires'. *22nd Symposium (International) on Combustion*, pp. 1299–1310. The Combustion Institute, Pittsburgh.

Wells, G. (1997). *Major Hazards and their Management*. Institution of Chemical Engineers, Rugby.

Welty, J.R., Wilson, R.E., and Wicks, C.E. (1976). *Fundamentals of Momentum, Heat and Mass Transfer*, 2nd Edition. John Wiley and Sons, New York.

White, A.G. (1925). 'Limits for the propagation of flame in inflammable gas/air mixtures. Part III. The effect of temperature on the limits'. *J. Chemical Society*, **127**, 672–684.

Whiteley, R.H. (1994). 'Short communication: some comments concerning the measurement of smoke'. *Fire and Materials*, **18**, 57–59.

Wichman, I.S. (1992). 'Review of flame spread'. *Progress in Energy and Combustion Science*, **18**, 553.

Wickstrom, U., Sundström, B., and Holmstedt, G. (1983). 'The development of a full-scale room fire test'. *Fire Safety Journal*, **5**, 191–197.

Williams, A. (1973). 'Combustion of droplets of liquid fuels: a review'. *Combustion and Flame*, **21**, 1–31.

Williams, F.A. (1965). *Combustion Theory*. Addison-Wesley.

Williams, F.A. (1974a). 'A unified view of fire suppression'. *J. Fire and Flammability*, **5**, 54–63.

Williams, F.A. (1974b). 'Chemical kinetics of pyrolysis', in *Heat Transfer in Fires* (ed. P.L. Blackshear), pp. 197–237. John Wiley & Sons, New York.

Williams, F.A. (1977). 'Mechanisms of fire spread'. *16th Symposium (International) on Combustion*, pp. 1281–1294. The Combustion Institute, Pittsburgh.

Williams, F.A. (1981). 'A review of flame extinction'. *Fire Safety Journal*, **3**, 163–175.

Williams, F.A. (1982). 'Urban and wildland fire phenomenology'. *Progress in Energy and Combustion Science*, **8**, 317–354.

Williamson, R.B., Revenaugh, A., and Mowrer, F.W. (1991). 'Ignition sources in room fire tests and some implications for flame spread evaluation'. *Proceedings of the 3rd International Symposium on Fire Safety Science*, pp. 657–666. Elsevier Science Publishers, Barking.

Wolfhard, H.G., and Simmons, R.F. (1955). 'Influence of methyl bromide on flames. Part I. Premixed flames'. *Transactions of the Faraday Society*, **51**, 1211–1217.

Wood, B.D., Blackshear, P.H., and Eckert, E.R.G. (1971). *Combustion Science and Technology*, **4**, 131.

Woodburn, P., and Drysdale, D.D. (1997). 'Fires in inclined trenches: the effects of trench and burner geometry on the critical angle'. *Proceedings of the 5th International Symposium on Fire Safety Science*, pp. 225–236. International Association for Fire Safety Science, Boston, MA.

Woolley, W.D. (1971). 'Decomposition products of PVC for studies of fires'. *British Polymer Journal*, **3**, 186–193.

Woolley, W.D. (1972). 'Studies of the dehydrochlorination of PVC in nitrogen and air'. *Plastics and Polymers*, pp. 203–208.

Woolley, W.D., and Ames, S.A. (1975). 'The explosion risk of stored foam rubber'. Building Research Establishment Current Paper CP 36/75.

Woolley, W.D., and Fardell, P.J. (1982). 'Basic aspects of combustion toxicology'. *Fire Safety Journal*, **5**, 29–48.

Woolley, W.D., Ames, S.A., and Fardell, P.J. (1979). 'Chemical aspects of combustion toxicology of fires'. *Fire and Materials*, **3**, 110–120.

Woolley, W.D., Ames, S.A., Pitt, A.I., and Murrell, J.V. (1975). 'Fire behaviour of beds and bedding materials'. *Fire Research Note No. 1038*.

Woolley, W.D., Raftery, M.M., Ames, S.A., and Murrell, J.V. (1979/80). 'Smoke release from wall linings in full scale compartment fires'. *Fire Safety Journal*, **2**, 61–72.

Yao, C. (1976). 'Development of large-drop sprinklers'. *FMRC Technical Report Serial No. 22476*. Factory Mutual Research Corporation.

Yao, C. (1980). 'Application of sprinkler technology'. Presented at Workshop on 'Engineering Applications of Fire Technology'. (National Bureau of Standards).

Yih, C.S. (1952). 'Free convection due to a point source of heat'. *Proceedings of the 1st US National Congress in Applied Mechanics*, pp. 941–947.

Yokoi, S. (1960). 'Study of the prevention of fire spread caused by hot upward currents'. Building Research Institute, Report No. 34, Tokyo.

You, H.Z., and Faeth, G.M. (1979). 'Ceiling heat transfer during fire plume and fire impingement'. *Fire and Materials*, **3**, 140–147.

You, H.Z., and Kung, H.C. (1985). 'Strong buoyant plumes of growing rack storage fires'. *20th Symposium (International) on Combustion*, pp. 1547–1554. The Combustion Institute, Pittsburgh.

Yule, A.J., and Moodie, K. (1992). 'A method of testing the flammability of sprays of hydraulic fluid'. *Fire Safety Journal*, **18**, 273–302.

Zabetakis, M.G. (1965). 'Flammability characteristics of combustible gases and vapours'. US Bureau of Mines, *Bulletin 627*.

Zabetakis, M.G., and Burgess, D.S. (1961). 'Research on the hazards associated with the production and handling of liquid hydrogen', US Bureau of Mines RI5707, Pittsburgh, PA.

Zalosh, R.G. (1995). 'Explosion protection', in *SFPE Handbook of Fire Protection Engineering*, 2nd Edition (eds P.J. Di Nenno *et al.*), pp. 3.312–3.329. Society of Fire Protection Engineers, Boston.

Zukoski, E.E. (1975). 'Convective Flows Associated with Room Fires', Semi Annual Progress Report, National Science Foundation Grant No. GI 31892 X1, Institute of Technology, Pasadena, California.

Zukoski, E.E. (1986). 'Fluid dynamic aspects of room fires'. *Proceedings of the 1st International Symposium on Fire Safety Science*, pp. 1–30. Hemisphere Publishing Corporation, Washington.

Zukoski, E.E. (1995). 'Properties of fire plumes', in *Combustion Fundamentals of Fire* (ed. G. Cox), pp. 101–219. Academic Press, London.

Zukoski, E.E., Kubota, T., and Cetegen, B. (1981a). 'Entrainment in fire plumes'. *Fire Safety Journal*, **3**, 107–121.

Zukoski, E.E., Kubota, T., and Cetegen, B. (1981b). 'Entrainment in the near field of a fire plume'. National Bureau of Standards, GCR-81–346, US Department of Commerce, Washington, DC.

Zukoski, E.E., Morehart, J.H., Kubota, T., and Toner, S.J. (1991). 'Species production and heat release rates in two-layered natural gas fires'. *Combustion and Flame*, **83**, 325–332.

Author Index

Subject Index